Lecture Notes in Computer Science

Lecture Notes in Artificial Intelligence 15244
Founding Editor

Jörg Siekmann

Series Editors

Randy Goebel, *University of Alberta, Edmonton, Canada*
Wolfgang Wahlster, *DFKI, Berlin, Germany*
Zhi-Hua Zhou, *Nanjing University, Nanjing, China*

The series Lecture Notes in Artificial Intelligence (LNAI) was established in 1988 as a topical subseries of LNCS devoted to artificial intelligence.

The series publishes state-of-the-art research results at a high level. As with the LNCS mother series, the mission of the series is to serve the international R & D community by providing an invaluable service, mainly focused on the publication of conference and workshop proceedings and postproceedings.

Dino Pedreschi · Anna Monreale ·
Riccardo Guidotti · Roberto Pellungrini ·
Francesca Naretto
Editors

Discovery Science

27th International Conference, DS 2024
Pisa, Italy, October 14–16, 2024
Proceedings, Part II

Editors
Dino Pedreschi
University of Pisa
Pisa, Italy

Anna Monreale
University of Pisa
Pisa, Italy

Riccardo Guidotti
University of Pisa
Pisa, Pisa, Italy

Roberto Pellungrini
Scuola Normale Superiore (SNS)
Pisa, Italy

Francesca Naretto
University of Pisa
Pisa, Italy

ISSN 0302-9743   ISSN 1611-3349 (electronic)
Lecture Notes in Artificial Intelligence
ISBN 978-3-031-78979-3   ISBN 978-3-031-78980-9 (eBook)
https://doi.org/10.1007/978-3-031-78980-9

LNCS Sublibrary: SL7 – Artificial Intelligence

© The Editor(s) (if applicable) and The Author(s), under exclusive license to Springer Nature Switzerland AG 2025
Chapters "ETIA: Towards an Automated Causal Discovery Pipeline", "An Attention-based CNN Approach to Detect Forest Tree Dieback Caused by Insect Outbreak in Sentinel-2 Images", and "Ensemble Counterfactual Explanations for Churn Analysis" are licensed under the terms of the Creative Commons Attribution 4.0 International License (http://creativecommons.org/licenses/by/4.0/). For further details see license information in the chapter.

This work is subject to copyright. All rights are solely and exclusively licensed by the Publisher, whether the whole or part of the material is concerned, specifically the rights of translation, reprinting, reuse of illustrations, recitation, broadcasting, reproduction on microfilms or in any other physical way, and transmission or information storage and retrieval, electronic adaptation, computer software, or by similar or dissimilar methodology now known or hereafter developed.
The use of general descriptive names, registered names, trademarks, service marks, etc. in this publication does not imply, even in the absence of a specific statement, that such names are exempt from the relevant protective laws and regulations and therefore free for general use.
The publisher, the authors and the editors are safe to assume that the advice and information in this book are believed to be true and accurate at the date of publication. Neither the publisher nor the authors or the editors give a warranty, expressed or implied, with respect to the material contained herein or for any errors or omissions that may have been made. The publisher remains neutral with regard to jurisdictional claims in published maps and institutional affiliations.

This Springer imprint is published by the registered company Springer Nature Switzerland AG
The registered company address is: Gewerbestrasse 11, 6330 Cham, Switzerland

If disposing of this product, please recycle the paper.

Preface

Discovery Science 2024 conference provides an open forum for intensive discussions and exchange of new ideas among researchers working in the area of Discovery Science. The conference focus is on the use of Artificial Intelligence, Data Science and Big Data Analytics methods in science. Its scope includes the development and analysis of methods for discovering scientific knowledge, coming from machine learning, data mining, intelligent data analysis, and big data analytics, as well as their application in various domains. The 27th International Conference on Discovery Science (DS 2024) was held in Pisa, Italy, during October 14–16, 2024.

This was the fourth time the conference was organized as a stand-alone physical event. Indeed, for its first 20 editions, DS was co-located with the International Conference on Algorithmic Learning Theory (ALT). In 2018 it was co-located with the 24th International Symposium on Methodologies for Intelligent Systems (ISMIS 2018). Then starting from 2019, it has been a stand-alone event. DS 2020 and DS 2021 were online-only events while DS 2022 and DS 2023 were located in Montpellier, France and Porto, Portugal, respectively.

DS 2024 received 121 international submissions, of which 25 were for the SoBigData++ track, a track dedicated to the usage of data and data science in science celebrating the conclusion of the SoBigData++ project, a sponsor of DS 2024. For each track, each submission was reviewed by at least two Program Committee (PC) members (some PC acted for both tracks) in a single-blind review process using the Microsoft CMT system. The PC decided to accept 45 papers for the regular research track of DS 2024 and 9 papers for the SoBigData++ track, for a total of 54 papers. This resulted in an overall acceptance rate of 45%. One paper was withdrawn two weeks before the beginning of the conference. Thus, during the conference, 53 papers were presented and included in these volumes.

The conference included three keynote talks: Roberto Navigli (Sapienza University of Rome and Babelscape) contributed a talk titled "What Is Missing in Today's Large Language Models?"; Carlos Castillo (ICREA and Universitat Pompeu Fabra) gave a presentation titled "Human Factors and Algorithmic Fairness"; and Francesca Toni (Imperial College London) contributed a talk titled "Bridging Explainable AI and Contestability". Abstracts of the invited talks are included in the front matter of these proceedings. Besides the presentation of the regular research papers and SoBigData++ papers in the main program, the conference offered two poster sessions titled "Late Breaking Contributions" and "Doctoral Consortium", featuring posters of very recent research results and PhD theses on topics related to Discovery Science.

We are grateful to Springer for their continued long-term support. Springer publishes the conference proceedings, as well as a regular special issue of the Machine Learning journal on Discovery Science. The latter offers authors a chance of publishing in this prestigious journal significantly extended and reworked versions of their DS conference papers, while being open to all submissions on DS conference topics.

On the program side, we would like to thank all the authors of the submitted papers and the PC members for their efforts in evaluating the submitted papers, as well as the keynote speakers. On the organization side, we would like to thank all the members of the Organizing Committee, in particular Roberto Pellungrini, Francesca Naretto, Vittorio Romano, Francesco Spinnato, Lorenzo Mannocci, Daniele Fadda and Rosalba Lubino for the smooth preparation and organization of all conference-associated activities. We are also grateful to the people behind Microsoft CMT for developing the conference organization system which has proved to be an essential tool in the paper submission and evaluation process.

October 2024

Dino Pedreschi
Anna Monreale
Riccardo Guidotti
Roberto Pellungrini
Francesca Naretto

Organization

General and Program Chairs

Dino Pedreschi	University of Pisa, Italy
Anna Monreale	University of Pisa, Italy
Riccardo Guidotti	University of Pisa, Italy

Special Session Chair

Roberto Trasarti — ISTI-CNR Pisa, Italy

Poster Session Chair

Francesca Naretto — University of Pisa, Italy

Doctoral Consortium Chair

Roberto Pellungrini — Scuola Normale Superiore, Italy
Fosca Giannotti — Scuola Normale Superiore, Italy

Social Media and Publicity Chair

Vittorio Romano — ISTI-CNR Pisa, Italy

Local Organization Committee

Francesco Spinnato — University of Pisa, Italy
Lorenzo Mannocci — University of Pisa, Italy
Daniele Fadda — ISTI-CNR Pisa, Italy

Steering Committee Chair

Michelangelo Ceci   University of Bari, Italy

Program Committee

Ad Feelders
Adriano Rivolli
Alberto Cano
Albrecht Zimmermann
Alexander H. Gower
Andreas Nuernberger
Andrew Aquilina
Angelica Liguori
Annalisa Appice
Anne Laurent
Antonio Mastropietro
Apostolos N. Papadopoulos
Arnaud Soulet
Bernard Zenko
Bernhard Pfahringer
Blaz Zupan
Brian Mac Namee
Bruno Cremilleux
Bruno E. Martins
Bruno Veloso
Carlo Metta
Carlos Soares
Caterina Senette
Chamalee Nisanala Wickrama Arachchi
Chetraj Pandey
Chiara Renso
Claire Nédellec
Clara R. T. Puga
Claudio Giovannoni
Cristiano Landi
Daniele Gambetta
Dariusz Brzezinski
Davide Vega
Dino Ienco
Domen Šoberl
Domenico Talia
Donato Malerba
Dragan Gamberger
Dragi Kocev
Elio Masciari
Eric Scott
Esther-Lydia Silva-Ramírez
Fabien FP Poirier
Fabio Fassetti
Fabrizio Angiulli
Fabrizio Marozzo
Federico Mazzoni
Florent Masseglia
Francesca Bugiotti
Francesca Naretto
Francesca Alessandra Lisi
Francesco Marcelloni
Francesco S. Pisani
Francesco Spinnato
Georgios Vardakas
Gianvito Pio
Giuseppe Manco
Giuseppina Andresini
Gizem Gezici
Günce Keziban Orman
Gustau Camps-Valls
Haimonti Dutta
Henrik Bostrom
Hoang-Anh Ngo
Howard J. Hamilton
Inês Dutra
Isacco Beretta
Ivan Reis Filho
Jaka Kokošar
Jelena Joksimovic
Joana Santos
Joao Mendes-Moreira
Johannes Fürnkranz
Jörg Wicker

José Luis Seixas Jr.
Juan G. Colonna
Julian Martin Rodemann
Julian Marvin Joers
Julio C. Muñoz-Benítez
Kai Puolamäki
Katharina Dost
Kaustubh Patil
Kohei Hatano
Larisa Soldatova
Larissa Andrade Silva
Lorenzo Mannocci
Lubos Popelínský
Luca Corbucci
Luca Ferragina
Maguelonne Teisseire
Maik Büttner
Marcilio de Souto
Marco Javier Suárez Barón
Marek Herde
Margarida A. Costa
Marina Sokolova
Mário Antunes
Marko Pranjifá
Marta Marchiori Manerba
Marta Moreno
Martin Atzmueller
Martin Holena
Martin Špendl
Martina Cinquini
Masaaki Nishino
Massimo Guarascio
Mathieu Roche
Matteo Zignani
Mattia Cerrato
Mattia Setzu
Maurizio Parton
Maximilien Servajean
Michael R. Berthold
Michele Fontana
Mirco Nanni
Myra Spiliopoulou
Nada Lavrafç
Nathalie Japkowicz
Nontokozo Mpofu
Nuno Silva
Panagiotis Papapetrou
Pance Panov
Pascal Poncelet
Paula Silva
Paulo Cortez
Pavel Brazdil
Pavlin G. Polifçar
Pedro C. Vieira
Pedro G. Ferreira
Pedro Henriques Abreu
Peter O. Koleoso
Peter van der Putten
Rafael Gomes Mantovani
Rafael Mamede
Rafael G. Teixeira
Raza Ul Mustafa
Reza Akbarinia
Ricardo Cardoso Pereira
Ricardo Cerri
Riccardo Cantini
Rita D. Nogueira
Robert Bossy
Roberto Corizzo
Roberto Interdonato
Roberto Pellungrini
Ross King
Rui Jorge Gomes
Sabrina Gaito
Salvatore Ruggieri
Saso Dzeroski
Silvia Corbara
Simona Nisticò
Simone Piaggesi
Sónia Teixeira
Takayasu Fushimi
Thiago Andrade
Van Anh Huynh-Thu
Vincent Labatut
Vincenzo Lagani
Vincenzo Pasquadibisceglie
Xuan Zhao
Yiping Tang
Zahra Donyavi
Zakaria Farou

Program Committee SoBigData++ Track

Albert Ali Salah
Antinisca Di Marco
Chiara Boldrini
Clara Punzi
Claudio Giovannoni
Fabrizia Auletta
Fabrizio Lillo
George Papastefanatos
Giovanni Stilo
Josep Domingo-Ferrer

Mark E. Cote
Marzio Di Vece
Maurizio Parton
Michela Natilli
Paolo Bellavista
Rami Haffar
Roberto Trasarti
Ruggero G. Pensa
Tom Emery
Valerio Grossi

Keynote Talks

What Is Missing in Today's Large Language Models?

Roberto Navigli

Sapienza University of Rome and Babelscape, Italy

Large Language Models (LLMs) like GPT-4 have demonstrated remarkable capabilities in generating human-like text, understanding context, and performing a wide range of tasks across various domains. However, despite their impressive performance, LLMs exhibit critical limitations that restrict their applicability and reliability in real-world scenarios. This talk delves into the key areas where LLMs fall short, including the lack of true understanding and reasoning, susceptibility to biases, difficulties with long-term context retention, and challenges in generating accurate outputs in non-predominant domains. I will also touch upon research directions that can provide potential solutions to enhance the capabilities and trustworthiness of future LLMs, such as performing lexical and sentence-level semantics, intersecting knowledge and results obtained from different tasks, improving training data diversity, and supporting factuality.

Human Factors and Algorithmic Fairness

Carlos Castillo

ICREA and Universitat Pompeu Fabra, Spain

In this talk, we present ongoing research on human factors of decision support systems that has consequences from the perspective of algorithmic fairness. We study two different settings: a game and a high-stakes scenario. The game is an exploratory "oil drilling" game, while the high-stakes scenario is the prediction of criminal recidivism. In both cases, a decision support system helps a human make a decision. We observe that in general users of such systems must thread a fine line between algorithmic aversion (completely disregarding the algorithmic support) and automation bias (completely disregarding their own judgment). The talk presents joint work led by David Solans and Manuel Portela.

Bridging Explainable AI and Contestability

Francesca Toni

Imperial College London, UK

AI has become pervasive in recent years, and the need for explainability is widely agreed upon as crucial towards safe and trustworthy deployment of AI systems. However, state-of-the-art AI and eXplainable AI (XAI) approaches mostly neglect the need for AI systems to be contestable, as advocated instead by AI guidelines (e.g. by the OECD) and regulation of automated decision-making (e.g. GDPR in the EU and UK). In this talk I will advocate forms of contestable AI that can (1) interact to progressively explain outputs and/or reasoning, (2) assess grounds for contestation provided by humans and/or other machines, and (3) revise decision-making processes to redress any issues successfully raised during contestation. I will then explore how contestability can be achieved computationally, starting from various approaches to explainability, including some drawn from the field of computational argumentation. Specifically, I will overview a number of approaches to (argumentation-based) XAI for neural models and for causal discovery and their uses to achieve contestability.

Contents – Part II

Tree-Based Models and Causal Discovery

Splitting Stump Forests: Tree Ensemble Compression for Edge Devices 3
  Fouad Alkhoury and Pascal Welke

Faithfulness of Local Explanations for Tree-Based Ensemble Models 19
  Amir Hossein Akhavan Rahnama, Pierre Geurts, and Henrik Boström

Random Forests for Heteroscedastic Data 34
  Hugo Bellamy and Ross D. King

Learning Deep Rule Concepts as Alternating Boolean Pattern Trees 50
  Florian Beck, Johannes Fürnkranz, and Van Quoc Phuong Huynh

ETIA: Towards an Automated Causal Discovery Pipeline 65
  Konstantina Biza, Antonios Ntroumpogiannis, Sofia Triantafillou, and Ioannis Tsamardinos

Security and Anomaly Detection

FedGES: A Federated Learning Approach for Bayesian Network Structure Learning 83
  Pablo Torrijos, José A. Gámez, and José M. Puerta

Enhancing Industrial Control Systems Security: Real-Time Anomaly Detection with Uncertainty Estimation 99
  Ermiyas Birihanu, Ayyoub Soullami, and Imre Lendák

ITERADE - ITERative Anomaly Detection Ensemble for Credit Card Fraud Detection 115
  Bahar Emami Afshar, Paula Branco, Tolga Kurt, Utku Gorkem Ketenci, and Hikmet Mazmanoglu

Purifying Adversarial Examples Using an Autoencoder 134
  Thijs van Weezel, Famke van Ree, Tychon Bos, Patrick Bastiaanssen, and Sibylle Hess

Approximate Compression of CNF Concepts 149
  Sieben Bocklandt, Vincent Derkinderen, Angelika Kimmig, and Luc De Raedt

Computer Vision and Explainable AI

Explainable AI in Time-Sensitive Scenarios: Prefetched Offline
Explanation Model 167
    Fabio Michele Russo, Carlo Metta, Anna Monreale,
    Salvatore Rinzivillo, and Fabio Pinelli

An Attention-Based CNN Approach to Detect Forest Tree Dieback
Caused by Insect Outbreak in Sentinel-2 Images 183
    Vito Recchia, Giuseppina Andresini, Annalisa Appice,
    Gianpietro Fontana, and Donato Malerba

Explaining Image Classifiers with Visual Debates 200
    Avinash Kori, Ben Glocker, and Francesca Toni

Resource-Constrained Binary Image Classification 215
    Sean Park, Jörg Wicker, and Katharina Dost

Towards a Multimodal Framework for Remote Sensing Image Change
Retrieval and Captioning 231
    Roger Ferrod, Luigi Di Caro, and Dino Ienco

Classification Models

Improving the Performance of Already Trained Classifiers Through
an Automatic Explanation-Based Learning Approach 249
    Andrea Apicella, Salvatore Giugliano, Francesco Isgrò,
    and Roberto Prevete

A Simple Method for Classifier Accuracy Prediction Under Prior
Probability Shift 267
    Lorenzo Volpi, Alejandro Moreo, and Fabrizio Sebastiani

Pairwise Difference Learning for Classification 284
    Mohamed Karim Belaid, Maximilian Rabus, and Eyke Hüllermeier

SoBigData++: City for Citizens and Explainable AI

Explaining Urban Vehicle Emissions in Rome 303
    Matteo Bohm, Patricio Reyes, Mirco Nanni, and Luca Pappalardo

Interpretable Machine Learning for Oral Lesion Diagnosis Through
Prototypical Instances Identification 316
   Alessio Cascione, Mattia Setzu, Federico A. Galatolo,
   Mario G. C. A. Cimino, and Riccardo Guidotti

Ensemble Counterfactual Explanations for Churn Analysis 332
   Samuele Tonati, Marzio Di Vece, Roberto Pellungrini,
   and Fosca Giannotti

This Sounds Like That: Explainable Audio Classification via Prototypical
Parts 348
   Andrea Fedele, Riccardo Guidotti, and Dino Pedreschi

TETRA: TExtual TRust Analyzer for a Gricean Approach to Social
Networks 364
   Federico Mazzoni, Simona Mazzarino, and Giulio Rossetti

SoBigData++: Societal Debates and Misinformation Analysis

What's Real News Today? A Multimodal, Continual-Learning Approach
for Detecting Fake News Over Time 381
   Luca Maiano, Martina Evangelisti, Silvia Bianchini,
   and Aris Anagnostopoulos

Beyond the Horizon: Using Mixture of Experts for Domain Agnostic Fake
News Detection 396
   Carmela Comito, Massimo Guarascio, Angelica Liguori,
   Giuseppe Manco, and Francesco Sergio Pisani

Quantifying Attraction to Extreme Opinions in Online Debates 411
   Davide Perra, Andrea Failla, and Giulio Rossetti

Structure-Attribute Similarity Interplay in Diffusion Dynamics on Social
Networks 425
   Salvatore Citraro, Valentina Pansanella, and Giulio Rossetti

Author Index 441

Tree-Based Models and Causal Discovery

Splitting Stump Forests: Tree Ensemble Compression for Edge Devices

Fouad Alkhoury[1,3](✉) and Pascal Welke[2]

[1] University of Bonn, Bonn, Germany
alkhoury@iai.uni-bonn.de
[2] TU Wien, Vienna, Austria
pascal.welke@tuwien.ac.at
[3] Lamarr Institute for Machine Learning and Artificial Intelligence, Dortmund, Germany

Abstract. We introduce Splitting Stump Forests – small ensembles of weak learners extracted from a trained random forest. The high memory consumption of random forest ensemble models renders them unfit for resource-constrained devices. We show empirically that we can significantly reduce the model size and inference time by selecting nodes that evenly split the arriving training data and applying a linear model on the resulting representation. Our extensive empirical evaluation indicates that Splitting Stump Forests outperform random forests and state-of-the-art compression methods on memory-limited embedded devices.

Keywords: Ensemble compression · Random forests · Edge devices

1 Introduction

The global count of Internet of Things (IoT) devices is expected to reach approximately 20 billion units by 2025 [41]. Many IoT devices require real-time decisions and therefore include limited computing capabilities [28]. Running machine learning models directly on embedded devices is increasingly popular due to improvements in reliability, affordability, and energy efficiency [17]. Local models also reduce or even eliminate the need for transferring data to cloud servers when connectivity, bandwidth consumption, communication costs, network latency, and privacy are significant concerns [20,30].

Ensemble models can outperform the predictive performance of individual classifiers in many machine learning tasks [13,37]. However, ensemble models combine multiple base models, resulting in high memory consumption. The model size is also a primary determinant of inference time, an aspect equally important as model accuracy in real-time applications. IoT devices, however, typically contain slow microcontrollers with limited flash memory, ranging from a few Kbytes to a maximum of a few Mbytes, as shown in Table 1. As a result, the best-performing ensemble model is often too large and too slow to be deployed for real-time applications in IoT devices. This work presents small tree ensemble models that provide responsive and highly accurate predictions. Decision trees efficiently represent sequences of if-else conditions and can be compiled to run efficiently on embedded devices [8,9]. We revisit these models and propose a lossy compression scheme for their ensembles, random forests. A random forest reduces the variance of the predictions compared to a single decision tree [5]. However, having only a few small trees in a random forest can hinder predictive performance, while large random forests pose a significant challenge for resource-limited devices. Our experiments show that high-performing random forests often exceed 100K nodes. As a result, these models may demand more than 5–10 Mbytes of memory even in very compact implementations [9].

We propose a novel method to create a new, *compressed* ensemble from a large random forest model, often comprising hundreds of thousands of nodes. To enable compression, the approach extracts a subset of test nodes from a trained random forest to build a smaller ensemble of *splitting stumps* with a total size of only a few Kbytes. Our approach combines the supervised selection of splits in the training of random forests with an unsupervised measure of balance on the training data. We argue that this reduces the tendency to overfit the training data. The final model transforms the input data into multi-hot encoding and trains a linear classifier to map the novel representation to the target domain. Our extensive experimental evaluation on various datasets shows the superiority of splitting stump forests over random forests and state-of-the-art competitive ensemble compression techniques in terms of compression rate, inference time, and predictive performance. Moreover, our experiments demonstrate that the selected test nodes are *informative* and not accidental.

This article proceeds as follows: We review related work in Sect. 2. Section 3 provides a detailed description of our method. Section 4 describes our empirical evaluation before Sect. 5 concludes. Code for the splitting stump forests is available on github.

Table 1. Flash Memory on different microcontroller units [1,2,4]

Microcontroller Unit	Flash Memory
ATmega169P	16 KB
Arduino Nano	26–32 KB
Arduino Uno	32 KB
Atmel ATSAM3S2AA-AU	64 KB
Arduino Mega	256 KB

2 Background

Our proposed random forest compression technique can be alternatively viewed as model compression or representation learning approach. We now review related work in these fields.

Ensemble Compression. The existing methods for ensemble size reduction can be categorized into blackbox and whitebox approaches. Blackbox approaches do not assume any particular model architecture. Instead, they work on the set of models in an ensemble without changing individual base models. White box approaches update base models, e.g., by pruning individual nodes of decision trees in a random forest. We focus our discussion on whitebox approaches.

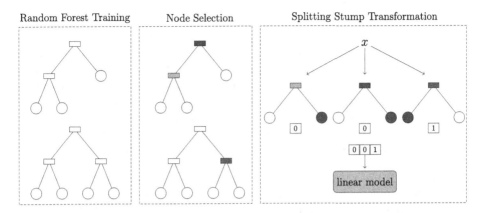

Fig. 1. Splitting stump forests at a glance. Given a random forest (left), selected nodes are selected (middle) and used as stumps (right). A training example is then transformed into a binary vector by the stumps and fed to a linear model.

Seeking an optimal sub-ensemble within large random forests is often impractical. In general, identifying an optimal subset of classifiers with the best generalization performance is an NP-complete problem [33]. Thus, most approaches identify a sub-ensemble with near-optimal performance. Several studies have been conducted on the pruning of machine learning models [25]. Moreover, considerable efforts concentrated on identifying subsets of random forest nodes that can match the original forest's accuracy. Peterson and Martinez [34] introduced a post-training technique that stores unique subtrees and combines redundant nodes into "parallel nodes" while maintaining the overall behavior. Buschjäger and Morik [10] introduced an innovative method that integrates regularization into the leaf-refinement process. Their proposed algorithm jointly prunes and refines trees, thereby enhancing the performance of tree ensembles. Prior research has indicated that high diversity among ensemble models can enhance their generalization performance. Li et al. [26] select classifiers that both minimize empirical error and lead to greater ensemble diversity. Ranking-based strategies sort individual models by their associated prediction error and select a few highly ranked members to compose the sub-ensemble [22]. Nakamura and Sakurada [31] reduce the number of distinct split conditions by sharing a common condition among multiple nodes which allows for practical model size reductions. Other studies have found that removing low-impact nodes from a decision tree can simplify it while preserving accuracy [15]. In contrast, our approach constructs a new ensemble that contains more but smaller trees than the original ensemble.

Representation Learning. Decision trees and random forests can be interpreted as representation learning [3], with studies exploring the use of a pre-trained random forest's transformed space as input for linear models. Ren et al. [36] enhance the fitting power of a tree ensemble model by global leaf value refinement using linear regression. Through global optimization, the approach iteratively merges insignificant pairs of adjacent leaves, effectively using complementary information from multiple trees and reducing model size. Likewise, Nakano et al. [32] integrate a representation learning component into the random forest methodology. Their method treats ensemble nodes as clusters formed by instances during recursive partitioning. Then it creates a binary vector where each element corresponds to a cluster node and is set to 1 if a training instance traverses the node. This newly created tree embedding is then combined with the original feature set. Welke et al. [43] identify frequent subtrees in a trained random forest and train a linear model on the resulting multi-hot leaf representation. Vens and Costa [42] use the encoding of nodes visited by data instances. The final feature encoding is obtained by concatenating the binary vectors of all trees in the forest. Estruch et al. [16] leverage the common components within decision tree ensembles. In this structure, the rejected splits are not discarded but stored as suspended nodes. This allows these nodes to be further explored, allowing the generation of new models.

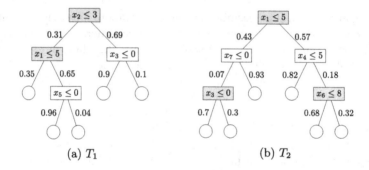

Fig. 2. A random forest F consisting of two decision trees T_1 and T_2. Round nodes represent leaves, rectangular nodes represent test nodes. The left (right) edge label represents the fraction of training instances evaluating to true (false). Selected nodes in $T[p]$ for filtering threshold $p = 0.2$, are colored in blue. (Color figure online)

3 The Splitting Stump Forests Method

In this section, we provide a description of our *splitting stump forests* method to extract a compact model from a potentially large random forest. See, e.g., Breiman [5] or Quinlan [35] for a background on random forest and decision tree algorithms. Given a trained random forest, we (1) compute a score for each node and select nodes with high scores (Sect. 3.1). Our selection technique prioritizes split values that yield balanced subtrees. Subsequently, we (2) construct one decision tree per selected node and form an ensemble (Sect. 3.2). Finally, our approach (3) integrates a representation learning module by using the transformed space derived from the constructed ensemble as input for a linear model

Algorithm 1. Splitting Stump Forest Transformation

Input: Random forest F, training points X_{train}, threshold parameter $p \in (0, 0.5]$
Output: Splitting stump forest F' and training data representation X'_{train}.

1: $F' \leftarrow \emptyset$
2: $X'_{train} \leftarrow \emptyset$
3: **for each** tree $T \in F$ **do**
4:    **for each** node $v \in T$ **do**
5:       X_v^t(resp. X_v^f)$\leftarrow$ set of training points in v that are True (resp. False)
6:       $score(v) \leftarrow \frac{min(|X_v^t|, |X_v^f|)}{|X_v^t| + |X_v^f|}$
7:       **if** $score(v) \geq p$ **then**
8:          add v to $T[p]$
9: **for each** $v \in T[p]$ **do**
10:    $T'_v \leftarrow$ construct a splitting stump by linking leaves to the node v
11:    add T'_v to F'
12: **for each** $x \in X_{train}$ **do**
13:    **for each** $T' \in F'$ **do**
14:       $f_{T'}(x) \leftarrow$ assign an encoding for x in T'
15:       add $f_{T'}(x)$ to $f_{F'}(x)$
16:    add $f_{F'}(x)$ to X'_{train}
17: **return** F' and X'_{train}

(Sect. 3.3). Figure 1 shows the pipeline of the three primary steps, which we will describe in turn. The pseudo-code of the first two steps is presented in Algorithm 1. In Sect. 3.4 we discuss balanced splits in more detail.

In what follows, we consider a supervised learning problem where the instance space is $X \subseteq \mathbb{R}^d$ and the target space is $Y \subseteq \mathbb{R}$. Each data point $x \in X$ is a d-dimensional vector described by a set of features and mapped to the corresponding label $y \in Y$. The goal is to find a function $f : X \to \mathbb{R}$ such that the difference between $f(x)$ and the true label y is minimal for all $x \in X$. Labeled instances $X_{train} \subseteq X$ are provided during training, while unlabeled instances $X_{test} \subseteq X$ are provided during testing. In this paper, we apply our approach to classification tasks, but an extension to regression problems is possible.

We call the root and all internal nodes of a decision tree test nodes. Test nodes are labeled with a split condition $x_a \leq s_v$ for a given attribute a and a split value $s_v \in \mathbb{R}$. In this work, test nodes have exactly two children, called left and right. When the split condition of node v for an instance $x \in X$ evaluates to True, the instance passes to the left child of v. A random forest classifier $F = \{T_j | j \in [1, t]\}$ consists of a set of t decision trees and a method to combine individual predictions, e.g., majority vote.

3.1 Splitting Node Selection

The aim of the first step is to select a subset of balanced splits from a trained random forest F. Technically, we propose a post-hoc selection criterion that favors split conditions that lead to balanced splits. In particular, for all $T \in F$ and for all nodes v of T, we count the number of incoming training points that evaluate to true (resp. false) using the split condition at node v (Line 5 of Algorithm 1). To qualify as balanced, v should attain a score that meets or exceeds a predetermined threshold p (Lines 6–7). Subsequently, we define $T[p]$ as the set of all nodes v with $score(v) \geq p$ (Line 8). In deep random forests, we can limit the size of $T[p]$ by arranging the scores in descending order and selecting a specific number of nodes with the highest scores. In a binary decision tree, $score(v) \in [0, 0.5]$ and higher values indicate a better division of training samples into two sub-samples of comparable sizes. Figure 2 shows the selection process on a small random forest. For efficient computation of $score(v)$, we store the count of training examples traversing tree edges during training. Alternatively, the counts can be derived by a single pass over an independent dataset that does not require labeling. Once these numbers are accessible, scores can be computed in constant time and high-scoring nodes can be selected by a single sweep across the random forest. Duplicate split conditions can be efficiently removed using an appropriate set data structure for $T[p]$.

Note that a scoring function for decision stump learning with a similar formula was introduced by Iba and Langley [21]. In contrast to our work, however, their score replaces e.g. the Gini index in decision stump learning and directly compares against the class label, while here we score a node in a decision tree based on the balance of training samples between its left and right branches.

3.2 Splitting Stump Transformation

$T[p]$, the result of the previous step, is a set of isolated vertices, each consisting of a feature and split value, and is hence not a random forest by our definition. By attaching two leaves to each node $v \in T[p]$, we transform v into the root of a decision tree T'_v of depth one (line 10). These decision trees can be viewed as learning a new data representation that maps a data point to a set of leaves. For each decision tree T'_v in the new ensemble F' of length k, we define a mapping function $f_{T'_v} : \mathbb{R}^d \to \{0, 1\}$ that returns one if and only if the split condition in node v evaluates to true on x (See line 14). For F', we thus construct a function $f_{F'} : \mathbb{R}^d \to \{0, 1\}^k$ which maps x to a new feature vector $\{f_{T'_{vi}}(x)\}_{i=1}^k$. That is, each training point x is *embedded* into the concatenation of features (ones and zeros) resulting from the stumps (line 16). Using $f_{T'_v}$ instead of the two leaf features of T'_v is sufficient due to the perfect correlation between the two leaf features resulting from each stump T'_v.

3.3 Training of Splitting Stump Forests

To enable deployment to devices with limited resources, we use a linear model to combine the individual predictions of the decision trees in F'. We apply logistic regression to model the relationship between the new feature vectors $f_{F'}(x)$ and the target variable. The resulting model is both resource-efficient and easy to interpret. Conceptually, this step can be seen as simultaneously learning the leaf

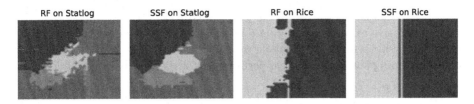

Fig. 3. Example of decision boundaries in data classification between random forests (RF) and the proposed splitting stump forests (SSF) on two-dimensional projections of two datasets. SSF achieves a comparable accuracy using only 0.002 of the total nodes employed by the RF method.

node assignments of all decision stumps and the voting scheme of the resulting random forest. Following the pruning of the random forest, similar post-training approaches have been shown to work well [10, 36]. Consider a single splitting stump T'_v: Training a logistic regression classifier on one-hot encoded representations $f_{T'_v}$ assigns a weight to each of the two dimensions that, for binary classification, corresponds to the likelihood of belonging to the target class. A similar approach works for regression tasks using a linear regression learner.

3.4 Further Details on Splitting Node Selection

The most common way to train random forests is based on bagging the training data and then using a recursive algorithm with a Gini-index or mutual information based split criterion selection. This reduces the variance of the predictions of the resulting model, but tends to increase the complexity of the model. Figure 3 shows the decision boundaries of trained random forests on two-dimensional feature-subsets of the statlog and rice datasets. In our two examples, the focus on pure splits in combination with voting results in the partition of the feature space into rather small and discontinuous regions. We argue that this may be detrimental to generalization and that simpler models may be found that yield similar performance at smaller sizes.

Small regions can arise when split criteria cut off a relatively small portion of the training data with pure labels. When the split condition of a node v evaluates to true (or false) for most incoming training points $X_v \subseteq X_{train}$, it leads to imbalanced subsets at the deeper level of the tree. These imbalanced subsets consist of a nearly pure and relatively small subset, thus facilitating good prediction on the training set, but also may cause overfitting and can increase sensitivity to noise and outliers. Conversely, the other branch typically contains a large subset. This results in deeper trees, longer inference times, and reduced human readability [24, 40]. Moreover, this behavior can negatively impact prediction accuracy as the algorithm prioritizes outliers or errors less relevant in the generalization process creating overly specific rules based on limited information. In this study, we investigate the effect of choosing attribute-value combinations that result in balanced splits. These combinations of balanced splits empirically facilitate good data partitioning, thereby helping learners avoid overfitting [7]. Figure 3 shows the decision boundaries of our corresponding splitting stump forests. In these illustrative examples, selecting nodes that lead to balanced splits for SSF increases the sizes of the continuous regions while reducing overall model size and maintaining similar predictive performance of the resulting model.

4 Experiments

Datasets. We experiment on 13 benchmark classification datasets with varying properties, primarily from the UCI repository (Adult, Letter Recognition, MAGIC, Spambase, Statlog, Waveform) [14], ALOI [19], Bank [29], Credit Card [45], Dry Bean [23], Rice [11], Room [39] and Shoppers [38]. This diverse selection enables the evaluation across varying complexities.

Competitive Methods. To evaluate the splitting stump forests approach (SSF), we perform a comparative analysis against the random forest (RF), the baseline cost complexity pruning (CCP) [6], and four state-of-the-art compression methods: the global refinement approach (LR) [36][1], the joint leaf refinement and ensemble pruning (LR+L1) [10][2], the diversity regularized ensemble pruning method (DREP) [26], and the individual error pruning method (IE) [22][3].

Table 2. Average rankings of methods across 13 datasets based on accuracy, compression ratio, and inference time for each depth. Here, rank one is assigned to the best-performing method, two to the second-best, and so on. The last column shows the global ranking across all depths and datasets.

Method		$d=5$	$d=10$	$d=15$	Global
RF	Acc.	5	4.69	3.23	4.31
	Size	7	7	7	7
	Inf.	6.08	5.46	5.46	5.67
CCP	Acc.	6.46	6.62	6.85	6.64
	Size	4	2.38	1.85	2.74
	Inf.	4.69	3.54	3.46	3.90
DREP	Acc.	4.23	3.62	3.69	3.85
	Size	3.62	4.69	5	4.44
	Inf.	3.07	3.08	2.85	3
IE	Acc.	4	3.23	3.38	3.54
	Size	3.69	4.54	5.08	4.43
	Inf.	2.92	3.23	2.92	3.03
LR	Acc.	2.69	3.15	4.38	3.41
	Size	5.23	4.77	4.85	4.95
	Inf.	4.92	6.77	6.69	6.13
LR+L1	Acc.	2.23	2.77	2.69	2.56
	Size	3.38	3.46	3	3.28
	Inf.	6	4.85	5.15	5.33
SSF	Acc.	**2**	**2.38**	**2.39**	**2.26**
	Size	**1.23**	**1.15**	**1.30**	**1.23**
	Inf.	**1.69**	**1.69**	**1.92**	**1.77**

Experimental Setting. We train random forests, using the Gini index reduction for splitting, by varying the maximum depth d of individual decision trees among $d \in \{5, 10, 15\}$ and the number t of decision trees among $t \in \{16, 32, 64\}$. The same d and t values are adopted as in the SSF method for each approach being compared. To determine optimal parameters, we conduct a grid search for each method. We perform 5-fold cross-validation and report

---

[1] available at github.com/gereleth/kaggle-telstra.
[2] available at github.com/sbuschjaeger/leaf-refinement-experiments.
[3] both available at github.com/sbuschjaeger/PyPruning.

the average accuracy achieved on test data for each d. The threshold parameter p is set to values $\{0.05, 0.1, ..., 0.4, 0.45\}$. Code for SSF and all experiments is accessible online.[4]

Comparison of Methods. We report the average test accuracy, the number of nodes (test nodes and leaves), and the prediction time, also referred to as inference time. Table 2 presents a summary of this experiment, displaying the average ranking achieved by each method across the 13 datasets. This mean ranking demonstrates the consistent superiority of the SSF approach, with respective global rankings of 2.26, 1.23, and 1.77 concerning predictive performance, model size compression, and inference time. In terms of predictive performance, SSF outperforms original random forests (RF) and state-of-the-art methods in 22 out of 39 runs (across 13 datasets, each with three maximum depth values). In most other runs, SSF accuracy is within 1% of the top-performing model. SSF significantly reduces model size by two to three orders of magnitude compared to other methods and achieves the best compression ratio in 32 out of 39 runs, achieving a global ranking of 1.23. Considering the mean ranking, SSF exhibits a marginal improvement in predictive performance compared to the LR+L1 method, while consistently excelling in reducing model size. Inference time for SSF is faster than the best-performing models of competing methods in 31 out of 39 runs, achieving a global ranking of 1.77. Scoring and selecting nodes in SSF training is efficiently done through a preorder tree traversal, with a time complexity of $O(n)$ where n is the number of nodes in the random forest if we store a record of the training examples that traverse edges during random forest training. Looking at Fig. 4, we note that the SSF outperforms competing methods in model size and inference time across all datasets while achieving the best or second-best levels of accuracy.

4.1 Predictive Performance on a Space Budget

Motivated by space constraints on small embedded devices, we analyze the performance of SSF and the competitive ensemble pruning methods in a space budget. To that end, we explore various random forest configurations with $d \in \{5, 10, 15\}$, $t \in \{8, 16, 32, 64\}$, and the corresponding parameters for each competitive method, evaluating predictive performance and model size (node count) for each configuration. To accommodate devices with limited storage capacity, we select the best models that can fit within 32 KB or 16 KB of memory. Such models are suitable for deployment on microcontroller units like the Arduino Uno and ATmega169P. We estimate the model size using the baseline implementation of decision trees established by Buschjäger and Morik [9] and applied in subsequent studies [10]. This implementation indicates that each node requires $17 + 4C$ bytes of memory, where C represents the number of classes.

Figure 5 shows that SSF models outperform RF models across all datasets. Notably, this improvement exceeds 3% in five datasets. Particularly in multi-classification tasks like the letter recognition dataset (26 classes) and the dry

---
[4] https://github.com/FouadAlkhoury/SplittingStumpsForests/.

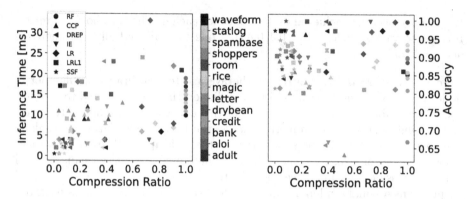

Fig. 4. The figure shows the highest test accuracy achieved with a maximum depth $d = 5$, along with its associated compression ratio and inference time, for each method and dataset.

bean dataset (7 classes), the improvement is significant, due to the complexity of multi-class problems [44]. In these tasks, deep random forests excel in capturing the complex decision boundaries necessary for reasonable accuracy. However, the best random forest model for the letter dataset requires 4 MB of memory, exceeding the IoT device's budget. Thus, we recommend using SSF models (of size below 10 KB in most datasets) for multi-classification tasks. Then, we employ the post-hoc Friedman test methodology as outlined by Demšar [12] for 32 KB and 16 KB budgets to check for statistically significant performance differences among the seven examined methods. We formulate the null hypothesis as all methods perform equally well without significant differences. The Friedman test ranks the methods for each dataset and each of 5 runs, assigning the top-performing method a rank of 1, the second-best a rank of 2, and so forth. This test determines whether the average ranks significantly deviate from the expected mean rank of 4. Average ranks provide a useful comparison of the methods, as illustrated in Table 3. Notably, the computed p-values for both 32 and 16 KB scenarios are 5.8×10^{-40} and 2.14×10^{-41} respectively, leading to the rejection of the null hypothesis at a highly significant level. As statistical significance is revealed, we apply a post-hoc procedure for multiple comparisons as proposed by García et al. [18]. Using the Conover Test, we conduct 21 pairwise comparisons among the seven methods, at confidence levels of 95%, 99%, and 99.9%. Figure 6 demonstrates that both SSF and LR+L1 significantly deviate from the other 5 methods at the highest confidence level p-value < 0.001 in both the 32 and 16 KB scenarios. Moreover, the computed p-value between the SSF and LR+L1 is 1.1×10^{-2} in the 16 KB scenario, indicating that SSF outperforms LR+L1 with 95% confidence on memory-limited devices.

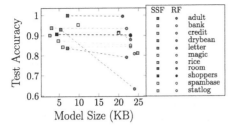
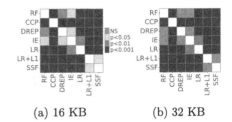

Fig. 5. The plot shows the best model attained by RF and SSF with a final model size below 32 KB.

Fig. 6. Pairwise comparisons through a Conover test between the top-performing models under 16 KB (left) and 32 KB (right).

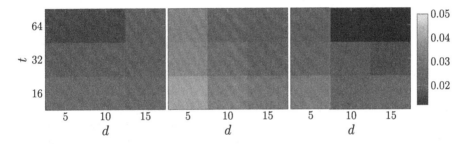

Fig. 7. The plot shows the compression ratio achieved in the datasets adult, shoppers, spambase (left to right) while permitting a 2% accuracy drop.

4.2 Compression on a Performance Budget

As a complementary experiment, we explore compression while tolerating a slight drop in accuracy. We identify the smallest SSF model within a 2% accuracy margin compared to the RF model with varying values of d and t. For a relatively small RF with $d = 5$ and $t = 16$, we achieved an average compression rate of 0.04 across all datasets. Moreover, as the random forest size increased, so did the compression rate for most datasets. Notably, the SSF method yielded compression values of $0.012, 0.02, 0.017, 0.025, 0.037$, and 0.005 for the datasets spambase, shoppers, adult, room, rice, and bank, respectively; see Fig. 7.

4.3 Optimizing Accuracy Vs. Compression

To validate our assumption regarding the informativeness of nodes with highly balanced branches, and as our problem involves balancing a trade-off between the predictive performance and model compression, we investigate the impact of varying filtering thresholds p on the trade-off between the two objectives. In particular, we examine the Pareto frontier which enables us to concentrate on the set of efficient choices of p known as non-dominated solutions [27].

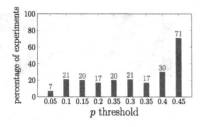
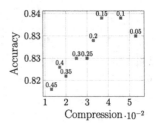

Fig. 8. The figure shows Pareto Frontier results for the min/max objectives compression ratio and accuracy. The left plot shows the percentage of experiments in which p is non-dominated by another p' in various problem settings $t \in \{16, 32, 64\}$, $d \in \{5, 10, 15\}$. Right, we exemplarily show the non-dominated thresholds in red, and dominated ones in blue on adult, $t = 64$, $d = 10$. (Color figure online)

A non-dominated filtering threshold p^* is where we cannot find another p that improves accuracy without sacrificing compression, or vice versa. We determine the set of non-dominated points for each dataset and for each RF setting $t \in \{16, 32, 64\}$ and $d \in \{5, 10, 15\}$. Next, we compute the frequency of each threshold p in the non-dominated set, focusing on data points achieving accuracy within 2% of the original RF accuracy. Omitting this step would result in any data point with the maximum threshold of 0.45 being incorrectly classified as non-dominated, given the monotonically increasing nature of the compression function. The findings, as shown in Fig. 8, indicate that the threshold $p = 0.45$ exhibits non-dominance in 71% of the experiments, while $p = 0.40$ demonstrates non-dominance in 30% of the cases. The other thresholds are non-

Table 3. The table shows the best accuracy with a model size below 16 KB and 32 KB. Bold entries indicate the best method for each dataset. The last line shows the average ranking of the accuracy of each method across all datasets.

Dataset	16 KB							32 KB						
	RF	CCP	DREP	IE	LR	LRL1	SSF	RF	CCP	DREP	IE	LR	LRL1	SSF
adult	82	81.9	80.5	82.9	84.4	85.7	**86.1**	85.1	82.3	83.4	84.3	85.9	**86.2**	86.1
	±0.4	±0.3	±0.2	±0.6	±0.4	±0.3	±0.3	±0.3	±0.4	±0.4	±0.5	±0.3	±0.3	±0.3
aloi	96.7	96.2	96.9	96.9	96.1	97.0	**97.1**	96.8	96.2	96.9	97	96.1	**97.1**	**97.1**
	±0.4	±0.1	±0.1	±0.1	±0.1	±0.1	±0.1	±0.1	±0.1	±0.1	±0.1	±0.1	±0.1	±0.1
bank	89.9	88.9	89.8	89.7	90.0	**90.2**	90.1	90	88.9	89.8	89.7	90.1	**90.4**	90.1
	±0.1	±0.5	±0.1	±0.1	±0.3	±0.3	±0.2	±0.1	±0.3	±0.1	±0.1	±0.2	±0.2	±0.1
credit	81	80.7	80.8	81.1	**81.4**	81.1	81.1	80.9	80.7	81	81.3	**81.4**	80.9	81.2
	±0.2	±0.3	±0.1	±0.2	±0.3	±0.1	±0.2	±0.1	±0.2	±0.1	±0.1	±0.2	±0.1	±0.1
dry bean	88.1	85.4	89.1	88.7	89.2	**91.8**	91.4	87.9	87.4	89.3	89.9	89.4	**91.8**	91.4
	±0.7	±0.3	±0.1	±0.3	±0.3	±0.1	±0.4	±1.0	±0.4	±0.5	±0.6	±0.5	±0.3	±0.4

continued

Table 3. continued

Dataset	16 KB							32 KB						
	RF	CCP	DREP	IE	LR	LRL1	SSF	RF	CCP	DREP	IE	LR	LRL1	SSF
letter	62.5 ±2.2	64.3 ±1.1	62.6 ±0.9	62.1 ±1.2	62.9 ±0.9	76.3 ±0.9	**93.0** ±1.7	63.3 ±3.6	61.2 ±1.4	65.9 ±1.0	65.8 ±1.7	78.2 ±1.5	71.4 ±1.4	**93.0** ±1.9
magic	84.9 ±0.5	83.2 ±0.9	83.7 ±0.5	84.2 ±0.5	84.9 ±0.01	85.8 ±0.2	**86.3** ±0.7	86 ±0.7	83.5 ±0.7	83.8 ±0.3	83.9 ±0.5	86.1 ±0.6	**86.5** ±0.2	86.3 ±0.4
rice	92.5 ±0.6	93.7 ±0.7	93.1 ±0.3	0.93 ±0.4	93.2 ±0.7	93.5 ±0.1	**93.8** ±0.1	93.4 ±0.7	93.7 ±0.5	93.6 ±0.2	93.5 ±0.3	93.2 ±0.6	93.5 ±0.1	**93.8** ±0.1
room	99.2 ±0.3	97.0 ±0.4	99.5 ±0.1	99.6 ±0.2	99.2 ±0.6	99.8 ±0.2	**99.9** ±0.2	99.5 ±0.1	99.3 ±0.2	99.7 ±0.1	**99.9** ±0.2	99.2 ±0.3	99.8 ±0.1	**99.9** ±0.1
shoppers	87.2 ±1.5	86.9 ±0.6	90.1 ±0.3	91.0 ±0.2	**91.6** ±0.4	91.3 ±0.3	90.7 ±0.4	90.2 ±1.2	86.9 ±0.2	90.3 ±0.4	91.2 ±0.1	91.6 ±0.4	**91.3** ±0.2	90.7 ±0.2
spambase	90.8 ±0.6	91.3 ±0.4	90.9 ±0.2	92.4 ±1.0	91.6 ±0.5	92.7 ±0.2	**95.3** ±0.4	91.1 ±0.6	91.7 ±0.6	92.9 ±0.6	92.4 ±1.2	94.3 ±0.4	93.2 ±0.2	**95.3** ±0.2
statlog	85.2 ±1.6	84.1 ±1.0	84.7 ±0.4	84.8 ±0.4	84.9 ±0.8	86.5 ±0.2	**87.4** ±0.3	85.1 ±0.8	84.9 ±0.6	84.8 ±0.7	84.7 ±0.6	87.1 ±0.5	**87.9** ±0.2	87.4 ±0.1
waveform	96.9 ±0.2	95.1 ±0.2	96.9 ±0.1	96.9 ±0.1	96.9 ±0.3	97.0 ±0.1	**97.1** ±0.2	96.6 ±0.1	95.1 ±0.1	97 ±0.1	97 ±0.1	96.9 ±0.2	**97.2** ±0.1	97.1 ±0.1
avg rank	4.79	5.86	4.79	4.17	3.94	2.09	**1.54**	4.72	6.32	4.45	4.28	3.51	2.1	**1.88**

dominant in about 20%, except for $p = 0.05$ in only 7% of runs. Our findings support our assumption that test nodes with high splitting power, like those with $p = 0.45$, provide more information than nodes with low splitting power. These high-scoring nodes represent only a small fraction of the entire nodes set in the random forest, producing well-balanced branches, and resulting in a highly accurate and compact model.

We validate the informativeness of our selected nodes by comparing their predictive performance to that of a randomly selected sample of the same size. For a given dataset D, we report the accuracy and number of selected nodes n using the parameters: $d = 15$, $t = 64$, $p = 0.4$. Then we randomly sample n nodes from the entire node set i.e. this case corresponds to $p = 0.0$. These nodes are then transformed into splitting stumps, and we proceed to train a linear model using their data representation. To ensure experiment

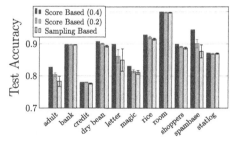

Fig. 9. Comparison of the predictive performance of the splitting stumps when $p = 0.4$, $p = 0.2$, and sampling-based splitting stumps of all scores. We report the mean and standard deviation of 10 random samples.

validity, we repeat sampling ten times and calculate the mean and standard deviation. We also sample n nodes that achieve a score better than $p = 0.2$. Comparing the predictive performance of score-based splitting stumps with $p = 0.4$ against sampling-based stumps, we find that score-based stumps tend to perform better across most datasets, as shown in Fig. 9. These high-scoring stumps also outperform an equivalent-sized set of lower-scoring nodes of $p = 0.2$, reinforcing our assumption that nodes with higher splitting power provide more information.

5 Conclusion

We introduced Splitting Stump Forests, an approach that extracts nodes from a trained random forest based on their splitting capabilities. Subsequently, we constructed decision trees for high-scoring nodes and trained a linear model over the derived data representations. Our extensive empirical tests indicate significant reductions in model size and improved inference speed without sacrificing accuracy across diverse datasets. We conducted a comprehensive comparison with competing methods and an ablation study of our split criterion. Our encouraging experimental findings revealed our method's superiority in model size compression and inference time acceleration while maintaining a comparable level of predictive performance. These outcomes raise interesting directions for future research. In particular, to develop practical deployment strategies, ensuring that the benefits of model compression can be fully realized in real-world applications following the ongoing integration of machine learning models in edge devices.

Acknowledgement. FA has been funded by the Federal Ministry of Education and Research of Germany and the state of North Rhine-Westphalia as part of the Lamarr Institute for Machine Learning and Artificial Intelligence, LAMARR24B. PW is financed by the Vienna Science and Technology Fund (WWTF) project StruDL (ICT22-059).

References

1. Atmel: ATMEGA169P 8-bit AVR microcontroller with 16k bytes in-system datasheet (2016). www.alldatasheet.com
2. Atmel: ATSAM3S2AA-AU-MC 32bit 64kb LQFP-48 (2016). distrelec.de
3. Bengio, Y., Courville, A., Vincent, P.: Representation learning: a review and new perspectives. Trans. Pattern Anal. Mach. Intell. **35**(8), 1798–1828 (2013)
4. Branco, S., Ferreira, A.G., Cabral, J.: Machine learning in resource-scarce embedded systems, FPGAs, and end-devices: a survey. Electronics **8**(11), 1289 (2019)
5. Breiman, L.: Random forests. Mach. Learn. **45**(1), 5–32 (2001)
6. Breiman, L., Friedman, J., Olshen, R., Stone, C.: Classification and Regression Trees. Chapman and Hall, New York, NY, USA (1984)
7. Bringmann, B., Zimmermann, A.: The chosen few: on identifying valuable patterns. In: International Conference on Data Mining (2007)
8. Buschjäger, S., Morik, K.: Decision tree and random forest implementations for fast filtering of sensor data. Trans. Circ. Syst. I: Regul. Pap. **65**(1), 209–222 (2017)

9. Buschjäger, S., Morik, K.: Improving the accuracy-memory trade-off of random forests via leaf-refinement. arXiv preprint arXiv:2110.10075 (2021)
10. Buschjäger, S., Morik, K.: Joint leaf-refinement and ensemble pruning through L1 regularization. Data Min. Knowl. Disc. **37**(3), 1230–1261 (2023)
11. Cinar, I., Koklu, M.: Classification of rice varieties using artificial intelligence methods. Int. J. Intell. Syst. Appl. Eng. **7**(3), 188–194 (2019)
12. Demšar, J.: Statistical comparisons of classifiers over multiple data sets. J. Mach. Learn. Res. **7**, 1–30 (2006)
13. Dietterich, T.G.: Ensemble methods in machine learning. In: International Workshop on Multiple Classifier Systems, pp. 1–15 (2000)
14. Dua, D., Graff, C.: UCI machine learning repository (2017)
15. Esposito, F., Malerba, D., Semeraro, G., Kay, J.: A comparative analysis of methods for pruning decision trees. Trans. Pattern Anal. Mach. Intell. **19**(5), 476–491 (1997)
16. Estruch, V., Ferri, C., Hernández-Orallo, J., Ramírez-Quintana, M.J.: Bagging decision multi-trees. In: Multiple Classifier Systems: International Workshop (2004)
17. Filho, C.P., et al.: A systematic literature review on distributed machine learning in edge computing. Sensors **22**(7), 2665 (2022)
18. García, S., Fernández, A., Luengo, J., Herrera, F.: Advanced nonparametric tests for multiple comparisons in the design of experiments in computational intelligence and data mining: experimental analysis of power. Inf. Sci. **180**(10), 2044–2064 (2010)
19. Geusebroek, J., Burghouts, G., Smeulders, A.: The Amsterdam library of object images. Int. J. Comput. Vis. **61**, 103–112 (2005)
20. Hua, H., Li, Y., Wang, T., Dong, N., Li, W., Cao, J.: Edge computing with artificial intelligence: a machine learning perspective. ACM Comput. Surv. **55**(9), 1–35 (2023)
21. Iba, W., Langley, P.: Induction of one-level decision trees. In: International Workshop on Machine Learning, pp. 233–240 (1992)
22. Jiang, Z., Liu, H., Fu, B., Wu, Z.: Generalized ambiguity decompositions for classification with applications in active learning and unsupervised ensemble pruning. In: AAAI Conference on Artificial Intelligence (2017)
23. Koklu, M., Ozkan, I.A.: Multiclass classification of dry beans using computer vision and machine learning techniques. Comput. Electron. Agric. **174**, 105507 (2020)
24. Leroux, A., Boussard, M., Dès, R.: Information gain ratio correction: improving prediction with more balanced decision tree splits. arXiv preprint arXiv:1801.08310 (2018)
25. Li, H., Kadav, A., Durdanovic, I., Samet, H., Graf, H.P.: Pruning filters for efficient convnets. In: International Conference on Learning Representations (2017)
26. Li, N., Yu, Y., Zhou, Z.H.: Diversity regularized ensemble pruning. In: European Conference on Machine Learning and Knowledge Discovery in Databases, pp. 330–345 (2012)
27. Lin, J.G.: Three methods for determining pareto-optimal solutions of multiple-objective problems. In: Directions in Large-Scale Systems: Many-Person Optimization and Decentralized Control (1976)
28. Merenda, M., Porcaro, C., Iero, D.: Edge machine learning for AI-enabled IoT devices: a review. Sensors **20**(9), 2533 (2020)
29. Moro, S., Cortez, P., Rita, P.: A data-driven approach to predict the success of bank telemarketing. Decis. Support Syst. **62**, 22–31 (2014)

30. Murshed, M.S., Murphy, C., Hou, D., Khan, N., Ananthanarayanan, G., Hussain, F.: Machine learning at the network edge: a survey. ACM Comput. Surv. **54**(8), 1–37 (2021)
31. Nakamura, A., Sakurada, K.: An algorithm for reducing the number of distinct branching conditions in a decision forest. In: European Conference on Machine Learning and Knowledge Discovery in Databases, pp. 578–589 (2019)
32. Nakano, F.K., Pliakos, K., Vens, C.: Deep tree-ensembles for multi-output prediction. Pattern Recogn. **121**, 108211 (2022)
33. Partalas, I., Tsoumakas, G., Vlahavas, I.: Pruning an ensemble of classifiers via reinforcement learning. Neurocomputing **72**(7-9) (2009)
34. Peterson, A.H., Martinez, T.R.: Reducing decision tree ensemble size using parallel decision DAGs. Int. J. Artif. Intell. Tools **18**(04), 613–620 (2009)
35. Quinlan, J.R.: Induction of decision trees. Mach. Learn. **1**, 81–106 (1986)
36. Ren, S., Cao, X., Wei, Y., Sun, J.: Global refinement of random forest. In: Conference on Computer Vision and Pattern Recognition (2015)
37. Sagi, O., Rokach, L.: Ensemble learning: a survey. Wiley Interdisc. Rev. Data Min. Knowl. Discov. **8**(4), e1249 (2018)
38. Sakar, C.O., Polat, S.O., Katircioglu, M., Kastro, Y.: Real-time prediction of online shoppers' purchasing intention using multilayer perceptron and LSTM recurrent neural networks. Neural Comput. Appl. **31**(10), 6893–6908 (2019)
39. Singh, A.P., Jain, V., Chaudhari, S., Kraemer, F.A., Werner, S., Garg, V.: Machine learning-based occupancy estimation using multivariate sensor nodes. In: IEEE Globecom Workshops, pp. 1–6 (2018)
40. Thakur, D., Markandaiah, N., Raj, D.S.: Re optimization of ID3 and C4.5 decision tree. In: International Conference on Computer and Communication Technology, pp. 448–450 (2010)
41. Vailshery, L.: Number of internet of things (IoT) connected devices worldwide from 2019 to 2030, by vertical (2023). www.statista.com
42. Vens, C., Costa, F.: Random forest based feature induction. In: International Conference on Data Mining, pp. 744–753 (2011)
43. Welke, P., Alkhoury, F., Bauckhage, C., Wrobel, S.: Decision snippet features. In: International Conference on Pattern Recognition, pp. 4260–4267 (2021)
44. Yan, J., Zhang, Z., Lin, K., Yang, F., Luo, X.: A hybrid scheme-based one-vs-all decision trees for multi-class classification tasks. Knowl.-Based Syst. **198**, 105922 (2020)
45. Yeh, I.C., Lien, C.H.: The comparisons of data mining techniques for the predictive accuracy of probability of default of credit card clients. Expert Syst. Appl. **36**(2), 2473–2480 (2009)

Faithfulness of Local Explanations for Tree-Based Ensemble Models

Amir Hossein Akhavan Rahnama[1]($\boxtimes$)[id], Pierre Geurts[2][id], and Henrik Boström[1][id]

[1] KTH Royal Institute of Technology, Stockholm, Sweden
{amiakh,bostromh}@kth.se
[2] University of Liège, Liège, Belgium
p.geurts@uliege.be

Abstract. Local explanation techniques provide insights into the predicted outputs of machine learning models for individual data instances. These techniques can be model-agnostic, treating the machine learning model as a black box, or model-based, leveraging access to the model's internal properties or logic. Evaluating these techniques is crucial for ensuring the transparency of complex machine-learning models in real-world applications. However, most evaluation studies have focused on the faithfulness of these techniques in explaining neural networks. Our study empirically evaluates the faithfulness of local explanations in explaining tree-based ensemble models. In our study, we have included local model-agnostic explanations of LIME, KernelSHAP, and LPI, along with local model-based explanations of TreeSHAP, Sabaas, and Local MDI for gradient-boosted trees and random forests models trained on 20 tabular datasets. We evaluate local explanations using two perturbation-based measures: Importance by Preservation and Importance by Deletion. We show that model-agnostic explanations of KernelSHAP and LPI consistently outperform model-based explanations from TreeSHAP, Saabas, and Local MDI when gradient-boosted tree and random forest models. Moreover, LIME explanations of gradient-boosted tree and random forest models consistently demonstrate low faithfulness across all datasets.

Keywords: Explainable AI · Explainable Machine Learning · Interpretable Machine Learning · Transparency in Machine Learning · Local Explanations

1 Introduction

Tree-based ensemble models have been shown to frequently have competitive predictive performance on tabular datasets, e.g., over neural network models [12]. Because of this, they are widely used across numerous applications [30]. Decision tree models are considered interpretable under certain assumptions, but tree-based ensemble models, typically consisting of several hundred, often conflicting, trees, are black boxes. Furthermore, these models are growing increasingly complex with the complexity of datasets. Therefore, understanding the logic of their predictions becomes an important part of the modeling process. Local explanation techniques are designed to help understand such complex models. These techniques provide information from the machine learning model without sacrificing accuracy [14,20].

Local explanation techniques are categorized into different types, e.g., Counterfactual, Prototype-based, or Feature attribution techniques [14,22]. Due to their flexibility, feature attribution explanations are arguably the most popular type of local explanations and they are therefore the focus of this study.

Local feature attribution explanation techniques[1] allocate real-valued importance scores to each feature in the explained data instance that shows its importance to the predicted output of black-box models. These techniques can be model-based, i.e., where they can only explain a particular class of model, or model-agnostic, i.e., where they can explain any machine learning model [22]. Some studies show that model-based explanation techniques not only have a computational advantage but also provide more faithful explanations[2] than their model-agnostic counterparts [1,15]. The success of local model-based explanation techniques can be because they have access to the internal structure of explained models or they can exploit prior assumptions about the model they explain [22,23].

Even though many consider these explanation techniques useful, numerous studies have presented cases in which these techniques fail [28,31]. However, the criticism has not been conclusive and has focused on a few examples. Instead of dismissing these techniques based on a few cases, more conclusive evaluation studies are needed to understand and correct their potential failures.

Local explanations provide transparency without the loss of accuracy. However, they come with a caveat: evaluating local explanations is inherently a complex and open research problem. This is because there is no unified approach to evaluating local explanations [34]. So far, two categories of methods for evaluating local explanations are proposed in the literature: human and functionally grounded methods [7]. In human evaluation methods, human subjects evaluate the quality of local explanations by replicating the prediction of models having only access to the instance explained and its corresponding local explanation [24]. The design and implementation of such methods are both time-consuming and costly and prone to inherent bias in human subjects. On the other hand, functionally grounded methods [8,23] use proxy measures to evaluate different aspects of explanations and can complement human evaluation methods. These evaluation techniques typically require less effort and can be helpful when iteratively evaluating explanation techniques, e.g., during their early

---

[1] In our study, we may refer to local feature attribution explanations as local explanations or simply explanations for brevity.

[2] Similar to [1], we use the terms faithful explanations or faithfulness of explanations instead of the term explanation accuracy. We avoid using the term accuracy in this context since our evaluation is not based on ground truth importance scores. Later in this section, we will provide more details.

development phase. Because of these advantages, our study focuses on the functionally grounded evaluation of local feature attribution explanation techniques.

Evaluating local explanations via functionally grounded evaluation measures has its challenges, one of which being that the ground truth feature importance scores for explanations are only available when explaining simpler models [2,26] or in cases where the models are trained on synthetic and simplistic datasets [19]. These types of evaluation measures, hence, rely on simplifying the data or the model we explain. However, another set of evaluation measures, called Importance by Preservation and Importance by Deletion, aims to evaluate local explanations without the need for ground truth or simplifying models and data [16,20,23]. In this study, we use these measures to evaluate the faithfulness of local explanations for tree ensemble models.

There is an important gap in the literature on explainable machine learning. Local explanations are arguably most useful in high-stake domains, such as health and judicial domains [7,13], and in such domains, tabular datasets are frequently used [33]. However, the majority of studies that evaluate the faithfulness of local explanations have focused solely on the explanations of neural network models [3,18,23]. We believe there is a need to evaluate the faithfulness of local explanations when explaining tree-based ensemble models. Moreover, no study compares the faithfulness of model-based versus model-agnostic explanations in the context of tree-based ensemble methods. The only available evaluation is in the work of [20], where the explanations are evaluated on three medical datasets that are not publicly available. We argue that for a fair comparison, the evaluations should include many explanation techniques and publicly available datasets.

In this study, we fill this gap by evaluating a wide range of local explanation techniques for tree-based ensemble models. We have included both model-agnostic explanations such as LIME [27], LPI [5], KernelSHAP[3] [21], along with model-based techniques, such as TreeSHAP (Observation and Interventional) [20], Local MDI [32] and Saabas [29]. Along with these techniques, we have included two explanation baselines in our evaluation: (Global) Impurity Feature Importance Scores [4] and random explanation. We have evaluated these techniques when explaining gradient-boosted trees and random forest models on 20 tabular datasets based on Importance by Preservation and Deletion as evaluation measures.

The main contributions of the study are:

- The first large-scale empirical evaluation of local explanations of tree-based ensemble models, spanning six explanation techniques, 20 datasets, two learning algorithms, and two evaluation measures. For reproducibility, the code for the experiments is publicly available[4].
- Key findings on the faithfulness of local explanation techniques, including:

---

[3] For brevity, we may refer to KernelSHAP as SHAP.
[4] The code repository: https://github.com/amir-rahnama/faithfulness_local_tree_ensemble_explanations/.

- The model-agnostic techniques KernelSHAP and LPI tend to be more faithful than their model-based counterparts.
- Interventional TreeSHAP is consistently more faithful than the Observational TreeSHAP across all datasets in explaining tree-based ensemble models.
- Saabas is *consistently* more faithful in explaining Random Forest models across all datasets than other model-based explanation techniques.
- The faithfulness of LIME in explaining tree-based ensemble models is consistently lower than our baselines, (global) impurity-based feature importance scores, and random explanations in numerous datasets.
- The overall agreement of local explanations positively correlates with the random forest models' accuracy and the number of features in the datasets for the gradient-boosted tree explanations.

2 Related Work

There are only a few studies that have focused on evaluating the local explanations of tree-based ensemble models. In [20], the authors proposed TreeSHAP and provided a minimal empirical evaluation. The authors concluded that TreeSHAP explanations are the most faithful for both gradient-boosted trees and random forests based on numerous variations of Importance by Deletion and Preservation measures. However, the findings in the study were limited to three medical datasets that are not publicly available.

In [32], Local MDI is evaluated against TreeSHAP and Saabas on six tabular datasets using pairwise similarity. The authors show a strong correlation between LocalMDI, TreeSHAP, and Saabas based on Spearman rank and Pearson correlation.

Other evaluation measures that are not based on perturbation, such as evaluating explanations with ground truth on synthetic data [13], and evaluating explanations with simpler white-box models [26] are available in the literature on explainable AI; however, due to the aforementioned limitations (Sect. 1), they are outside the scope of our study. An extensive survey of evaluation measures for local explanations is presented in [34].

3 Background

In this section, we first provide the notation used in this paper. In Sect. 3.2, we briefly introduce the explanation techniques used in our study. Finally, we describe the employed evaluation measures in Sects. 3.3 and 3.4.

3.1 Notation

In this study, we focus on classification problems with numerical features only. Let $X \in \mathbb{R}^{N \times M}$ be the dataset (N the number of samples and M the number of features) and $y \in \mathbb{R}^N$ be the output labels where $y_i \in \{1, ..., C\}$ and C is the total number of classes. The model $f : X \to y$ is trained on the (X, y) pairs. The vector $\Phi^c_{x,f} = [\phi_1, ..., \phi_M] \in \mathbb{R}^M$ is the local explanation of explained instance x with respect to the underlying model f for a given designated class c.

In tabular datasets, features are usually divided into important and unimportant based on their absolute importance scores from $\Phi^c_{x,f}$ given a threshold K. The set S where $|S| = K$ indicates the subset of features with the top-K largest absolute importance scores, from $\Phi^c_{x,f}$. We refer to this set as the **Top-K important features**. Let S' be the complement set of S, which indicates the subset of features with the smallest absolute scores. Note that $|S'| = M - K$.

3.2 Explanation Techniques

In this section, we briefly describe the employed explanation techniques. For full details, we refer the reader to the original studies of each explanation technique.

LIME and KernelSHAP. LIME [27] and KernelSHAP [21] explain individual instances by approximating the black-box model with an interpretable surrogate model. First, they create binary interpretable representations of the explained instance x [10]. For a sample size T, set by the user, different variations of x are generated where a random subset of interpretable features in x are nullified in each sample. The surrogate model is trained on $(x', f(x'))$ pairs. The distance between these samples x' and the original explained instance x is calculated using a kernel function k and is used as sample weights for training the surrogate model. The main difference between LIME and KernelSHAP is the choice of the kernel function k. LIME uses an exponential kernel, while KernelSHAP uses a combinatorics-based kernel function.

TreeSHAP, Saabas, and Local MDI. TreeSHAP [20] is a local model-based explanation technique that aims to explain the predicted output of tree model f by estimating $E[f(x)|x_S]$ where x_S denotes the explained instance x that only has feature values in the subset S, i.e. feature values not included in S are nullified.

TreeSHAP has access to the internal structure of the tree model. To calculate $E[f(x)|x_S]$, TreeSHAP simulates the missing features not in S by replacing their values with a baseline value. To replace these values, TreeSHAP uses two different approaches. **Interventional sampling** (I) replaces the values from a background dataset, e.g., training or validation set. In contrast, **Observational** (O) sampling, or **Tree Path Dependent**, follows the decision tree and uses the feature values from the instances in each leaf. The latter is most useful when the training data is unavailable. Our study uses both Interventional and Tree Path dependent sampling, where the former uses the training set as input.

Local MDI [32] is another model-based explanation technique that uses the structure of the tree models to allocate importance scores ϕ to features. Formally,

$\phi_j = \frac{1}{N_T} \sum_T \sum_{v(s_t)=j, x \in t} i(t) - i(t_j)$ where N_T is the number of trees in the ensemble, $v(s_t) = j$ denotes all nodes in tree t that use feature j for splitting s_t, t_j is the successor of node t in the tree and $i(.)$ is the impurity function.

Saabas is a model-based explanation technique that explains the predicted output $f(x)$ by a sum that includes the bias of each tree, c_{full} and the sum of the feature contributions $f(x) = c_{\text{full}} + \sum_{j=1}^{M} \phi_j$ where ϕ_j is a positive or negative feature importance value. In this technique, the importance scores of feature j are the magnitude in which splitting on feature j changes the predicted scores $f(x)$ from its value at the root note, c_{full} along the decision path.

LPI. Local permutation importance (LPI) [5] is a local model-agnostic explanation technique that can be considered the local variation of the (global) Permutation Importance algorithm of [4]. In LPI, instead of permuting all feature values in the dataset, each feature value in the explained instance is replaced with unique values for that feature in the dataset. As before, the change in the predicted output of the black-box model is recorded before and after this replacement.

3.3 Perturbation-Based Evaluation

Evaluations of local explanations based on perturbation are mainly performed by two measures: Importance by Deletion and Importance by Preservation. In **Importance By Deletion** (IBD) [11,23], we create instance x' from explained instance x by nullifying the K most important features from its corresponding local explanation. We then measure the change in the predicted output of the black-box model f before and after this nullification by $\epsilon = |f(x) - f(x')|$ where $f(x)$ is the predicted probability score for a designated class c. The designated class is often the predicted class for x by f. Faithful local explanations are expected to have large values of ϵ. Importance By Deletion (IBD) is sometimes referred to as Prediction Gap on Important feature perturbation (PGI) [1] or Selectivity [23] in the literature.

Importance By Preservation. (IBP) works similarly to Importance By Deletion (IDB). We create x' from the explained instance x' by nullifying the least k important features from its corresponding local explanations. We then measure the change in the predicted output of the black-box model before and after this nullification by $\epsilon = |f(x) - f(x')|$. Faithful local explanations are expected to have low values of ϵ. Importance By Preservation (IBD) is sometimes referred to as Prediction Gap on Unimportant feature perturbation (PGU) [1].

The **nullification** method for measuring Importance by Deletion and Preservation is usually performed in two ways. The first approach is to replace the values of each important (unimportant) feature with a **single** baseline value, such as the average value of that feature in the dataset. This can create a bias in cases where explained instances have feature values close to the average values in the dataset [2,25,26]. Because of this, another approach for nullification is to perform **re-sampling**. In this approach, we repeatedly replace the feature values of the nullified feature by sampling uniformly from the set of T unique values of that feature in the dataset[5]. At each step t, we obtain a new variation of the explained instance, x, which we call x'_t. We then calculate the average change in the predicted output of the model before and after this nullification process: $\epsilon = \sum_{t=1}^{T} \frac{|f(x)-f(x'_t)|}{T}$. As described before, selecting the most and least important features relies on the hyper-parameter K, i.e., the number of selected features. To make the evaluation not dependent on a specific choice of K, we marginalize this choice by progressively increasing the value of K and calculate the AUC of each measure as suggested in [16]. Naturally, we expect AUC values of Importance by Deletion (Preservation) to be large (small) for faithful explanations.

3.4 Pairwise Rank Correlation

In some cases, two local explanation techniques can generate similar explanations or, in other words, their explanations *agree* with one another [1,18]. There are numerous similarity metrics to measure this agreement. We use Spearman's rank correlation for measuring the similarity, similar to proposals of [18,26]. This is largely because different explanation techniques have different objectives, and their importance scores can have semantic differences. Using this measure, we compare the similarity of explanations only based on the features' ranks and not absolute scores.

4 Experiments

In this section, after briefly describing the setup for datasets, models and explanations in our study (Sects. 4.1 and 4.2), we present the results from comparing the local explanation techniques concerning Rank Pairwise Similarity in Sect. 4.3. Finally, the main results are discussed in Sect. 4.4.

4.1 Dataset and Models

In Table 1, the information about the datasets used in the study is presented. The table includes information about the OpenML identifier of the dataset, training and test sample sizes, the number of features, and the accuracy of our explained models. We have used the Scikit-Learn implementation of gradient-boosted tree (GB) and random forest (RF) models. The default hyper-parameters were used for both models[6].

A 30 % randomly selected hold-out set was used for testing unless a separate validation or test set was provided in the repository.

---

[5] In our study, we sample these values randomly without replacement.
[6] For gradient boosted trees, we use 100 estimators, a learning rate of 0.1, and for random forest models, we use 100 estimators, and max feature is set to "sqrt".

Table 1. Information about the datasets and models in our study. The ID column represents the ID of the dataset on the OpenML platform. Gradient-Boosted Tree (GB) and Random Forest (RF) accuracy is measured on the test set.

Dataset	ID	Training Size	Test Size	Number of Features	GB Accuracy	RF Accuracy
Banknote Auth	1462	919	453	4	0.98	0.98
Blood Transfusion	1464	501	247	4	0.79	0.79
Covertype	293	13400	6600	54	0.78	0.86
Diabetes	37	514	254	8	0.74	0.76
Electricity	151	30359	14953	7	0.83	0.88
Eye Movements	1044	7327	3609	24	0.72	0.73
Higgs	42769	13400	6600	28	0.71	0.72
House 16H	821	15265	7519	16	0.88	0.9
Ionosphere	59	235	116	34	0.9	0.92
KDD IPUMS (Small)	44124	3475	1713	20	0.89	0.88
Magic Telescope	1120	12743	6277	10	0.87	0.88
MiniBooNE	41150	13400	6600	50	0.93	0.93
Mozilla4	1046	10415	5130	4	0.94	0.95
Phoneme	1489	3620	1784	5	0.86	0.9
Pol	722	10050	4950	48	0.97	0.98
Qsar-Biodeg	1494	706	349	41	0.88	0.89
Spambase	44	3082	1519	57	0.95	0.96
Steel Plates Fault	1504	1300	641	33	1	0.99
WDBC	1510	381	188	30	0.97	0.96
Wisconsin	15	468	231	9	0.94	0.96

4.2 Explanation Techniques

The sample size for LIME and SHAP is set to 5000, and the entire training set is used as background datasets for these techniques, as suggested in the original studies and implementations. For TreeSHAP explanations, we obtain the explanations for probability scores. We generate uniformly distributed values between -1 and 1 for all features for random baseline explanations. As a second baseline, we also included global feature importance scores based on impurity in the study. We obtain local explanations for the predicted class of all explained instances in the test dataset.

4.3 Pairwise Rank Similarity

In Fig. 1, the similarity of the explanations of all test instances averaged over all datasets is provided. Note that SHAP and TreeSHAP (I and O) are more similar for Gradient Boosted Tree than for Random Forest explanations. Moreover, Interventional and Observation TreeSHAP explanations show equal agreement across both ensemble models.

In Fig. 2, we can see that the agreement between all explanations in a dataset is positively correlated with model accuracy for explanations of random forests and the number of features in a dataset for explanations of gradient-boosted trees.

4.4 Evaluation by Perturbation

In this section, we begin our analysis of the faithfulness of local explanations based on Importance by Preservation and Deletion using average as baselines in Sect. 4.4 as this is the most common approach in evaluating local explanations [1]. In Sect. 4.4, we provide the analysis based on re-sampling techniques.

Average Baseline. In Fig. 3, we show the results from measuring the Importance of the Preservation and Deletion for the Spambase dataset. As mentioned in Sect. 3.3, we progressively nullify the top-K percentage of important (unimportant) features from local explanation into its corresponding explained instance. Then, we measure the change in the predicted probability scores for that instance before and after this nullification process. In this figure, we have averaged the results of overall test instances to provide a dataset-level perspective. We expect faithful explanations to have large (small) Importance by Deletion (Preservation) values, specifically more faithful than the random baseline

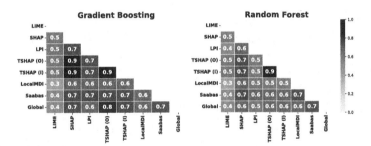

Fig. 1. Average Rank Similarity between explanations

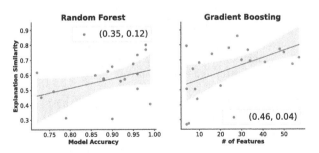

Fig. 2. Regression plots with highlighted error intervals between the average pairwise similarity of all explanations and factors such as model accuracy and the number of features in the dataset. Pearson correlation and corresponding p-values are reported in the legend.

explanations. In this example, the LIME explanation does not satisfy this expectation at some cutoff points. We calculate the AUC of these charts to provide an overview summary of this process.

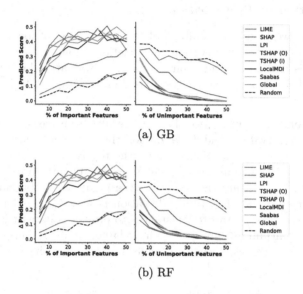

Fig. 3. Importance by Deletion (Left) and Preservation (Right) for explanations of Gradient Boosting (a) and Random Forest (b) Models. The line chart depicts the values over test instances at % cutoff of nullified features. Faithful explanations must show larger (smaller) Importance by Deletion (Preservation) values.

In Table 2, the average AUC for the Importance of Preservation and Deletion of explanations are shown[7]. SHAP provides the most faithful explanations based on both measures on average across all datasets. Even though LIME provides a local explanation and is expected to be more faithful, its results are at times less faithful than the global baseline.

Table 2. Faithfulness based on Importance by Preservation (IBP) and Importance by Deletion (IBD) averaged over all datasets. Large (small) values of Importance by Deletion (Preservation) indicate higher faithfulness.

Model	Measure	LIME	SHAP	LPI	TSHAP (O)	TSHAP (I)	LocalMDI	Saabas	Global	Random
GBT	IBP	2.45	**0.95**	0.95	1.05	0.98	1.54	1.27	1.26	4.53
	IBD	5.79	**7.53**	7.19	7.04	7.07	6.67	6.99	6.92	3.83
RF	IBP	3.31	**1.84**	2	2.31	2.35	2.21	2.01	2.35	4.83
	IBD	5.61	**6.9**	6.56	6.28	6.48	6.66	6.74	6.33	4.23

---

[7] The AUC results for each dataset can be seen in our repository.

In Figs. 4 and 5, we summarize the rank of all explanations based on their faithfulness across all datasets with boxplot visualizations. Lower ranks indicate more faithful explanations in both charts. These figures help us understand which local explanations *consistently* outperform other techniques. This consistency is a desirable attribute of a faithful explanation technique. Otherwise, explanation techniques that provide faithful explanations in one dataset can provide explanations with low faithfulness in other datasets.

These charts show that KernelSHAP is consistently the most faithful explanation across all measures for both ensemble models. After KernelSHAP, LPI provides the second most consistent faithfulness across all datasets. On the other hand, LIME explanations show very low faithfulness across all datasets for explanations of both ensemble models. LIME's faithfulness is worse than the global and random baseline explanations in many cases.

Sabaas shows more faithfulness in explaining both models among the model-based explanation techniques, except for the Importance by Preservation of Gradient Boosting explanations.

Observation TreeSHAP is less faithful than the Interventional TreeSHAP explanations based on our studied evaluation measures. Our empirical results agree with findings in [17] that interventional Shapley values are more faithful because this way of sampling breaks the relationship between the predicted output and the features before computing their feature importance. On the other hand, our empirical results disagree with findings in [9] where authors proposed that observational TreeSHAP is a more reliable approach to sampling. In this study, authors claim that Interventional sampling generates instances outside the original data distribution used in training the explained model.

We find it surprising that model-based explanations are less faithful than model-agnostic ones. These techniques have access to the internal structure of tree ensemble models' properties, such as their decision paths, which encode a lot of information about the internal logic of the prediction of tree ensemble models. Moreover, our results do not agree with the findings in [20], where the authors showed that TreeSHAP provided the most faithful explanation of Gradient Boosting and Random Forest models.

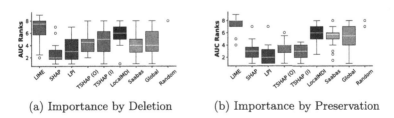

(a) Importance by Deletion   (b) Importance by Preservation

Fig. 4. Boxplots of the ranks of Gradient Boosting explanations based on Preservation and Deletion across all datasets. Lower ranks indicate higher fidelity.

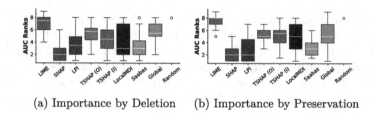

(a) Importance by Deletion (b) Importance by Preservation

Fig. 5. Boxplots of the ranks of Random Forest model explanations based on Preservation and Deletion across all datasets. Lower ranks indicate higher robustness.

Sampling Baseline. So far, to calculate Importance by Deletion and Preservation, we have used the average values as the baseline for the nullified features. In Tables 3 and 4, we show the median rank of explanations calculated based on the sampling baseline and different sampling sizes. Based on the results, LPI provides the most faithful explanations for explanations of both models. In numerous cases, except in Importance by Preservation in Gradient-Boosted Tree explanations, SHAP provides close or equally faithful explanations to LPI. The sample size does not significantly affect the median rank of explanations.

Table 3. Gradient Boosting: median rank of explanations across all datasets based on their faithfulness by Importance by Perturbation measures (with resampling baseline). Lower ranks indicate more faithfulness. Bold values indicate the most faithful explanations for each measure and corresponding sample size.

Model	Sample Size	LIME	SHAP	LPI	TSHAP (O)	TSHAP (I)	LocalMDI	Saabas	Global	Random
IBP	1	8	3	**1**	4	4	6	6	5	9
	10	8	3	**1**	4	3	6	6	5.25	9
	20	8	3	**1**	4	3	6	6	5	9
	50	7.75	3	**1**	4	3.5	6	6	5	9
IBD	1	7.25	**2**	**2**	4	4	5.5	5	5.5	9
	10	8	**2**	**2**	4	3.75	5.75	5	6	9
	20	8	2	**1.75**	4	3.75	6	5	6	9
	50	8	**2**	**2**	4	3	6.5	5	6	9

The improvement in the faithfulness of LPI based on the sampling baseline is expected and can be explained. As mentioned in Sect. 3.2, LPI replaces feature values of the explained instance with alternative unique values of that feature in the dataset to allocate importance scores. This process is similar to the way Importance By Preservation (Deletion) works, particularly when a sampling baseline is used. In other words, LPI is directly optimizing for Importance by Deletion and Perturbation measures in its internal logic.

Table 4. Random Forest: median rank of explanations across all datasets based on their faithfulness by Importance by Perturbation measures (with nullification using resampling). Lower ranks indicate more faithfulness. Bold values indicate the most faithful explanations for each measure and corresponding sample size.

Model	Sample Size	LIME	SHAP	LPI	TSHAP (O)	TSHAP (I)	LocalMDI	Saabas	Global	Random
IBP	1	7.5	2	**1**	6	5	4.5	4	7	9
	10	8	2	**1**	6	5	4.5	3	6.5	9
	20	8	2	**1**	6	5	4	3	7	9
	50	8	2	**1**	6	5	5	3	6.5	9
IBD	1	7	**2**	2.25	6	5	4.5	4	6.25	9
	10	8	2	**1**	6	4.75	3.5	3	7	9
	20	8	2	**1**	6	5	3	3.5	7	9
	50	7.5	2	**1**	6	5	3.5	4	7	9

5 Concluding Remarks

Our study empirically evaluates the faithfulness of local (feature attribution) explanations when explaining tree-based ensemble models, using Importance by Preservation and Importance by Deletion as evaluation measures. The investigation included TreeSHAP, Sabaas, Local MDI, LIME, KernelSHAP, and LPI for explaining gradient-boosted trees and random forests on 20 tabular datasets.

In the experiments, no single local explanation technique could show faithfulness across all datasets and measures. However, some local explanation techniques outperform others with more consistency. KernelSHAP was shown to be the most faithful explanation on average across measures when explaining both ensemble models, using the average baseline value for the nullification. When instead using a re-sampling-based approach to calculate the evaluation measures, LPI was shown to be the most faithful explanation on average across both measures. Across all measures, the faithfulness of LIME explanations when explaining tree ensemble models is consistently less faithful than the Global Impurity Feature importance baseline. Investigating the failures of LIME in explaining tree-based ensemble models can be a potential future study.

Given that model-based explanations have access to the internal logic of local explanations, we expected that they would provide more faithful explanations than their model-agnostic counterpart. However, our study showed empirical evidence against this. Future studies can highlight the strengths and limitations of model-based local explanations compared to local model-agnostic explanations for tree ensemble models.

The study can be further extended in several ways. In the current investigation, we considered datasets with numerical features only. This choice was primarily based on the fact that categorical variables are handled in different ways in different implementations of both tree-based ensembles and explanation techniques, e.g., using one-hot encoding vs. partition-based splits. An extension of this study could look into the effect of the handling of categorical features on faithfulness. Another direction concerns also investigating very high-dimensional datasets, e.g., gene datasets as investigated in [6]. Finally, it should be noted that functionally grounded evaluation of local explanation techniques, which has been the focus of this study, is not a replacement for human-grounded evaluation methods. Even for explanations that are considered faithful in our study, human studies are needed to evaluate these local explanations further before they can be deployed in high-stake domains.

References

1. Agarwal, C., et al.: OpenXAI: towards a transparent evaluation of model explanations. Adv. Neural. Inf. Process. Syst. **35**, 15784–15799 (2022)
2. Akhavan Rahnama, A.H.: The blame problem in evaluating local explanations and how to tackle it. In: Nowaczyk, S., et al. (eds.) Artificial Intelligence. ECAI 2023 International Workshops. ECAI 2023. Communications in Computer and Information Science, pp. 66–86. Springer Nature Switzerland, Cham (2024). https://doi.org/10.1007/978-3-031-50396-2_4
3. Alvarez-Melis, D., Jaakkola, T.S.: On the robustness of interpretability methods. In: ICML Workshop on Human Interpretability in Machine Learning (2018)
4. Breiman, L.: Random forests. Mach. Learn. **45**(1), 5–32 (2001)
5. Casalicchio, G., Molnar, C., Bischl, B.: Visualizing the feature importance for black box models. In: Berlingerio, M., Bonchi, F., Gärtner, T., Hurley, N., Ifrim, G. (eds.) Machine Learning and Knowledge Discovery in Databases: European Conference, ECML PKDD 2018, Dublin, Ireland, September 10–14, 2018, Proceedings, Part I, pp. 655–670. Springer International Publishing, Cham (2019). https://doi.org/10.1007/978-3-030-10925-7_40
6. Chen, X., Ishwaran, H.: Random forests for genomic data analysis. Genomics **99**(6), 323 329 (2012)
7. Doshi-Velez, F., Kim, B.: Towards a rigorous science of interpretable machine learning. arXiv preprint arXiv:1702.08608 (2017)
8. Fong, R.C., Vedaldi, A.: Interpretable explanations of black boxes by meaningful perturbation. In: Proceedings of the IEEE International Conference on Computer Vision, pp. 3429–3437 (2017)
9. Frye, C., de Mijolla, D., Begley, T., Cowton, L., Stanley, M., Feige, I.: Shapley explainability on the data manifold. In: The International Conference on Learning Representations (ICLR) (2021)
10. Garreau, D., Luxburg, U.: Explaining the explainer: a first theoretical analysis of lime. In: International Conference on Artificial Intelligence and Statistics, pp. 1287–1296. PMLR (2020)
11. Ghorbani, A., Abid, A., Zou, J.: Interpretation of neural networks is fragile. In: Proceedings of the AAAI Conference on Artificial Intelligence, vol. 33, pp. 3681–3688 (2019)
12. Grinsztajn, L., Oyallon, E., Varoquaux, G.: Why do tree-based models still outperform deep learning on typical tabular data? Adv. Neural. Inf. Process. Syst. **35**, 507–520 (2022)
13. Guidotti, R.: Evaluating local explanation methods on ground truth. Artif. Intell. **291**, 103428 (2021)

14. Guidotti, R., Monreale, A., Ruggieri, S., Turini, F., Giannotti, F., Pedreschi, D.: A survey of methods for explaining black box models. ACM Comput. Surv. (CSUR) **51**(5), 1–42 (2018)
15. Hsieh, C.-Y., et al.: Evaluations and methods for explanation through robustness analysis. arXiv preprint arXiv:2006.00442 (2020)
16. Hsieh, C.-Y., et al.: Evaluations and methods for explanation through robustness analysis (2021)
17. Janzing, D., Minorics, L., Blöbaum, P.: Feature relevance quantification in explainable AI: a causal problem. In: International Conference on artificial intelligence and statistics, pp. 2907–2916. PMLR (2020)
18. Krishna, S., et al.: The disagreement problem in explainable machine learning: a practitioner's perspective. arXiv preprint arXiv:2202.01602 (2022)
19. Liu, Y., Khandagale, S., White, C., Neiswanger, W.: Synthetic benchmarks for scientific research in explainable machine learning. arXiv preprint arXiv:2106.12543 (2021)
20. Lundberg, S.M., et al.: From local explanations to global understanding with explainable AI for trees. Nat. Mach. Intell. **2**(1), 56–67 (2020)
21. Lundberg, S.M., Lee, S.-I.: A unified approach to interpreting model predictions. In: Advances in Neural Information Processing Systems, vol. 30 (2017)
22. Molnar, C.: Interpretable Machine Learning, 2 edition (2022)
23. Montavon, G., Samek, W., Müller, K.-R.: Methods for interpreting and understanding deep neural networks. Digit. Signal Process. **73**, 1–15 (2018)
24. Poursabzi-Sangdeh, F., Goldstein, D.G., Hofman, J.M., Vaughan, J.W.W., Wallach, H.: Manipulating and measuring model interpretability. In: Proceedings of the 2021 CHI Conference on Human Factors in Computing Systems, pp. 1–52 (2021)
25. Akhavan Rahnama, A.H., Boström, H.: A study of data and label shift in the lime framework. In: Neurip 2019 Workshop on Human-Centric Machine Learning (2019)
26. Akhavan Rahnama, A.H., Bütepage, J., Geurts, P., Boström, H.:. Can local explanation techniques explain linear additive models? Data Min. Knowl. Discov., 1–44 (2023)
27. Ribeiro, M.T., Singh, S., Guestrin, C.: Model-agnostic interpretability of machine learning. In: ICML Workshop on Human Interpretability in Machine (2016)
28. Rudin, C.: Please stop explaining black box models for high stakes decisions. Stat **1050**, 26 (2018)
29. Saabas, A.: treeinterpreter. (2015)
30. Sharma, H., Kumar, S.: A survey on decision tree algorithms of classification in data mining. Int. J. Sci. Res. (IJSR) **5**(4), 2094–2097 (2016)
31. Sundararajan, M., Najmi, A.: The many Shapley values for model explanation. In: International Conference on Machine Learning, pp. 9269–9278. PMLR (2020)
32. Sutera, A., Louppe, G., Huynh-Thu, V.A., Wehenkel, L., Geurts, P.: From global to local mdi variable importances for random forests and when they are Shapley values. Adv. Neural Inf. Process. Syst. **34**, 3533–3543 (2021)
33. Wang, C., Han, B., Patel, B., Rudin, C.: In pursuit of interpretable, fair and accurate machine learning for criminal recidivism prediction. J. Quant. Criminol. **39**(2), 519–581 (2023)
34. Zhou, J., Gandomi, A.H., Chen, F., Holzinger, A.: Evaluating the quality of machine learning explanations: a survey on methods and metrics. Electronics **10**(5), 593 (2021)

Random Forests for Heteroscedastic Data

Hugo Bellamy[✉] and Ross D. King

University of Cambridge, Cambridge, UK
hpb32@cam.ac.uk

Abstract. Random forests are a popular machine learning technique that are effective across a range of scientific problems. We extend the standard algorithm to incorporate the uncertainty information that arises in heteroscedastic data - datasets where the amount of noise in the target value varies between datapoints. We consider datasets where the relative amount of measurement noise in different datapoints is known. This is not the standard scenario, but does commonly exist in real data, as we illustrate on 10 drug design datasets. Utilising this uncertainty information can lead to significantly better predictive performance. We introduce three random forest variations to learn from heteroscedastic data: parametric bootstrapping, weighted random forests and variable output smearing. All three can improve model performance, demonstrating the adaptability of random forests to heteroscedastic data and thus expanding their applicability. Additionally, variations in the relative performance of the three methods across datasets provides insight into the mechanisms of random forests and the purpose of the different random elements within the model.

Keywords: Random forests · Noise

1 Introduction

Since their introduction by Breiman [4], random forests have been successfully deployed across a wide range of scientific problems including drug design [5,20], agriculture [25,28], environmental modelling [13] and bioinformatics [16,23]. The popularity of random forests stems from their ability to make good predictions without the need for extensive training time or hyperparameter tuning. They can generalise well from small datasets [1,2] and are easily scalable, making them appropriate for use on large datasets [10]. This study investigates the application of random forests to regression datasets with heteroscedastic noise present in the target values - a useful extension of random forests due to their effectiveness in low signal to noise environments [17].

Random forests ensemble the predictions of individual decision trees to generate a final prediction. For the ensemble process to be useful, the individual decision trees must differ from one another; this is achieved by adding random elements into the decision trees during model fitting. Early random forests algorithms tried many sources of randomness including random feature selection, bagging [4], output smearing [3], random split selection [7] and extremely randomized trees [11]. Breiman noted that "various combinations of randomness can be added to see what works the best" [4]. Most modern implementations of random forest use the bootstrap and random feature selection as the only sources of randomness. We investigate modifying different randomisation techniques to exploit knowledge about heteroscedastic uncertainties in the data.

Heteroscedastic data are characterized by datapoints having non-constant variance. It can arise in real datasets when datapoints are measured multiple times and averaged, combined from multiple sources or when the measurement process itself introduces heteroscedastic uncertainties. Work on machine learning with heteroscedastic noise generally assumes the noise is unknown and adapts the fitting procedure to account for heteroscedasticity. Models are simultaneously fit for the mean and variance of a function and a joint loss function minimised. Examples include neural networks [27], Gaussian processes [14,15] and Bayesian decision trees [22]. Additionally, least squares regression can be generalised to work with known heteroscedastic noise in both inputs and outputs [29]. However, none of these approaches transfer easily to a random forest because they rely on modifying a loss function with a probabilistic interpretation (generally negative log likelihood of the training data) - a feature not present in random forests. This makes adapting random forests for heteroscedastic data a more difficult problem because the tree building procedure itself must be adapted. Hence, we investigate using random forests when uncertainty information is known; aiming to find effective ways of including uncertainty information in the random forest algorithm without the additional complexity of simultaneously modelling variance. These methods are useful on their own, as we demonstrate on drug design data, and can serve as a starting point for future work on generalising random forests to the unknown noise scenario.

We introduce three methods for incorporating uncertainty information into random forests.

1. **Parametric bootstrap** - In the bootstrap step of the random forest, select samples with lower variance measurements more often.
2. **Weighted random forest** - Build a random forest with weighted decision trees [12]. These weight the splits of the decision tree so datapoints with lower variance have greater influence over the splits.
3. **Variable Output smearing** - Output smearing is an alternative randomisation method in a random forest that adds Gaussian noise to each sample. With variable output smearing, high variance datapoints are smeared more.

The remainder of this paper looks at how these methods affect performance on noisy datasets. We focus on regression problems but both the parametric bootstrap and weighted random forest methods can be applied to classification problems. Section 2 gives an overview on the related literature. In Sect. 3 we introduce each of our three methods in detail. In Sect. 4, we consider the problem of method selection by comparing optimised versions of each method on simulated and real datasets. The work finishes with a discussion of the results and how they can be understood within the context of the random forest algorithm (Sect. 5).

2 Related Work

We are not aware of any previous study investigating the use of random forests with heteroscedastic noise in the outputs. Shy et al. [26] used random forests for classification problems with known heteroscedastic noise present in the features. The method they used, variable input smearing, mirrors the variable output smearing we implement but applies the smearing to the features rather than the target values. Shy et al. noted that the addition of noise in this way is a general technique and can be applied to any machine learning method, even those not typically ensembled. While variable input smearing did not improve prediction accuracy, it allowed for better quantification of prediction uncertainty. An approach very similar to input smearing was effectively applied to missing data imputation by Ramosaj and Pauly [24].

Hashemi and Karimi [12] introduced weighted versions of various machine learning methods including a weighted version of a decision tree. On all the methods they used the weighted versions were more effective on a classification task with known heteroscedastic uncertainties. We extend their weighted decision trees to build a weighted random forest.

A related method for learning with heteroscedastic noise is multi-fidelity modelling, which divides data based on quality, usually a small high-fidelity set and a large low fidelity set. Typically, these methods combine a low-fidelity model with a high-fidelity model to maximize data utility [8]. While these methods are related to the problems we consider, they do not generally extend to the case of a continuous quality measure.

3 Methods

3.1 Parametric Bootstrap

In a random forest, individual trees are built using a bootstrap sample of the original data. The bootstrap chooses n samples with replacement, usually the value of n is chosen to be the size of the original dataset, but smaller values can also be used. The parametric bootstrap instead selects samples (with replacement) from a specified probability distribution.

We assume that for each sample we have a set of features (x_i), target value (y_i) and variance (σ_i^2). To use the parametric bootstrap with data of this form we convert the variances to an unnormalised probability distribution using:

$$z_i = e^{-\alpha \sigma_i^2} \tag{1}$$

where z_i is the unnormalised probability and α is a hyperparameter that controls how much more often low variance points are sampled than high variance points. The parametric bootstrap then selects points with probability proportional to z_i. We use this method for generating probabilities because it can tune the relatively probabilities based on α and if $\alpha = 0$ it becomes a standard random forest. Additionally, it can work with any measure of datapoint quality instead of the variance. We tested this method on a simulated dataset with heteroscedastic noise following:

$$y_i = N(f(x_i), \sigma_i^2) \qquad (2)$$

The function, $f(x_i)$, used was *make_friedman1* from the sci-kit learn package for python [9,21]; the values of σ_i^2 were generated from a uniform distribution between 0 and a fixed maximum noise value equal to $2\gamma \text{var}(y)$, here γ sets the ratio between the mean of the noise added to a datapoint and the variance of the noise free target values. Models were built on a training set containing 200 samples and tested using a set of 2000 noise free samples. Figure 1 shows how the coefficient of determination (R^2) for a random forest with parametric bootstrap changes as the hyperparameter α is varied, using the dataset we have described with $\gamma=3$. We use the coefficient of determination on the test set as a performance metric throughout this study because it is easily interpretable and informative of model performance [6] (except in Table 2, where we use mean squared error, as here the mean squared error allows for a comparison of how additional noise effects model performance).

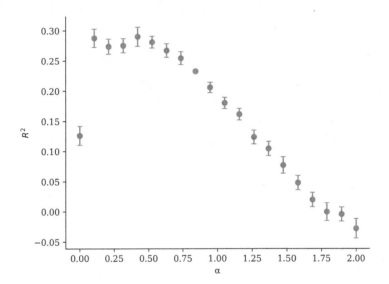

Fig. 1. Coefficient of determination with different α values for a random forest with a parametric bootstrap. Experiments run on a simulated dataset containing known heteroscedastic noise, results show the mean and standard deviation of 10 runs.

The parametric bootstrap with $\alpha < 1.25$ outperforms a standard bootstrap ($\alpha=0$) on this dataset. The optimal value on this dataset is around 0.25 and leads to a large increase in the coefficient of determination. Exploring the method further we repeated the experiment with different amounts of noise present (γ values) in the training data. These results are shown in Fig. 2, with the left hand axis showing the best R^2 value at each noise level and the right hand axis showing the corresponding α values for these datapoints.

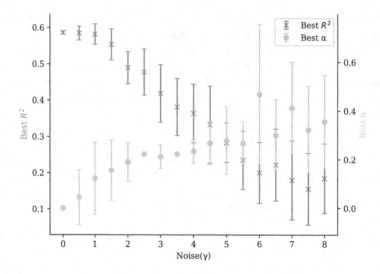

Fig. 2. Coefficient of determination and the corresponding best α values for random forest with a parametric bootstrap with different levels of noise(γ) present. Experiments run on simulated datasets containing known heteroscedastic noise, results show the mean and standard deviation of 10 runs.

As the amount of noise increases the optimal value of α increases - with more noise present greater variability in the sampling probabilities of points is required to get the best performance. Additionally, with more noise present the model performance decreases. The parametric bootstrap was useful even the smallest amount of heteroscedastic noise tested. More results for the parametric bootstrap and a full comparison to other methods are given in Sect. 4.1.

3.2 Weighting Splits

A weighted random forest is built using weighted decision trees [12], which differ from standard decision trees by weighting samples when choosing splits and assigning leaf values. In our experiments, by growing all trees until each leaf contains only a single sample, we look only at the effect of weighting splits and not of weighting leaves. This decision is discussed in Sect. 5.2. We used the z_i values from Eq. 1 as weights. Figure 3 shows the effect of changing α on the effectiveness of weighting splits using the same simulated dataset as used for Fig. 1.

Figure 3 shows that introducing weighting improves performance initially but further increasing α above small values gives only small changes in performance. Following our experiments with the parametric bootstrap, we tested weighted

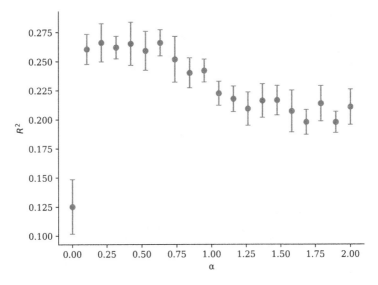

Fig. 3. Coefficient of determination with different α values for random forest with weighted splits. Experiments run on a simulated dataset containing known heteroscedastic noise, results show the mean and standard deviation of 10 runs.

random forests with different amounts of noise present and plotted how the best value of α and the resulting R^2 changed, these results are shown in Fig. 4. The results follow the same pattern as for the parametric bootstrap; as the amount of noise increases the best value of α increases and the resulting R^2 decreases. On a weighted random forest the increase in best α value was rapid around $\gamma = 3$ and flatter elsewhere, as opposed to the parametric bootstrap where the best α value increased gradually. Additionally, the best α values were larger at higher noise levels with a weighted random forest.

3.3 Variable Output Smearing

In our experiments variable output smearing was used as a replacement to random feature selection. It adds random noise to the outputs during fitting of the decision trees. This means the bootstrap is still used and every tree uses perturbed values for the training data target values. These perturbations follow a normal distribution, the target value used for sample i in tree j is sampled from:

$$y_{i,j} \sim N(\bar{y}_i, \frac{\beta}{z_i}) \qquad (3)$$

where $y_{i,j}$ is the value of training point i used in tree j, $\bar{y}_i$ is the observed value of training point i, β is a hyperparameter that controls the amount of output smearing, z_i values are from Eq. 1. Any quality measure can be used and then appropriate values for α (Eq. 1) and β chosen. It can be easier to replace β using:

$$\beta = \beta_2 var(y) \frac{1}{n} \Sigma_i(z_i) \qquad (4)$$

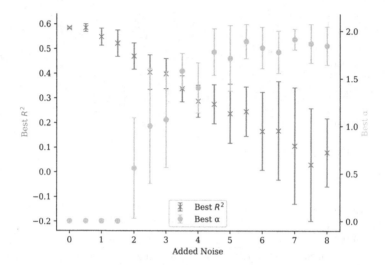

Fig. 4. Coefficient of determination and the corresponding best α values for random forest with weighted splits with different levels of noise(γ) present. Experiments run on simulated datasets containing known heteroscedastic noise, results show the mean and standard deviation of 10 runs.

Where $var(y)$ is the variance of the target values in the training set and n the number of points in the training set. Here β_2 controls the ratio of the mean variance of added noise to the variance of the target values in the original dataset. This approach is used in our code to make selecting values of β easier because it causes the optimal values to be of similar magnitudes across all datasets.

On the dataset from Figs. 1 and 3 but with less noise (we used γ=1 instead of γ=3), we tried variable output smearing with a fixed β value ($\beta = 2$) and varying α. These results are shown in Fig. 5. On this dataset, with uniform noise, variable output smearing makes the model performance worse. However, variable output smearing can improve model performance in some cases. As an example Fig. 6 shows the results for an experiment on the same dataset where noise is distributed as a function of the y value(input dependent noise):

$$\sigma_i^2 \propto \begin{cases} 1 & \text{if } 0.6 < X[0] < 1 and 0.2 < X[1] < 0.6 \\ 5 & \text{otherwise} \end{cases} \quad (5)$$

The values of σ_i^2 were normalised so that the mean amount of added noise was controlled by a parameter γ, as in the case of adding uniform noise.

When noise is a function of the feature values output smearing is useful. Performance improves for all non zero values of α. This shows the distribution of noise is an important factor to determine if variable output smearing is useful. The next section explores this further by running hyperparameter optimisation with both noise distributions on two datasets at multiple noise levels.

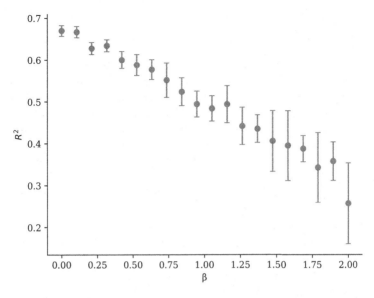

Fig. 5. Coefficient of determination with different α values for random forest with variable output smearing. Experiments run on a simulated dataset containing known heteroscedastic noise, results show the mean and standard deviation of 10 runs.

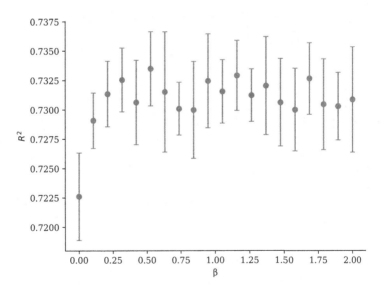

Fig. 6. Coefficient of determination with different α values for random forest with variable output smearing using a dataset where the amount of noise is a function of X. Experiments run on a simulated dataset containing known, output dependent, heteroscedastic noise, results show the mean and standard deviation of 10 runs.

4 Results

4.1 Model Selection

We used grid search cross validation to search for the most effective model hyperparameters for datasets with both uniformly distributed heteroscedastic noise and output dependant noise (Eq. 4). Four random forest models are compared; a standard random forest, parametric bootstrap, weighted random forests and variable output smearing. Additionally, on the linear simulated dataset, we compare to weighted and unweighted ridge regression because weighted ridge regression it is an established method that can use known heteroscedacity information [18,29]. The methods and the corresponding hyperparameters that were optimised are detailed in Table 1. Mtry is the fraction of the features that are considered at each split, the bootstrap size is the size of the bootstrap sample as a fraction of the total number of training points. The α values are those from Eq. 1, for variable output smearing the β value refers to the β from Eq. 3.

Table 1. Methods tested in hyperparameter optimisation and the hyperparameters optimised in the search.

Method	Hyperparameters Optimised
standard	mtry and bootstrap size
parametric bootstrap	mtry, bootstrap size and α
weighted	mtry, bootstrap size and α
variable output smearing	bootstrap size, α and β
least squares	regularisation parameter
weighted least squares	regularisation parameter

Here we give the results for 8 datasets, corresponding to two different functions from the scikit learn package [21](*make_friedman1* [9] and *make_regression*), at two different levels of noise ($\gamma = 1$ and $\gamma = 3$) and the two different noise generation mechanisms - uniform and output dependent, as described in sections 3–5. The *friedman1* dataset was used following Breimans original random forest paper [4] and the linear *make_regression* dataset was used to give a best case scenario for the linear model baseline. The use of simple functions should not effect our results otherwise because random forests have been shown to have similar performance across regression functions if the signal to noise ratio is constant [17,30].

We report the result on a withheld test-set after training the model using the best hyperparameters from the corresponding cross validation. The performance of the optimised models is given in Table 2. We use a paired t-test to test the alternative hypothesis that the new methods have a lower squared error on the test set than a standard random forest. In the Table we mark all values that are significant at the 5% significance level.

Table 2. Mean squared error of the random forest variations after hyperparameter optimisation. Results are shown for two different generating functions (friedman1 and a linear model), two different distributions of noise and two different noise levels. Values marked with an asterisk are those that are significantly better than a standard forest at a 1% significance level. The best results in each row are printed in bold.

dataset	noise type	γ	Model standard	parametric bootstrap	weighted	output smearing	weighted least squares	least squares
friedman1	uniform	1	6.60	5.07*	5.81*	5.85*	-	-
		3	14.5	7.45*	11.6*	11.3*	-	-
	input dep.	1	11.4	8.50*	9.88*	11.5	-	-
		3	23.3	14.9*	19.2*	27.1	-	-
Linear	uniform	1	0.47	0.47	0.47	0.46*	0.66	0.87
		3	1.07	1.07	1.08	1.06	10.0	13.4
	input dep.	1	0.55	0.45*	0.44*	0.56	1.22	2.91
		3	0.88	0.66*	0.67*	0.83	4.79	6.49

The parametric bootstrap is the most consistent method, performing significantly better than a standard random forest on 6 of the 8 datasets and was the best method on 4 datasets. Weighted least squares performed better than least squares but worse than the random forests, particularly at the higher noise level. This is due to the effectiveness of random forests on noisy datasets.

4.2 Real Datasets

We collected drug activity datasets from PubChem [19] which contained heteroscedastic noise in the form of repeat measurements. The number of repeat measurements in these datasets varies because we combine the results from screening assays - where large number of molecules are tested once at a single drug concentration - and confirmatory assays - where a much smaller number of molecules are tested at a range of concentrations. Combining these gives a datasets where datapoints have different numbers of measurements because the confirmatory assays repeat the concentration from the initial screening. We use the reciprocal of the number of measurements as a confidence metric for each point because this is proportional to the variance of the uncertainty in the mean of the measurements, assuming the measurement error is unbiased with constant variance. Dataset details are given in Table 3.

The target value for machine learning was the measured variable in the experiments - usually % inhibition of the target protein. Chemicals were represented using 1024 bit Morgan fingerprints with radius 2 generated using the RdKit python package. We used 10% of the dataset as a training set and 10% as a test set. Splits were made ensuring that the proportion of molecules with repeat measurements was equal to the average in the entire dataset (training and test sets each contained 10% of the molecules with repeat measurements).

Table 4 shows the results. We again used a paired t-test to test if the average squared error on the random forest variations was lower than the results with a

Table 3. Sources for real datasets. We give the PubChem aid for the relevant bioassays and give the original source that provided the data to PubChem. Data can be accessed online at "https://pubchem.ncbi.nlm.nih.gov/bioassay/" + AID. The datasets were originally from the Scripps Research Institute Molecular Screening Center (Scripps) or the Burnham Center for Chemical Genomics(Burnham).

Dataset name	Screening AID	Confirmitory AID	Orginal Source
K7	652039	686949	Scripps
DAX1	652010	652134	Scripps
ADAM10	720582	743254	Scripps
PLC	720704	743261	Scripps
TAAR1	624467	651783	Scripps
PAFAH	492953	493034	Scripps
scp1	493091	540281	Burnham
Gli-SUFU	588413	602428	Burnham
CRF-R2	588743	602473	Burnham
EBI2	651636	651997	Burnham
PFG	504690	504753	Burnham

standard random forest. Models that are significant at the 5% significance level are marked with an asterisk.

Use of the parametric bootstrap leads to a significant improvement on 6 of the 11 datasets tested and the weighted random forest gave a significant improvement of 3 of the datasets. On the datasets where there is not a significant performance improvement these methods generally perform as well or slightly better than standard. The R^2 for both these methods was either better than or within 0.001 of standard on all datasets. Output smearing does not generally perform well compared to the other methods on these real datasets and is sometimes worse than a standard random forest.

The number of repeat measurements varied between 0.02% and 2.5% of the total number of datapoints. Of the 3 datasets where this was under 1%, only 1 had a significant improvement from our new methods. The other 8 datasets were all over 2% repeat measurments and of these 6 had a significant improvement. Overall, our results demonstrate that incorporating the heteroscedacity information that exists in real datasets can give significant performance improvements and, if using the parametric or weighted methods, is unlikely to make predictions worse.

Table 4. Coefficient of determination of random forest models on real datasets. Models that had a significantly lower mean squared error than the standard random forest algorithm and marked with an asterisk.

Dataset Name	Standard	Parametric	Weighted	Output Smearing
K7	0.119	0.159*	0.117	0.104
DAX1	0.078	0.078	0.077	0.052
ADAM10	0.309	0.310	0.310	0.295
PLC	0.328	0.337*	0.337*	0.332
TAAR1	0.077	0.086*	0.076	0.078
PAFAH	0.148	0.160*	0.149	0.145
scp1	0.093	0.094	0.092	0.089
Gli-SUFU	0.131	0.132	0.131	0.132
CRF-R2	0.119	0.122	0.125*	0.107
EBI2	0.098	0.106*	0.101*	0.102*
PFG	0.136	0.141*	0.136	0.135

5 Discussion

5.1 Weighting Methods

A regression random forest makes prediction by averaging the values of each decision tree in the forest. These trees values are the mean value of the training points in the corresponding node. This means that each prediction is a weighted average of all training points. A random forest prediction can therefore be rewritten as:

$$y_{pred} = \sum_i w_i y_i \qquad (6)$$

where y_{pred} is the new prediction and w_i is the weight corresponding to training set sample y_i. These weights are not calculated directly by the random forest algorithm, but they could be found for any test point using the fitted model.

The parametric bootstrap affects the model in two ways. Firstly, it decreases the weights of datapoints with higher variance by making them less likely to appear in the original bootstrap sample. If a point is not in the bootstrap sample, it cannot be in any nodes for that tree - this decreases the average weights of higher variance datapoints. Secondly, the parametric bootstrap will also affect how the model splits occur. Because it is likely that in each bootstrap sample low variance points are sampled multiple times and high variance points are likely to only be sample once (or not at all), the splits will be weighted towards what is best for datapoints with lower variance. The combination of these two factors reduces the impact of high variance data whilst still allowing them to impact the model. Proper tuning of the hyperparameter α can control this extent of this impact and allow for better predictions.

A weighted random forest weights the splits directly and then weights points within a node when making the trees prediction. In the version of random forests we used, each leaf node contained only one sample. So, the within node weighting had no effect and the weighted random forest cannot reduce the weights of more noisy points. This feature can prevent points with very large uncertainties from affecting predictions.

5.2 Why Weight Before?

The parametric bootstrap and weighted random forests are similar methods as described above. Our implementation of weighted random forests is disadvantaged because it cannot weight the sample node as we grow trees to full depth. This is standard with many random forest packages and for many real datasets pruning does not improve forest performance [30]. To get the full advantage of weighted random forests a pruning hyperparameter would have to be optimised simultaneously, making hyperparameter selection more complicated. Additionally, whilst a weighted random forest chooses splits using the full data distribution, the weighting in the leaf nodes (if they contain multiple points) would be a relative weighting with the other points that share the leaf node. If the node contained a mix of qualities, this weighting effect would be similar to the parametric bootstrap. But if the weights of points in a node are similar the weights will only make a small difference when applied. Conversely, the parametric bootstrap weights the points directly but the weights used in the splits are only an approximation to the input weights.

Both weighted random forests and the parametric bootstrap act to decrease the weights of high variance points in the final prediction and to limit the effect that high variance points have on splits. Whichever one of these two things is done first makes the second more difficult, but the parametric bootstrap can achieve both effects with one less hyperparameter. Further work could look for a method that can combine these two effects without any downside.

5.3 Output Smearing

Figures 5 and 6 show that the effectiveness of variable output smearing can vary as the distribution of noise changes. This could be explained be considering that the randomness added to a random forest acts as regularisation [17]. Variable output smearing means higher variance datapoints are smeared more. If noise is uniformly distributed, these high variance points will be uniformly distributed and these points will then be the main source of variation between trees. It is relatively likely that these high variance points are outliers to the rest of the dataset because they are smeared the most. As a squared error criterion is used to decide on splits, the high variance datapoints will have a large influence on where the data splits. This is the opposite of what the parametric bootstrap and weighted methods do as it is gives high variance datapoints more influence on the model, the result of this is seen in Fig. 5, where increasing α decreases performance.

In the output dependent noise case the high variance points are grouped. With variable output smearing the splitting in these high variance areas is much more random than in low variance areas of the data. Output smearing could be adaptively regularising the model. Giving improved performance because the data is fit more accurately in low variance regions without overfitting in the high variance regions. However, the performance improvement is small and the use of variable output smearing was generally less effective than the other random forest variations.

6 Conclusions

We have introduced three methods for dealing with known heteroscedastic noise in random forests and have demonstrated their effectiveness on both simulated and real datasets. The most consistent method was the parametric bootstrap, which performed significantly better than a standard random forest on 6 out of 8 simulated datasets and 6 out of 11 real datasets. Similarly, weighted random forests can improve performance on heteroscedastic data, but across our experiments, they were less effective than the parametric bootstrap. Both of these methods generally improve performance and, even when they do not enhance performance, they do not perform worse than a standard random forest. Variable output smearing can be useful, but its effectiveness depends on the distribution of noise. Additionally, on some datasets, it performed worse than a standard random forest. Overall, our results demonstrate that random forests can be adapted to exploit uncertainty information, improving predictive performance.

The main limitation of our approach is that it requires data where the relative quality of each datapoint is known beforehand which, as discussed in the introduction, is uncommon. Additionally, using the method requires the user to select an additional parameter. Unlike the other parameters in a random forest this value needs be carefully tuned based on the dataset and cannot be left at default values. Further work could look at applying these methods to more real datasets, particularly in fields where random forests are already among the best performing models. Careful analysis of data from different domains may reveal known heteroscedacity information being available. Other avenues for further research are to extend our approach to the more general case of unknown heteroscedastic noise or to explore further variations of the methods themselves; for examples, removing the default of growing trees to full depth or trying to combine the parametric bootstrap within a weighted random forest.

Acknowledgments. The authors thank BASF for providing the funding that allowed this work to be completed (grant number G113369).

Code Availability. The full code for this work is available on github.

Disclosure of Interests. There are no competing interests.

References

1. Ali, J., Khan, R., Ahmad, N., Maqsood, I.: Random forests and decision trees. Int. J. Comput. Sci. Issues (IJCSI) **9**(5), 272 (2012)
2. Biau, G., Scornet, E.: A random forest guided tour. TEST **25**, 197–227 (2016)
3. Breiman, L.: Randomizing outputs to increase prediction accuracy. Mach. Learn. **40**, 229–242 (2000). https://doi.org/10.1023/A:1007682208299
4. Breiman, L.: Random forests. Mach. Learn. **45**, 5–32 (2001). https://doi.org/10.1023/A:1010933404324
5. Cano, G., et al.: Automatic selection of molecular descriptors using random forest: application to drug discovery. Expert Syst. Appl. **72**, 151–159 (2017)
6. Chicco, D., Warrens, M.J., Jurman, G.: The coefficient of determination r-squared is more informative than SMAPE, MAE, MAPE, MSE and RMSE in regression analysis evaluation. PeerJ Comput. Sci. **7**, e623 (2021). https://doi.org/10.7717/peerj-cs.623
7. Dietterich, T.G.: Ensemble methods in machine learning. In: Multiple Classifier Systems, pp. 1–15. Springer, Berlin, Heidelberg (2000). https://doi.org/10.1007/3-540-45014-9_1
8. Fernández-Godino, M.G., Park, C., Kim, N.H., Haftka, R.T.: Review of multifidelity models. arXiv preprint arXiv:1609.07196 (2016). https://doi.org/10.48550/arXiv.1609.07196
9. Friedman, J.H.: Multivariate adaptive regression splines. Ann. Stat. **19**(1), 1–67 (1991). https://doi.org/10.1214/aos/1176347963
10. Genuer, R., Poggi, J.M., Tuleau-Malot, C., Villa-Vialaneix, N.: Random forests for big data. Big Data Res. **9**, 28–46 (2017)
11. Geurts, P., Ernst, D., Wehenkel, L.: Extremely randomized trees. Mach. Learn. **63**, 3–42 (2006)
12. Hashemi, M., Karimi, H.A.: Weighted machine learning. Stat. Optim. Inf. Comput. **6**, 497–525 (2018). https://doi.org/10.19139/soic.v6i4.479
13. He, S., Wu, J., Wang, D., He, X.: Predictive modeling of groundwater nitrate pollution and evaluating its main impact factors using random forest. Chemosphere **290**, 133388 (2022)
14. Kersting, K., Plagemann, C., Pfaff, P., Burgard, W.: Most likely heteroscedastic gaussian process regression. In: Proceedings of the 24th international conference on Machine learning, pp. 393–400 (2007). https://doi.org/10.1145/1273496.1273546
15. Liu, H., Ong, Y.S., Cai, J.: Large-scale heteroscedastic regression via gaussian process. IEEE Trans. Neural Netw. Learn. Syst. **32**(2), 708–721 (2020)
16. Mehrmohamadi, M., Mentch, L.K., Clark, A.G., Locasale, J.W.: Integrative modelling of tumour DNA methylation quantifies the contribution of metabolism. Nat. Commun. **7**(1), 13666 (2016)
17. Mentch, L., Zhou, S.: Randomization as regularization: a degrees of freedom explanation for random forest success. J. Mach. Learn. Res. **21**(1), 6918–6953 (2020)
18. Montgomery, D.C., Peck, E.A., Vining, G.G.: Introduction to linear regression analysis. John Wiley & Sons (2021)
19. National Center for Biotechnology Information: Pubchem bioassay record for aid 651739. https://pubchem.ncbi.nlm.nih.gov/bioassay/651739 (2023). Accessed 10 Oct 2023
20. Olier, I., et al.: Meta-QSAR: a large-scale application of meta-learning to drug design and discovery. Mach. Learn. **107**, 285–311 (2018)

21. Pedregosa, F., et al.: Scikit-learn: machine learning in python. J. Mach. Learn. Res. **12**, 2825–2830 (2011)
22. Pratola, M.T., Chipman, H.A., George, E.I., McCulloch, R.E.: Heteroscedastic BART via multiplicative regression trees. J. Comput. Graph. Stat. **29**(2), 405–417 (2020)
23. Qi, Y.: Random forest for bioinformatics. Ensemble Mach. Learn. Methods Appl., 307–323 (2012)
24. Ramosaj, B., Pauly, M.: Who wins the miss contest for imputation methods? Our vote for miss BooPF. arXiv preprint arXiv:1711.11394 (2017)
25. Sakamoto, T.: Incorporating environmental variables into a MODIS-based crop yield estimation method for united states corn and soybeans through the use of a random forest regression algorithm. ISPRS J. Photogramm. Remote. Sens. **160**, 208–228 (2020)
26. Shy, S., Tak, H., Feigelson, E.D., Timlin, J.D., Babu, G.J.: Incorporating measurement error in astronomical object classification. Astron. J. **164**(1), 6 (2022). https://doi.org/10.3847/1538-3881/ac6e64
27. Stirn, A., Wessels, H., Schertzer, M., Pereira, L., Sanjana, N., Knowles, D.: Faithful heteroscedastic regression with neural networks. In: International Conference on Artificial Intelligence and Statistics, pp. 5593–5613. PMLR (2023)
28. Tariq, A., Yan, J., Gagnon, A.S., Riaz Khan, M., Mumtaz, F.: Mapping of cropland, cropping patterns and crop types by combining optical remote sensing images with decision tree classifier and random forest. Geo-Spatial Inf. Sci. **26**(3), 302–320 (2023)
29. York, D.: Least squares fitting of a straight line with correlated errors. Earth Planet. Sci. Lett. **5**, 320–324 (1968)
30. Zhou, S., Mentch, L.: Trees, forests, chickens, and eggs: when and why to prune trees in a random forest. Stat. Anal. Data Min. ASA Data Sci. J. **16**(1), 45–64 (2023)

Learning Deep Rule Concepts as Alternating Boolean Pattern Trees

Florian Beck[1(✉)], Johannes Fürnkranz[1,2], and Van Quoc Phuong Huynh[2]

[1] LIT Artificial Intelligence Lab, Linz, Austria
{juffi,fbeck}@faw.jku.at
[2] Institute for Application-oriented Knowledge Processing (FAW), Johannes Kepler University Linz, Altenberger Straße 66b/69, 4040 Linz, Austria
vqphuynh@faw.jku.at

Abstract. In classification problems, trees are mostly known in the form of decision trees. In the simplest case, the classifier consists of a single tree whereby each class can be described as a set of paths leading to its prediction—usually depicted as a set of conjunctive rules. An alternative usage are pattern trees, consisting of logical operators as inner nodes and input features as leaf nodes. Each such tree thus encodes a Boolean function on the input features. While pattern trees were introduced a few years ago in a fuzzy context, we propose a greedy algorithm learning pattern trees limited to alternating the Boolean operators AND and OR, which can also be transformed into compact deep rule concepts. Our experiments on UCI data sets indicate that the learned alternating Boolean pattern trees achieve similar performance as state-of-the-art rule learners. In particular on categorical data, they outperform fuzzy pattern trees in both accuracy and efficiency, which, on the other hand, have more effective operators for dealing with numeric attributes. We also demonstrate how these pattern trees may be interpreted as deep rule sets, which do not directly link input features to a prediction but do so via automatically formed intermediate concepts.

Keywords: pattern trees · constructive induction · rule learning

1 Introduction

While most of the current work in inductive rule learning focuses on learning conjunctive rules which are then directly combined to predictive rule models [11], recent work in this area has also focused on learning deeper structures, which are able to define intermediate concepts in very much the same way the hidden layers in a neural network are able to develop subconcepts that are helpful for the overall classification task. Algorithms that follow this line of work either try to optimize a given network structure [3, 19], or form a network by iterative, layer-wise application of a rule learning algorithm [4]. In this work, we investigate an alternative approach that aims at constructing a Boolean network node by node.

In particular, we consider an algorithm for the top-down construction of alternating Boolean pattern trees. Thus, instead of learning local and therefore special rule concepts that are combined into a global model, we start with a global model and refine it in every iteration by expanding the pattern tree. It is important to note that an expansion of the pattern tree does not necessarily correspond to a specialization of the model, unlike in decision tree and rule learners, where only conjunctions are allowed. Instead, in every iteration, both specializations (conjunctions) and generalizations (disjunctions) of the pattern tree are evaluated, resulting in an more flexible learning process than in a layer-wise construction of a network, while still maintaining the strictly alternating structure of conjunctive and disjunctive layers, which is needed for a rule-based interpretation of the model. Furthermore, the proposed top-down approach is an anytime algorithm, which provides a valid pattern tree after every iteration. The granularity of the model can be chosen depending on the available computing power or time. Finally, the resulting pattern trees are more compact, whereas in a layer-wise, bottom-up approach many concepts defined in earlier layers turn out to be redundant or irrelevant so that they are not used in subsequent layers, thereby bloating the learned model.

We start our discussion with a brief summary of relevant related work, most notably in deep rule learning Sect. 2, before we introduce alternating Boolean pattern trees and our iterative algorithm for learning them Sect. 3. Section 4 describes an extensive experimental evaluation of algorithms, which allows us to draw several conclusions, which are summarized in Sect. 5.

2 Deep Rule Learning and Related Work

Current rule learning algorithms are limited to learning conjunctive rules that directly map the input features to the desired outcome. Multiple individual rules can be combined into rule models—typically rule sets [16,18], decision lists [1,23], or hybrids [6]—which are sufficiently expressive to encode any Boolean function. While traditional algorithms focus on learning individual rules and combining them later, typically using a variant of the covering algorithm [9], modern algorithms often integrate both via sequential loss minimization [7,22] or directly aim at learning rule sets or lists [25,26].

Sets of conjunctive rules may be combined into a single logical expression which corresponds to a logical formula in disjunctive normal form (DNF). In previous work [3], we related such DNF expressions to shallow neural networks. The connections between the input and the hidden layer are conjunctive (the rule bodies), whereas the connections between the hidden and the output layer are disjunctive (one rule in the set has to fire in order to make a positive prediction). The fact that every logical formula can be represented as such a DNF expression and thus as a shallow rule model can be related to the universal approximation theorem [12] for neural networks, which analogously states that, in principle, a single hidden layer is sufficient to encode any continuous function. Nevertheless, we have seen that deep neural networks have many practical advantages [17].

For this reason, work in learning deep logical structures has received some attention. Work in this area can be roughly divided into two different approaches:

Neural Network Approach: This line of work draws upon the idea of optimizing the parameters of a fixed (logical) network structure. Binary or ternary neural networks [20], which train a neural network and then round the weights to values 1, 0, and/or -1 may be viewed in this context, even though they are typically not interpreted as logical expressions. In previous work [3], we attempted to optimize Boolean networks with successive conjunctive and disjunctive layers using a randomized gradient ascent algorithm, and were able to demonstrate the utility of deeper structures in artificial datasets, even though their advantage on conventional benchmark data sets was considerably less pronounced. An interesting variant is DIFFLOGIC [19], which does not consider the weights as the parameters of the network, but instead aims at finding the best Boolean operator in each node of a binary network.

Rule Learning Approach: An alternative approach has been taken in [4], where a deep rule network is constructed with a conventional rule learning algorithm [14], by learning alternative conjunctive and disjunctive layers in a framework that allows to flexibly exploit the duality of learning DNF and CNF expressions. In this way, the network structure does not have to be pre-defined, but can be constructed during learning.

In this work, we take a different route. While the rule learning approach effectively forms a deep rule network layer-by-layer, we aim at constructing the network node-by-node, by incrementally forming so-called pattern trees, as discussed below in Sect. 3. These are analogous to fuzzy pattern trees, which have been proposed in [13] and refined in [24]. However, while fuzzy pattern trees excel in numerical data and also include special operators such as averaging, we restrict our trees to combinations of the crisp logical operators AND and/or OR. Also, algorithmically, our approach differs from the top-down construction proposed in [24] where the pattern trees are only refined at the leafs, whereas our algorithm also allows the tree to grow at the root and at interior nodes.

Note that our work aims at learning in propositional logic only, unlike many of the neuro-symbolic approaches which aim at combining representations in first-order logic with neural network architectures. Our work is thus also related to Boolean function synthesis [5], where the goal is to derive a Boolean function from its desired input/output behavior. However, these works have a stronger focus on finding compact representations, and typically do not focus on generalization to unseen data [21]. A notable exception is the work of Aoga et al. [2] on Boolean circuits, where constraint programming techniques are applied on machine learning problems. However, their results are not directly comparable to ours because they use slightly modified versions of the UCI data sets (e.g. treating multi-class problems as binary classification problems).

3 Alternating Boolean Pattern Trees

In this section, an iterative method to learn alternating Boolean pattern trees is presented, and its operation is demonstrated on a small example data set.

3.1 Theory and Iterative Creation

Given is a standard supervised learning task with instances $x \in \mathbb{X} = \mathbb{X}_1 \times \mathbb{X}_2 \times \cdots \times \mathbb{X}_m$ consisting of features of m attributes, and classes $y \in \mathbb{Y}$, combined to training examples $(x, y) \in \mathbb{X} \times \mathbb{Y}$.

To apply any Boolean operators on features x_i, $i \in \{1 \ldots m\}$, it is required that these are binary. Thus, any numeric attributes have to be discretized and any nominal attributes converted to attribute-value-pairs in a pre-processing step. In the following, we denote the resulting binary features with letters $\{a, b, c, d, e \ldots\}$ for easier readability. Similarly, we will also typically operate in a concept learning setting, where we aim at learning the definition of a concept from positive and negative training examples. Multi-class problems are tackled by considering each class value $y \in \mathbb{Y}$ as a separate concept.

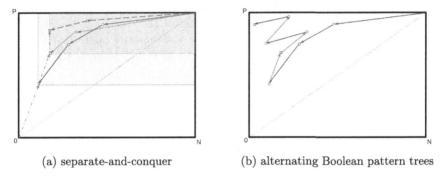

(a) separate-and-conquer   (b) alternating Boolean pattern trees

Fig. 1. Incremental construction of Boolean expressions in coverage space.

The task of learning an *alternating Boolean pattern tree (ABPT)* is quite similar to learning a rule: For a specific class $y_j \in \mathbb{Y}$, a tree $t : y_j \leftarrow B$ is learned, where B is a logical expression defined over the input features. In rule learning, this expression B usually only consists of features and conjunctions for a single rule, and multiple expressions of different rules can afterwards be connected by disjunctions to form a rule set, so that eventually B is an expression in *disjunctive normal form (DNF)*. Figure 1a illustrates the iterative construction of a DNF expression in coverage space [10], a non-normalized version of ROC space: conjunctive rules are formed by iteratively adding conditions to single rules (the red paths originating in the upper right corner, which represents a model that covers all positive and negative examples), and these are then disjunctively combined (the green path that adds one rule at a time to an initially empty set that covers none of the examples). Note that this is quite inflexible, so that, e.g., the target model in the upper left corner (covering all the positive and none of the negative examples) can no longer be reached after the first rule has been added, because covered negative examples can no longer be removed by adding more rules (unless additional mechanisms such as rule weights and voting are employed that go beyond the learning of purely logical expressions).

For ABPTs, the construction of the Boolean expression B is much more flexible. Binary features can be connected by conjunctions and disjunctions in any arbitrary order, so that potentially shorter expressions can be found, as is illustrated in Fig. 1b. This also complicates the iterative learning of B: For rules, the expression can be considered as a set; a feature can only be added once and its position in B does not matter. In contrast, for pattern trees, the insertion position of a new feature is crucial if conjunctions and disjunctions are mixed: Inserting a disjunction with c in $B = a \wedge b$ can result in $c \vee (a \wedge b)$, $(a \vee c) \wedge b$ or $a \wedge (b \vee c)$, which are all logically different (see also Fig. 2).

Obviously, the placement of parentheses is important as well. This becomes even more clear when the expression is visualized in form of a tree: The features are the leaves of the tree (in a fixed order) but can be combined in different orders by the interior nodes AND and OR. In this representation we can distinguish between three[1] ways to insert a new feature into the tree of size n:

- *addParent*: Inserts a new feature at the top of the tree. If $n > 1$, the root node of the tree is an operator. We choose the opposite operator (AND for OR or OR for AND) as the new root of the tree and connect the previous root with the newly added feature.
- *addChild*: Inserts a new feature in the interior of the tree as a new child of an existing operator. Since the operator is not restricted to binary operations, we can just connect it to the new feature.
- *addSibling*: Inserts a new feature as a sibling of an existing leaf. The leaf is replaced by the opposite operator of the previous parent of the leaf, connecting the leaf with the new feature.

The size n of the tree increases by 1 for *addChild* and by 2 for *addParent* and *addSibling*, since a new operator has to be added along with the new feature. Since for each node of the tree there is either an *addChild* or *addSibling* expansion possible, plus one global *addParent* expansion, there are in total $n+1$ possible expansions of the tree per feature. Note that by applying these expansions, the pattern tree always maintains a strict alternation of conjunctive and disjunctive layers; all subsequent equal operator nodes are implicitly merged by the *addChild*-method. In Fig. 2, all expansion methods are illustrated.

To build a classifier, one pattern tree for each class $y \in \mathbb{Y}$ has to be learned, typically using a one-against-all approach. When predicting the class of a test example in a fuzzy pattern tree classifier, all pattern trees are evaluated and the highest probability decides about the class prediction. In the Boolean context, the output of the Boolean expression represented by the pattern tree can only be true or false for the features of the test example, which leads to ties if a test example is matched by multiple pattern trees. An easy way to break these ties is to order the pattern trees in a list. The class of the first pattern tree in the list that matches the given example is predicted; if none of the trees matches, a default class is predicted.

---

[1] Actually, a fourth operation, *splitChildren*, is possible after *addChild* but is omitted here because it is quite expensive to compute.

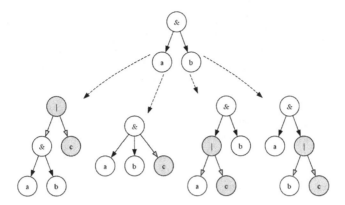

Fig. 2. Four possible expansions of a sample tree $a \wedge b$ with feature c. From left to right, the tree is expanded by adding c as a (a) parent of $\wedge$, (b) child of $\wedge$, (c) sibling of a, (d) sibling of b. The respective nodes and arrows added to the tree are highlighted in blue. (Color figure online)

3.2 Algorithm

Based on the iterative approach presented in the previous section, we develop a simple greedy algorithm. For every class, it learns a pattern tree by starting with an empty tree and adding exactly one feature in every iteration, by applying one of the *addParent-*, *addChild-* or *addSibling-*methods.

The main task of the algorithm is to select the best expansion. For this, we need to determine how the tested expansions influence the prediction of the training instances, i.e., whether the instances are covered by the new Boolean expression encoded by the tree, and whether they belong to the class the pattern tree was learned for. In order to use this information for estimating the quality of a candidate pattern tree, we make use of rule learning heuristics. These are typically based on the number of examples covered by a pattern, i.e., the number of true positives p, and the number of false positives n. There have been many proposals in the literature for how to combine these two measures into a single evaluation heuristic, we refer to [10] for a theoretical discussion.

In order to reduce the danger of making a wrong, myopic choice, and to ensure a broader variety of trees, we optimize the expansion regarding multiple different metrics and store multiple candidates for the next iteration. While the algorithm is designed to be general enough to make use of various heuristics, in our experiments we use the following metrics, which implement different cost trade-offs between true and false positives:

Accuracy: $h_{acc}(t) = \frac{p+N-n}{P+N}$.

Accuracy computes the percentage of correctly classified examples. Maximizing accuracy is equivalent to maximizing the difference $p-n$, thus giving equal weight to covering positive examples and not covering negative examples.

Linear cost metric: $h_{lc}(t) = c \cdot p - (1-c) \cdot n$.

The *linear cost* metric h_{lc} allows to assign different weights c and $1 - c$ ($c \in [0, 1]$) to true positives and false positives. If c is close to 0, trees with a small n are preferred, if is close to 1, those with a high p. Obviously, the h_{acc} is equivalent to the special case where $c = 0.5$.

While we use multiple parametrizations of h_{lc} for expanding the candidates tree, the final selection for which pattern tree to include in the classifier will be made by the main metric h_{acc}.

To efficiently compute the number of true positives p and false positives n for a given Boolean expression B, we make use of *n-Lists* and *PPC-Trees* [8], two data structures data have their origins in the field of frequent item set mining, but have recently been adopted for classification rule learning [14].

Algorithm 1: ABPT: Learning an alternating Boolean pattern tree

Input: Set of classes $\mathbb{Y}$, number of iterations k, set of features F, list of metrics M, empty tree t_0, main metric m_0
Output: Ordered list of pattern trees TM

1 $TM \leftarrow [\,]$;
2 **for** $y \in \mathbb{Y}$ **do**
3 $\quad t^* \leftarrow t_0$;
4 $\quad h^* \leftarrow evaluate(t^*, y, m_0)$;
5 $\quad i \leftarrow 0$;
6 $\quad T_0 \leftarrow [t_0]$;
7 $\quad$ **while** $i < k$ **do**
8 $\quad\quad C \leftarrow \emptyset$;
9 $\quad\quad$ **for** $t \in T_i$ **do**
10 $\quad\quad\quad$ **for** $f \in F$ **do**
11 $\quad\quad\quad\quad$ **if** $t.contains(f)$ **then**
12 $\quad\quad\quad\quad\quad$ *continue*;
13 $\quad\quad\quad\quad C \leftarrow C \cup t.addParent(f)$;
14 $\quad\quad\quad\quad$ **for** $node \in t$ **do**
15 $\quad\quad\quad\quad\quad$ **if** $node \in leaves(t)$ **then**
16 $\quad\quad\quad\quad\quad\quad C \leftarrow C \cup t.addSibling(node, f)$;
17 $\quad\quad\quad\quad\quad$ **else**
18 $\quad\quad\quad\quad\quad\quad C \leftarrow C \cup t.addChild(node, f)$;
19 $\quad\quad T_{i+1} \leftarrow \emptyset$;
20 $\quad\quad$ **for** $m \in M$ **do**
21 $\quad\quad\quad T_{i+1} \leftarrow T_{i+1} \cup \arg\max_{c \in C} evaluate(c, y, m)$;
22 $\quad\quad\quad$ **if** $m = m_0 \wedge \max_{c \in C} evaluate(c, y, m) > h^*$ **then**
23 $\quad\quad\quad\quad t^* \leftarrow \arg\max_{c \in C} evaluate(c, y, m)$;
24 $\quad\quad\quad\quad h^* \leftarrow \max_{c \in C} evaluate(c, y, m)$;
25 $\quad\quad i \leftarrow i+1$;
26 $\quad TM.insert(t^*, h^*)$;
27 **return** TM;

Algorithm 1 shows the described learning algorithm in a more formal way. For each class y, we learn the tree t^* with the best heuristic b^* according to the main metric m_0. These values are initialized with the empty tree t_0 (lines 3–4), which is also the only tree to be expanded in the first iteration (l. 6).

In lines 8 to 15, we generate a set of candidate trees C by expanding each optimal tree of the previous iteration. The candidates are all addParent-, addChild- and addSibling-expansions with features that are not yet part of the tree[2]. From lines 20 to 24, C is then reduced to a smaller set T, which stores the optimal trees regarding the tested metrics M, and updates t^* and h^* if needed.

This procedure is repeated until the maximum number of iterations k (= number of features in the pattern tree) is reached (l. 7). Finally, the best found pattern tree t^* is inserted to the tree model (l. 26), which is sorted by heuristic in descending order. The algorithm is demonstrated in the following Example 1.

Example 1. Given a data set with 7 binary attributes $\{a, ..., g\}$ (with all $2^7 = 128$ possible combinations as examples) and a binary class attribute. The examples of the positive class are defined by the concept

$$+ \leftarrow (a \vee b) \wedge (c \vee d) \wedge [e \vee (f \wedge g)]. \tag{1}$$

Obviously, this expression can also be interpreted as a pattern tree with six interior nodes for the operators (three for each of $\wedge$ and $\vee$) and seven leaf nodes for the features, which is indeed learned by ABPT as Fig. 3 shows.

For this, we used the main metric *accuracy* along with the two linear cost metrics for $c = 0$ (lc_0) and $c = 1$ (lc_1) for additional refinements. Figure 3 shows the best candidate tree after each iteration; on the left side for lc_0, in the center for *accuracy* and on the right side for lc_1, whereby duplicate trees are only shown in the center. For example, for all three metrics, the best tree after the first iteration consists of the feature e, so that only one candidate tree is kept for the next iteration. This tree is then expanded by "$\wedge\ a$" (best for lc_0 and *accuracy*) or by "$\vee\ f$" (best for lc_1) resulting in two candidates for the next iteration. The maximum number of three candidates is only used in the third and fourth iteration. Overall, ABPT finds the optimal tree after seven iterations, highlighted by the path of red arrows. Note that along *accuracy* the metric lc_0 was needed to not get stuck in a local optimum.

In contrast to this, a conventional rule learner like RIPPER would not be able to learn the compact representation shown in Eq. 1, but would have to find the corresponding minimal DNF which has 28 literals distributed over 8 rules:

$$\begin{aligned}+ \leftarrow\ &(a \wedge c \wedge e) \vee (a \wedge c \wedge f \wedge g) \vee (a \wedge d \wedge e) \vee (a \wedge d \wedge f \wedge g) \vee \\ &(b \wedge c \wedge e) \vee (b \wedge c \wedge f \wedge g) \vee (b \wedge d \wedge e) \vee (b \wedge d \wedge f \wedge g)\,.\end{aligned} \tag{2}$$

As to be expected, this representation is harder to learn if training data is incomplete, as confirmed by the resulting predictive accuracy in a 10-fold cross-validation (96.88% for ABPT vs. 92.19% for RIPPER).

---

[2] This slightly improves the run time and prevents learning redundant expressions.

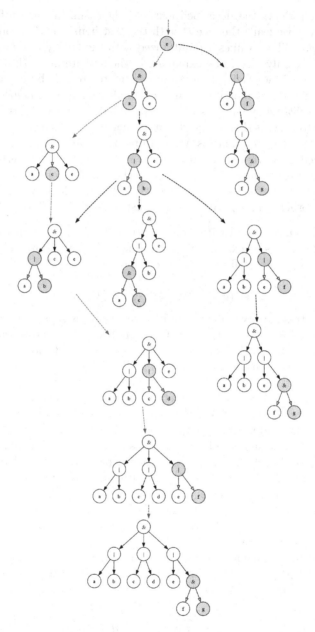

Fig. 3. Learning of the alternating Boolean pattern tree for the positive class of data set of Example 1. This sample application uses three metrics (linear cost with $c = 0$ on the left, accuracy in the center, linear cost with $c = 1$ on the right), whereby duplicate trees are only shown in the center. The nodes and arrows added to the tree in a growing step are highlighted in blue. The correct Eq. 1 is found after seven iterations—the learning path is shown by red arrows. (Color figure online)

4 Experiments

In this section we compare the presented algorithm ABPT[3] with the fuzzy pattern tree learner PTTD [24] and the two state-of-the-art rule learners RIPPER [6] and LORD [14], which has compared favorably to various other modern rule learning algorithms such as interpretable decision sets [16]. The comparison is made based on 29 UCI data sets [15] with varying number of instances, classes and attributes. In particular, the selected data sets differ with regard to the distribution of nominal and numeric attributes; half of them contain both types of attributes while the other half consists of purely nominal and purely numeric (excluding the class attribute) data sets. Numeric attributes are handled by the native methods of the respective algorithms. All algorithms are applied on the same folds of a 10-fold-cross-validation.

4.1 Hyperparameters

For ABPT, we use seven metrics for choosing the candidate trees: *accuracy* as the main metric and *linear cost* with six values $c \in [0.01, 0.1, 0.25, 0.75, 0.9, 0.99]$ as additional metrics.

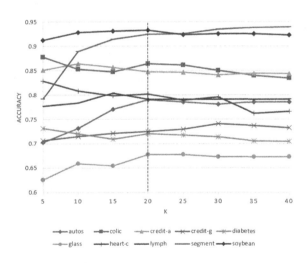

Fig. 4. Predictive accuracy of ABPT on 10 UCI data sets for various values of k.

Furthermore, the choice of the maximum number of iterations k is crucial. Figure 4 shows the accuracy on 10 representative data sets for eight different values $[5, 10, \ldots, 40]$. For some data sets (*autos*, *glass*, *segment*) a high number of k is needed to fit the data, others (*colic*, *diabetes*, *heart-c*) overfit and lose accuracy for higher k. For our experiments, we chose $k = 20$ as a trade-off

---

[3] The code of the ABPT algorithm can be found at https://github.com/f-beck/ABPT.

between these two cases. No further stopping criteria are defined, so that—except for the rare case of running out of features even earlier—the algorithm keeps growing the candidate trees for the full 20 iterations even if in intermediate steps no metric could be improved.

All other algorithms were used in their default settings, except that the default configuration of PTTD resulted in excessive run times, so that we had to limit the number of iterations to 20. Hence, both ABPT and PTTD use the same number of iterations, however, PTTD's more powerful set of operators (it includes, e.g., the possibility of averaging numerical values in a node) still allow PTTD to learn more flexible models at the cost of a much longer run time than ABPT (cf. Sect. 4.3 for further details).

4.2 Accuracy

Table 1 summarizes the predictive accuracies of the four learners on the 29 UCI data sets. The best accuracy for each data set is highlighted in bold and is almost evenly distributed among the four algorithms: For every algorithm, there are at least six data sets where it performs better than all its competitors. This also leads to similar average ranks between 2.33 and 2.74, resulting in no statistical difference between the algorithms. Hereby, ABPT ranks below LORD (-0.17) and RIPPER (-0.07), but above PTTD ($+0.24$).

However, looking at the average ranks for specific subsets of the data sets, we notice some clear trends: Most notably, PTTD outperforms all algorithms on numeric data. Despite the small number of numeric data sets ($N = 8$), the difference between PTTD and ABPT is even statistically significant when a Friedman and post-hoc Nemenyi test is applied. This indicates that in this case PTTD clearly benefits from the additional numerical operations it uses. RIPPER also performs much better than LORD and ABPT on numeric datasets, which we also assume is due to a more flexible treatment of numeric attributes during the learning phase. LORD and ABPT both have to rely on pre-discretization, which does not appear to be the best strategy.

On the other hand, however, PTTD performs much worse than all other algorithms on both nominal and mixed data. On these 21 data sets with at least one nominal attribute, a Friedman and post-hoc Nemenyi test shows a statistically significantly worse performance of PTTD when compared with ABPT or LORD. It seems that if nominal data is included, crisp Boolean operators are not only sufficient, but even preferable over fuzzified or numerical operators to learn models with good predictive accuracy.

4.3 Model Properties and Run Time

Besides the predictive accuracy of the algorithms, we also took a closer look into the models learned. Arguably, the model size should not be too large to maintain interpretability. Since LORD finds a locally optimal rule for each training example, its model is much larger than the models of the other three learners. While these individual rules provide good explanations for a specific test example, the

Table 1. Predictive accuracies of the pattern tree learners ABPT and PTTD and the rule learners RIPPER and LORD on 29 UCI data sets. The corresponding rank when compared with the other learners is denoted in parantheses. The average rank is computed for all 29 data sets as well as for the subsets of pure nominal, mixed and pure numeric data sets. The number of nominal and numeric attributes in each data set is shown along with number of instances and classes on the right side of the table.

	ABPT	PTTD	RIPPER	LORD	#inst	#cls	#num	#nom
anneal	**99.78 (1)**	97.66 (4)	97.88 (3)	98.78 (2)	898	5	6	32
audiology	**79.20 (1)**	75.66 (4)	77.88 (3)	78.81 (2)	226	24	0	69
autos	**79.02 (1)**	66.83 (4)	75.61 (3)	76.52 (2)	205	6	15	10
balance-scale	74.56 (4)	**90.08 (1)**	81.44 (3)	82.39 (2)	625	3	4	0
breast-cancer	72.73 (2)	71.68 (3)	**73.08 (1)**	71.33 (4)	286	2	0	9
breast-w	94.85 (4)	**96.42 (1)**	95.71 (2)	95.13 (3)	699	2	9	0
colic	**86.41 (1)**	84.51 (4)	85.33 (2)	85.04 (3)	368	2	7	15
credit-a	84.78 (4)	85.51 (3)	86.09 (2)	**86.81 (1)**	690	2	6	9
credit-g	72.50 (4)	73.30 (2)	72.70 (3)	**74.30 (1)**	1000	2	7	13
diabetes	72.01 (3)	**76.82 (1)**	73.83 (2)	71.49 (4)	768	2	8	0
glass	67.76 (3)	**69.16 (1)**	68.22 (2)	65.00 (4)	214	6	9	0
heart-c	80.20 (3.5)	80.86 (2)	80.20 (3.5)	**83.10 (1)**	303	2	6	7
heart-h	80.27 (2)	79.93 (3.5)	79.93 (3.5)	**81.31 (1)**	294	2	6	7
heart-statlog	73.70 (4)	**81.85 (1)**	79.63 (2)	76.30 (3)	270	2	13	0
hepatitis	81.94 (2)	80.65 (3)	80.00 (4)	**83.92 (1)**	155	2	6	13
hypothyroid	98.54 (2)	92.42 (4)	**99.39 (1)**	97.51 (3)	3772	4	6	22
iris	94.67 (2)	**96.00 (1)**	94.00 (3.5)	94.00 (3.5)	150	3	4	0
kr-vs-kp	98.03 (3)	94.34 (4)	**99.00 (1)**	98.72 (2)	3196	2	0	36
labor	**85.96 (1)**	43.86 (4)	80.70 (2)	80.67 (3)	57	2	8	8
lymph	79.05 (2)	**83.78 (1)**	77.03 (4)	78.29 (3)	148	4	3	15
mushroom	**100.00 (2)**	98.52 (4)	**100.00 (2)**	**100.00 (2)**	8124	2	0	22
primary-tumor	38.05 (3)	42.77 (2)	37.76 (4)	**45.43 (1)**	339	21	0	17
segment	92.42 (3)	90.95 (4)	**94.37 (1)**	93.59 (2)	2310	7	19	0
sick	97.72 (2)	94.04 (4)	**98.36 (1)**	95.15 (3)	3772	2	6	22
soybean	**93.27 (1)**	88.14 (4)	91.22 (3)	91.50 (2)	683	19	0	35
vote	94.48 (3)	**95.63 (1)**	95.40 (2)	94.04 (4)	435	2	0	16
vowel	63.23 (3)	44.75 (4)	70.20 (2)	**75.86 (1)**	990	11	10	3
yeast	55.86 (4)	57.48 (2)	**58.49 (1)**	55.93 (3)	1484	10	8	0
zoo	89.11 (2)	88.12 (3)	87.13 (4)	**90.18 (1)**	101	7	1	16
Avg. rank	2.5000	2.7414	2.4310	**2.3276**				
– nominal	**2.1429**	3.1429	2.2857	2.4286	(7 datasets)			
– mixed	2.1786	3.2500	2.7143	**1.8571**	(14 datasets)			
– numeric	3.3750	**1.5000**	2.0625	3.0625	(8 datasets)			

whole model is usually too large to be considered interpretable and is therefore omitted in the following comparisons.

When comparing ABPT with PTTD, we note that the average number of features in the pattern tree model is slightly lower for ABPT (57.53) than for PTTD (64.74) for the presented experiments using $k = 20$. We conclude that ABPT achieves a similar performance for the same model size without using any other operators other than $\wedge$ and $\vee$.

RIPPER, however, only uses 20.21 features in the learned rule model—a.o. due to the separate-and-conquer approach. Only for 4 out of 29 data sets, ABPT learns a smaller model than RIPPER; one of them being the *mushroom* data set, where both algorithms achieve a perfect accuracy of 100%. On average, ABPT only uses a model with 9 features while RIPPER learns a larger model consisting of 12.1 features. Figure 5 compares the two models learned for (class=p).

c1 ← (stalk-surface-below-ring=y).
c1 ← (population=c).
c2 ← (stalk-surface-above-ring=k).
c2 ← (gill-size=n).
c ← (odor=f).
c ← (gill-color=b).
c ← (odor=p).
c ← (odor=c).
c ← (spore-print-color=r).
c ← c1 ∧ c2.

(a) ABPT model with 9 features.

c ← (odor=f).
c ← (gill-size=n) ∧ (gill-color=b).
c ← (gill-size=n) ∧ (odor=p).
c ← (odor=c).
c ← (spore-print-color=r).
c ← (stalk-surface-above-ring=k) ∧ (gill-spacing=c).
c ← (habitat=l) ∧ (cap-color=w).
c ← (stalk-color-above-ring=y).

(b) RIPPER model with 12 features.

Fig. 5. Models learned by ABPT and RIPPER on the *mushroom* data set for class=p (poisonous). For a better comparison, the tree and rule list are transformed into a uniform representation as a set of conjunctive rules c ← B, i.e. a Prolog-like backwards implication that concept c is fulfilled if the conjunctive expression B—consisting of attribute-value-combinations—is true. As the last row in model (a) shows, intermediate concepts can be reused in subsequent expressions B.

We notice two reasons why ABPT is able to find a more compact model than RIPPER. First, because RIPPER chooses the feature greedily, leading to two rules using the feature (gill-size=n) even if it is not needed. Second, RIPPER is unable to find models deeper than the two layers of DNF. As the top part of the ABPT model in Fig. 5a shows, only four instead of five features are needed to describe the remaining examples. To achieve this, two intermediate rule concepts c1 and c2 are defined and afterwards combined.

Finally, the run time on the used machine (Core i7-8665U 1.90GHz) differs considerably between the algorithms. While RIPPER and LORD learn the rule sets in less than a second, ABPT needs on average 19 s to learn the pattern tree model, and PTTD even 1,064 s.

5 Conclusion

In this paper, we proposed alternating Boolean pattern trees as a new machine learning model which can be translated into deep rule concepts, and presented a greedy algorithm ABPT that iteratively learns such trees with a maximum size k. The experimental evaluation shows that ABPT can compete with state-of-the-art rule learners, and slightly outperforms a fuzzy pattern tree learner PTTD in terms of predictive accuracy. While being significantly worse than PTTD on numeric data sets, ABPT is significantly better than PTTD on nominal and mixed data sets. Furthermore, ABPT learns model of the same size on average 30 times faster than PTTD, and thanks to the limitation to conjunctive and disjunctive operations, arguably, they are even easier to interpret, since the alternating structure of the Boolean trees can be directly transformed into rule concepts.

References

1. Angelino, E., Larus-Stone, N., Alabi, D., Seltzer, M.I., Rudin, C.: Learning certifiably optimal rule lists for categorical data. J. Mach. Learn. Res. **18**, 234:1–234:78 (2017)
2. Aoga, J.O., Nijssen, S., Schaus, P.: Modeling pattern set mining using Boolean circuits. In: Principles and Practice of Constraint Programming: 25th International Conference, CP 2019, Stamford, CT, USA, September 30–October 4, 2019, Proceedings 25, pp. 621–638, Springer (2019). https://doi.org/10.1007/978-3-030-30048-7_36
3. Beck, F., Fürnkranz, J.: An empirical investigation into deep and shallow rule learning. Front. Artif. Intell. **4**, 145 (2021)
4. Beck, F., Fürnkranz, J., Huynh, V.Q.P.: Layerwise learning of mixed conjunctive and disjunctive rule sets. In: International Joint Conference on Rules and Reasoning, pp. 95–109, Springer (2023). https://doi.org/10.1007/978-3-031-45072-3_7
5. Brayton, R.K., Hachtel, G.D., McMullen, C.T., Sangiovanni-Vincentelli, A.L.: Logic Minimization Algorithms for VLSI Synthesis. Kluwer Academic Publishers (1984)
6. Cohen, W.W.: Fast effective rule induction. In: Prieditis, A., Russell, S. (eds.) Proceedings of the 12th International Conference on Machine Learning (ML-95), pp. 115–123, Morgan Kaufmann, Lake Tahoe, CA (1995)
7. Dembczyński, K., Kotłowski, W., Słowiński, R.: ENDER: a statistical framework for boosting decision rules. Data Min. Knowl. Disc. **21**(1), 52–90 (2010). https://doi.org/10.1007/s10618-010-0177-7
8. Deng, Z.H., Lv, S.L.: PrePost+: an efficient n-lists-based algorithm for mining frequent itemsets via children-parent equivalence pruning. Expert Syst. Appl. **42**(13), 5424–5432 (2015)
9. Fürnkranz, J.: Separate-and-conquer rule learning. Artif. Intell. Rev. **13**(1), 3–54 (1999)
10. Fürnkranz, J., Flach, P.A.: ROC 'n' rule learning - towards a better understanding of covering algorithms. Mach. Learn. **58**(1), 39–77 (2005)
11. Fürnkranz, J., Gamberger, D., Lavrač, N.: Foundations of Rule Learning. Springer-Verlag (2012). ISBN 978-3-540-75196-0

12. Hornik, K.: Approximation capabilities of multilayer feedforward networks. Neural Netw. **4**(2), 251–257 (1991)
13. Huang, Z., Gedeon, T.D., Nikravesh, M.: Pattern trees induction: a new machine learning method. IEEE Trans. Fuzzy Syst. **16**(4), 958–970 (2008)
14. Huynh, V.Q.P., Fürnkranz, J., Beck, F.: Efficient learning of large sets of locally optimal classification rules. Mach. Learn. (2023). https://doi.org/10.1007/s10994-022-06290-w, ISSN 1573-056
15. Kelly, M., Longjohn, R., Nottingham, K.: The UCI machine learning repository (2024). https://archive.ics.uci.edu
16. Lakkaraju, H., Bach, S.H., Leskovec, J.: Interpretable decision sets: a joint framework for description and prediction. In: Proceedings of the 22nd ACM SIGKDD International Conference on Knowledge Discovery and Data Mining (KDD-16), pp. 1675–1684, ACM, San Francisco, CA (2016)
17. Mhaskar, H., Liao, Q., Poggio, T.A.: When and why are deep networks better than shallow ones? In: Singh, S.P., Markovitch, S. (eds.) Proceedings of the 31st AAAI Conference on Artificial Intelligence, pp. 2343–2349, AAAI Press, San Francisco, California, USA (2017)
18. Michalski, R.S.: On the quasi-minimal solution of the covering problem. In: Proceedings of the 5th International Symposium on Information Processing (FCIP-69), vol. A3 (Switching Circuits), pp. 125–128, Bled, Yugoslavia (1969)
19. Petersen, F., Borgelt, C., Kuehne, H., Deussen, O.: Deep differentiable logic gate networks. In: Koyejo, S., Mohamed, S., Agarwal, A., Belgrave, D., Cho, K., Oh, A. (eds.) Advances in Neural Information Processing Systems 35 (NeurIPS-22). LA, USA, New Orleans (2022)
20. Qin, H., Gong, R., Liu, X., Bai, X., Song, J., Sebe, N.: Binary neural networks: a survey. Pattern Recogn. **105** (2020)
21. Rai, S., Neto, W.L., Miyasaka, Y., et al.: Logic synthesis meets machine learning: Trading exactness for generalization. In: Proceedings of the Conference on Design, Automation and Test in Europe (DATE), pp. 1026–1031. IEEE, Grenoble, France (2021)
22. Rapp, M., Mencía, E.L., Fürnkranz, J., Nguyen, V.-L., Hüllermeier, E.: Learning gradient boosted multi-label classification rules. In: Hutter, F., Kersting, K., Lijffijt, J., Valera, I. (eds.) Machine Learning and Knowledge Discovery in Databases: European Conference, ECML PKDD 2020, Ghent, Belgium, September 14–18, 2020, Proceedings, Part III, pp. 124–140. Springer International Publishing, Cham (2021). https://doi.org/10.1007/978-3-030-67664-3_8
23. Rivest, R.L.: Learning decision lists. Mach. Learn. **2**(3), 229–246 (1987). https://doi.org/10.1007/BF00058680
24. Senge, R., Hüllermeier, E.: Top-down induction of fuzzy pattern trees. IEEE Trans. Fuzzy Syst. **19**(2), 241–252 (2010)
25. Wang, T., Rudin, C., Doshi-Velez, F., Liu, Y., Klampfl, E., MacNeille, P.: A Bayesian framework for learning rule sets for interpretable classification. J. Mach. Learn. Res. **18**, 70:1–70:37 (2017)
26. Yu, J., Ignatiev, A., Stuckey, P.J., Bodic, P.L.: Learning optimal decision sets and lists with SAT. J. Artif. Intell. Res. **72**, 1251–1279 (2021)

ETIA: Towards an Automated Causal Discovery Pipeline

Konstantina Biza[1,3], Antonios Ntroumpogiannis[1,3], Sofia Triantafillou[2,3], and Ioannis Tsamardinos[1,3(✉)]

[1] Computer Science Department, University of Crete, Rethimno, Greece
{kbiza,droubo,tsamard}@csd.uoc.gr
[2] Department of Mathematics and Applied Mathematics, University of Crete, Rethimno, Greece
sof.triantafillou@uoc.gr
[3] Institute of Applied and Computational Mathematics, FORTH, Heraklion, Greece

Abstract. We introduce the concept of Automated Causal Discovery (AutoCD), defined as any system that aims to fully automate the application of causal discovery and causal reasoning methods. AutoCD's goal is to deliver all causal information that an expert human analyst would provide and answer user's causal queries. To this goal, we introduce ETIA, a system that performs dimensionality reduction, causal structure learning, and causal reasoning. We present the architecture of ETIA, benchmark its performance on synthetic data sets, and present a use case example. The system is general and can be applied to a plethora of causal discovery problems.

Keywords: Automated learning · Causal discovery

1 Introduction

Causal Discovery is a field of machine learning and statistics aiming to induce causal knowledge from data [23,36]. There is a large corpus of algorithms and methodologies in the field, spanning tasks like learning causal models, estimating causal effects, and determining optimal interventions. While there are several public libraries of algorithms for these tasks, combining and applying them to any given problem is a challenging endeavor that requires extensive knowledge of the methods and a deep understanding of the theory to interpret results.

In this paper, we introduce the concept of Automated Causal Discovery (**AutoCD**), defined as the effort to fully automate the application of causal discovery. AutoCD's goals are to deliver not just the optimal causal model that fits the data, but answers to queries, visualizations, interpretations, explanations and all other information that a human expert analyst would give. AutoCD is meant to make an esoteric field and its methodologies accessible to non-experts. These methods produce results whose interpretation requires a deep understanding of the causal modeling theory. Just like in the term AutoML (automated machine learning), the term "Auto" implies that an optimization of the causal learning pipeline is taking place.

Working towards this goal, we implement and introduce ETIA (meaning cause in Greek), an AutoCD architecture (Fig. 1), which is capable of dealing with high-dimensional and mixed-type data. It reduces the dimensionality of the problem and optimizes the causal discovery algorithm and its hyper-parameters to learn the causal model that best fits the data. It then employs the Markov Equivalence class of the model to answer user-defined causal queries and visualize the results. We also perform several experiments to illustrate the performance of our methods on synthetic data.

The rest of this document is organized as follows. In Sect. 2, we present the basic concepts of causal discovery that are needed to understand the paper. In Sect. 3, we discuss related work and in Sect. 4, we describe our proposed architecture. In Sect. 5, we use a case-study to highlight ETIA's functionalities. In Sect. 6, we provide an experimental benchmarking of our framework on synthetic data. Finally, in Sect. 7, we discuss a real-world application of ETIA, the limitations, and the future work.

2 Background

Two types of graphs are commonly used to represent the causal relationships, the Directed Acyclic Graph (DAG) [23] and the Maximal Ancestral Graph (MAG) [29]. When a DAG is annotated with conditional probability densities, it becomes a quantitative model, namely a Bayesian Network (BN) [23]. A DAG contains only directed edges and if it is interpreted causally, a direct edge represents direct causality, i.e., if $X \longrightarrow Y$ then X is a direct cause of Y (direct in the context of the observed variables). A MAG contains both directed and bi-directed edges. MAGs admit and represent latent confounding; they are used when common causes (confounders) are not measured. We refer to such cases as causally insufficient systems. The interpretation of edges in a MAG is not as straightforward as in DAG models. In a MAG, an edge $X \longrightarrow Y$ denotes that X causes Y, but the causal relation may not be direct, and it may be confounded by latent variables (see [7,29,39] for a deeper look into the interpretation of MAG edge semantics). An edge $X \longleftrightarrow Y$ in a MAG means that X does not cause Y and Y does not cause X, and the two share a latent common cause.

Both DAGs and MAGs capture the independence relationships between the observed variables [23, 29, 36]. Different graphs over the same set of variables can entail the same conditional independencies. These graphs are called Markov Equivalent graphs and represent different causal theories that equally well fit the data (their conditional independences to be specific). A Completed Partially Directed Acyclic Graph (CPDAG) [9] represents a class of Markov Equivalent DAGs. Similarly, a Partial Ancestral Graph (PAG) [36] represents a class of Markov Equivalent MAGs. Both CPDAGs and PAGs contain only the edges (ignoring direction) shared by all members of the equivalence class. The endpoints of the edges in PAGs can be either arrows, tails or circles. An arrow and a tail denote that the edge has the specific direction shared by all members of the class, while a circle indicates disagreement between members (could be an arrow or a tail). For example, the edge $A \circ \rightarrow B$ in a PAG denotes that there exist members of the class with the edge $A \rightarrow B$ as well as $A \leftrightarrow B$. The interpretation of the edges is that A is causally affecting B (possibly indirectly and possibly the relation is also confounded by a latent variable), or, neither variable is causing each other and there exist a latent confounding variable [29,44].

A path between X and Y is a sequence of distinct nodes, in which every pair of successive nodes (V_i, V_{i+1}) is adjacent (i.e., connected with any type of edge). In any of the aforementioned graphs, a *directed path* (also called *causal path*) from a node X to Y contains only directed edges pointing away from X and towards Y. If there is at least one arrowhead towards X, we say that it is a *non-causal path* [24,35]. A *potentially directed path* in a PAG is a path that could be oriented into a directed path by changing the circles on the path into appropriate tails or arrowheads [44]. A node Z is a collider on a path p, if both adjacent edges to Z are towards Z. A path p between X and Y is a *collider path* if all nodes (except X and Y) are colliders on p [24].

In a causal graph, the Markov boundary (Mb) of a node X consists of its adjacent nodes and the nodes that are reachable through a collider path [8]. The $\mathbf{Mb}(X)$ has the following important properties: *it is the minimal set that renders X conditionally independent of any other node; it is the solution to the feature selection problem when trying to predict X under certain broad conditions* [40]. It is shown that in distributions faithful to a BN (intuitively all and only conditional independences in the distribution are shown in the BN, see [36] page 31 for a definition), the Markov boundary is unique. Hence, under faithfulness, different equivalent MAGs/DAGs may disagree in the direction of causal edges, but agree on all Markov boundaries (all Mb are invariant).

A wide range of causal discovery algorithms have been proposed in the literature to learn a causal structure from observational data. These algorithms utilize various methodologies (e.g., conditional independence testing, scoring functions, neural networks), are proposed for different data types (e.g., cross-sectional or temporal), and are suitable under different assumptions (e.g., linear or non-linear relationships). The algorithms that assume causal sufficiency estimate a single DAG or its corresponding CPDAG, and those that allow causal insufficiency induce a MAG or its corresponding PAG. For a comprehensive review of causal discovery algorithms, see the surveys [15,16].

3 Related Work

Various libraries of causal analysis-related algorithms exist (e.g., Tetrad[1]), but they do not constitute automated analysis systems. The three works most relevant to ETIA are PyWhy[2], OpportunityFinder [21], and CausalMGM [14].

---

[1] https://github.com/cmu-phil/tetrad
[2] http://pywhy.org.

PyWhy is a system for causal learning and inference with several libraries, including causal-learn [45] for causal discovery, DoWhy [33] for causal effect estimation and what-if questions, EconML [4] for conditional causal effect estimation, and CausalTune [13] for selecting the optimal approach for causal effect estimation. OpportunityFinder [21] is a causal inference framework, which aims to automate the estimation of the causal effect, inspired by the AutoML community. Both PyWhy and OpportunityFinder mainly focus on the automation of causal inference, a term commonly used when our goal is to study the effect of an intervention to the system. On the contrary, ETIA currently focus on the automation of causal discovery (a term used when our goal is to find the causal structure) and the information that can be extracted from the causal graph. A tuning method is the common feature in the above works, specialized for causal discovery in ETIA and causal effect estimation in PyWhy and OpportunityFinder.

CausalMGM [14] is a web-based tool for causal discovery that performs feature selection, causal discovery, and visualization. These three modules appear also in ETIA, however, we point out the following differences: (a) CausalMGM applies a single feature selection algorithm, while ETIA has an automated pipeline to select the optimal feature selection method among an extendable set of algorithms. (b) ETIA uses causal tuning [6] to automate causal discovery, selecting from a variety of causal discovery algorithms and hyper-parameters, while CausalMGM uses the StEPS [31] tuning procedure to select optimal hyper-parameters of a single algorithm. (c) ETIA allows for causally insufficient systems, while CausalMGM assumes no latent confounding, and (d) in ETIA we include additional information, such as edge and path confidence estimations, using the estimated causal structure. On the other hand, CausalMGM has an interactive web platform that allows a non-expert user to apply causal discovery and visualize the results without any prior knowledge. A web-based tool is one of the future goals for ETIA.

4 ETIA's Architecture

Figure 1 shows our proposed methodology. The architecture consists of three modules: Automated Feature Selection (AFS), Causal Learning (CL), and Causal Reasoning and Visualization (CRV). In the first step, AFS reduces the dimensionality of the problem, by filtering in variables that are necessary for optimizing an optional set of user-defined outcomes. In the second step, CL learns a causal model over the selected features, by optimizing the causal analysis pipeline. In the third step, CRV visualizes, summarizes, and interprets the estimated causal structure.

4.1 Automated Feature Selection

The purpose of the AFS module is to reduce the number of features required to be considered for subsequent causal modeling (Fig. 1a). Specifically, in high-dimensional datasets, the user can optionally specify one or more outcomes of

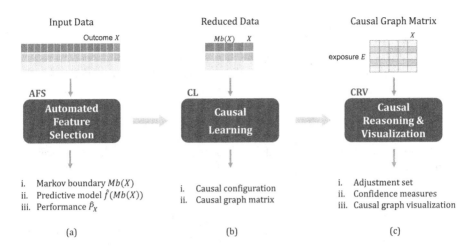

Fig. 1. The proposed architecture of ETIA.

interest; AFS then seeks to identify their Mb. The union of all Mb returned by AFS will be used in the second module that learns a causal model. This guarantees that direct parents of each outcome will not be filtered out. However, we note that filtering out features and learning the causal model of the marginal distribution, may result in missed opportunities for learning the directionality of causal edges or causal relationships in general (e.g., may not show the causal path from X to Y, when X is an indirect cause of Y).

The input to AFS is (a) a dataset and (b) one or more outcomes of interest. The output of AFS are (i) estimates of the Markov boundary of each outcome (feature selection results), (ii) a predictive model for each outcome using its Mb, and (iii) an out-of-sample estimate of the predictive performance of the model with respect to each outcome.

AFS is implemented using an automated machine learning platform for predictive modeling, equipped with feature selection. The automated machine learning approach implies the machine learning pipeline (called a **configuration**, hereafter) is tuned and optimized for the data and outcome at hand. This is an important factor in achieving high performance in terms of selecting the best approximations of the Mb. AFS's configurations consist of two steps: feature selection, and modeling. We follow the design principles of JADBio (Just Add Data Bio, see Fig. 6 in [41]), a commercial AutoML tool. A Configurator Generator (CG) generates the configurations to try within the configuration space. Contrary to JADbio, the configuration space in AFS is always the same for all datasets, and includes a smaller set of predictive learning and feature selection algorithms. For feature selection, AFS employs the FBED, and SES algorithms [8,19] from the MXM R Package [19]. These algorithms scale up to tens of thousands of features, apply on several types of outcomes (nominal, ordinal, continuous, time-to-event), and return multiple statistically equivalent solutions, i.e., multiple possible Mb whose predictive performance is statistically indistinguishable. The algorithms also have well-defined causal properties. Specifically, FBED is guaranteed to return the Mb asymptotically, in faithful distributions, assuming perfect tests of conditional independence, while SES is guaranteed to return the causal neighbors (direct causes and direct effects) of the outcome, under the same conditions. Regarding modeling algorithms, the current implementation of AFS optimize over hyper-parameters of the Random Forests and linear regression. The estimation protocol in AFS is always 5-fold cross-validation. Next, the Configuration Evaluator (CE) uses the selected estimation protocol to determine the predictive performance of a configuration. The AUROC, the average AUROC of each class-vs-rest, and the coefficient of determination R^2 are the performance metrics of choice to optimize, for binary, multi-class, and continuous outcomes, respectively. We note that these measures are used for observational predictions. Applying the winning configuration on all available data returns the final selection of features (Mb) and produces the final model, expected to be the best performing on average.

Table 1. Causal Discovery Algorithms in CL

	Algorithms	Latent	Tests/Scores	Data Type
Tetrad	PC variants [11,28,36]	✗	FisherZ, BIC	C
	FGES [10,27]	✗	Chi2, G2	D
	FCI variants [12,26,28,36]	✓	BDeu [9,17]	D
	GFCI [22]	✓	CG [1]	M
	Direct-LiNGAM [34]	✗	DG [2]	M

4.2 Causal Learning

The CL module is analogous in design to the AFS, in the sense that it searches in the space of configurations to identify the one with the highest performance (Fig. 1b). The notable difference is that the configuration space includes configurations (causal discovery algorithms and their hyper-parameters) that produce causal graphs and not predictive models. The input to CL is a dataset over the outcomes of interest and the Markov boundaries returned by AFS. The output is (i) the selected causal configuration and (ii) the matrix of the causal graph.

As in AFS, the causal Configurator Generator (CG) decides which algorithms to apply to a given problem and the ranges of their hyper-parameters, and is currently implemented as a simple grid search. Table 1 presents the currently available algorithms in the system and whether they admit the presence of latent confounders (*Latent*). *Tests/Scores* column shows the supported conditional independence tests and scoring functions employed internally by the algorithms. These indicate the type of assumptions made by the algorithm (e.g., linearity) and the data type to which they apply. The data type is explicitly shown in the column *Type*, where C, D, and M stand for continuous, discrete, and mixed, respectively. CL currently employs the causal discovery implementations from the Tetrad project. These support prior knowledge, and so CL can also take as input information about (a) edges that should or should not appear and (b) nodes that have no incoming edges or are ancestors of another set of nodes.

The causal CE system is responsible for evaluating the configurations and identifying the winner. Unlike AFS, *the problem of fitting a causal model is unsupervised*. This means that a standard cross-validation procedure cannot be applied directly. Several methodologies have been devised for causal model selection and/or causal configuration optimization [20,25,31]. In this work, we employ the OCT (Out-of-sample Causal Tuning) algorithm proposed in [6], which outperformed other methods and is applicable to mixed data types and models admitting latent confounding variables. OCT exploits the fact that the true causal graph induces the correct Mb for each variable and hence can lead to optimally predictive models for each node given its Mb. The true causal graph is expected to exhibit the highest average predictive performance over all nodes, under faithfulness assumption, in the large sample limit. OCT uses an out-of-sample protocol to estimate the predictive performance of a configuration and penalizes causal models with large Mb. To enable averaging the performances over mixed data types, OCT uses the mutual information between the outcome and the predictions of the predictive model.

4.3 Causal Reasoning and Visualization

Given the estimated causal structure, CRV module currently supports the following functionalities: (i) identification of an adjustment set, (ii) computation of a confidence measure on the causal findings, and (iii) visualization of the causal graph and subgraphs of interest (Fig. 1c).

Adjustment Set Identification. In many disciplines, the researchers are often interested to predict the effect of an intervention on a variable E (exposure) to a target variable T. Under assumptions, this causal effect can be identifiable using pre-interventional data. The identification relies on finding an appropriate set of covariates to control confounding (adjustment set) [23,35]. Several graphical criteria have been proposed to identify an adjustment set, if such exists [23,24,43]. In the current version, we use the R packages pcalg [18] and dagitty [37] to find the canonical and the minimal adjustment sets. Currently, ETIA includes the identification of the adjustment set as part of the causal information that can be obtained from the estimated causal structure. Causal effect estimation is one of the future tasks of ETIA.

Confidence Calculations on Causal Findings. CRV uses bootstrapping to calculate the confidence of causal findings in the optimal estimated causal graph $\hat{G}$. Specifically, we create bootstrap versions of the data and learn a population of corresponding causal graphs, using the selected causal configuration. For each causal feature of interest in $\hat{G}$, e.g., the presence of a directed edge $A \longrightarrow B$, we measure the percentage of graphs in the population containing the specific edge, named as **edge discovery frequency**. Notice that one bootstrap sample may lead to the finding $A \longrightarrow B$ and another to $A \circ \to B$. The first edge is a special case of the second. Both edges are *consistent* with each other since their semantics can both be true in the same PAG. In other words, if the edge $A \longrightarrow B$ is found 40% of the time, and $A \circ \to B$ is found another 40% of the time, then we are %80 confident in the edge $A \circ \to B$. In this case, we report the **edge consistency frequency** for each edge in the selected causal graph, that is the percentage an edge is found in all of its consistent forms. We extend the above confidence metrics to paths (i.e., the **path discovery frequency** and the **path consistency frequency**). In this case, for a given path, we measure the percentage of graphs in the population containing the specific path, or a *consistent path*, respectively. A *consistent path* is a path that contains only consistent edges, as defined above. We note that for all cases, *the computed percentage reflects the uncertainty of the specific causal finding due to the finite sample size of the real dataset; it should not be interpreted as the probability the finding is true irrespective of the causal discovery algorithm used.*

Graph Visualization. In case of large datasets, the representation of the above results can be quite challenging. We provide a visualization library to report the causal findings and facilitate the interpretation of the output causal graphs (especially the dense graphs). For this task, we utilize Cytoscape, an open source software platform for visualizing complex networks [32]. Other works that use Cytoscape are the Tetrad project, the CausalMGM [14] and the CausalPath [3] Apart from the above, Dagitty [37] is also a web-based environment for visualizing DAGs and reporting causal information. In this paper, we visualize the results either through the Cytoscape platform or with a jupyter notebook (example in our github repository). In the case of large and dense causal graphs, we focus on the causal neighborhood of the nodes of interest. The library (i) highlights specific nodes (e.g., target, exposure, adjustment set), (ii) creates subnetworks of interest (e.g., shows only the paths up to length k between two nodes of interest), (iii) exports and highlights the paths with specific causal characteristics, and (iii) reports edge and path confidences (either as weights above the edges or controlling the width of the edges). To provide an insight to a non-expert user, we group the paths, given two nodes X and Y, in five categories: (a) the directed path from X to Y, that is responsible for the causal effect of X to Y, (b) the potentially directed path from X to Y, that may be responsible for the causal effect in an equivalent MAG of the estimated PAG, (c) the *blocking path*, that has at least one collider and is responsible for 'blocking' the causal information and making X and Y unconditionally independent, (d) the *confounding path*, that has one common cause (confounder), does not have any collider, and contributes to the unconditionally dependency of X and Y, (e) the non-causal path, that is neither directed nor potentially directed and cannot be reported as blocking or confounding (e.g., due to unoriented edges). We note that blocking and confounding paths are special cases of the non-causal paths.

5 Motivation and Case Study

In many real-world problems, a variety of dataset characteristics, such as the high-dimensionality, might pose challenges to the subsequent causal analysis. In ETIA, our first goal is to reduce the dimensionality of the problem, utilizing AutoML strategies inside the AFS module. In addition, the researchers often need to optimize one or more variables of interest, such as the expression levels of a protein or the product sales, given an appropriate variable set for interventions (exposures). In this case, a local causal structure may be adequate to answer their questions. We propose to apply the CL module over a causal neighborhood around their Markov boundaries, as indicated by AFS. We then apply the CRV module, to provide a more detailed view of the causal relationships around the variables of interest, and further restrict the space, highlighting important paths (e.g., causal paths) and sets of nodes (e.g., adjustment set).

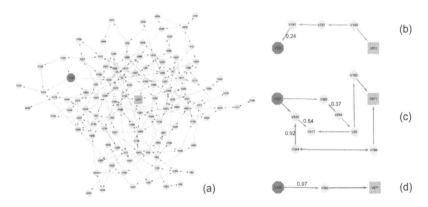

Fig. 2. An example of ETIA's application. (a) The estimated PAG over a subset of variables. We highlight the exposure variable (red), the target (green) and the estimated minimal adjustment set (yellow). (b–d) Subgraphs showing three categories of paths found up to length 5 between exposure and target. There are one confounding path (b), three blocking paths (c) and one directed path (d). We report the corresponding path consistencies as weights above the paths. (Color figure online)

To illustrate the application of ETIA, we use a synthetic dataset over $\mathbf{V}$ ($|\mathbf{V}| = 1000$), assuming that there is a target variable T and one exposure variable E. We initially apply the AFS module to reduce the dimensionality of the problem. We search for a subset $\mathbf{C} \in \mathbf{V}$, which contains the Mb of the target, $\mathbf{Mb}(T)$, the Mb of the exposure, $\mathbf{Mb}(E)$, and the Mb of each node in $\mathbf{Mb}(T)$. We then apply the CL module over the set $\mathbf{C}$ ($|\mathbf{C}| = 117$) to estimate an optimal causal graph. Since we excluded the rest of the variables in **V**, we need to assume that latent variables exist and apply the corresponding causal discovery algorithms (Table 1). Next, we apply the CRV module to extract causal information from the estimated structure. We identify the adjustment set from E to T and we compute the edge consistency and discovery frequencies for all estimated edges. CRV then uses the above information to visualize the results. In Fig. 2a, we show the estimated PAG over **C**. We report the target variable (green), the exposure variable (red) and the identified adjustment set (yellow). Each edge is equipped with two weights, the edge consistency frequency and the edge discovery frequency. Even though the estimated causal structure is smaller, the graph is still difficult to interpret. Given the two nodes of interest, E (red) and T (green), we extract all the paths between them, up to length k (k is set by the user, here is 5). For each path category, we then create the corresponding subgraphs and report the path consistency frequency (Fig. 2b–d).

6 Experiments

Experimental Setup. We create synthetic data of 100, 200, 500 and 1000 nodes and 3 average node degree. For the data generation, we assume linear relations and additive normal Gaussian noise. We use the Tetrad project for the construction of the random DAGs and the data generation. We create 10 datasets of 2000 samples for each network size. In each DAG, we randomly select a target variable (T) and one exposure variable (E) that is connected to the target with at least one directed path. For the following experiments, the true predictive model for these datasets is a linear regression model denoted as f. For each repetition and network size, we also simulate hold-out data of 500 samples for the estimation of the predictive performance in the AFS module.

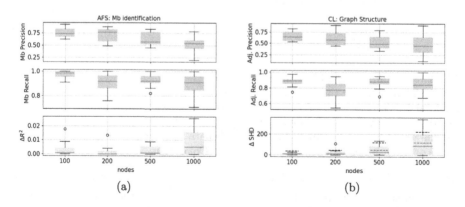

Fig. 3. Evaluation of the AFS and CL modules

AFS: Evaluate the Mb Identification. In the AFS module, we search over twelve predictive configurations, consisting of two predictive learning algorithms

(RF, linear regression), two feature selection algorithms (FBED, SES) and three significance levels (0.01, 0.05, 0.1). As in the case study, we search for the Mb of T, $\mathbf{Mb}_{est}(T)$, the Mb of E, $\mathbf{Mb}_{est}(E)$, and the Mb of each node in $\mathbf{Mb}_{est}(T)$. We denote the union of the above sets as $\mathbf{Mb}_{est}$. The corresponding set $\mathbf{Mb}_{true}$ is determined by $\mathcal{G}_{true}$. In Fig. 3a, we plot the precision and recall of the Mb identification and the difference between the predictive performances (as measured by R^2), called ΔR^2, between the fitted model $f(\mathbf{Mb}_{est}(T))$ returned by AFS and the optimal model $f(\mathbf{Mb}_{true}(T))$ of the gold standard. The larger the difference, the worse the predictive model by AFS. Precision and recall are high for smaller network sizes, but precision decreases as we increase the number of nodes. Although we obtain many false positive nodes, AFS does not miss many nodes that can be important in the next steps of the analysis (recall is above 0.8 even for 1000 nodes). The difference ΔR^2 shows that we obtain optimal predictive performance for the target, regardless the network size.

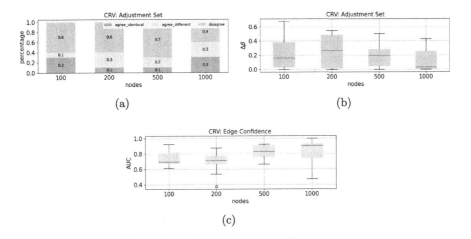

Fig. 4. Evaluation of the CRV module

CL: Evaluate the Output Causal Structure. The CL module returns the selected causal graph $\mathcal{G}_{est}^{Mb}$, where the superscript indicates that it is learned only over the variables returned by AFS. We compare this graph with $\mathcal{G}_{true}^{Mb}$, which is the marginal of the true graph over the variables of the true Mb (see [29] for theoretically computing the marginal). The OCT tuning method in the CL module searches over six causal configurations, consisting of two causal discovery algorithms (FCI, GFCI) and three significance levels (0.01, 0.05, 0.1), and returns the selected graph $\mathcal{G}_{est}^{Mb}$. In the first two rows of Fig. 3b, we show the precision and recall of the adjacencies (i.e., edges ignoring orientation) in the output $\mathcal{G}_{est}^{Mb}$. As we increase the network size, adjacency precision decreases but recall remains high. This is in accordance with the previous results on Mb identification. In the last row (Fig. 3b), we evaluate the tuning performance of

OCT, and we plot the difference in SHD between the optimal and the selected causal configuration. SHD counts the number of steps needed to reach the true PAG from the estimated PAG [38]. As a result, SHD reflects both the adjacency and the orientation errors. For the sake of comparison, we also show the median ΔSHD of a random choice (blue line) and the worst choice (black line). OCT can select an optimal configuration in many cases. We note that ΔSHD is low for small networks, but increases for larger networks. This appears because of the larger SHD differences among the causal configurations.

CRV: Evaluate the Adjustment Set Identification. The CRV module takes as input the estimated causal graph and the selected causal configuration. Here our goal is to compare the minimal adjustment sets $\mathbf{Z}_{true}$ and $\mathbf{Z}_{est}$, in the true DAG and estimated PAG, respectively. We evaluate the above sets by reporting two measures: (a) the percentage of agreement between $\mathbf{Z}_{true}$ and $\mathbf{Z}_{est}$ and (b) how well we can estimate the causal effect of the exposure to the target. In the first case (Fig. 4a), we report the percentages of the following cases: (i) **Agree-Identical**: same conclusion about identifiability, and same sets if identifiable, (ii) **Agree-Different**: same conclusion about identifiability, but different sets if identifiable, (iii) **Disagree**: different conclusion about identifiability. While different conclusions are common in smaller networks ($\sim 60\%$), this is not the case for the 1000 nodes. We note that in this experiment, different conclusions include only the cases where $\mathbf{Z}_{true}$ is identifiable but $\mathbf{Z}_{est}$ is not. Based on our previous results, the false positive nodes and the false positive edges in the graph may affect the adjustment set identification, accordingly. Our second evaluation is based on the metric, Causal Mean Square Error [42], which measures the squared difference between the true and the estimated causal effect. We note that this metric assumes conditional linear Gaussian distributions and so can be applied in our experimental setting. We fit two regression models $T = \beta_0 + \beta_E E + \beta_Z \mathbf{Z}_{true}$ and $T = \hat{\beta}_0 + \hat{\beta}_E E + \hat{\beta}_Z \mathbf{Z}_{est}$. We then measure the difference $\Delta \beta = \sqrt{(\beta_E - \hat{\beta}_E)^2}$ for each network. As with CMSE, if either $\mathbf{Z}_{true}$ or $\mathbf{Z}_{est}$ is not identifiable, we set the corresponding coefficient to 0. In Fig. 4b, we plot the computed $\Delta \beta$, which are in accordance with the results in Fig. 4a. The different conclusions regarding identifiability are not unexpected; for all network sizes we estimate a PAG over only $\sim 20\%$ (on average) of the input nodes. This makes adjustment set identification quite challenging. In the future, we aim to study larger causal neighborhoods, starting from the AFS module.

CRV: Evaluate the Confidence on Causal Findings. We evaluate the calculation of the edge consistency frequency using the bootstrapping approach. For each dataset, we apply the selected causal configuration on 100 bootstrapped samples. Using the estimated bootstrapped graphs $\{\mathcal{G}i_{est}^{Mb}\}$, we compute the edge consistency frequency for each pair of nodes in the union $\mathbf{Mb}_{true}$ and $\mathbf{Mb}_{est}$. We note that we measure our confidence not only for an edge that appears in $\mathcal{G}_{est}^{Mb}$, but also for a missing edge in $\mathcal{G}_{est}^{Mb}$. We use the AUC metric to compare these estimates with $\mathcal{G}_{true}^{Mb}$ (e.g., a consistent edge in $\mathcal{G}_{true}^{Mb}$ has label 1). In Fig. 4c, we show the AUC values, which are above 0.6, regardless the network size.

7 Discussion

In this work, we propose ETIA, a framework that aims to automate causal discovery for non-experts users. We introduce an architecture to facilitate the learning procedure and the interpretation of the results, and we evaluate our methods on synthetic data. The code of ETIA is available on Github[3].

Real-World Application. ETIA has been applied to a real-world telecommunication problem, to learn the causal relations among time-series measurements. A detailed description of the dataset, the challenges, and the results are described in the preliminary release of ETIA [5]. For our analysis, we used a time-lagged dataset, where each row represents a number of time points (time-lags) for each temporal variable (e.g., $t-2, t-1, t$). We then applied ETIA's modules, treating the time-lagged dataset as atemporal, and created a time-lagged causal graph around the target variable of interest (network throughput). We also evaluated AFS, CL and CRV modules on resimulated data, that were constructed using the telecommunication dataset and the causal-based simulation technique [6]. The time-lagged resimulated data had 123 variables and various sample sizes. The results are close to the results on synthetic data: high precision and recall and optimal predictive performance in the AFS module, lower adjacency recall and optimal tuning performance in the CL module, and similar edge confidences in the CRV module. However, we note that common constraints for time-series analysis, such as stationarity [30], were not imposed in our pipeline. In the future, we aim to enrich ETIA's modules to include such constraints for temporal data.

Limitations and Future Work. Despite our effort to automate the process of causal discovery, a user of ETIA may face the following challenges. Real-world data often require a preprocessing step, to handle missing values or transform the variables. In addition, the functional relationships might be challenging (e.g., non-linear), and so suitable conditional independence tests should be used in the AFS and CL modules. ETIA currently cannot handle the above cases. Based on our experiments, ETIA can scale up to 1000 nodes, requiring approximately 2.5 h. We have already included parallel programming for some tasks, however, we note that the computational cost and the ability to scale strongly depend on the algorithms that are included in the corresponding modules. In addition, they depend on the density of the underlying graph, and the type of variables (continuous, discrete, or mixed). We also note that the dimensionality reduction, performed by AFS, effectively reduces the computational cost for causal discovery in the CL module.

In the next version of ETIA, we aim to address the above challenges. In particular, we aim to (a) search for the optimal data representation in the predictive configuration space of the AFS module, (b) handle several data types (such as non-linear and temporal), and (c) implement more functionalities in parallel. Furthermore, our goal is to add more methods in the CRV module, such as (i) causal effect estimation, (ii) optimization of the outcome of interest using the

---

[3] https://github.com/mensxmachina/AutoCD.

estimated causal graph, and (iii) evaluation of the validity of the causal estimations. Finally, an important next step is to construct a web-based platform to facilitate a non-expert user.

Acknowledgments. We thank Gregory Cooper and Nikolaos Gkorgkolis for their helpful comments on the manuscript. Funded by the European Union (ERC, AUTOCD, project number 101069394). Views and opinions expressed are however those of the author(s) only and do not necessarily reflect those of the European Union or the European Research Council Executive Agency. Neither the European Union nor the granting authority can be held responsible for them. The study was part of the project "National research network to elucidate the genetic basis of Alzheimer's and Parkinson's neurodegenerative diseases, detect reliable biomarkers, and develop innovative computational technologies and therapeutic strategies underpinning precision medicine", Brain Precision - TAEDR-0535850 that is Funded by the European Union- Next Generation EU, Greece 2.0 National Recovery and Resilience plan, National Flagship Initiative "Health and Pharmaceuticals".

References

1. Andrews, B., Ramsey, J., Cooper, G.F.: Scoring Bayesian networks of mixed variables. Int. J. Data Sci. Anal. **6**(1), 3–18 (2018). https://doi.org/10.1007/s41060-017-0085-7
2. Andrews, B., Ramsey, J., Cooper, G.F.: Learning high-dimensional directed acyclic graphs with mixed data-types. In: The 2019 ACM SIGKDD Workshop on Causal Discovery, pp. 4–21. PMLR (2019)
3. Babur, Ö., et al.: Causal interactions from proteomic profiles: molecular data meet pathway knowledge. Patterns **2**(6) (2021)
4. Battocchi, K., et al.: EconML: a Python package for ML-based heterogeneous treatment effects estimation, version 0.x (2019). https://github.com/py-why/EconML
5. Biza, K., Ntroumpogiannis, A., Triantafillou, S., Tsamardinos, I.: Towards automated causal discovery: a case study on 5G telecommunication data. arXiv preprint arXiv:2402.14481 (2024)
6. Biza, K., Tsamardinos, I., Triantafillou, S.: Out-of-sample tuning for causal discovery. IEEE Trans. Neural Netw. Learn. Syst. **35**, 4963–4973 (2022)
7. Borboudakis, G., Triantafillou, S., Tsamardinos, I.: Tools and algorithms for causally interpreting directed edges in maximal ancestral graphs. In: Sixth European Workshop on Probabilistic Graphical Models (2012)
8. Borboudakis, G., Tsamardinos, I.: Forward-backward selection with early dropping. J. Mach. Learn. Res. **20**(8), 1–39 (2019)
9. Chickering, D.M.: Learning equivalence classes of Bayesian-network structures. J. Mach. Learn. Res. **2**, 445–498 (2002)
10. Chickering, D.M.: Optimal structure identification with greedy search. J. Mach. Learn. Res. **3**, 507–554 (2002)
11. Colombo, D., Maathuis, M.H.: Order-independent constraint-based causal structure learning. J. Mach. Learn. Res. **15**, 3741–3782 (2014)
12. Colombo, D., Maathuis, M.H., Kalisch, M., Richardson, T.S.: Learning high-dimensional directed acyclic graphs with latent and selection variables. Ann. Stat. 294–321 (2012)

13. Debono, T., et al.: CausalTune: a Python package for Automated Causal Inference model estimation and selection (2022). https://github.com/py-why/causaltune
14. Ge, X., Raghu, V.K., Chrysanthis, P.K., Benos, P.V.: CausalMGM: an interactive web-based causal discovery tool. Nucleic Acids Res. **48**, W597–W602 (2020)
15. Glymour, C., Zhang, K., Spirtes, P.: Review of causal discovery methods based on graphical models. Front. Genet. **10**, 524 (2019)
16. Hasan, U., Hossain, E., Gani, M.O.: A survey on causal discovery methods for IID and time series data. Trans. Mach. Learn. Res. (2023)
17. Heckerman, D., Geiger, D., Chickering, D.M.: Learning Bayesian networks: the combination of knowledge and statistical data. Mach. Learn. **20**, 197–243 (1995)
18. Kalisch, M., Mächler, M., Colombo, D., Maathuis, M.H., Bühlmann, P.: Causal inference using graphical models with the R package PCALG. J. Stat. Softw. **47**(11), 1–26 (2012)
19. Lagani, V., Athineou, G., Farcomeni, A., Tsagris, M., Tsamardinos, I.: Feature selection with the R package MXM: discovering statistically equivalent feature subsets. J. Stat. Softw. **80**(7), 1–25 (2017)
20. Maathuis, M., Kalisch, M., Buhlmann, P.: Estimating high-dimensional intervention effects from observational data. Ann. Stat. **37**, 3133–3164 (2009)
21. Nguyen, H., Grover, P., Khatwani, D.: Opportunityfinder: a framework for automated causal inference. arXiv preprint arXiv:2309.13103 (2023)
22. Ogarrio, J.M., Spirtes, P., Ramsey, J.: A hybrid causal search algorithm for latent variable models. In: Conference on Probabilistic Graphical Models, pp. 368–379. PMLR (2016)
23. Pearl, J.: Causality: Models, Reasoning and Inference, 2nd edn. Cambridge University Press, Cambridge (2009)
24. Perković, E., Textor, J., Kalisch, M., Maathuis, M.H.: A complete generalized adjustment criterion. In: Proceedings of the Thirty-First Conference on Uncertainty in Artificial Intelligence. UAI'15, pp. 682–691 (2015)
25. Raghu, V.K., Poon, A., Benos, P.V.: Evaluation of causal structure learning methods on mixed data types. In: Proceedings of 2018 ACM SIGKDD Workshop on Causal Discovery, pp. 48–65. PMLR (2018)
26. Raghu, V.K., et al.: Comparison of strategies for scalable causal discovery of latent variable models from mixed data. Int. J. Data Sci. Anal. **6**(1), 33–45 (2018). https://doi.org/10.1007/s41060-018-0104-3
27. Ramsey, J., Glymour, M., Sanchez-Romero, R., Glymour, C.: A million variables and more: the fast greedy equivalence search algorithm for learning high-dimensional graphical causal models, with an application to functional magnetic resonance images. Int. J. Data Sci. Anal. **3**, 121–129 (2017)
28. Ramsey, J., Spirtes, P., Zhang, J.: Adjacency-faithfulness and conservative causal inference. In: Proceedings of the Twenty-Second Conference on Uncertainty in Artificial Intelligence. UAI'06, pp. 401–408 (2006)
29. Richardson, T., Spirtes, P.: Ancestral graph Markov models. Ann. Stat. **30**(4), 962–1030 (2002)
30. Runge, J.: Causal network reconstruction from time series: from theoretical assumptions to practical estimation. Chaos: Interdisc. J. Nonlinear Sci. **28**(7) (2018)
31. Sedgewick, A.J., Shi, I., Donovan, R.M., Benos, P.V.: Learning mixed graphical models with separate sparsity parameters and stability-based model selection. BMC Bioinform. **17**, 307–318 (2016)
32. Shannon, P., et al.: Cytoscape: a software environment for integrated models of biomolecular interaction networks. Genome Res. **13**(11), 2498–2504 (2003)

33. Sharma, A., Kiciman, E.: Dowhy: an end-to-end library for causal inference. arXiv preprint arXiv:2011.04216 (2020)
34. Shimizu, S., et al.: DirectLiNGAM: a direct method for learning a linear non-Gaussian structural equation model. J. Mach. Learn. Res. **12**, 1225–1248 (2011)
35. Shpitser, I., VanderWeele, T.J., Robins, J.M.: On the validity of covariate adjustment for estimating causal effects. In: Conference on Uncertainty in Artificial Intelligence (2010)
36. Spirtes, P., Glymour, C., Scheines, R.: Causation, Prediction, and Search. MIT Press, Cambridge (2001)
37. Textor, J., Van der Zander, B., Gilthorpe, M.S., Liśkiewicz, M., Ellison, G.T.: Robust causal inference using directed acyclic graphs: the R package 'dagitty'. Int. J. Epidemiol. **45**(6), 1887–1894 (2016)
38. Triantafillou, S., Tsamardinos, I.: Score-based vs constraint-based causal learning in the presence of confounders. In: CFA@UAI (2016)
39. Triantafillou, S., Tsamardinos, I.: Constraint-based causal discovery from multiple interventions over overlapping variable sets. J. Mach. Learn. Res. **16**(1), 2147–2205 (2015)
40. Tsamardinos, I., Aliferis, C.F.: Towards principled feature selection: relevancy, filters and wrappers. In: International Workshop on Artificial Intelligence and Statistics, pp. 300–307. PMLR (2003)
41. Tsamardinos, I., et al.: Just add data: automated predictive modeling for knowledge discovery and feature selection. NPJ Precis. Oncol. **6**(1), 38 (2022)
42. Tsirlis, K., Lagani, V., Triantafillou, S., Tsamardinos, I.: On scoring maximal ancestral graphs with the max-min hill climbing algorithm. Int. J. Approx. Reason. **102**, 74–85 (2018)
43. Van der Zander, B., Liskiewicz, M., Textor, J.: Constructing separators and adjustment sets in ancestral graphs. In: CI@ UAI, pp. 11–24 (2014)
44. Zhang, J.: Causal reasoning with ancestral graphs. J. Mach. Learn. Res. **9**(7) (2008)
45. Zheng, Y., et al.: Causal-learn: causal discovery in python. J. Mach. Learn. Res. **25**(60), 1–8 (2024)

Open Access This chapter is licensed under the terms of the Creative Commons Attribution 4.0 International License (http://creativecommons.org/licenses/by/4.0/), which permits use, sharing, adaptation, distribution and reproduction in any medium or format, as long as you give appropriate credit to the original author(s) and the source, provide a link to the Creative Commons license and indicate if changes were made.

The images or other third party material in this chapter are included in the chapter's Creative Commons license, unless indicated otherwise in a credit line to the material. If material is not included in the chapter's Creative Commons license and your intended use is not permitted by statutory regulation or exceeds the permitted use, you will need to obtain permission directly from the copyright holder.

Security and Anomaly Detection

FedGES: A Federated Learning Approach for Bayesian Network Structure Learning

Pablo Torrijos[1,2]([✉]) , José A. Gámez[1,2] , and José M. Puerta[1,2]

[1] Instituto de Investigación en Informática de Albacete (I3A), Universidad de Castilla-La Mancha, Albacete 02071, Spain
[2] Departamento de Sistemas Informáticos, Universidad de Castilla-La Mancha, Albacete 02071, Spain
{Pablo.Torrijos,Jose.Gamez,Jose.Puerta}@uclm.es

Abstract. Bayesian Network (BN) structure learning traditionally centralizes data, raising privacy concerns when data is distributed across multiple entities. This research introduces Federated GES (FedGES), a novel Federated Learning approach tailored for BN structure learning in decentralized settings using the Greedy Equivalence Search (GES) algorithm. FedGES uniquely addresses privacy and security challenges by exchanging only evolving network structures, not parameters or data. It performs collaborative model development, using structural fusion to combine the limited models generated by each client in successive iterations. A controlled structural fusion is also proposed to enhance client consensus when adding any edge. Experimental results on various BNs from bnlearn's BN Repository validate the effectiveness of FedGES, particularly in high-dimensional (a large number of variables) and sparse data scenarios, offering a practical and privacy-preserving solution for real-world BN structure learning.

Keywords: Federated learning · Bayesian Network structure learning · Bayesian Network fusion/aggregation

1 Introduction

Bayesian Network (BN) structure learning [4,5,8] is a significant challenge in machine learning due to BNs' succinct depiction and interpretation of intricate probabilistic relationships [10], being powerful tools for uncovering (in)dependence relationships in complex datasets. The recent interest in causal models [33] and the growing demand for explainable models [2] have led to a widespread exploration of their applications and research.

These characteristics have allowed BNs to be adopted in areas such as agriculture [7], healthcare [13], and renewable energy [3]. However, the computational complexity of learning BN structures becomes a significant challenge as the number of variables increases, as it is an NP-hard problem [6].

In response, distributed learning efforts [14,15,27] have divided the learning process among multiple nodes or clients. However, this approach raises privacy concerns, requiring a central node to access the entire dataset. In response, Federated Learning (FL) [17,18,34] has emerged as a transformative paradigm, offering a collaborative framework for training machine learning models across privacy-sensitive environments. FL allows entities or clients to collaboratively learn a global model while maintaining data privacy locally, sharing only parameters, statistics, or model updates with a central server, which, if appropriately configured, do not compromise privacy.

Federated Learning's rapid development is evident in its applications in sectors facing data-sharing constraints, particularly in fields like healthcare [25,29] and fraud detection [32], where privacy concerns necessitate decentralized data processing. It also finds utility in scenarios involving client devices with limitations, such as low-powered Internet of Things (IoT) devices or mobile phones [16,18,22]. Additionally, it supports real-time systems where relying solely on server-sent models is impractical, as seen in applications like energy demand prediction [26].

While FL has predominantly been applied to Neural Networks (NNs) [17,34], this study focuses on BNs, which offer interpretability lacking in the inherent black-box nature of NNs. Despite its potential, Federated Learning's application in Bayesian Network structure learning remains relatively unexplored. Existing literature in this domain typically employs horizontal data partitioning[1]. Methods range from adaptations of continuous optimization techniques [21] to approaches using regret-based learning [19,20], and federated independence tests [31]. These algorithms yield a global model comprising an unparameterized BN structure suitable for symbolic reasoning, such as relevance analysis. However, they often encounter challenges such as high execution times, generation of suboptimal BNs, or compromises in data privacy. In response, this paper proposes a novel Federated Learning BN structural learning paradigm based on the state-of-the-art Greedy Equivalence Search (GES) algorithm [5] and Bayesian Network fusion techniques [24].

Contributions. We present Federated GES (FedGES), a novel approach for Federated Bayesian Network structural learning in horizontally partitioned data settings. Our contributions include:

- We present an iterative approach that combines (fuses) locally generated limited BNs from individual clients to construct a unified global structure. Our method ensures data security by exchanging only network structures or lists of (in)dependencies, thereby not exposing sensitive parameters (probabilities, statistics, etc.) between clients and the server.

---

[1] Horizontal partitioning divides data instances across clients, where each client possesses complete records but for different samples or segments. In contrast, vertical partitioning splits data attributes across clients, with each retaining all instances but only for specific attributes or features.

- Our approach maintains the same theoretical properties as GES (identifying the optimal structure given sufficient and faithful data). FedGES includes mechanisms to ensure convergence and control the complexity of the fused structures, which are essential when these properties are challenged.
- We validate the efficacy of the proposed FedGES method through comprehensive experiments on various BNs from the bnlearn's Bayesian Network Repository [28]. We focus on the final DAG obtained at the server, the global model, after the iterative process.
- The implementation of our algorithms is provided to ensure reproducibility and foster future research on this topic.

Organization of the Paper. Sect. 2 provides the necessary background and the related works. Section 3 describes our proposed FedGES method in detail. Section 4 presents an experimental evaluation of our method using several benchmark BNs. Finally, Sect. 5 concludes the paper and discusses future research directions.

2 Preliminaries

2.1 Bayesian Network

A Bayesian Network (BN) [10], denoted as $\mathcal{B} = (\mathcal{G}, \mathcal{P})$, is a probabilistic graphical model with two main components. On the one hand, a Directed Acyclic Graph (DAG), represented as $\mathcal{G} = (\mathcal{X}, \mathcal{E})$, encapsulates the network structure, where $\mathcal{X} = X_1, \ldots, X_n$ denotes the problem domain variables and $\mathcal{E} = \{X_i \rightarrow X_j \mid X_i, X_j \in \mathcal{X} \land X_i \neq X_j\}$ encode the (in)dependency relationships between $\mathcal{X}$ through directed edges. On the other hand, a set of Conditional Probability Tables (CPTs), denoted by $\mathcal{P}$, factorizes the joint probability distribution $P(\mathcal{X})$ using the graphical structure $\mathcal{G}$ and the Markov's condition:

$$P(\mathcal{X}) = P(X_1, \ldots, X_n) = \prod_{i=1}^{n} P(X_i \mid pa_{\mathcal{G}}(X_i)), \quad (1)$$

where $pa_{\mathcal{G}}(X_i)$ denotes the set of parents of X_i in $\mathcal{G}$. In this paper, as usual in BN literature, we only consider discrete variables.

2.2 Structure Learning of Bayesian Networks

Given a problem domain $\mathcal{X} = \{X_1, \ldots, X_n\}$ and a dataset defined over it $\mathcal{D} = \{(x_1^i, \ldots, x_n^i)\}_{i=1}^{m}$, the process of Bayesian Network structure learning [4,5,8] involves the derivation of a DAG $\mathcal{G} \in \mathcal{G}^n$, i.e. the space of DAGs defined over n variables, such that $\mathcal{G}$ captures the relationships of (in)dependence among the variables in $\mathcal{X}$ supported by $\mathcal{D}$. Notably, BN structural learning poses an NP-hard problem [6], necessitating the utilization of heuristic methods. Two primary approaches, namely constraint-based and score+search methods, are commonly employed in this context.

Constraint-based approaches, exemplified by the PC algorithm [30], aim to identify the skeleton of the underlying undirected graph by discerning dependencies among variables. This is accomplished through conditional independence tests, systematically removing edges inconsistent with the observed relations.

This work falls in the score+search approach, which seeks the optimal DAG $\mathcal{G}^* = \mathrm{argmax}_{\mathcal{G} \in \mathcal{G}^n} f(\mathcal{G} : \mathcal{D})$, where $f(\mathcal{G} : \mathcal{D})$ is a scoring function that quantifies how well the DAG fits the given data $\mathcal{D}$. We employ the Bayesian Dirichlet equivalent uniform (BDeu) score [9], a decomposable, score-equivalent metric that exhibits local and global consistency, characteristics for which it has been widely used over the years [1,4,5,8,15]. Several efficient local search-based algorithms, including the state-of-the-art algorithm, Greedy Equivalence Search (GES) [5], leverage these properties during the search process.

Greedy Equivalence Search Algorithm. The Greedy Equivalence Search (GES) algorithm [5] stands out as a BN structural learning approach, showcasing high efficiency in practical applications. GES performs a greedy search within the space of equivalence classes of BN structures, consisting of two key phases: Forward Equivalence Search (FES) and Backward Equivalence Search (BES). During the FES stage, edges are incrementally added until a local maximum is reached. Conversely, edges are systematically deleted in the BES stage until a local optimum is attained. When coupled with a locally and globally consistent metric, the GES algorithm provides theoretical guarantees for identifying the optimal equivalence class given the data under sufficient and faithful data assumptions. However, practical considerations arise in the presence of certain substructures, necessitating modifications to reduce computational complexity while preserving the algorithm's theoretical properties [1]. These adaptations enhance the algorithm's applicability and efficiency in real-world scenarios.

2.3 Bayesian Network Structural Fusion

Bayesian Network Structural Fusion aims to construct a unified BN structure from a set of BNs that share the same variables. The objective is to synthesize input BNs by emphasizing their common independence relationships. Solving BN Fusion is a challenging task, being NP-hard, necessitating the utilization of heuristics for practical solutions. Heuristic approaches attempt to approximate the fusion of networks by relying on a common variable ordering σ that may not be optimal [23]. A recent contribution introduces a practical and efficient greedy heuristic method for BN Fusion [24]. The proposed Greedy Heuristic Order (GHO) efficiently determines a suitable order σ to guide the fusion process. The fusion process uses this order to obtain a minimal directed independence map[2] $\mathcal{G}_i^\sigma$ following σ for each input DAG $\mathcal{G}_i \in \{\mathcal{G}_1, \ldots, \mathcal{G}_k\}$. Once the graphs are compatible with a common ordering, the consensus DAG is computed as the union of all the edges in $\{\mathcal{G}_1^\sigma, \ldots, \mathcal{G}_k^\sigma\}$. Authors in [24] show that the fusion obtained by using the heuristic ordering is almost identical to the optimal solution but requires, by far, less time.

2.4 Federated Learning

Since its introduction [18], Federated Learning (FL) [17,34] has rapidly gained significance in scenarios involving privacy-sensitive and data-sharing constraints. Federated Learning can be defined as the process of learning a global model $\mathcal{M}$ through the collaboration of k clients $\mathcal{C} = \{\mathcal{C}_1, \ldots, \mathcal{C}_k\}$, each possessing its dataset $\{\mathcal{D}_1, \ldots, \mathcal{D}_k\}$ that is not shared with the other clients. If each database $\mathcal{D}_i$ contains the same set of n variables $\mathcal{X} = \{X_1, \ldots, X_n\}$, we would be talking about horizontal FL; otherwise, it would be vertical FL.

Due to the high accuracy of deep learning approaches, most developments in Federated Learning involve deep neural networks. In this context, it is commonly assumed that the network structure is the same across all clients and the server; hence, only parameters (weights) are shared between the server and clients. The server aggregates the received parameters and dispatches the results to the clients, who then use the updated model alongside their own data to refine or update the model (weights). In the following, we will say that models are exchanged between the different nodes since, with the structure being common, the parameters differentiate the networks (models).

To summarize, the standard FL process is organized in a series of rounds. At each round: (1) The server $\mathcal{S}$ sends its model (initially random, empty, etc.) $\mathcal{M}$ to each client $\mathcal{C}_i$; (2) Each client $\mathcal{C}_i$ trains its local model $\mathcal{M}_i$ starting from $\mathcal{M}$ and using its respective dataset $\mathcal{D}_i$, and send it to the server; finally (3) the server aggregates all the received client's models $\{\mathcal{M}_1, \ldots, \mathcal{M}_k\}$ to create a new global model $\mathcal{M}$, which becomes the starting point for the next iteration.

2.5 Related Works

There has been limited exploration of BNs in the FL domain. Featured FL approaches, like federated averaging [18], mainly focus on continuous optimization, typically applied to learning NNs. In this context, the NOTEARS-ADMM [21] algorithm emerges, adapting advances in continuous optimization to BN structure learning, specifically addressing horizontally partitioned data.

---

[2] The minimal DAG $\mathcal{G}^\sigma$ that being compatible with σ preserves as much as possible of the conditional independences in $\mathcal{G}$, although the number of arcs considerably increases.

Furthermore, recent developments[3] have introduced methods based on both score+search algorithms, such as GES [5], and constraint-based algorithms, such as PC [30]:

- In the score+search category, notable contributions include the Regret-based Federated Causal Discovery (RFCD) [19] and its successor PERI [20]. Both algorithms employ a regret-based search, where each client initially discovers its best-fitting local model using a score-based algorithm. The distinction lies in the search strategy, with RFCD utilizing a basic beam-search and PERI employing the GES algorithm. Following this, the server proposes networks, and clients return regret values relative to the proposed models. Utilizing these regrets, a global model is learned by minimizing the worst-case regret from all clients in a privacy-preserving manner.
- On the constraint-based side, the FEDC^{2}SL algorithm [31] stands out by introducing a federated framework for BN structure learning with a federated conditional independence χ^2 test. This ensures an interaction with data that preserves privacy, enabling secure statistical evaluations of conditional independence between variables without sharing private data. By incorporating this federated framework, the FEDPC algorithm emerges as an innovative approach to BN structure learning, extending the capabilities of the PC algorithm to the FL paradigm.

3 FedGES

This section introduces Federated GES (FedGES), a novel approach to federated Bayesian Network structural learning. FedGES addresses the challenge of BN structure learning in privacy-sensitive environments by leveraging the Federated Learning paradigm. The algorithm follows a client-server framework, where the server orchestrates the fusion of client BNs into a global structure, ensuring privacy by exchanging only network structures (DAGs), not parameters. It is essential to highlight that since no parameters or statistics are shared, a malicious agent cannot reconstruct or sample data from the Directed Acyclic Graph (DAG), even with access to it.

The FedGES scheme is outlined in Algorithm 1. Let $\{\mathcal{D}_1, \ldots, \mathcal{D}_k\}$ represent k unique datasets, one for each of the k clients, defined on the same set of variables $\mathcal{X} = \{X_1, \ldots, X_n\}$. The server starts with an empty DAG, $\mathcal{M} = \mathcal{G} = (\mathcal{X}, \emptyset)$, and the models $\{\mathcal{M}_1, \ldots, \mathcal{M}_k\} = \{\mathcal{G}_1, \ldots, \mathcal{G}_k\}$ represent the specific DAG of each client, initially also empty.

The process begins with each client fusing the received global DAG $\mathcal{G}$ with its DAG $\mathcal{G}_i$, obtaining $\mathcal{G}'_i$. Any fusion can be used for this, or even the incoming

---

[3] It is essential to highlight that, despite the terminology referring to structure learning of Causal Networks, in the three aforementioned contributions, we can use this term interchangeably with Bayesian Networks. This is attributed to their exploration of the space of Markov equivalence classes rather than the space of DAGs, highlighting their emphasis on equivalent causal structures.

Algorithm 1: FedGES

Data: k clients; $\mathcal{D} = \{\mathcal{D}_1, \ldots, \mathcal{D}_k\}$ datasets defined over $\mathcal{X} = \{X_1, \ldots, X_n\}$ variables; $maxIt$, the number of FL rounds; l, the limit of edges that GES can add at each client in each FL round; $fusionClients$ and $fusionServer$, the types of BN fusion the clients and the server execute, respectively.

Result: $\mathcal{G} = (\mathcal{V}, \mathcal{E})$, the resulting DAG of the server; $\{\mathcal{G}_1, \ldots, \mathcal{G}_k\} = \{(\mathcal{X}, \mathcal{E}_1), \ldots, (\mathcal{X}, \mathcal{E}_k)\}$, the resulting client's DAGs.

1 $\mathcal{G} \leftarrow (\mathcal{X}, \emptyset)$; // Initialize server DAG
2 each client $i = 1, \ldots, k$ do in parallel
3 $\quad \mathcal{G}_i \leftarrow (\mathcal{X}, \emptyset)$; // Initialize client-specific DAGs to the empty network
4 for $(round = 1, \ldots, maxIt)$ do
5 $\quad$ each client $i = 1, \ldots, k$ do in parallel
6 $\quad\quad \mathcal{G}'_i \leftarrow fusionClient(\mathcal{G}_i, \mathcal{G})$; // Client-Side Fusion
7 $\quad\quad \mathcal{G}_i \leftarrow GES(init = \mathcal{G}'_i, data = \mathcal{D}_i, limit = l)$; // Obtain clients DAGs
8 $\quad$ if $convergenceCheck()$ then
9 $\quad\quad$ break ; // Convergence Verification
10 $\quad \mathcal{G} = fusionServer(\{\mathcal{G}_1, \ldots, \mathcal{G}_k\})$; // Server-Side Fusion

DAG can overwrite the local one ($\mathcal{G}'_i = \mathcal{G}$). At each round of the FL cycle, each client obtains $\mathcal{G}_i$ running a BN learning algorithm which takes $\mathcal{G}'_i$ as initial state and is *constrained to add a limited number of edges, l*, thus facilitating a more gradual learning process that avoids excessively complex fusions in the server. In this paper, the GES algorithm with the BDeu score (Sect. 2.2) is employed as the locally used structural learning algorithm. Then, the locally learned models $\{\mathcal{G}_1, \ldots, \mathcal{G}_k\}$ are then sent to the server, which fuses them into a single model $\mathcal{G}$ by using the fusion method proposed in [24]. Therefore, uniform fusion is used in this approach, and after the fusion, the same (global) model is sent to all the clients. This iterative process continues until the maximum number of rounds ($maxIt$) or when none of the new local networks $\mathcal{G}_i$ differs from the previous iteration. The latter criterion can be used when the BN learning algorithm is deterministic, as with GES.

The proposed approach offers several advantages. Firstly, the framework is highly customizable, allowing adjustments to the base structural learning algorithm or its score function and modifications to the fusion process on both the client and server sides. Secondly, this approach maintains the same good theoretical properties as the GES algorithm when using a globally and locally consistent scoring criterion as the BDeu score [5]. It asymptotically converges to the gold-standard BN ($\mathcal{G}^*, \theta^*$) under the same assumptions. This convergence is demonstrated by the fact that given sufficient data faithful to the probability distribution θ^* encoded by the optimal BN, the GES algorithm for each client i returns a learned model $\mathcal{G}_i$ equal to $\mathcal{G}^*$. Consequently, the subsequent server fusion, $\mathcal{G}$, also converges to $\mathcal{G}^*$. This holds when using a fusion method that, given a list of equal DAGs $\{\mathcal{G}, \ldots, \mathcal{G}\}$, always results in $\mathcal{G}$ (as is the case with all the methods used in this paper).

The convergence of each GES run on clients is guaranteed by employing a consistent scoring criterion, ensuring that each $\mathcal{G}_i$ score is greater than or equal to the initial $\mathcal{G}'_i$ score. However, the overall convergence of FedGES is not assured because the resulting $\mathcal{G}'_i$ from fusing the last $\mathcal{G}_i$ with the global $\mathcal{G}$ may obtain a lower score than $\mathcal{G}_i$. While this can potentially enhance the quality of generated Bayesian Networks (BNs), as incorporating information from other networks may lead to a score decrease, it can also result in cyclical behavior. Some clients might repeat the same DAGs every certain number of iterations. This fact can be addressed by modifying the convergence function to check whether any of the $\mathcal{G}_i$ generated in the current iteration has not been created earlier in previous iterations rather than checking only for changes with respect to the previous iteration.

4 Experimental Evaluation

This section presents the practical evaluation of our proposal compared to alternative hypotheses. We discuss the domains and algorithms involved, detail the experimental approach, and present the obtained results.

4.1 Algorithms

The algorithms evaluated in this study include[4]:

- The RFCD algorithm [19].
- The FEDC2SL algorithm [31] instantiated in FEDPC. We also include in the comparison their baseline algorithm PC-Voting [31].
- An enhanced version of the GES algorithm [5] described in [1]. Although not initially designed as a federated algorithm, we learn a network for each client and fuse them using the procedure outlined in [24], obtaining a baseline for the global ($\mathcal{G}$) model. Notice that in this method, GES is run at each client until convergence; that is, no limit on the number of edges is set. This baseline method can be viewed as a one-shot FL algorithm because only a single round of communication is needed.
- The proposed FedGES algorithm (see Sect. 3), configured with a limit of $l = 10$ edges added by GES at each round. Three fusion strategies are tested on the server side: The canonical BN fusion described in [24], denoted as UNION, which adds the edges of all the BNs once converted to the same σ ancestral order; and C25 and C50, which only add those edges that appear in at least 25% or 50% of input DAGs transformed to σ, respectively. This ensures a minimum consensus among different clients when adding each arc while also limiting the complexity (number of arcs) of the resulting fused model.

---

[4] We ran algorithms for which the source code is publicly available. NOTEARS-ADMM was not included in our tests, as previous works [19,20] have demonstrated its inferior performance compared to the other methods we used to evaluate FedGES.

4.2 Methodology

To assess the validity of FedGES, we have selected 14 BNs from the bnlearn's Bayesian Network Repository[5] [28], generating 10 samples (datasets) with 5000 instances for each one. The focus is on BNs of Medium, Large, and Very Large sizes, described in Table 1. Small BNs are excluded intentionally. The selected BNs constitute a commonly used benchmark in the BN learning literature.

In the experimental evaluation, each sample of 5000 instances is distributed among the varying number of clients used in each algorithm (5, 10, or 20). Consequently, each client is allocated 1000, 500, or 250 instances, respectively. The data is distributed independently and identically across clients (IID), ensuring that each client's dataset retains the same statistical properties as the overall dataset. However, when the number of clients increases, and each client receives fewer instances, the datasets can vary more significantly between clients, which introduces additional challenges for the learning process. This design allows us to study scenarios with limited data, where Federated Learning becomes more valuable in real-world applications. Future work will explore the impact of non-IID data distributions on the performance of FedGES.

Table 1. Bayesian Networks used in the experimental evaluation.

Network	Features					
	NODES	EDGES	PARAMETERS	MAX PARENTS	EMPTY	SMHD
CHILD	20	25	230	4		30
INSURANCE	27	52	1008	3		70
WATER	32	66	10083	5		123
MILDEW	35	46	540150	3		80
ALARM	37	46	509	4		65
BARLEY	48	84	114005	4		126
HAILFINDER	56	66	2656	4		99
HEPAR2	70	123	1453	6		158
WIN95PTS	76	112	574	7		225
PATHFINDER	109	195	72079	5		208
ANDES	223	338	1157	6		626
DIABETES	413	602	429409	2		819
PIGS	441	592	5618	2		806
LINK	724	1125	14211	3		1739

To evaluate the networks obtained by using the tested methods, this article only compares the structure of the discovered networks because using learning scores as BDeu only makes sense on the client's side, as the server has no access

---
[5] https://www.bnlearn.com/bnrepository/.

to any data. In particular, the Structural Moralized Hamming Distance (SMHD) [12,14,15] is used to assess the similarity of the learned BNs to the original one from which the data were sampled. This metric is akin to the literature's widely used Structural Hamming Distance (SHD) [11], but it compares the moralized graphs to avoid considering different equivalent (in)dependencies. As the SMHD compares the structure of the BNs, we can use this metric to assess the quality of networks obtained by the clients and also by the server.

4.3 Reproducibility

To ensure reproducibility, we implemented the GES algorithm and the federated framework from scratch using Java (OpenJDK 20) and the Tetrad 7.1.2-2[6] causal reasoning library. For comparisons with FEDC^{2}SL and RFCD, we use the Python implementation of both algorithms as provided by the authors of FEDC^{2}SL, available on GitHub[7], and executed them with Python 3.10.8. All experiments were conducted on machines equipped with Intel Xeon E5-2650 8-Core Processors and 64 GB of RAM per execution.

Furthermore, for transparency and accessibility, we have created a shared repository on OpenML[8] containing the 10 datasets sampled for each of the 14 BNs, with references to their original papers included in their descriptions. These databases, along with all source code, are also available on GitHub[9].

4.4 Results

In this section, we assess the performance of FedGES by evaluating the global DAG ($\mathcal{G}$) obtained after the federated process.

The SMHD scores of the global BNs $\mathcal{G}$ generated by the server for each algorithm and configuration (#clients, fusion) are presented in Fig. 1[10]. Additionally, Table 2 provides the count of added edges in these networks.

The obtained results lead to several key conclusions:

– FedGES consistently outperforms other algorithms in all scenarios, except for the PATHFINDER network, where, in specific instances, the PC-Voting and RFCD algorithms achieve slightly better results. This divergence can be attributed to the unique characteristics of the PATHFINDER network, which has a semi-naive Bayes structure[11], where improvements in the BDeu score do not necessarily translate to better SMHD [15]. As a result, neither algorithm enhances the SMHD of an empty network (208). When analyzing the number

---

[6] https://github.com/cmu-phil/tetrad/releases/tag/v7.1.2-2.
[7] https://github.com/wangzhaoyu07/FedC2SL.
[8] https://www.openml.org/search?type=data&uploader_id=%3D_33148&tags.tag=bnlearn.
[9] https://github.com/ptorrijos99/BayesFL.
[10] With five clients, C25 and UNION fusion are equivalent ($\lfloor 5 \cdot 0.25 \rfloor = \lfloor 1.25 \rfloor = 1$), involving the addition of all edges present in the DAGs.
[11] https://www.bnlearn.com/bnrepository/discrete-verylarge.html#pathfinder.

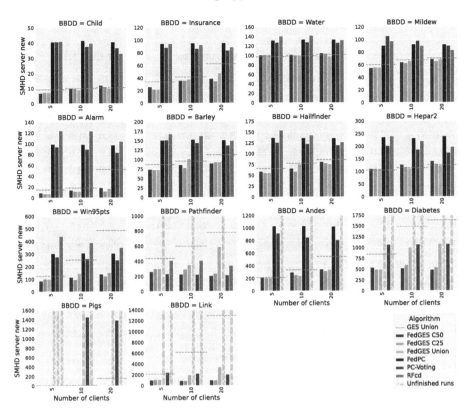

Fig. 1. Mean SMHD of the final global BNs $\mathcal{G}$. Comparison with baselines. "GES Union" lines correspond to running the GES algorithm with no iteration limit on all clients and fusing the resulting networks with a UNION. The "Unfinished run" bars correspond to algorithms that cannot finish the execution in a reasonable amount of time.

of added edges, it is evident that all algorithms achieve a better SMHD when adding fewer edges. Furthermore, this count decreases with an increasing number of clients. This dynamic favors PC-Voting, returning a nearly empty network. In all other networks, the performance of FEDPC, PC-Voting, and RFCD is inferior, with PC-Voting surprisingly outperforming FEDPC in virtually all BNs. Also noteworthy is the case of PIGS, wherein the two cases where PC-Voting terminates (10 and 20 clients), it obtains very poor results in SMHD. At the same time, the three FedGES configurations practically attain the optimal network.
- The two competing algorithms specifically designed for FL, FEDPC and RFCD, fail to run on large BNs. This fact highlights the relevance of FedGES, especially in scenarios involving large networks.
- The application of Federated Learning with FedGES improves the results of merging BNs generated by non-federated GES. This improvement becomes more pronounced as the size of the BN increases and when more clients (each

Table 2. Mean number of edges of the final networks generated by the server.

Network	5 Clients				10 Clients				20 Clients			
	C50	C25	Union	GES	C50	C25	Union	GES	C50	C25	Union	GES
Child	21.6	21.8	21.6	20.0	20.9	23.0	22.4	17.0	18.0	23.0	25.9	
Insurance	37.5	45.0	48.5	32.0	43.0	48.0	57.4	29.0	32.0	54.0	68.2	
Water	25.0	31.0	35.9	20.0	23.0	31.0	40.8	18.0	19.0	31.0	40.2	
Mildew	28.0	29.0	32.0	23.0	27.0	29.0	31.9	17.0	21.0	26.0	26.8	
Alarm	44.0	48.5	55.0	42.0	46.0	53.0	61.4	38.5	44.0	53.5	90.8	
Barley	49.0	52.0	62.2	40.0	47.0	52.0	58.1	35.5	47.0	47.0	62.6	
Hailfinder	62.0	68.0	81.9	46.0	59.0	73.0	84.1	35.0	44.0	58.7	74.4	
Hepar2	43.5	51.0	54.6	30.0	39.5	50.5	52.4	16.0	27.0	51.3	57.2	
Win95pts	108.0	136.0	176.4	87.0	109.0	150.0	262.5	70.0	92.0	150.0	392.4	
Pathfinder	135.5	170.0	298.2	103.0	155.0	200.0	414.4	75.5	116.2	305.0	534.1	
Andes	255.5	273.0	267.8	216.0	242.0	295.0	339.0	193.0	220.3	330.0	437.5	
Diabetes	561.0	620.0	980.6	499.0	577.5	646.2	1451.9	462.2	533.0	846.0	1463.7	
Pigs	592.0	601.0	600.5	593.0	600.0	600.0	607.5	592.0	594.0	622.5	764.4	
Link	856.0	1012.0	2108.7	778.2	959.0	655.5	3947.1	642.0	819.8	2211.0	6386.4	

Network	5 Clients				10 Clients				20 Clients			
	FedPC	PC-Voting	RFcd	FedPC	PC-Voting	RFcd	FedPC	PC-Voting	RFcd			
Child	27.2	24.7	28.8	27.0	21.3	25.5	26.0	21.4	16.2			
Insurance	46.8	38.7	47.5	48.2	36.4	45.5	48.0	27.1	39.8			
Water	31.6	19.0	55.8	32.2	18.1	47.8	34.2	17.0	35.8			
Mildew	20.2	42.7	31.0	23.4	33.3	22.5	22.6	23.1	12.8			
Alarm	54.2	47.1	86.0	53.2	42.0	85.5	52.0	32.1	62.5			
Barley	56.0	51.9	74.2	58.8	41.0	70.7	57.2	30.4	48.3			
Hailfinder	56.0	42.9	71.5	54.8	38.1	58.0	55.4	34.4	37.0			
Hepar2	95.4	53.9	103.2	92.0	35.9	73.5	101.4	21.3	56.0			
Win95pts	105.2	69.7	264.5	108.0	48.1	209.5	104.8	33.6	156.0			
Pathfinder	-	41.4	237.0	-	25.8	227.0	-	14.3	149.0			
Andes	430.8	310.9	-	429.0	240.7	-	423.8	194.4	-			
Diabetes	-	260.0	-	-	268.0	-	-	276.5	-			
Pigs	-	-	-	-	664.0	-	-	592.5	-			
Link	-	704.5	-	-	504.7	-	-	304.3	-			

one with less data) are involved. It is logical since clients with smaller datasets introduce more variability in the networks they generate. As a result, the fusion of all these networks may contain numerous unnecessary edges, resulting in very poor SMHDs. This effect is more significant in larger and more

challenging networks such as PATHFINDER, DIABETES, or LINK with more possible edges to add.
- Comparing the three fusions carried out in FedGES (C50, C25, and UNION), it is evident that the more the BNs generated by the clients diverge among them, whether through an increase in the number of clients or through the creation of larger BNs (as seen in the previous case), the more advisable it becomes to utilize C50 or C25. This is logical since, in small BNs and with a large amount of data, the BNs generated by each client will not differ enough for a C50 or C25 strategy to be noticeable. Furthermore, with five clients, C25 is equivalent to UNION, and C50 actually adds each edge if it appears in 2 of the five networks, which is quite likely. However, in the most complex BNs, we can clearly see how the use of C50 or C25 produces very good results when the number of clients increases. Both UNION and GES increase the number of edges added, and so the SMHD value, which in some cases is even worse than that of PC-Voting or RFCD.

Table 3. Total execution time (seconds) normalized by the number of clients.

Network	5 Clients							10 Clients							20 Clients						
	C50	C25/UN	GES	FEDPC	PC-VOT	RFCD		C50	C25	UNION	GES	FEDPC	PC-VOT	RFCD	C50	C25	UNION	GES	FEDPC	PC-VOT	RFCD
CHILD	0.2	0.2	0.2	7.2	1.7	30.5		0.1	0.1	0.1	0.1	4.5	1.2	18.3	0.1	0.1	0.1	0.0	3.3	0.9	12.5
INSURANCE	0.2	0.3	0.2	11.2	3.2	106.3		0.2	0.4	0.7	0.1	6.9	2.3	56.5	0.3	0.3	11.0	0.1	4.7	1.5	34.8
WATER	0.3	0.3	0.2	2.8	0.9	140.8		0.1	0.4	0.2	0.1	1.7	0.7	78.5	0.1	0.1	0.3	0.1	1.1	0.5	43.4
MILDEW	1.0	1.0	0.4	172.1	11.3	270.9		1.4	1.2	19.3	0.2	130.8	5.4	176.5	1.1	0.9	7.6	0.1	102.3	3.0	96.9
ALARM	0.4	0.4	0.3	8.4	3.6	328.1		0.5	0.3	0.3	0.2	5.0	3.0	165.4	0.2	0.4	1.4	0.1	3.4	2.4	88.6
BARLEY	1.3	1.5	0.5	67.5	7.8	820.8		3.1	16.4	16.3	0.3	42.9	5.3	468.9	12.3	22.4	28.7	0.2	31.1	3.6	276.2
HAILFINDER	0.8	1.4	0.4	17.8	5.6	980.8		0.6	1.7	3.1	0.3	10.6	4.4	480.2	0.7	0.4	2.8	0.2	7.0	3.5	216.3
HEPAR2	0.6	1.4	0.5	34.0	6.1	2727.9		0.4	1.4	3.7	0.3	21.6	4.2	4609.1	0.2	0.3	2.9	0.2	15.1	3.2	1441.3
WIN95PTS	2.2	1.6	0.9	25.0	12.2	6248.5		1.1	1.5	3.1	0.7	14.9	10.1	2185.9	1.6	2.0	251.2	0.5	9.7	8.0	884.3
PATHFINDER	155.8	367.5	2.8	-	2564.2	13972.9		15.3	13.3	114.2	2.0	-	308.6	6625.7	7.6	24.0	37.2	1.5	-	81.5	2941.2
ANDES	15.7	12.6	16.3	215.4	75.0	-		14.7	15.1	13.9	13.8	132.0	57.3	-	17.6	22.3	73.2	10.9	92.4	47.7	-
DIABETES	786.5	746.2	154.3	-	6119.2	-		601.3	665.5	1313.8	148.2	-	2694.3	-	677.4	518.5	511.0	120.2	-	1614.9	-
PIGS	106.8	70.1	150.2	-	-	-		129.4	68.9	96.8	147.9	-	3320.6	-	131.5	90.7	96.9	143.6	-	699.3	-
LINK	3186.2	1811.7	656.0	-	2925.1	-		1343.6	1226.2	543.9	608.8	-	1845.9	-	608.5	1235.1	513.6	504.2	-	1070.1	-

Finally, Table 3 presents the total execution times (sum of all clients and the server) for various algorithms, normalized by the number of clients for clarity. It is evident that FEDPC, PC-Voting, and RFCD exhibit computational complexities several orders of magnitude higher than the FedGES configurations and GES, rendering them impractical for larger Bayesian Networks.

When comparing FedGES with different fusion strategies, we observe that with fewer clients (5), resulting in fewer and more consistent BNs due to increased data per client, the UNION fusion (equivalent to C25) generally outperforms C50 in larger BNs. As the number of clients increases (10, 20), more constrained fusions yield shorter execution times compared to UNION by restricting complex fusions that complicate subsequent FL rounds. GES maintains manageable execution times due to its one-shot strategy; the complexity would lie in initiating a new FL iteration from the intricate result generated by GES. Therefore, given that FedGES performs subsequent iterations, it is advantageous to limit the maximum number of edges it can add per iteration and opt for more restrictive fusions.

5 Conclusions

We introduce Federated GES (FedGES), a novel Federated Learning approach for Bayesian Network (BN) structure learning in decentralized settings. FedGES utilizes the Greedy Equivalence Search (GES) algorithm to iteratively fuse the limited locally generated BN structures from individual clients, creating a unified global structure. The proposed controlled structural fusion in FedGES enhances consensus among different BNs learned by clients, offering a valuable solution in scenarios with larger BNs or numerous clients. This approach ensures data privacy by exchanging only network structures, thus avoiding the exposure of sensitive parameters. Our practical demonstrations over 14 BNs highlight FedGES's real-world applicability, consistently outperforming various federated and non-federated algorithms while maintaining the privacy of local data.

In future research, we plan to investigate the impact of client heterogeneity on the performance of FedGES, and its ability to handle non-IID data distributions among clients would also be valuable. Different strategies for non-uniform fusion in the server will also be investigated. A second line of research will address the process of parameter learning for the network in a federated manner, adding a second phase to FedGES. On the other hand, we could explore additional security mechanisms, e.g., considering the application of obfuscation processes such as differential privacy to further secure the exchange of DAGs. Studying the response of FedGES to different types of malicious attacks in depth is also a promising line of research.

Acknowledgements. The following projects have funded this work: TED2021-131291B-I00 (MICIU/AEI/10.13039/501100011033 and European Union Next GenerationEU/PRTR), SBPLY/21/180225/000062 (Junta de Comunidades de Castilla-La Mancha and ERDF A way of making Europe), PID2022-139293NB-C32 (MICIU/AEI/10.13039/501100011033 and ERDF, EU), FPU21/01074 (MICIU/AEI/10.13039/501100011033 and ESF+); 2022-GRIN-34437 (Universidad de Castilla-La Mancha and ERDF A way of making Europe).

References

1. Alonso, J.I., de la Ossa, L., Gámez, J.A., Puerta, J.M.: Scaling up the greedy equivalence search algorithm by constraining the search space of equivalence classes. Int. J. Approximate Reasoning **54**(4), 429–451 (2013)
2. Barredo Arrieta, A., Díaz-Rodríguez, N., Del Ser, J., et al.: Explainable artificial intelligence (XAI): concepts, taxonomies, opportunities and challenges toward responsible AI. Inf. Fusion **58**, 82–115 (2020)
3. Borunda, M., Jaramillo, O., Reyes, A., Ibargüengoytia, P.H.: Bayesian networks in renewable energy systems: a bibliographical survey. Renew. Sustain. Energy Rev. **62**, 32–45 (2016)
4. de Campos, C.P., Ji, Q.: Efficient structure learning of bayesian networks using constraints. J. Mach. Learn. Res. **12**, 663–689 (2011)

5. Chickering, D.M.: Optimal structure identification with greedy search. J. Mach. Learn. Res. **3**, 507–554 (2002)
6. Chickering, D.M., Heckerman, D., Meek, C.: Large-sample learning of Bayesian networks is NP-hard. J. Mach. Learn. Res. **5**, 1287–1330 (2004)
7. Drury, B., Valverde-Rebaza, J., Moura, M.F., de Andrade Lopes, A.: A survey of the applications of Bayesian networks in agriculture. Eng. Appl. Artif. Intell. **65**, 29–42 (2017)
8. Gámez, J.A., Mateo, J.L., Puerta, J.M.: Learning Bayesian networks by hill climbing: efficient methods based on progressive restriction of the neighborhood. Data Min. Knowl. Disc. **22**(1), 106–148 (2011)
9. Heckerman, D., Geiger, D., Chickering, D.: Learning Bayesian networks: the combination of knowledge and statistical data. Mach. Learn. **20**, 197–243 (1995)
10. Jensen, F.V., Nielsen, T.D.: Bayesian Networks and Decision Graphs. Springer, New York (2007)
11. de Jongh, M., Druzdzel, M.J.: A comparison of structural distance measures for causal Bayesian network models. In: Klopotek, M., Przepiorkowski, A., Wierzchon, S.T., Trojanowski, K. (eds.) Recent Advances in Intelligent Information Systems, Challenging Problems of Science, Computer Science series, pp. 443 – 456. Academic Publishing House EXIT (2009)
12. Kim, G.H., Kim, S.H.: Marginal information for structure learning. Stat. Comput. **30**(2), 331–349 (2019)
13. Kyrimi, E., McLachlan, S., Dube, K., Neves, M.R., Fahmi, A., Fenton, N.: A comprehensive scoping review of Bayesian networks in healthcare: past, present and future. Artif. Intell. Med. **117**, 102108 (2021)
14. Laborda, J.D., Torrijos, P., Puerta, J.M., Gámez, J.A.: A ring-based distributed algorithm for learning high-dimensional Bayesian networks. In: Bouraoui, Z., Vesic, S. (eds.) Symbolic and Quantitative Approaches to Reasoning with Uncertainty, pp. 123–135. Springer Nature Switzerland, Cham (2024)
15. Laborda, J.D., Torrijos, P., Puerta, J.M., Gámez, J.A.: Parallel structural learning of Bayesian networks: iterative divide and conquer algorithm based on structural fusion. Knowl.-Based Syst. **296**, 111840 (2024)
16. Leroy, D., Coucke, A., Lavril, T., Gisselbrecht, T., Dureau, J.: Federated learning for keyword spotting. In: ICASSP 2019 - 2019 IEEE International Conference on Acoustics, Speech and Signal Processing (ICASSP). IEEE (2019)
17. Li, L., Fan, Y., Tse, M., Lin, K.Y.: A review of applications in federated learning. Comput. Ind. Eng. **149**, 106854 (2020)
18. McMahan, B., Moore, E., Ramage, D., Hampson, S., Arcas, B.A.y.: Communication-Efficient Learning of Deep Networks from Decentralized Data. In: Singh, A., Zhu, J. (eds.) Proceedings of the 20th International Conference on Artificial Intelligence and Statistics. Proceedings of Machine Learning Research, vol. 54, pp. 1273–1282. PMLR (2017)
19. Mian, O., Kaltenpoth, D., Kamp, M.: Regret-based federated causal discovery. In: Le, T.D., Liu, L., Kıcıman, E., Triantafyllou, S., Liu, H. (eds.) Proceedings of The KDD 2022 Workshop on Causal Discovery. Proceedings of Machine Learning Research, vol. 185, pp. 61–69. PMLR (2022)
20. Mian, O., Kaltenpoth, D., Kamp, M., Vreeken, J.: Nothing but regrets - privacy-preserving federated causal discovery. In: Ruiz, F., Dy, J., van de Meent, J.W. (eds.) Proceedings of The 26th International Conference on Artificial Intelligence and Statistics. Proceedings of Machine Learning Research, vol. 206, pp. 8263–8278. PMLR (2023)

21. Ng, I., Zhang, K.: Towards federated Bayesian network structure learning with continuous optimization. In: Camps-Valls, G., Ruiz, F.J.R., Valera, I. (eds.) Proceedings of The 25th International Conference on Artificial Intelligence and Statistics. Proceedings of Machine Learning Research, vol. 151, pp. 8095–8111. PMLR (2022)
22. Nguyen, D.C., Ding, M., Pathirana, P.N., Seneviratne, A., Li, J., Vincent Poor, H.: Federated learning for internet of things: a comprehensive survey. IEEE Commun. Surv. Tutorials **23**(3), 1622–1658 (2021)
23. Peña, J.: Finding consensus Bayesian network structures. J. Artif. Intell. Res. (JAIR) **42** (2011)
24. Puerta, J.M., Aledo, J.A., Gámez, J.A., Laborda, J.D.: Efficient and accurate structural fusion of Bayesian networks. Inf. Fusion **66**, 155–169 (2021)
25. Rieke, N., et al.: The future of digital health with federated learning. NPJ Digital Med. **3**(1) (2020)
26. Saputra, Y.M., et al.: Energy demand prediction with federated learning for electric vehicle networks. In: 2019 IEEE Global Communications Conference (GLOBECOM). IEEE (2019)
27. Scanagatta, M., Salmerón, A., Stella, F.: A survey on Bayesian network structure learning from data. Prog. Artif. Intell. **8**(4), 425–439 (2019)
28. Scutari, M.: Learning Bayesian networks with the bnlearn R Package. J. Stat. Softw. **35**(3), 1–22 (2010)
29. Silva, S., et al.: Federated learning in distributed medical databases: meta-analysis of large-scale subcortical brain data. In: 2019 IEEE 16th International Symposium on Biomedical Imaging (ISBI 2019). IEEE (2019)
30. Spirtes, P., Glymour, C., Scheines, R.: Causation, Prediction, and Search. The MIT Press (2001)
31. Wang, Z., Ma, P., Wang, S.: Towards practical federated causal structure learning, p. 351–367. Springer Nature Switzerland (2023)
32. Yang, W., et al.: FFD: A federated learning based method for credit card fraud detection, p. 18-32. Springer International Publishing (2019)
33. Yao, L., Chu, Z., Li, S., Li, Y., Gao, J., Zhang, A.: A survey on causal inference. ACM Trans. Knowl. Discov. Data **15**(5), 1–46 (2021)
34. Zhang, C., Xie, Y., Bai, H., Yu, B., Li, W., Gao, Y.: A survey on federated learning. Knowl. Based Syst. **216**, 106775 (2021)

Enhancing Industrial Control Systems Security: Real-Time Anomaly Detection with Uncertainty Estimation

Ermiyas Birihanu[(✉)], Ayyoub Soullami, and Imre Lendák

Data Science and Engineering Department, Faculty of Informatics, Eötvös Loránd University, Pázmány Péter str. 1/A, 1117 Budapest, Hungary
{ermiyasbirihanu,lendak}@inf.elte.hu

Abstract. Industrial Control Systems (ICS) are crucial for managing essential infrastructure like energy, healthcare, and water treatment. However, as ICS become increasingly complex and interconnected with more edge devices, they become more difficult to manage and vulnerable to new threats. The objective of this study is to proposed a real-time anomaly detection framework and assess the uncertainty associated with the proposed method. An anomaly detection framework for ICS data streams is used, employing Docker containers for scalable and manageable model deployment. Three machine learning models: Autoencoder (AE), LSTM Autoencoder (LSTM-AE), and Seq to Seq (Seq2Seq) were trained on the Secure Water Treatment (SWaT) and HIL- based Augmented ICS security (HAI) datasets to detect anomalies based on reconstruction error deviations. To enhance detection accuracy and assess prediction confidence, Monte Carlo (MC) Dropout and Bayesian Neural Networks (BNNs) were employed. This approach reduces false positives and enables adaptive thresholding for more focused investigations and faster responses to critical anomalies. The results indicate that the proposed models effectively learn normal patterns and accurately detect anomalies. The AE model provided consistent predictions with moderate uncertainty, while the LSTM-AE and Seq2Seq models successfully captured time-based patterns with varying confidence levels. This study highlights that integrating anomaly detection with uncertainty estimation enhances ICS reliability and security by measuring prediction confidence, preventing failures, and supporting resilient industrial processes. Our code is publicly available: https://github.com/Ermiyas21/Realtime-Anomaly-detection-for-ICS.

Keywords: Uncertainty estimation · Anomaly detection · Real time · Industrial Control Systems · latency · throughput

1 Introduction

Industrial Control Systems (ICS) are the backbone of critical infrastructure in sectors like energy, healthcare, and water treatment, essential for smooth industrial operations [1,2]. The increasing interconnectivity and complexity of these systems have made them a key focus within the cybersecurity landscape [3]. As ICS integrate with IT networks and the Internet, they present a larger attack surface and new vulnerabilities. They control essential services, making them high-profile targets for cyberattacks that can cause significant disruptions. Smart devices in ICS are increasingly used by hackers due to the rapid rise in vulnerabilities. In 2018, IoT vulnerabilities in industry grew 14.7% faster than overall network vulnerabilities, indicating that traditional cyber defense strategies may be insufficient for detecting attacks in ICS [4].

To handle abnormalities in ICS data, detecting unusual patterns is a common practice. Researchers employ various methods for this purpose, ranging from simple statistics to complex techniques like deep learning [5]. However, the massive amount of time-based data generated by edge devices in ICS has made this task much more challenging. The time-based characteristic of ICS data is a valuable tool for anomaly detection. Hence, various researchers have proposed real-time anomaly detection frameworks [4] [5]. In real-time anomaly detection, there is a need for analysis results to enable rapid responses and to effectively adjust to dynamic environments, such as those found in ICS. The approach by [6] uses real-time data analysis in ICS utilities to detect anomalies by learning normal behaviors and raising alarms for abnormalities. In [7] the authors introduce a flexible stream processing framework using Docker for online anomaly detection in Industrial Internet of Things (IIoT). It includes a prediction-based model combining batch training with real-time anomaly detection.

Moreover, the proposed methods for anomaly detection in ICS uses various deep learning models [8]. This approach leverages the strengths of multiple models to improve detection accuracy, enhance robustness, and adapt to different types of anomalies typically found in ICS environment. However, these models (DNNs) are known to sometimes make unexpected, incorrect, but overconfident predictions. This can have serious consequences in ICS. Uncertainty estimation aims to address this by estimating the confidence of DNN predictions, going beyond just performance metrics. In recent years, various Uncertainty estimation methods have been developed for DNNs [9]. Uncertainty estimation in deep learning models is classified into two types: data uncertainty (aleatoric) and model uncertainty (epistemic) [9]. Epistemic uncertainty, resulting from insufficient training data, reflects the model's knowledge limitations and decreases with more relevant, high-quality data. In contrast, aleatoric uncertainty stems from inherent randomness in the data itself and remains irreducible regardless of data volume [10]. Addressing these uncertainties is crucial for improving the reliability and accuracy of anomaly detection in ICS, ensuring robust and effective monitoring and predictive capabilities [11,12].

Real-time anomaly detection in ICS is essential for ensuring smooth operations and preventing sudden disruptions. These systems, overseeing critical industrial processes, are highly vulnerable to anomalies that could cause failures or safety hazards. By swiftly identifying and responding to irregular patterns, real-time anomaly detection helps to protect against cyber-physical threats and strengthens the resilience of industrial operations [13]. Developing a real-time anomaly detection framework for ICS environments is challenging due to several factors, including managing vast data dimensions, limited hardware resources, the diversity of normal behaviors, temporal dependencies, unreliable connectivity, and the absence of robust ML/DL techniques. With the rapid development of sensor and Internet of Things (IoT) technologies, huge amounts of time-series data are being generated very quickly. This large dataset is often utilized to make real-time decisions in areas such as IT operations, finance, and healthcare [14]. Given the dynamic nature and complexity of the ICS system, the quality of services (QoS) is crucial. Ensuring low latency, minimal packet-drop, and high reliability are essential for maintaining system integrity and operational efficiency [15].

Various approaches have been proposed to address real-time anomalies in ICS [4,5,7], and uncertainty estimation [16] for deep learning methods. However, these approaches often fall short in terms of performance, scalability for anomaly detection, measurement of real-time quality of service (latency and throughput), and uncertainty estimation for models proposed in ICS altogether.

In this study, we propose a real-time data stream processing framework that integrates machine learning techniques with Docker containers to enhance model performance and scalability for real-time data in ICS. By leveraging Autoencoders, Seq2Seq, and LSTM networks, we use reconstruction error to detect anomalies when data points deviate from the normal data distribution. Addressing uncertainty estimation methods, our approach aims to improve the reliability and robustness of ICS. This integration enhances the system's capability to detect and respond to anomalies, thereby reducing the risk of operational failures and cyber-physical threats.

The contributions of this study are as follows:

- This study introduces a real-time data stream processing framework that integrates machine learning techniques with Docker containers to enhance model performance and scalability in ICS.
- This study conducts comprehensive comparison of various machine learning models and datasets used in reconstruction-based anomaly detection for ICS.
- This study evaluates the proposed model uncertainty using Monte Carlo (MC) dropout simulations and Bayesian Neural Networks (BNNs), comparing their effectiveness to determine the better approach.
- This study examines the impact of Docker containers on quality of service by measuring improvements in data handling scalability, latency, and throughput metrics. These metrics are used to compare performance with and without containerization.

2 Background and Related Works

2.1 Hierarchical Management of ICS Processes

ICS manages critical physical processes, ensuring operational safety and reliability through sensor, actuator, and control system orchestration. According to the

Purdue model of ICS, the system is divided into levels, ranging from the physical process at Level 0, which has the strictest access control, to higher-level applications at Level 3, with less stringent access. At Level 0, sensors and actuators enable feedback and input to the physical process. At Level 1, programmable logic controllers (PLCs) interface directly with these sensors and actuators to automate the process. Level 2, which encompasses Supervisory Control and Data Acquisition (SCADA) systems, oversees multiple PLCs by collecting data and providing operators with an interface for controlling and analyzing the physical process [17]. These systems operated on private, dedicated networks. However, with the widespread adoption of remote management, ICS systems now communicate over open IP networks, such as the Internet. This transition exposes ICS systems to cyberspace, making them susceptible to cyber-attacks via the Internet [18].

2.2 Realtime Data Streaming Technologies

Data stream processing involves continuously ingesting, analyzing, and transforming data in real-time. It is crucial for ICS scenarios because it aids in monitoring, control, optimization, prediction, and decision-making. However, several challenges arise, such as handling the large volume, speed, variety, and reliability of data; limited resources and connectivity of edge devices; differences and distribution of data sources and processing nodes; and the complex needs of ICS applications. To address these issues, various stream processing technologies have been proposed or adopted to support data stream processing in ICS scenarios [7].

Some of the most popular streaming technologies are Apache Spark Streaming, Apache Flink, and Apache Kafka, which are used for ingestion, storage, query processing, scheduling, fault tolerance, and load balancing for data streaming [7]. There are also big data databases like InfluxDB, used for time-series data, and Elasticsearch, used for search and analytics. Additionally, visualization tools such as Grafana are available for streaming data, enabling users to create interactive and real-time dashboards. Together, these technologies provide a comprehensive solution for managing, analyzing, and visualizing data streams in real-time. All of these applications can be integrated using Docker, a well-known and popular open-source containerization framework in the software industry. Docker automates the development and deployment of application containers by allowing developers to bundle an application with its runtime requirements into a container image, which can then be executed on a host system [19]. In this study, we utilized containerized real-time streaming technologies for streaming the data.

2.3 Related Works

Machine learning approaches like statistical, distance-based, and density-based methods have been used for anomaly detection [20], but each has limitations in ICS scenarios. Statistical methods struggle with dynamic, contextual data streams due to their need for fixed distributions. Distance-based methods are computationally intensive, requiring distance calculations for all data pairs. Density-based methods are sensitive to noise and outliers, affecting density estimations. Some hybrid and deep learning methods, using techniques like CNNs and autoencoders, have been proposed to improve anomaly detection in IoT data streams [21]. However, hybrid methods can suffer from drawbacks like low accuracy and high latency since they depend on multiple methods [7].

Real-time anomaly detection is crucial for quickly spotting unusual behavior in systems. It addresses the drawbacks of the dynamic nature of ICS and the complex relationships between edge devices. Additionally, it improves the performance of the proposed method by being scalable for big data and ensuring low latency. Chen et al. [22] propose a real-time, fully distributed anomaly detection algorithm (GAD) for Networked Industrial Sensing Systems (NISS). Through extensive simulations, the effectiveness and scalability of this method for monitoring building structures and smart grids have been validated. While Chen et al. [22] GAD algorithm demonstrates effectiveness in building structure and smart grid monitoring, further advancements are necessary to address the complexities of broader industrial applications.

Bhatia et al. [23] introduce Midas, a real-time anomaly detection algorithm for dynamic graphs in edge streams. Midas excels at identifying microcluster anomalies, sudden bursts of similar edges often indicating denial-of-service attacks. It uses count-min sketches and chi-squared tests for efficient anomaly assessment. To further enhance accuracy, Midas-F is proposed, modifying the scoring function and adding a conditional merge step. Experiments show Midas-F achieves significant accuracy improvements (up to 62% in ROC-AUC) compared to recent baselines. However, further optimization is needed for handling very large and diverse data patterns.

The study in [24] explores real-time anomaly detection for network data streams in cloud data centers using PySpark. Their method monitors CPU load and memory usage changes across VMware environments. PySpark with flat-increment clustering establishes normal attributes, leading to lower processing latency compared to Storm and Scala-Spark. While achieving high efficiency and accuracy, the study suggests integrating machine learning for even better performance with complex data patterns.

In [25] the authors present a Seq2Seq anomaly detection algorithm for real-time network traffic analysis. Their encoder-decoder model learns normal TCP packet sequences and flags deviations as anomalies. Trained on attack-free data, it achieves 97% accuracy for anomaly detection and 100% accuracy for specific attacks like Port Sweep and Neptune DOS. However, further testing on diverse and complex datasets is recommended for real-world applicability.

Ahmad et al. [26] introduced unsupervised real-time anomaly detection in streaming data using Hierarchical Temporal Memory (HTM). This approach continuously monitors individual streams for unusual behaviors and adapts to concept drift without labeled data. Tested on real-world data with labeled anomalies (Numenta Anomaly Benchmark), it effectively detects both spatial and temporal anomalies. However, further work is needed for complex data patterns and real-world application integration.

The current related works lacks comprehensive performance analyses, including measurements of throughput, latency, and uncertainty estimation for the proposed method in dynamic and streaming ICS data. In this study, we address this gap by providing detailed performance analyses of system latency, throughput and stability, and uncertainty estimation for the proposed method.

3 Methodology

The goal of this study is to propose a real-time framework for identifying anomalies and measuring uncertainty estimation in ICS. The methodology involves a two-module approach, as depicted in Fig. 1: a containerized stack module and an anomaly detection module. The containerized stack module (Docker container) is responsible for data acquisition and preprocessing, streaming data from the source and preparing it for analysis through cleaning and normalization. The hypothesis is that containerization can address common challenges associated with big data transactions within ICS environments. The container stack used in this module includes Docker[1], Apache Kafka[2], Apache Spark[3], InfluxDB[4], and Grafana[5]. These technologies manage large amounts of data from different sources, ensuring swift and seamless processing.

The anomaly detection module utilizes a pre-trained model. Initially, the model is trained and stored in Keras format for subsequent use. This saved model is integrated into a Docker stack, enabling real-time analysis of streaming data. The model is deployed to assess and classify incoming data, determining whether anomalies are present.

Next, the anomaly detection module leverages preprocessed data to evaluate the performance of the proposed method in identifying anomalies, detecting deviations from normal patterns, and quantifying the uncertainty associated with prediction results. Together, these components form a comprehensive framework that enhances the robustness and trustworthiness of anomaly detection in ICS. This framework not only identifies anomalies but also provides a measure of the uncertainty in the detection process.

The process begins with data stored in a Comma-Separated Values (CSV) file, which is prepared and transformed for model training. The prepared data is ingested into a Kafka topic using a Kafka producer that reads the CSV file. PySpark, running within a Spark container, processes the data stream in real-time. A pre-trained model (such as Autoencoder, LSTM-AE, or Seq2Seq) for anomaly prediction, housed in the container, receives the processed data and generates predictions, detecting anomalies using reconstruction error.

---

[1] https://www.docker.com/.
[2] https://kafka.apache.org/.
[3] https://spark.apache.org/.
[4] https://www.influxdata.com/.
[5] https://grafana.com/.

The reconstruction error in system states within ICS are utilized to discover the presence of anomalies. First, system states $\tilde{X}$ over the previous m timesteps are collected from the edge device of ICS physical traffic, up to the current timestep t. Second, the pre-trained model is provided the system state sequence $(\tilde{X}_{t-m}, \tilde{X}_{t-m+1}, \ldots, \tilde{X}_t)$ and predicts the next system state $\tilde{X}'_{t+1}$. Third, the predicted and observed states are compared, and the reconstruction error $\tilde{e}_t$ is computed through the root mean-squared-error (RMSE): $\tilde{e}_t = \sqrt{(\tilde{X}'_t - \tilde{X}_t)^2}$. Lastly, the prediction $\tilde{y}_t$ is calculated over a sequence of reconstruction errors $(\tilde{e}_0, \tilde{e}_1, \ldots, \tilde{e}_t)$. Specifically, $\tilde{y}_t = 1$ when the reconstruction error exceeds a predetermined threshold T.

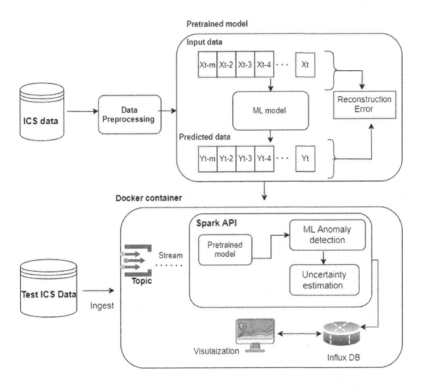

Fig. 1. Real Time anomaly detection frame work for ICS

The uncertainty of these prediction models is measured using methods MC dropout and BNNs to assess confidence levels. Incorporating MC dropout and BNN into LSTM, autoencoder, and Seq2Seq models estimates predictive uncertainty by using dropout during inference. Multiple forward passes provide a distribution of predictions, from which mean and variance quantify uncertainty. This is crucial for reliable anomaly detection and system monitoring in ICS time series prediction. Compute the mean and variance of the predictions to estimate the predictive uncertainty for LSTM, AE and Seq2Seq.

The mean prediction $\hat{\mathbf{y}}_t$ is calculated as the average of these K predictions: $\hat{\mathbf{y}}_t = \frac{1}{K}\sum_{k=1}^{K} \hat{\mathbf{y}}_t^k$. The variance $\text{Var}(\hat{\mathbf{y}}_t)$ is computed as the average squared deviation from the mean: $\text{Var}(\hat{\mathbf{y}}_t) = \frac{1}{K}\sum_{k=1}^{K}(\hat{\mathbf{y}}_t^k - \hat{\mathbf{y}}_t)^2$.

Predictions and uncertainty measures are saved to InfluxDB, a time-series database, and visualized in Grafana. Grafana dashboards monitor the model's predictions and uncertainty scores in real-time, triggering alerts if the uncertainty score exceeds a threshold T, even when the model suggests normal behavior. This continuous monitoring setup enhances the system's reliability and security by enabling timely detection and response to anomalies.

4 Experiment

In this section, we perform experiments to assess the real-time anomaly detection algorithm, the proposed data stream processing framework, and the uncertainty estimation.

4.1 Experiment Setup

We used the Secure Water Treatment (SWaT), and HIL-based Augmented ICS Security (HAI) datasets.

Secure Water Treatment (SWaT): It is a crucial resource designed to assist researchers in securing Cyber-Physical Systems (CPS)[6]. Data collection spanned 11 days: the first seven days captured normal operations to establish a strong baseline state of the system, while the last four days included various cyber and physical attacks. The dataset comprises two main types of data: physical properties data, recorded every second from sensors and actuators, and network traffic data, captured using equipment from Check Point Software Technologies. The 2015 A1 and A2 SWaT Dataset contains 53 features, with 496,800 training records from normal operations and 449,919 records from attack scenarios of the physical properties [27].

HIL- Based Augmented ICS security (HAI): It includes both normal and attack data, with each instance comprising 51 features. Each instance in the HAI dataset represents a set of control loop operations that monitor and adjust process variables. The simulation incorporates a HAI simulator, a turbine, a water treatment system, and a boiler. The dataset is split into two training files and two testing files, each containing different attack scenarios and varying data recording intervals. It consists of 63 features, both numerical and categorical, spanning a 10-day period for training (with only normal data) and 5.5 days for testing (including 38 attacks)[7].

---

[6] https://itrust.sutd.edu.sg/itrust-labs_datasets/dataset_info/.
[7] https://www.usenix.org/conference/cset20/presentation/shin.

Data Preprocessing: The Data Preprocessing for the SWaT and HAI datasets began with checking for missing values. Through our experiments, none were found, ensuring data integrity and eliminating the need for imputation or deletion. We then focused on identifying and removing consistent features-those having the same value across all samples. One-hot encoding was used to encode categorical features in the dataset. This process transforms categorical features into a numerical format suitable for machine learning algorithms. For numerical features, we used Min-Max scaling. In the proposed method, 70% of the training dataset is randomly chosen for training the ML model. The remaining 30%, referred to as the validation dataset, is used for validation.

The proposed method faced class imbalance in our dataset, with the attack class representing around 0.9% of the HAI sample and 5.8% of the SWaT samples. To address this, we used the Synthetic Minority Oversampling Technique for Nominal and Continuous (SMOTE-NC) method [28][8]. This technique generates synthetic data points for numerical features through linear interpolation and assigns categorical feature values based on the most common category among the nearest neighbors. SMOTE-NC ensures the minority class is adequately represented, creating a balanced dataset that improves the performance and reliability of our machine learning models.

The proposed method used GridSearch[9] for hyperparameter tuning to find the best parameters and enhance its performance. The best hyperparameters for training the proposed method on both datasets are listed in Table 1.

Table 1. Hyperparameters used for AE and LSTM

Model	Hyperparameter	SWaT	HAI
AutoEncoder	Batch Size	32	256
	Hiden_dim	8	8
	Epochs	50	50
LSTM	Batch Size	64	64
	learning_rate	0.001	0.001
	Epochs	10	50
	Lstm_Units	64	64

4.2 Experiment I-Baseline Method

Our baseline models included an Autoencoder with an 8-neuron hidden layer trained with Adam optimizer, MSE loss, and early stopping for dimensionality

---

[8] https://imbalanced-learn.org/dev/references/generated/imblearn.over_sampling.SMOTENC.html.
[9] https://scikit-learn.org/stable/modules/grid_search.html.

reduction, and an LSTM network with two LSTM layers and sigmoid activations trained similarly to serve as a benchmark for time series analysis. We denoted the baseline autoencoder and LSTM as b-AE and b-LSTM.

4.3 Experiment II Proposed Method

We built three anomaly detection models: an autoencoder with dropout for dimensionality reduction, an LSTM autoencoder for capturing temporal patterns, and a Seq2Seq model for understanding regular sequences. All models used dropout to prevent over fitting and early stopping to avoid over training. These methods use supervised learning techniques to train models on historical data and predict future values of data streams. We used RMSE to estimate reconstruction error and identify anomalies exceeding a predefined threshold. Finally, the proposed method estimates the uncertainty of the model.

After developing the anomaly detection model, accurately estimating uncertainty becomes essential. This quantifies the model's confidence and provides insights into its reliability and robustness. To measure the uncertainty surrounding the predictions made by the model, the MC Dropout and BNNs techniques are employed. This involves generating multiple predictions with dropout activated during inference, enabling a comprehensive assessment of the model's confidence in its predictions. Through MC Dropout, the model's uncertainty regarding its predictions is quantified, providing valuable insights into the reliability and robustness of the anomaly detection process. Following processing, the model's predictions and the calculated uncertainty measures are written to InfluxDB, which serves as a storage point for Grafana to pull data for visualization. Figure 2 shows the uncertainty estimation for real-time anomaly detection on test data using Grafana. The X-axis represents time, and the Y-axis represents the threshold values. When new data deviates from the threshold values, an alert is triggered for ICS engineers, indicating anomalies in the system.

We now present the approach used to estimate uncertainty in our real-time ICS anomaly detection framework. In Figs. 3 and 4, the X-axis represents the feature, and the Y-axis shows the predicted values. Additionally, the figures display the estimated predictive uncertainty in the two top-performing AE models for the HAI dataset and the LSTM-AE model for the SWaT dataset in the proposed anomaly detection method, using test data. In these figures, the red line indicates the mean of algorithm predictions. The gray interval represents the estimated prediction interval, indicating that each time step's prediction falls within this interval. When new data arrives, denoted by the blue line, if it falls outside the gray area, we would consider it an abnormal data point.

4.4 Result and Discussion

This section compares the performance of our proposed real-time anomaly detection models with baseline models and the latest related work. Table 2 presents the results for both the SWaT and HAI datasets, including those from the baseline methods, related work and our proposed real-time approach.

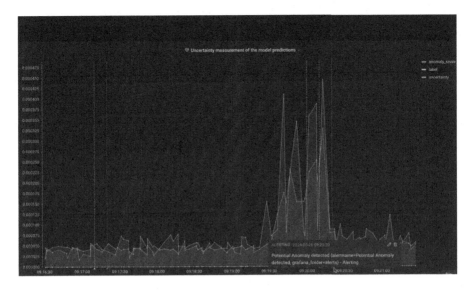

Fig. 2. Grafana: Uncertainty measurement for AE model

SWaT Dataset: The AE model demonstrated strong performance on the SWaT dataset with a recall of 0.89, indicating successful identification of 89% of actual anomalies. It achieved a precision of 0.91, accurately flagging 91% of detected anomalies as true anomalies, resulting in a high F1-score of 0.90, reflecting a favorable balance between precision and recall. Moreover, the AUC of 0.96 underscored the model's robust ability to discriminate between normal and anomalous instances. In contrast, the LSTM-AE exhibited a recall of 0.62, lower than the AE, indicating detection of fewer actual anomalies. However, it had a remarkably high precision of 0.99, implying that almost all detected anomalies were true anomalies. Despite this, its F1-score stood at 0.76, indicating a comparatively lower balance between precision and recall. The AUC of 0.93 suggested commendable overall performance, although slightly lower than that of the AE. The Seq2Seq model consistently performed with both recall and precision at 0.76, resulting in an F1-score of 0.76. Although its AUC of 0.91 indicated its capability to distinguish between normal and anomalous data, it fell slightly short compared to the AE. In summary, the AE and b-AE models demonstrated the best performance on the SWaT dataset, with high F1-scores and AUC values. The LSTM-AE and Seq2Seq models also performed well but were slightly less effective in comparison. The b-LSTM model had the weakest performance among the tested models.

HAI Data Set: The AE achieved a recall of 0.74, correctly identifying 74% of actual anomalies, with a precision of 0.70. This balanced performance resulted in an F1-score of 0.72 and an AUC of 0.79, indicating a solid ability to distinguish between normal and anomalous instances. The LSTM-AE achieved a slightly higher recall at 0.76, capturing more actual anomalies than the AE, but with

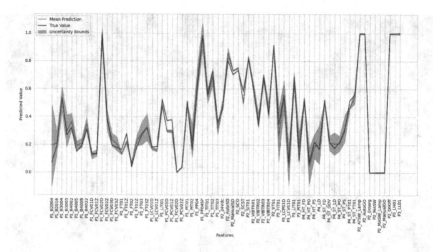

Fig. 3. AE predictions with uncertainty on HAI (Color figure online)

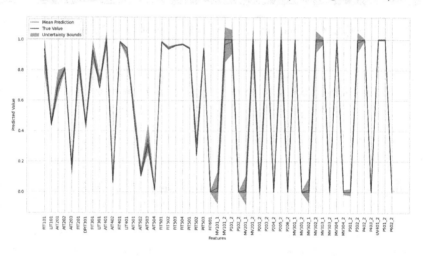

Fig. 4. LSTM-AE predictions with uncertainty on SWaT (Color figure online)

a lower precision of 0.63, resulting in an F1-score of 0.69 and an AUC of 0.76, showing competent but slightly less effective performance compared to the AE. The Seq2Seq model had a recall of 0.71 and a precision of 0.59, leading to an F1-score of 0.65 and an AUC of 0.71. While it could differentiate between normal and anomalous data, it was less reliable than the AE and LSTM-AE models. In summary, on the HAI dataset, the AE showed balanced and solid performance. The LSTM-AE and Seq2Seq models were competent but less effective. The b-AE had the highest recall but lower precision, while the b-LSTM had the weakest overall performance. In summary, on the HAI dataset, the AE showed balanced and solid performance. The LSTM-AE and Seq2Seq models were competent but less effective. The b-AE had the highest recall but lower precision, while the b-LSTM had the weakest overall performance.

Table 2. Performance comparison of different models on the SWaT and HAI datasets

	Model	Recall	Precision	F1-score	AUC
Hussain Nizam et al. [5]	XGBoost	0.77	0.67	0.72	-
	CNN	0.71	0.68	0.69	-
	Autoencoder	0.69	0.72	0.70	-
Proposed models on SWaT	Autoencoder	0.89	0.91	0.90	0.96
	LSTM-AE	0.62	**0.99**	0.76	0.93
	Seq2Seq	0.76	0.76	0.76	0.91
	b-AE	**0.91**	0.88	**0.91**	**1.00**
	b-LSTM	0.76	0.55	0.64	0.62
Proposed models on HAI	Autoencoder	0.74	**0.70**	**0.72**	**0.79**
	LSTM-AE	0.76	0.63	0.69	0.76
	Seq2Seq	0.71	0.59	0.65	0.71
	b-AE	**0.89**	0.55	0.68	0.66
	b-LSTM	0.58	0.55	0.56	0.58

This experiment was conducted using Colab Pro and a personal computer with Python 3.10.12, TensorFlow, and Keras. The hardware included an 11th Gen Intel Core i5-1135G7 CPU @ 2.40 GHz with 8 GB RAM and Intel Xeon CPUs @ 2.20 GHz with 50.99 GB RAM. Docker containers were used to orchestrate services like Kafka, Grafana, PySpark, Zookeeper, and InfluxDB for distributed data processing, monitoring, and visualization. Zookeeper handles coordination, Kafka manages messaging, and PySpark processes data. InfluxDB stores time-series data, and Grafana visualizes it. Services communicate via a bridge network, with persistent volumes ensuring data durability, even after restarts. The setup is modular for easy expansion.

ICS components require fast communication, but security protocols can slow it down [29]. Many SCADA systems communicate at very low rates, such as 19200, 1200, or even 300 bits per second [30]. When streaming data, it is critical to measure response time (latency) and throughput (data points transmitted per second). In our study, latency and throughput were evaluated using Autoencoder and LSTM-AE methods on the SWaT test data over the last 72 data points, as these methods achieved the highest performance scores. Figure 5 shows latency and throughput trends over time, with the Autoencoder method demonstrating lower latency and higher throughput compared to LSTM-AE. Both methods achieve efficient data transmission within 0–5 s, with a minimum throughput of 1907 data points per second. As shown in Fig. 5, the reduced throughput in LSTM-AE proposed model over time can be attributed to several factors. These include increasing latency buildup, computational demands, growing sequence lengths, and resource constraints.

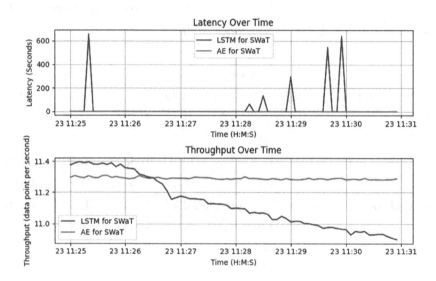

Fig. 5. Real time analysis for streaming data

5 Conclusion

Anomaly detection is a critical task in ICS environments due to its significant impact on normal system operations. Anomaly detection can occur in two ways: offline or online mode. In offline mode, models are constructed using large historical datasets to identify anomalies in batches, which involves looking back in time and is not suitable for real-time applications. On the other hand, online mode provides nearly real-time results, enabling quick responses and adaptation to changing conditions.

This study introduces a real-time anomaly detection framework that integrates machine learning techniques, uncertainty estimation, and reconstruction error analysis to enhance anomaly detection in time series data. The framework incorporates a pre-trained model utilizing AE, LSTM, and Seq2seq methods, deployed within a scalable Docker container environment to handle big data and streaming technologies like Kafka and PySpark. Evaluation of the proposed method demonstrates its effectiveness in addressing anomaly detection challenges in ICS, alongside assessments of performance of the proposed methods, latency and throughput for streaming data.

Future research will focus on addressing several limitations not covered in this study. These include incorporating data uncertainty, integrating domain knowledge feedback, applying federated learning (FL), and developing mitigation strategies.

References

1. ISACA. Introduction to ICS/OT systems and their role in critical infrastructure (2023). https://www.isaca.org/resources/news-and-trends/isaca-now-blog/2023/introduction-to-ics-ot-systems-and-their-role-in-critical-infrastructure. Accessed 19 May 2024
2. National Institute of Standards and Technology (NIST). Industrial control systems (ICS) (2024). https://www.nist.gov/itl/applied-cybersecurity/cybersecurity-and-privacy-applications/industrial-control-systems-ics. Accessed 19 May 2024
3. Birihanu, E., Barcsa-Szabó, Á., Lendák, I.: Proximity-based anomaly detection in securing water treatment. In: 2022 IEEE 2nd Conference on Information Technology and Data Science (CITDS), pp. 34–38. IEEE (2022)
4. Hao, W., Yang, T., Yang, Q.: Hybrid statistical-machine learning for real-time anomaly detection in industrial cyber-physical systems. IEEE Trans. Autom. Sci. Eng. **20**(1), 32–46 (2021)
5. Nizam, H., Zafar, S., Lv, Z., Wang, F., Xiaopeng, H.: Real-time deep anomaly detection framework for multivariate time-series data in industrial IoT. IEEE Sens. J. **22**(23), 22836–22849 (2022)
6. Clotet, X., Moyano, J., León, G.: A real-time anomaly-based ids for cyber-attack detection at the industrial process level of critical infrastructures. Int. J. Crit. Infrastruct. Prot. **23**, 11–20 (2018)
7. Wang, R., Qiu, H., Cheng, X., Liu, X.: Anomaly detection with a container-based stream processing framework for industrial internet of things. J. Industr. Inf. Integr. **35**, 100507 (2023)
8. Gawlikowski, J., et al.: A survey of uncertainty in deep neural networks. Artif. Intell. Rev. **56**(Suppl 1), 1513–1589 (2023)
9. He, W., Jiang, Z.: A comprehensive survey on uncertainty quantification for deep learning. ACM Comput. Surv. **37**(4) (2024)
10. iMerit. A comprehensive introduction to uncertainty in machine learning (2022). https://imerit.net/blog/a-comprehensive-introduction-to-uncertainty-in-machine-learningall-una/. Accessed 14 May 2024
11. Tyralis, H., Papacharalampous, G.: A review of predictive uncertainty estimation with machine learning. Artif. Intell. Rev. **57**(4), 94 (2024)
12. Legrand, A., Trannois, H., Cournier, A.: Use of uncertainty with autoencoder neural networks for anomaly detection. In: 2019 IEEE Second International Conference on Artificial Intelligence and Knowledge Engineering (AIKE), pp. 32–35. IEEE (2019)
13. Shehu, Y., Harper, R.: Efficient periodicity analysis for real-time anomaly detection. In: NOMS 2023-2023 IEEE/IFIP Network Operations and Management Symposium, pp. 1–6. IEEE (2023)
14. Sun, W., Li, H., Liang, Q., Zou, X., Chen, M., Wang, Y.: On data efficiency of univariate time series anomaly detection models. J. Big Data **11**(1), 1–31 (2024)
15. Ravikumar, G., Govindarasu, M.: Anomaly detection and mitigation for wide-area damping control using machine learning. IEEE Trans. Smart Grid (2020)

16. Ardestani, S.B.: Time series anomaly detection and uncertainty estimation using LSTM autoencoders (2020)
17. Fung, C., Srinarasi, S., Lucas, K., Phee, H.B., Bauer, L.: Perspectives from a comprehensive evaluation of reconstruction-based anomaly detection in industrial control systems. In: Atluri, V., Di Pietro, R., Jensen, C.D., Meng, W. (eds.) ESORICS 2022. LNCS, vol. 13556, pp. 493–513. Springer, Cham (2022). https://doi.org/10.1007/978-3-031-17143-7_24
18. Teixeira, M.A., Salman, T., Zolanvari, M., Jain, R., Meskin, N., Samaka, M.: Scada system testbed for cybersecurity research using machine learning approach. Future Internet **10**(8), 76 (2018)
19. Nandi, A., Xhafa, F., Kumar, R.: A docker-based federated learning framework design and deployment for multi-modal data stream classification. Computing **105**(10), 2195–2229 (2023)
20. Wang, J., Ma, Y., Zhang, L., Gao, R.X., Wu, D., Methods and applications: Deep learning for smart manufacturing. J. Manuf. Syst. **48**, 144–156 (2018)
21. Zhang, C., et al.: A deep neural network for unsupervised anomaly detection and diagnosis in multivariate time series data. In: Proceedings of the AAAI Conference on Artificial Intelligence, vol. 33, pp. 1409–1416 (2019)
22. Chen, P.-Y., Yang, S., McCann, J.A.: Distributed real-time anomaly detection in networked industrial sensing systems. IEEE Trans. Industr. Electron. **62**(6), 3832–3842 (2014)
23. Bhatia, S., Liu, R., Hooi, B., Yoon, M., Shin, K., Faloutsos, C.: Real-time anomaly detection in edge streams. ACM Trans. Knowl. Discov. Data (TKDD) **16**(4), 1–22 (2022)
24. Ranganathan, G.: Real time anomaly detection techniques using Pyspark frame work. J. Artif. Intell. **2**(01), 20–30 (2020)
25. Loganathan, G., Samarabandu, J., Wang, X.: Sequence to sequence pattern learning algorithm for real-time anomaly detection in network traffic. In: 2018 IEEE Canadian Conference on Electrical and Computer Engineering (CCECE), pp. 1–4. IEEE (2018)
26. Ahmad, S., Lavin, A., Purdy, S., Agha, Z.: Unsupervised real-time anomaly detection for streaming data. Neurocomputing **262**, 134–147 (2017)
27. Goh, J., Adepu, S., Junejo, K.N., Mathur, A.: A dataset to support research in the design of secure water treatment systems. In: Havarneanu, G., Setola, R., Nassopoulos, H., Wolthusen, S. (eds.) CRITIS 2016. LNCS, vol. 10242, pp. 88–99. Springer, Cham (2017). https://doi.org/10.1007/978-3-319-71368-7_8
28. Almeida, G., Bacao, F.: Umap-smotenc: a simple, efficient, and consistent alternative for privacy-aware synthetic data generation. Knowl.-Based Syst. 112174 (2024)
29. Solomakhin, R., Tsang, P., Smith, S.: High security with low latency in legacy SCADA systems. In: Moore, T., Shenoi, S. (eds.) ICCIP 2010. IAICT, vol. 342, pp. 63–79. Springer, Heidelberg (2010). https://doi.org/10.1007/978-3-642-16806-2_5
30. Wright, A.K., Kinast, J.A., McCarty, J.: Low-latency cryptographic protection for SCADA communications. In: Jakobsson, M., Yung, M., Zhou, J. (eds.) ACNS 2004. LNCS, vol. 3089, pp. 263–277. Springer, Heidelberg (2004). https://doi.org/10.1007/978-3-540-24852-1_19

ITERADE - ITERative Anomaly Detection Ensemble for Credit Card Fraud Detection

Bahar Emami Afshar[1](✉), Paula Branco[1], Tolga Kurt[2], Utku Gorkem Ketenci[2], and Hikmet Mazmanoglu[2]

[1] EECS, Faculty of Engineering, University of Ottawa, Ottawa, ON, Canada
bemam006@uottawa.ca
[2] H3M Analytics Inc., Montreal, QC, Canada

Abstract. Anti-money laundering (AML) efforts are critical not just for financial stability but also for global security, as money laundering supports various criminal activities like terrorism, human trafficking, and drug trade. Fraud detection, as part of AML compliance processes, is essential for companies to protect their customers' transactions and accounts from fraudulent activity. Fraud and Money Laundering intersect in two different areas: laundering of the proceeds of the fraud, or using fraudulent accounts for money laundering. Machine learning has shown promising results in multiple domains. However, in an AML setting, particularly in supervised learning, it faces important challenges associated with limited available labels, difficulty in detecting emerging criminal tactics, and scarce representation of suspicious cases. Despite the high volume of data available, the instances are typically unlabeled and unbalanced; the real suspicious cases (true positives) are rare, and the labeling process is costly, labor-intensive, and requires domain expertise. To address the above issues, this paper introduces ITERADE, a novel iterative anomaly detection ensemble that integrates dynamic clustering to isolate anomalous and normal cases effectively while allowing the end-user to set the desired budget of cases to be returned. In a completely unsupervised manner, ITERADE provides a strategic starting point for labeling and instances' inspection by domain experts. This budget-agnostic solution outperforms existing alternatives while offering flexibility in adjusting labeling budgets according to end-user preferences. ITERADE can select a subset of the entire dataset within a flexible budget, achieving a higher percentage of positive cases by a factor of 3 to 15. This capability is particularly beneficial for highly imbalanced datasets, especially when working with a smaller labeling budget.

Keywords: Financial Crime · Machine learning · Anomaly detection · Imbalanced domain · Fraud detection · Unsupervised learning

1 Introduction

With the rise of e-commerce and online transactions, the use of credit and debit cards has surged, leading to an increase in fraudulent activities. Detecting and preventing credit card fraud is crucial not only to avoid financial losses and data breaches but also to maintain productivity and customer trust. In 2021, the Federal Trade Commission (FTC) received nearly 390,000 reports of credit card fraud, highlighting its prevalence in the U.S. [1]. A significant portion of this fraud is card-not-present fraud, which includes online, phone, and mail-order transactions, resulting in an estimated $5.72 billion in losses in 2022 alone [2].

Machine learning (ML) techniques have shown promise in detecting credit card fraud [6,16,23,24]. However, several challenges persist. One of these challenges lies in labeling the data, an expensive and time-consuming task requiring substantial domain knowledge and expertise. Often, the data available for training ML models is insufficiently labeled, i.e., only a small number of cases are labeled, leading researchers to use alternative methods to supervised learning, such as semi-supervised and active learning strategies. Moreover, in areas such as fraud detection, the labeled data is not only scarce but also highly imbalanced, complicating the process further. In this paper, we focus on the common setting where there are no labels available, which is common for especially fintechs that are just entering the market and require fraud detection from day 1. This occurs when companies wish to deploy ML models but have not yet labeled any cases. This is a typical setting in fraud detection scenarios because there is a vast amount of data but only a few fraud cases. This raises the question of which cases to label first, i.e., how to start the costly, labor-intensive, and time-consuming process of labeling. If the cases are not correctly selected, there is a high probability of labeling only non-fraudulent cases as finding fraud cases among a vast majority of non-fraudulent instances is very hard. Thus, identifying the most suspicious transactions to label first is critical in these imbalanced datasets to maximize the detection of fraudulent activities.

In this study, we introduce ITERADE, an iterative anomaly detection ensemble with a dynamic clustering approach. ITERADE aims to separate anomalous cases from normal ones, reducing the imbalance ratio in a completely unsupervised manner. This method provides a valuable starting point for labeling experts and offers a flexible solution that outperforms existing baselines while accommodating various labeling budgets.

Contributions: Our main contributions are as follows:

– We provide a literature review focused on credit card fraud detection and we cover recent endeavors involving various categories of anomaly detection.
– We propose ITERADE, an unsupervised solution with an adjustable budget that selects a subset of the dataset with a higher percentage of fraud cases compared to the initial dataset in credit card fraud detection, making it easier for labeling experts to start identifying fraudulent transactions.

– We conduct a thorough experimental comparison on multiple datasets, with different imbalance ratios, showing the competitiveness of the proposed ITERADE method.

Organization: This paper is organized as follows. Section 2 provides background information on various types of anomaly detection, followed by a review of the most relevant related works. Section 3 presents our proposed algorithm, ITERADE. In Sect. 4, we detail our experimental configuration. The results from our experiments are presented and discussed in Sect. 5. Finally, Sect. 6 concludes the paper and offers insights for future research avenues.

2 Related Work

We initiate our exploration of related works and recent endeavors by categorizing anomaly detection into distinct types. Goldstein and Uchida [8], conducted a comprehensive survey, comparing 19 unsupervised anomaly detection algorithms across 10 datasets. Their study categorizes anomaly detection setups into three distinct categories based on available dataset labels.

– Supervised Anomaly Detection: Used when both normal and anomalous instances are fully labeled in the data. However, this approach is limited by imbalanced data and specific classifiers. It's less practical in real-world scenarios where anomalies aren't known in advance.
– Semi-supervised Anomaly Detection: Also known as "one-class" classification [Perera et al. [19]], it involves labeled training data with only normal instances, while anomalies appear in the test set. This method trains a model on normal data to predict anomalies based on deviations from the model.
– Unsupervised Anomaly Detection: This setup treats all data as a single set without distinction between training and test sets and requires no labels. The algorithms predict the anomalies based on inherent properties of the data, often using metrics such as distance and density to define the normal and anomalous instances.

In this literature review, we focus on related works in detecting anomalies through unsupervised approaches, as supervised methods are not suitable for our imbalanced domain.

Several works have explored using autoencoders in the fraud detection domain. As an example of one-class anomaly detection, focusing solely on training the model with non-fraudulent cases, is the work of Jiang et al. [11]. The authors introduced an unsupervised attentional framework employing autoencoders and GANs to address the credit card fraud detection problem. Through experiments on datasets such as the Kaggle Credit Card Fraud Detection Dataset[1] and IEEE-CIS Fraud Detection Dataset [9], they presented a method consisting of an initial autoencoder and a final encoder. To further enhance

---

[1] https://www.kaggle.com/datasets/mlg-ulb/creditcardfraud.

the learning process, a channel-wise feature attention layer is added between the autoencoder and the encoder to refine the reconstruction error. Also using autoencoders, Kennedy et al. [12] proposed an unsupervised iterative method for detecting fraudulent activities in highly imbalanced data. Their approach involves iteratively separating the majority class from the data. In the final iteration, an autoencoder is trained on the majority class. The reconstruction error is used to identify the minority class cases. However, the experiments in this work are limited to a single dataset, and the criteria for terminating the iterations are not explicitly defined. In extremely imbalanced scenarios, one-class classification methods, such as autoencoders, one-class SVMs, and their deep learning counterparts, can be effectively applied to the entire dataset. However, this approach operates under the assumption that, initially, the dataset consists entirely of normal cases, which is not always ensured. This method has been used in studies like the one by Zong et al. [27], where a deep autoencoder was employed to generate a low-dimensional representation of the data and compute the reconstruction error for each input data point. These features were then fed into a Gaussian Mixture Model for further analysis. Other works combine different alternative approaches. For instance, Rezapour [21] proposes using three unsupervised methods-autoencoder, one-class support vector machine, and robust Mahalanobis outlier detection-on a real credit fraud detection dataset. While autoencoders and one-class SVMs require labels for normal data, Mahalanobis outlier detection can be trained without any labels, aligning with the focus of our investigation. Overall, these methods rely on the assumption that the majority of instances are non-fraudulent, and thus the cases are used to model the patterns in the non-fraudulent class.

Other researchers have focused on using other unsupervised methods. For instance, Lepoivre et al. [13] combine Principal Component Analysis (PCA) and K-Means algorithms and expand the used data with spatial information of the transactions. In [20] an unsupervised Neural Network, an autoencoder, the Local Outlier Factor (LOF), the Isolation Forest (IForest) and the K-Means clustering are compared.

Another interesting approach to tackle fraud detection through cross-dataset analysis was followed by Zhu [26]. The proposed cross-outlier detection involves the analysis of customer behavior through distance metrics and uses clustering techniques to examine neighboring behaviors. This method identifies suspicious cases by investigating each transaction across two datasets.

Data resampling techniques have also been used. Liu et al. [15] introduced two new undersampling ensembles, EasyEnsemble and BalanceCascade, to deal with highly imbalanced datasets in the fraud detection domain. Both EasyEnsemble and BalanceCascade tackle the imbalance by undersampling the majority class. However, these methods are constrained by their reliance on labeled data.

Finally, some studies compare supervised and unsupervised learning methods in the context of credit card fraud detection (e.g., [5,18]). Generally, these works found that unsupervised learning has a lower performance being less interesting from this perspective. However, unsupervised learning is still considered suitable for credit card fraud detection because no balanced dataset is needed.

Overall, research in credit card fraud detection is still not well explored when considering using unsupervised learning techniques. This is a critical application domain where labeled data is often not available, and thus developing more effective solutions is crucial.

3 ITERADE: ITERative Anomaly Detection Ensemble

Identifying where to initiate the labeling process poses a significant challenge for domain experts when faced with vast volumes of unlabeled data where the majority is expected to be non-fraudulent. This process causes substantial costs, particularly in imbalanced domains where experts prioritize labeling the fraudulent group of cases. However, this often requires investigating a large volume of transactions. To address this issue, we introduce ITERADE, an iterative anomaly detection ensemble.

The key goal of ITERADE is to rebalance the dataset by iteratively identifying and isolating anomalous fraud cases while identifying safe instances that do not need to be further inspected. The main idea of ITERADE is to iteratively apply an anomaly detection ensemble to clusters within a dataset that is modified at each iteration. At each iteration, the optimal number of clusters to be formed is dynamically determined for the current dataset. After this step the clustering algorithm is applied and the anomaly detection ensemble is used inside each cluster obtained. The less suspicious cases are considered safe (non-fraud) and are removed from the dataset. The process then restarts with the calculation of the optimal number of clusters for the newly obtained dataset.

In more detail, initially, we apply an outlier preprocessing step with IForest to clean the data from any noise that might be present. This step is crucial, as eliminating noise enhances model accuracy, allowing the focus to be directed toward identifying fraud-related outliers. Then, in each iteration, the optimal number of clusters is determined using K-Means and the KneeLocator [22] library. The pseudo-code for this step is shown in the Algorithm 1. We compared K-Means with Gaussian Mixture Models (GMMs) to select the most suitable clustering algorithm. Although GMMs are more resource-intensive and time-consuming, they did not yield better results than K-Means based on the silhouette score of the clusters. K-Means also offer easier optimization, which is crucial for our iterative approach. By dynamically adjusting the number of clusters in each iteration, we ensure model flexibility and prevent biases associated with a static cluster number. The KneeLocator library helps identify the optimal number of clusters using the elbow method. Subsequently, K-Means clustering is applied using this determined number of clusters. We must highlight that this step is repeated in each iteration, allowing the fitting of different clusters as the algorithm moves forward in the iterations. This is important because each iteration modifies the dataset by removing cases considered safe in the previous iteration. For each cluster, an anomaly detection ensemble is employed to calculate scores

Algorithm 1. Find Optimal K

1: **procedure** FINDK(*df*)
2:    **Initialize** *d* as an empty dictionary
3:    **for** *k* **in** 5 to 20 **do**
4:       **Fit** K-Means with *k* clusters on *df*
5:       **Store** inertia in *d*[*k*]
6:    **end for**
7:    **Return** knee point of *d* values using KneeLocator
8: **end procedure**

for individual data points. Instances with scores above the upper threshold are flagged as anomalies and stored in a suspicious list, while those below the lower threshold are identified as safe cases and are removed from the dataset. Thus, the algorithm will be able to focus on the most problematic and potentially unsafe cases in the next iterations. Upon completing all iterations, the instances in the suspicious list that appear across all iterations are identified and labeled as anomalies. Once we calculate the common cases across all iterations, we have two options: we can either analyze all the anomalous common cases or concentrate on the cases that consistently appear in every iteration, which we refer to as Most Repeated (MR) cases. This strategy allows for greater flexibility in managing our solution's budget. These labeled instances can then be forwarded to domain experts for further investigation. The pseudo-code for ITERADE is depicted in Algorithm 2.

We employed two distinct ensembles within this algorithm: Ensemble 1 consists of DBSCAN [7], IForest [14], and HDBSCAN [17], and Ensemble 2 uses LOF [4], IForest, and HDBSCAN. We use the first ensemble when applying a hard voting mechanism to predict whether each instance is anomalous. The average of these predictions is computed, and instances are categorized as either suspicious or safe based on predefined thresholds. The second ensemble employs soft voting as a probabilistic anomaly detection ensemble. It predicts the degree of anomaly for each instance using parameters such as negative_outlier_factor_ for LOF, average anomaly score (decision_function) for IForest, and probabilities_ for HDBSCAN. These values are then scaled to a range between 0 and 1, and their sum is utilized as the instance anomaly score. This score is compared against the specified up and down thresholds to isolate anomalous and safe instances accordingly.

4 Experimental Settings

4.1 Datasets

We used two datasets specifically generated for the credit card fraud detection domain: the Credit Card Transactions (CCT) Dataset [3] and the Bank Account Fraud (BAF) Dataset [10]. The CCT Dataset includes over 20 million transactions generated from a multi-agent virtual world simulation performed by IBM,

Algorithm 2. ITERADE Algorithm

1: **procedure** ITERADE($X, y, up\_threshold, down\_threshold, voting, iterations$)
2:    **Initialize** variables
3:    $X, y \leftarrow$ CLEANDATA(X, y)
4:    **for** $i = 1$ **to** $iterations$ **do**
5:       $k \leftarrow$ FINDK(X)
6:       **Fit** K-Means with k clusters
7:       **Predict** clusters for X
8:       **if** $voting == "SoftVoting"$ **then**   ▷ Perform Soft Voting Option
9:          **for each** cluster **do**
10:            **Fit** IForest, LOF, and HDBSCAN on cluster data
11:            **Store** anomaly scores
12:          **end for**
13:          **Scale** anomaly scores using MinMaxScaler
14:       **else**   ▷ Perform Hard Voting Option
15:          **for each** cluster **do**
16:            **Fit** IForest, DBSCAN, and HDBSCAN on cluster data
17:            **Store** anomaly predictions
18:          **end for**
19:       **end if**
20:       **Calculate** vote score by adding the anomaly scores obtained for each instance
21:       **Remove** safe cases with vote score $\leq down\_threshold$ from X and y and store separately
22:       **Store** cases with vote score $> up\_threshold$ in the suspicious list
23:       **Update** variables and performance metrics
24:    **end for**
25:    **Calculate** common instances in the suspicious list
26:    **Save** metrics, common instances, and safe cases to CSV
27: **end procedure**

with approximately 33,000 fraudulent transactions. It covers 2,000 synthetic consumers residing in the United States who travel worldwide, spans decades of purchases, and includes multiple cards for many consumers. While synthetic, the dataset closely resembles real data in key aspects like fraud rates, purchase amounts, and Merchant Category Codes (MCCs), as claimed by the producers. The dataset contains three CSV files: one detailing transactions, one with card information, and one with customer information. For our study, we merged these files to gather all user and card details for each unique transaction.

In terms of the preprocessing steps, we dropped the "Errors?" column due to major missing values and imputed the remaining rows. The "Amount" feature was converted to an integer for model compatibility. Moreover, all categorical features were numerically encoded using hashing, and the entire dataset was normalized to have a mean of 0 and a standard deviation of 1. Additionally, we calculated the correlation between features and the target column, removing redundant, noisy, and uninformative features. Finally, for feature engineering, we

Table 1. Main characteristics of the datasets used.

Name	Size	#Features	#Categorical	IR	Fraud Cases	Non-Fraud Cases
CCT - Low	50571	34	14	0.21	10620	39951
CCT - Med				0.077	3894	46677
CCT - High				0.00077	39	50532
CCT - Original				0.0014	71	50500
BAF - Low	52000	30	5	0.21	10920	41080
BAF - Med				0.077	4004	47996
BAF - High				0.00077	40	51960
BAF - Original				0.01102	573	51427

introduced new aggregated features, such as the total amount of purchases from the same merchant, the total number of transactions from the same merchant, and the total number of cards used in transactions for each user.

The BAF suite of datasets was published at NeurIPS 2022 and includes six different synthetic bank account fraud tabular datasets. BAF provides a comprehensive and robust test bed for evaluating machine learning algorithms as it is realistic and based on a current real-world fraud detection dataset. For our study, we selected the base dataset, sampled to best represent the original dataset. This dataset contains 1 million records, with 11,029 fraudulent cases.

To preprocess the data, we encoded all categorical features numerically using hashing and normalized the dataset to have a mean of 0 and a standard deviation of 1. Feature selection was performed using feature importance results obtained from XGBoost and the features with zero importance were removed. For this dataset, it was not possible to calculate the extra aggregation features because the dataset does not contain the same information as the previous one.

The predictive task for both datasets is a binary classification to determine whether a transaction is fraudulent or not. To simulate real-world conditions, we received real imbalance scenarios from our industry partner, H3M Analytics, which typically deals with the following imbalance ratios (IRs)[2] in their projects: low imbalance (0.21), medium imbalance (0.077), and high imbalance (0.00077). To simulate these IRs, we selected four random subsets of the datasets and applied the mentioned preprocessing stages. We also carry out experiments with the original imbalance present in the datasets to observe how different IRs affect our framework's results. Table 1 contains the characteristics of each dataset used in our experiments.

---

[2] We use the following definition of Imbalance Ratio (IR): ratio between the number of fraudulent cases (transactions) and the total number of cases (transactions) in the dataset.

4.2 Evaluation Methodology

Given our imbalanced datasets and unsupervised approaches, we employed a combination of performance metrics to evaluate both minority and majority groups. For both groups, we calculated the precision and F1 score to assess the accuracy of identifying safe cases and the detection of isolated anomalous cases. Additionally, for the minority group, which is our main focus, we measured the improvement in the IR after the selection of potentially anomalous cases. We calculated this improvement as the imbalance ratio obtained in the sample of cases selected by an approach divided by the imbalance ratio of the original data.

To establish a starting point for the experts to inspect and label cases, we recognize the significance of budgetary considerations. Hence, an important metric to consider, which we also report, is the total number of anomalies detected, as these will subsequently be forwarded to experts for labeling purposes.

We considered the following baseline algorithms: DBSCAN, HDBSCAN, IForest, and LOF which were then integrated into ITERADE alongside Empirical Cumulative Distribution Functions (ECOD) from the PYOD [25] package. We compared ITERADE against these baselines using two main variants: hard voting (HV)and soft voting (SV). Each algorithm is fitted in the entire dataset and the cases detected as anomalies are labeled as positive (fraud).

Throughout our experiments, we worked with two sets of parameters: the internal hyperparameters of the baseline anomaly detection models and the external hyperparameters of the ITERADE framework. The internal hyperparameters of all baseline models-IForest, Local Outlier Factor, HDBSCAN, DBSCAN, and ECOD-were tuned in an unsupervised manner using grid search, with the silhouette score as the scoring function. After fine-tuning, these models were integrated into our hard-voting and soft-voting ensembles. This strategy allows us to explore the most suitable hyperparameters to use in the baselines and our proposed solution without any data leakage from the true class labels. The hyperparameters of the ensemble baselines explored during our tuning phase are detailed in Tables 2 and 3.

For our ITERADE framework, we introduced the up_threshold and down_threshold hyperparameters, which depend on the voting scheme (soft or hard) used in the corresponding ensemble. For the hard voting ensemble, we tested values of 3 and 2 for the up_threshold, aiming to detect instances predicted as anomalies by all 3 models and by more than 2 models, respectively. For the soft voting scheme of ITERADE, we investigated four combinations of thresholds for both up_threshold and down_threshold. Details of our experiments for ITERADE variants are presented in Table 4.

Table 2. Hyperparameters for Different Models and Configurations - CCT Dataset

Model	Searched Parameters		Best Parameter Values			
	Parameter	Range	Low	Medium	Original	High
DBSCAN	eps	range(0.1, 1.5, 0.1)	0.1	0.1	0.1	0.1
	min_samples	range(2, 10, 1)	2	2	2	2
HDBSCAN	min_cluster_size	range(5, 50, 5)	45	35	45	45
	min_samples	range(2, 10, 1)	7	8	6	4
IForest	n_estimators	range(50, 200, 50)	50	150	50	150
	max_samples	linspace(0.1, 1.0, 10)	0.6	0.2	0.2	0.5
	contamination	linspace(0.01, 0.3, 10)	0.01	0.01	0.01	0.01
LOF	n_neighbors	range(5, 50, 5)	25	30	30	40
	leaf_size	range(10, 50, 10)	10	10	10	10
ECOD	contamination	linspace(0.01, 0.3, 10)	0.01	0.01	0.01	0.01

Table 3. Hyperparameters for Different Models and Configurations - BAF Dataset

Model	Searched Parameters		Best Parameter Values			
	Parameter	Range	Low	Medium	Original	High
DBSCAN	eps	range(0.1, 1.5, 0.1)	1.1	0.7	0.5	1.3
	min_samples	range(2, 10, 1)	3	2	2	4
HDBSCAN	min_cluster_size	range(5, 50, 5)	10	5	30	30
	min_samples	range(2, 10, 1)	2	2	2	2
IForest	n_estimators	range(50, 200, 50)	150	100	100	150
	max_samples	linspace(0.1, 1.0, 10)	0.6	0.9	0.3	0.2
	contamination	linspace(0.01, 0.3, 10)	0.01	0.01	0.01	0.01
LOF	n_neighbors	range(5, 50, 5)	35	15	25	30
	leaf_size	range(10, 50, 10)	10	10	10	10
ECOD	contamination	linspace(0.01, 0.3, 10)	0.01	0.01	0.01	0.01

Table 4. Experimental configurations of ITERADE Algorithm.

Voting	Ensembles	Variant	up_threshold	down_threshold
Hard Voting (HV)	HDBSCAN; IForest; DBSCAN	HV 1.1	3	0
		HV 1.2	2	0
Soft Voting (SV)	HDBSCAN; IForest; LOF	SV 2.1	2.7	1.5
		SV 2.2	2.5	1.2
		SV 2.3	2.7	1.2
		SV 2.4	2.5	1.5

5 Results and Discussion

5.1 Experimental Results

To ensure a fair comparison between ITERADE and the baselines, we defined budget buckets based on the ranges obtained from our baseline models. Our main experimental results on the eight datasets, evaluated across multiple performance metrics, are detailed in Tables 5 and 6 and Fig. 1. We present two types of results: in the tables, we provide the average results of our proposed solution across all iterations in each budget bucket, while in the figure we show the results of the iteration where the proposed solution achieved the best result. By showing the average across all iterations in each budget bucket for ITERADE our goal was to provide a less optimistic view of our solution. This handicap that is inputted into our results shows that our algorithm works well not only at a given particular iteration (best result) but also has an overall good performance. In all our experiments, after calculating the common cases from all iterations, we can either examine all the anomalous common cases retrieved or focus on the cases that have repeated in every iteration, which we denote as MR (Most Repeated) in the tables. This approach allows us to adjust our budget effectively. For instance, in some variants like HV 1.1.2, the total common cases are usually above 5,000, while the most repeated cases can be as few as 50 to 500, typically resulting in a better ratio of frauds to total cases retrieved and a more realistic scenario for labeling experts.

From Tables 5 and 6, we observe an overwhelming advantage of using ITERADE across all budgets. Our algorithm is always able to provide more fraud (positive) cases and typically outputs a lower number of anomalies (two last columns in the tables). This is observed for both datasets on all imbalance ratios and budgets considered, except for the BAF-High dataset at a 600 budget, where our algorithm's performance was comparably close to the baseline ECOD. Looking in more detail at the results of ITERADE, we see that different ITERADE variants, based on the thresholds used in the experiments, are suited for different types of budgets. Overall, we observe that variants with lower thresholds for the anomalous cases (up_threshold) signal a higher number of anomalies, being thus more suitable for cases with higher budgets. The variants with higher anomalous case thresholds will be more adequate for lower budgets, outputting a smaller set of anomalous cases. For example, SV 2.2, SV 2.4, and HV 1.2 generally detect many cases as anomalies and are more appropriate for high budgets. However, they can also be used with low budgets by focusing on their most repeated cases. On the other hand, HV 1.1, SV 2.1, and SV 2.3 are more suitable for lower budgets. For illustration, the number of anomalies detected in each iteration, including the most repeated anomalies, and the number of frauds with the most repeated frauds in each iteration for the CCT original dataset, are plotted in Fig. 2. These plots include horizontal lines indicating the budget buckets, providing a visual representation of the possible solutions for each budget bucket. Furthermore, the best values achieved within each budget category

Table 5. Experimental results for CCT datasets across different imbalance ratios, with average performance reported for each budget bucket.

Dataset	Budget	Algorithm	Performance		Improv.	Average Anomalies	
			Precision	F1 Score		TP	Anom. Det.
CCT-Original	≤600	IForest	0.00198	0.00395	1.414	1	506
		ECOD	0.00988	0.01957	7.057	5	506
		ITERADE - SV 2.4 - MR	0.02104	0.04121	**15.029**	2.5	118.8
		ITERADE - SV 2.3 - MR	0.00567	0.01128	4.05	0.53	93.46
		ITERADE - HV 1.1	0.00184	0.00367	1.314	0.6	325.26
	>600 & ≤4000	HDBSCAN	0.0011	0.0022	0.786	4	3647
		ITERADE - SV 2.1 - MR	0.00972	0.01925	**6.943**	12	1234
		ITERADE - HV 1.2 -MR	0.00328	0.00654	2.343	6.5	1983
		ITERADE - SV 2.3	0.00622	0.01236	4.443	18.33	2947.66
	>4000 & ≤8000	LOF	0.00148	0.00296	1.057	11	7457
		ITERADE - SV 2.1	0.00416	0.00829	2.971	26.75	6426.75
		ITERADE - SV 2.3	0.00424	0.00844	**3.029**	27	6369.75
	>8000 & ≤20000	DBSCAN	0.0026	0.00519	1.857	64	24652
		ITERADE - SV 2.4	0.00294	0.00586	2.1	46.77	15893.22
		ITERADE - HV 1.2	0.00419	0.00835	**2.993**	44.12	10535.52
CCT-Low	≤600	IForest	0.14822	0.25817	0.706	75	506
		ECOD	0.24111	0.38854	1.148	122	506
		ITERADE - HV 1.1	0.18583	0.31342	0.885	55.1	296.5
		ITERADE - SV 2.1 - MR	0.39202	0.56324	**1.867**	46	117.34
		ITERADE - SV 2.4 - MR	0.36214	0.53172	1.724	180.2	497.6
	>600 & ≤4000	HDBSCAN	0.13736	0.24154	0.654	500	3640
		ITERADE - SV 2.3 - MR	0.29443	0.45492	**1.402**	495.75	1683.75
		ITERADE - SV 2.2 - MR	0.20561	0.34109	0.979	363.88	1769.74
		ITERADE - SV 2.4 - MR	0.26303	0.41651	1.253	420.3	1597.91
	>4000 & ≤8000	LOF	0.17967	0.30461	0.856	1414	7870
		ITERADE - SV 2.1	0.30022	0.4618	**1.43**	1856.2	6182.8
		ITERADE - SV 2.2 - MR	0.29194	0.45194	1.39	1516.5	5194.5
	>8000 & ≤20000	DBSCAN	0.28431	0.44274	1.354	8378	29468
		ITERADE - SV 2.1	0.29944	0.46088	**1.426**	3565	11905.62
		ITERADE - HV 1.2	0.25215	0.40275	1.201	2592.95	10283.43
		ITERADE - SV 2.4	0.25321	0.4041	1.206	4244.5	16762.62
CCT-Med	≤600	IForest	0.10672	0.19286	1.386	54	506
		ECOD	0.16798	0.28764	2.182	85	506
		ITERADE - HV 1.2 - MR	0.17935	0.30415	2.329	43.29	241.37
		ITERADE - SV 2.1 - MR	0.18113	0.30671	2.352	6.74	37.21
		ITERADE - SV 2.3 - MR	0.19489	0.32621	**2.531**	18.38	94.31
	>600 & ≤4000	HDBSCAN	0.0508	0.09669	0.66	206	4055
		ITERADE - HV 1.2 - MR	0.12787	0.22675	1.661	238	1861.33
		ITERADE - SV 2.1	0.13519	0.23818	**1.756**	425.25	3145.5
	>4000 & ≤8000	LOF	0.07498	0.1395	0.974	574	7655
		ITERADE - SV 2.3	0.12184	0.21721	1.582	818	6713.9
		ITERADE - SV 2.1	0.13192	0.23309	**1.713**	914.42	6931.5
	> 8000 & ≤ 20000	DBSCAN	0.12853	0.22778	**1.669**	3455	26880
		ITERADE - SV 2.1	0.12789	0.22678	1.661	1079.41	8440.25
		ITERADE - SV 2.2	0.10738	0.19394	1.395	1790.8	16676.86
CCT-High	≤600	IForest	0.00395	0.00787	5.13	2	506
		ECOD	0.00593	0.01179	7.701	3	506
		ITERADE - HV 1.1	0.00592	0.01177	7.688	1.5	253.5
		ITERADE - HV 1.2 - MR	0.0005	0.001	0.649	0.1	199
		ITERADE - SV 2.4 - MR	0.00625	0.01242	**8.117**	1.58	253
	>600 & ≤4000	HDBSCAN	0.00059	0.00118	0.766	2	3362
		ITERADE - SV 2.1	0.00389	0.00775	5.052	12.5	3215
		ITERADE - SV 2.2 - MR	0.00197	0.00393	2.558	3.46	1755.5
		ITERADE - SV 2.3 - MR	0.00493	0.00981	**6.403**	6.33	1283.83
	>4000 & ≤8000	LOF	0.00118	0.00236	1.532	9	7610
		ITERADE - SV 2.2 - MR	0.00323	0.00644	**4.195**	15	4651
		ITERADE - SV 2.1	0.00235	0.00469	3.052	16.7	7115
	>8000 & ≤20000	DBSCAN	0.00141	0.00282	1.831	35	24756
		ITERADE - HV 1.2	0.00239	0.00477	**3.104**	24.5	10231.43
		ITERADE - SV 2.2	0.00172	0.00343	2.234	25.9	15082.1

Table 6. Experimental results for BAF datasets across different imbalance ratios, with average performance reported for each budget bucket.

Dataset	Budget	Algorithm	Performance		Improv.	Average Anomalies	
			Precision	F1 Score		TP	Anom. Det.
BAF-Original	≤600	LOF	0	0	0	0	25
		HDBSCAN	0.01266	0.025	1.149	2	158
		IForest	0.01538	0.03029	1.396	8	520
		ECOD	0.02885	0.05608	2.618	15	520
		ITERADE - HV 1.1	0.01538	0.03029	1.396	10.33	671.5
		ITERADE - HV 1.2 - MR	0.04348	0.08334	**3.946**	4.8	110.4
		ITERADE - SV 2.4 - MR	0.02373	0.04636	2.153	14	590
	>600 & ≤4000	ITERADE - HV 1.1	0.01581	0.03113	1.435	12.81	810.22
		ITERADE - SV 2.4	0.02441	0.04766	**2.215**	33	1352
		ITERADE - HV 1.2 - MR	0.01352	0.02668	1.227	47.5	3513
	>4000 & ≤20000	DBSCAN	0.01102	0.0218	1	573	51998
		ITERADE - HV 1.2	0.01316	0.02598	**1.194**	285	21657
		ITERADE - SV 2.4	0.01167	0.02307	1.059	149.33	12798.33
BAF-Low	≤600	LOF	0.22222	0.36363	1.058	6	27
		HDBSCAN	0.31797	0.48251	1.514	69	217
		IForest	0.26923	0.42424	1.282	140	520
		ECOD	0.32885	0.49494	1.566	171	520
		ITERADE - HV 1.1 - MR	0.32171	0.48681	1.532	1.66	5.16
		ITERADE - HV 1.2 - MR	0.31582	0.48004	1.504	30.42	96.32
		ITERADE - SV 2.3 - MR	0.35681	0.52595	**1.699**	42	117.71
	>600 & ≤4000	ITERADE - HV 1.2 - MR	0.26714	0.42164	1.272	978	3661
		ITERADE - SV 2.2	0.2866	0.44552	**1.365**	517.66	2085.33
	>4000 & ≤20000	DBSCAN	0.21001	0.34712	1	10920	51997
		ITERADE - SV 2.2	0.23119	0.37556	**1.101**	1423	6155
		ITERADE - HV 1.2	0.21838	0.35848	1.04	1534.48	7026.51
BAF-Med	≤2000	HDBSCAN	0	0	0	0	0
		LOF	0.16667	0.28572	2.165	6	36
		IForest	0.13654	0.24027	1.773	71	520
		ECOD	0.18462	0.31169	2.398	96	520
		ITERADE - HV 1.1	0.10548	0.19083	1.37	15.4	146
		ITERADE - SV 2.3	0.05204	0.09893	0.676	17.16	329.76
		ITERADE - SV 2.4	0.19074	0.32037	**2.477**	84.5	443
	>600 & ≤4000	ITERADE - HV 1.2	0.11457	0.20559	**1.488**	323.91	2827.16
		ITERADE - SV 2.2	0.09632	0.17572	1.251	155.8	1617.47
	>4000 & ≤20000	LOF	0.07498	0.1395	0.974	574	7655
		ITERADE - SV 2.3	0.12184	0.21721	1.582	818	6713.9
		ITERADE - SV 2.1	0.13192	0.23309	**1.713**	914.42	6931.5
BAF-High	≤600	HDBSCAN	0	0	0	0	150
		LOF	0	0	0	0	34
		IForest	0	0	0	0	520
		ECOD	0.00192	0.00383	**2.494**	1	520
		ITERADE - SV 2.1 - MR	0.00141	0.00282	1.831	2	1414
	>2000 & ≤20000	DBSCAN	0.00077	0.00154	1	40	51992
		ITERADE - SV 2.1	0.00098	0.00196	**1.273**	9.38	9600.3
		ITERADE - SV 2.4	0.00043	0.00086	0.558	4	9332
		ITERADE - SV 2.2	0.00064	0.00128	0.831	12	18732

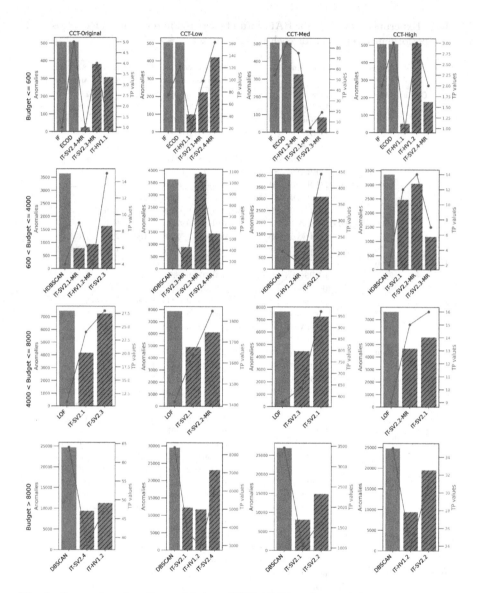

Fig. 1. Comparison baselines with best ITERADE results on CCT dataset for the 4 budgets considered. Each row shows a different budget in decreasing order from top to bottom. Each column shows one of the variants of the CCT dataset - from left to right: Original, Low, Med, and High. (Blue bars: anomalies detected in baseline; dashed blue bars: anomalies detected in ITERADE; the red line shows, on the right y-axis, the number of positive (fraud cases) contained in the signaled cases). (Color figure online)

for each algorithm are listed in Figs. 1, which may present an overly optimistic estimation.

In each budget bucket, ITERADE outperforms the baselines in terms of the number of anomalous cases detected and the precision of fraud cases within the anomalous cases. The improvement of each solution is calculated by dividing the final imbalance ratio (precision of the minority group) by the initial imbalance ratio. The improvement results are displayed in Tables 5 and 6 in column "Improv.". While the baselines show lower improvements or degradation, the ITERADE algorithm shows consistently higher improvements, even when considering the average across all iterations of the algorithm for each budget range.

Another notable observation is that as datasets become more imbalanced and the budget decreases, the improvements achieved by ITERADE become more significant. We achieved an average of 15 times improvement in the fraud cases signaled for the "CCT-Original" dataset with a budget of less than 600, using ITERADE - SV 2.4 - MR. For budgets between 600 and 4000, our ITERADE - SV 2.1 - MR outperformed other algorithms, providing an average 6.94 times improvement in the fraud cases present in the anomalies detected. As the imbalance decreases to low and medium levels, improvements become more modest, averaging 1.4 to 2.5 times improvement. However, we must highlight that, in these cases, the baselines are showing a much worse performance.

Furthermore, we found that the performance of anomaly detection baselines impacts the performance of the ensemble. For example, in the BAF dataset with high imbalance, all four baselines used in the ensemble failed to detect any anomalies, and DBSCAN labeled every instance as an anomaly. This affected the performance of the ensembles, resulting in less significant improvements. Despite these challenges, our algorithm still demonstrated superiority, achieving an average of 1.2 to 1.8 times improvement in the fraud cases detected out of the anomalies signaled.

Looking at the results of Fig. 1, we observe that in the vast majority of cases, the baselines signal a larger set of anomalous cases (clear blue bars represent the number of anomalies detected by the baselines and the dashed blue bars correspond to the ITERADE). Simultaneously, we observe that the number of fraud cases contained in the signaled cases is many times higher than the baselines (red lines in the figure). This shows that ITERADE provides a lower number of cases to inspect as potential anomalies. Moreover, in those cases signaled, more correspond to actual fraud cases.

Due to space constraints, all extra plots for all imbalance ratios and extra tables containing additional performance metrics for both the majority and minority groups are available in the repository associated with this paper at https://github.com/beafshar/ITERADE-Results.

Our study demonstrates the effectiveness of ITERADE in addressing the labeling problem in credit card fraud detection. By leveraging the benefits of various anomaly detection models and an iterative approach with two voting schemes and using dynamic clustering, we provide an algorithm with a flexible

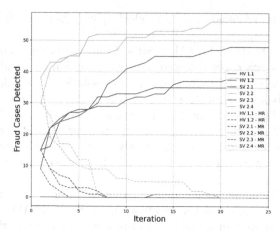

(a) Number of fraud cases and most repeated fraud cases detected per iteration for the CCT original imbalance dataset.

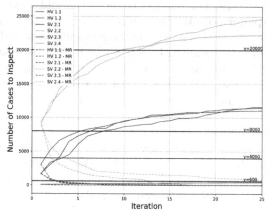

(b) Number of anomalies and most repeated anomalies detected per iteration for the CCT original imbalance dataset. Horizontal lines indicate budget buckets, showing solutions for each budget range.

Fig. 2. ITERADE results across all iterations for CCT original imbalance dataset.

budget that delivers significant improvements, especially as the datasets become more imbalanced.

Finally, we found that ITERADE and this study still present some challenges that need to be further explored. In this study, we discovered that the performance of the ensemble heavily depends on the effectiveness of the baseline algorithms it employs. The primary challenge affecting the performance of these baselines is the unsupervised hyperparameter tuning. Without labels or a validation set, achieving optimal hyperparameter tuning is extremely difficult. Moreover, we did not use any information regarding the imbalance in the dataset to not bias the parameter selection. Experiments can be improved by expanding the search space and using larger steps in the grid search for hyperparameter tuning. Also, other unsupervised solutions for hyperparameter tuning should be investigated.

Additionally, the flexible budget is influenced by the up and down thresholds used in the voting scheme. To provide more budget buckets with diverse solutions, experiments can be conducted using various combinations of thresholds for both voting schemes.

6 Conclusions

In this study, we proposed ITERADE as a solution for unsupervised anomaly detection in fraud scenarios, demonstrating its effectiveness across various imbalanced domains. Our framework offers an adjustable budget that can be tailored based on the preferences of domain experts. Our experiments demonstrate that ITERADE outperforms the baseline methods across all datasets and imbalance domains, effectively increasing the fraud to anomalies detected ratio and resulting in a less imbalanced dataset. We observed that ITERADE can select a subset of the entire dataset within a flexible budget, achieving a higher percentage of positive cases by a factor of 3 to 15. This capability is particularly advantageous for highly imbalanced datasets, especially when working with a limited labeling budget. Moreover, our solution provides as a side product a set of safe instances for which no further inspection is needed.

In the future, we plan to integrate our approach with a human-in-the-loop system that can provide actual labels for instances flagged by ITERADE. With enough labeled instances, the unsupervised problem can transition into an active learning or semi-supervised learning scenario, enabling the model to learn from limited data and incrementally increase its labeled dataset. We aim to test our framework on realistic datasets obtained from banking institutions and integrate it with active or semi-supervised learning to enhance performance and expand the labeled dataset. Additionally, further experiments can be conducted to determine the optimal point of transition—the imbalance ratio threshold—from unsupervised to active learning.

References

1. Federal trade commission: Consumer sentinel network data book 2021 (2022). https://www.ftc.gov/system/files/ftc_gov/pdf/CSN%20Annual%20Data%20Book%202021%20Final%20PDF.pdf. Accessed 09 June 2024
2. Largest banks in the us - list (2024). https://www.emarketer.com/insights/largest-banks-us-list/. Accessed 16 June 2024
3. Altman, E., Nitsure, A., Mroueh, Y.: Credit card transactions fraud detection and other analyses (2019). https://www.kaggle.com/datasets/ealtman2019/credit-card-transactions. Accessed 20 June 2024
4. Breunig, M.M., Kriegel, H.P., Ng, R.T., Sander, J.: LOF: identifying density-based local outliers. In: Proceedings of the 2000 ACM SIGMOD International Conference on Management of Data, pp. 93–104 (2000)

5. Caroline Cynthia, P., Thomas George, S.: An outlier detection approach on credit card fraud detection using machine learning: a comparative analysis on supervised and unsupervised learning. In: Intelligence in Big Data Technologies-Beyond the Hype: Proceedings of ICBDCC 2019, pp. 125–135. Springer (2021)
6. Cheng, D., Ye, Y., Xiang, S., Ma, Z., Zhang, Y., Jiang, C.: Anti-money laundering by group-aware deep graph learning. IEEE Trans. Knowl. Data Eng. **35**(12), 12444–12457 (2023). https://doi.org/10.1109/TKDE.2023.3272396
7. Ester, M., Kriegel, H.P., Sander, J., Xu, X., et al.: A density-based algorithm for discovering clusters in large spatial databases with noise. In: KDD, vol. 96, pp. 226–231 (1996)
8. Goldstein, M., Uchida, S.: A comparative evaluation of unsupervised anomaly detection algorithms for multivariate data. PLoS ONE **11**(4), e0152173 (2016)
9. Howard, A., Bouchon-Meunier, B., IEEE CIS, Inversion, John Lei, L., Abbass, P.H.: IEEE-CIS fraud detection (2019). https://kaggle.com/competitions/ieee-fraud-detection
10. Jesus, S., et al.: Turning the tables: biased, imbalanced, dynamic tabular datasets for ML evaluation. In: Advances in Neural Information Processing Systems (2022)
11. Jiang, S., Dong, R., Wang, J., Xia, M.: Credit card fraud detection based on unsupervised attentional anomaly detection network. Systems **11**(6), 305 (2023)
12. Kennedy, R.K., Salekshahrezaee, Z., Khoshgoftaar, T.M.: A novel approach for unsupervised learning of highly-imbalanced data. In: 2022 IEEE 4th International Conference on Cognitive Machine Intelligence (CogMI), pp. 52–58. IEEE (2022)
13. Lepoivre, M.R., Avanzini, C.O., Bignon, G., Legendre, L., Piwele, A.K.: Credit card fraud detection with unsupervised algorithms. J. Adv. Inf. Technol. **7**(1) (2016)
14. Liu, F.T., Ting, K.M., Zhou, Z.H.: Isolation forest. In: 2008 Eighth IEEE International Conference on Data Mining, pp. 413–422. IEEE (2008)
15. Liu, X.Y., Wu, J., Zhou, Z.H.: Exploratory undersampling for class-imbalance learning. IEEE Trans. Syst. Man Cybern. Part B (Cybern.) **39**(2), 539–550 (2008)
16. Lu, M., et al.: Bright-graph neural networks in real-time fraud detection. In: Proceedings of the 31st ACM International Conference on Information & Knowledge Management, pp. 3342–3351 (2022)
17. McInnes, L., Healy, J., Astels, S., et al.: HDBSCAN: hierarchical density based clustering. J. Open Source Softw. **2**(11), 205 (2017)
18. Niu, X., Wang, L., Yang, X.: A comparison study of credit card fraud detection: supervised versus unsupervised. arXiv preprint arXiv:1904.10604 (2019)
19. Perera, P., Oza, P., Patel, V.M.: One-class classification: a survey. arXiv preprint arXiv:2101.03064 (2021)
20. Rai, A.K., Dwivedi, R.K.: Fraud detection in credit card data using unsupervised machine learning based scheme. In: 2020 International Conference on Electronics and Sustainable Communication Systems (ICESC), pp. 421–426. IEEE (2020)
21. Rezapour, M.: Anomaly detection using unsupervised methods: credit card fraud case study. Int. J. Adv. Comput. Sci. Appl. **10**(11) (2019)
22. Satopaa, V., Albrecht, J., Irwin, D., Raghavan, B.: Finding a "kneedle" in a haystack: detecting knee points in system behavior. In: 2011 31st International Conference on Distributed Computing Systems Workshops, pp. 166–171. IEEE (2011)
23. Xiang, S., et al.: Semi-supervised credit card fraud detection via attribute-driven graph representation. In: Proceedings of the AAAI Conference on Artificial Intelligence, vol. 37, pp. 14557–14565 (2023)

24. Zhang, Y.L., et al.: A framework for detecting frauds from extremely few labels. In: Proceedings of the Sixteenth ACM International Conference on Web Search and Data Mining, pp. 1124–1127 (2023)
25. Zhao, Y., Nasrullah, Z., Li, Z.: PyOD: a python toolbox for scalable outlier detection. J. Mach. Learn. Res. **20**(96), 1–7 (2019). http://jmlr.org/papers/v20/19-011.html
26. Zhu, T.: An outlier detection model based on cross datasets comparison for financial surveillance. In: 2006 IEEE Asia-Pacific Conference on Services Computing (APSCC 2006), pp. 601–604. IEEE (2006)
27. Zong, B., et al.: Deep autoencoding gaussian mixture model for unsupervised anomaly detection. In: International conference on learning representations (2018)

Purifying Adversarial Examples Using an Autoencoder

Thijs van Weezel[✉] [ID], Famke van Ree [ID], Tychon Bos [ID], Patrick Bastiaanssen [ID], and Sibylle Hess [ID]

Mathematics and Computer Science Department, Eindhoven University of Technology, Eindhoven, The Netherlands
{t.g.g.v.weezel,f.v.ree,t.m.j.bos,p.c.a.bastiaanssen}@student.tue.nl,
s.c.hess@tue.nl

Abstract. One of the most prominent security challenges to neural networks are adversarial examples - inputs with often barely perceptible perturbations causing misclassification. In this study, we propose a defense mechanism that uses an autoencoder to restore adversarial examples before classification. That is, the autoencoder purifies input data points from potential adversarial perturbations. The method is titled Autoencoder-based Adversarial Purification (AAP). We demonstrate the effectiveness of AAP on multiple datasets, attack methods, and perturbation levels. While certain limitations exist, this research offers valuable insights and a promising direction for robust defense mechanisms in adversarial deep learning.

Keywords: adversarial attack · autoencoder · purification

1 Introduction

The rapid advancements in Neural Networks (NNs) have transformed the field of image classification, enabling machines to perform tasks surpassing human capabilities [1]. These advancements have led to the widespread adoption of NNs across various domains, ranging from everyday applications like social media filters to more critical uses in healthcare, autonomous vehicles, and security systems.

With these advancements comes the growing realization of the vulnerabilities within these systems, particularly through adversarial attacks. Adversarial examples are inputs with slight perturbations, often unrecognizable to the human eye, causing a NN to predict the wrong class with high confidence [2]. An example is displayed in Fig. 1. The existence of adversarial examples prohibits an application of NNs in sensitive domains, as it shows that we cannot trust their predictions. Understanding and mitigating the risks posed by adversarial attacks is essential to fully realize the potential of NNs in image classification, especially in domains where the cost of errors is high.

Adversarial attacks can be divided into white-box and black-box attacks based on the amount of knowledge they leverage. White-box attacks have full or almost full knowledge of the target model's parameters and the data. In contrast, black-box attacks require less knowledge and their applications are therefore more versatile [3].

Clean image   Pertubation   Adversarial Example

Fig. 1. A demonstration of the generation of an adversarial example. Although the clean image can be correctly classified, after application of the perturbation, it is misclassified. ϵ is usually a small value in $[0, 1]$.

Defense mechanisms against adversarial attacks can be classified into three main strategies: detection, robustness, and purification. Detection strategies aim to identify adversarial examples within the input [4]. Robustness strategies, such as adversarial training [5], incorporate adversarial examples in training, aiming to make the model more robust. Purification strategies aim to restore the adversarial example to the original [6]. The current state-of-the-art purification techniques show promising results [7–9], but several challenges, such as generalizability across diverse attacks, remain unaddressed [10].

Therefore, in this paper, we propose a new defense mechanism making the following key contributions:

- **Novel Defense Mechanism**: We propose a defense method founded on the theoretical rationale for the efficacy of employing an autoencoder (AE) as a comprehensive solution for the purification of adversarial examples prior to their classification.
- **Benchmarking**: Measured on varying data and across multiple experimental setups, our defense mechanism typically demonstrates capacities exceeding those established by current state-of-the-art techniques.

2 Related Work

In recent years, there has been significant research focused on enhancing the robustness of NNs, particularly in the face of adversarial examples. Various approaches have been explored to mitigate the vulnerabilities of these models. An overview is presented below of attacks and defense mechanisms.

2.1 Autoencoders

In this paper, we will use an autoencoder to defend against adversarial examples. Autoencoders (AEs) are unsupervised neural networks designed to learn efficient codings of their input [11]. AEs transform data into a low-dimensional latent space Z while minimizing the error resulting from reconstructing the input using Z. An AE consists of an encoder $f_{\theta_E} : \mathbb{R}^d \to \mathbb{R}^Z$ and a decoder $f_{\theta_D} : \mathbb{R}^Z \to \mathbb{R}^d$, conjointly represented by $f_{\theta_{AE}} : \mathbb{R}^d \to \mathbb{R}^d$, parameterized by θ_E, θ_D, and θ_{AE}, respectively. The objective is usually to minimize the loss function in Equation (1), in which X is the set of inputs.

$$\mathcal{L}_{AE} = \sum_{x \in X} ||x - f_{\theta_{AE}}(x)||^2. \tag{1}$$

2.2 Adversarial Attacks

In the following, a neural network classifier is parameterized by θ, and trained to minimize a loss function $\frac{1}{n} \sum_{i=1}^{n} J(\theta, x_i, y_i)$, with x_i representing the input data, y_i the class label and n the number of training examples. Out of the plethora of attack methods that exist [12], we choose to focus on three classical attacks. A commonly used white-box attack method is the Fast Gradient Sign Method (FGSM) [5]. FGSM adds a perturbation of size ϵ to every pixel in the direction that maximizes the loss function, as shown in 2.

$$\hat{x} = x + \epsilon \cdot \text{sign}(\nabla_x J(\theta, x, y)). \tag{2}$$

This method can result in a misclassification rate of 89.4% of the MNIST dataset [13]. An extension of FGSM also deployed in this paper is Iterative FGSM (iFGSM) [14], which iteratively conducts a series of FGSM attacks with a step size α, commonly set to 0.01. Starting with $\hat{x}_0 = x$, iFGSM iteratively performs the update rule expressed in 3, in which t represents the iteration. The infinity norm is implemented by constraining the perturbations to a range.

$$\hat{x}_{t+1} = \text{clip}_{x,\epsilon}(\hat{x}_t + \alpha \cdot \text{sign}(\nabla_x J(\theta, \hat{x}_t, y))). \tag{3}$$

Compared to FGSM, which crafts the examples in a single step, iFGSM applies FGSM multiple times, making it more effective, as the iterative process helps find an adversarial example that is more likely to result in misclassification while keeping the perturbations to a minimum [2,14].

Carlini Wagner (CW) attack uses an iterative optimization approach to find adversarial examples. In here, the process of finding an adversarial example is optimized, by minimizing the perturbation added to the original input while ensuring misclassification by the model. As the objective is directly related to minimizing the distances between the original and the adversarial example, the perturbations are smaller and less detectable and therefore more likely to evade detection and cause misclassification [15].

2.3 Adversarial Defenses

The development of defenses chases the development of attacks, and vice versa. A common issue of defense methods is that they do not generalize over multiple attacks. Defense strategies can be divided into methods to detect adversarial examples, methods to create robustness in models, and methods to purify the adversarial examples [6].

Adversarial example detection is frequently done with AEs [16–18]. These methods focus on the reconstruction error, in which the assumption is made that an AE's reconstruction error is small if it is similar to the training data. Hence, a higher reconstruction error is interpreted as an indication of an adversarial example. While these detection methods can be effective at identifying adversarial examples, they do not provide comprehensive protection to the network, as detection alone does not enable the network to further utilize the adversarial input.

An approach to increase the robustness of models is adversarial training, in which the model is trained on both adversarial examples and clean input. Advantages of this approach is that it is effective at increasing robustness, easily scalable, and can be trained effectively against various attacks [19]. However, adversarial training significantly decreases the accuracy of clean samples, increases the training time, and does not generalize well to attacks it hasn't been trained on [20].

In comparison, purifying adversarial examples has the advantage that retraining of the targeted model is not necessary. Purification methods use a separate model to reconstruct clean images from the adversarial examples. Such reconstructed – or purified – images can be used straightforwardly for subsequent processing. Purifying adversarial examples can be performed with different approaches.

PixelDefend [9] is one such approach. Using the PixelCNN architecture [21], this method finds a new image by evaluating and adjusting the probability of each pixel in the adversarial image given its neighboring pixels. The model estimates the probability density of the training data distribution, which can be used to identify whether an image is likely to be part of the training data distribution and therefore detect adversarial examples [9]. The results of PixelDefend show success on simple attacks like FGSM. However, against more advanced attacks like CW, this method necessitates deeper networks, which can suffer from disadvantages such as vanishing gradients. As a result, the authors propose to use this approach to defend against simple attacks, while combining it with adversarial training otherwise.

An approach that can defend against advanced attacks is MagNet, which trains an AE to purify the adversarial samples [22]. It consists of two components: The detector and the reformer. The detector is an AE that indicates whether the input is suspected to be adversarially perturbed. If that is the case, the reformer tries to produce an image that approximates the adversarial image of which the representation in the manifold is moved to that of clean images. The framework does not require specific knowledge on the type of adversarial attack, and is therefore well generalizable. A downside of this approach is that the effectiveness of the detector is depended on how well it can learn the manifold of clean data. Due to it's setup, it is highly dependent on well-trained AE's as both the detector and the reformers are autoencoders [22]. In addition, the approach is vulnerable to white-box attacks because the detector and reformer leave their parameters exposed for the attacker to use. Therefore, white-box attacks can succeed on these components as well as the classifier.

Generative Adversarial Networks (GANs) [23] and Wasserstein GANS (WGANs) [24,25] employed for purification of adversarial images can solve this problem, as demonstrated by Defense-GAN [8]. GANs in general suffer from time efficiency problems, making them unsuitable for real-time applications. A WGAN learns to generate images that are similar to clean images by minimizing the Wasserstein distance between the generated and clean images. Once trained, the WGAN can purify the adversarial examples by transforming them to resemble clean images. This approach is able to defend against advanced attacks like CW, does not need retraining of the classification model and can be used against any attack without prior knowledge of the attack beforehand.

A more time-efficient solution is proposed with PuVAE [7], which is based on a Variational AE (VAE) [26], an autoencoder in which the latent space is constrained to a prior distribution. The VAE is trained to reconstruct an image using as input the concatenation of a clean image and its corresponding label. During inference, the image is reconstructed with each label, to find the reconstruction that results in the lowest reconstruction error. This reconstruction is the purified sample, and is fed to the classifier. This method achieves a better performance on various attacks than aforementioned methods. Another main contribution is the method's improved tractability. However, the process by which the purified sample is selected is in itself a form of classification, which makes it a potential target for the adversarial attack.

The disadvantages of the existing methods of purification can thus be summarized as either being time inefficient, failing to achieve its objective, or not being generalizable to multiple attack methods. Our method aims to overcome these issues by being effective, generalizable and straightforward to implement.

3 Methodology

We propose a novel method for defending against adversarial examples that focuses on the shortcomings of alternatives discussed in Sect. 2.3. This method is hereafter referred to as Autoencoder-based Adversarial Purification (AAP). AAP attempts to overcome shortcomings such as inference speed and generalizability by using a cooperative training of an AE and a classification CNN. AAP defends against adversarial attacks on the latter model. For this, white-box attacks are chosen due to their properties of linearity. AE-based models are renowned for being relatively quick generative models [27] and therefore suitable for real-time applications. They furthermore are generally difficult to attack [28], especially given a separation of their parameters from the targeted model as elaborated below. This makes their use as defense mechanism potentially effective against various attack methods. In the remainder of this section, all inputs are assumed to be images.

3.1 Algorithm

Considering that white-box adversarial attacks access the parameters of a neural network to calculate the required perturbations to mislead it, they are specific only to the targeted network, and thus not necessarily effective against other networks.

Building on this premise, we can construct an encoder network with parameters distinct from those of a classification network targeted by an adversarial attack. This encoder aims to map adversarial examples to a latent space representation that a decoder network can meaningfully interpret. Given that the decoder is trained to reconstruct clean images, the latent space is expected to be interpreted as a representation of clean images, rather than adversarial ones. Thus, the encoder and decoder, jointly forming an AE, take adversarial examples as input. These are represented as a latent variable in an intermediate step, and from this, clean images are ultimately produced.

3.2 Training

A classification CNN $f_{\theta_C} : \mathbb{R}^d \to \mathbb{R}^c$, parameterized by θ_C, is considered the target of an adversarial attack. The following describes the training process of f_{θ_C} and an AE $f_{\theta_{AE}} : \mathbb{R}^d \to \mathbb{R}^d$ with parameters θ_{AE}. For all samples in the dataset X, adversarial examples, denoted as $\hat{x}$, are produced using $FGSM(\theta_C, x, \epsilon)$, with x being a minibatch and ϵ a random value in the range $[0, 0.4)$. The adversarial minibatch is inputted to $f_{\theta_{AE}}$ to produce $f_{\theta_{AE}}(\hat{x})$, a reconstruction of x. This reconstruction is hereafter referred to as the purification of $\hat{x}$. f_{θ_C} receives $f_{\theta_{AE}}(\hat{x})$ as input and must attempt to predict the corresponding correct labels y. In addition to these purifications, f_{θ_C} classifies x, to ensure that it is able to handle clean input. $\hat{x}$ is not used as input for f_{θ_C}, as the effect of adversarial training should not be considered, so as to obtain an accurate evaluation of the method introduced here. The training procedure is elaborated in pseudocode in Algorithm 1 and the inference process is schematically represented in Fig. 2.

We concatenate the encoder and the decoder to form $f_{\theta_{AE}}$. Its loss function is stated in Eq. (4), is an adaptation of the loss function specifically designed for AEs introduced in [29] and based on the Structural Similarity Index (SSIM) developed by [30], in which α is set to 0.84 as proposed by the authors. See Algorithm 1 for its use during training. SSIM seeks to approximate the perceived quality of a reconstructed image by comparing local statistics to those of the original image. The equation describing the SSIM between 11×11 local regions x and y of two images is given in Eq. (5). c_1 and c_2 are constants to stabilize the division in case of a weak denominator.

$$\mathcal{L}_{AE} = \alpha \cdot -SSIM(x, f_{\theta_{AE}}(\hat{x})) + (1-\alpha) \cdot ||x - f_{\theta_{AE}}(\hat{x})||_1. \qquad (4)$$

$$SSIM(x,y) = \frac{(2\mu_x\mu_y + c_1) + (2\sigma_{xy} + c_2)}{(\mu_x^2 + \mu_y^2 + c_1)(\sigma_x^2 + \sigma_y^2 + c_2)}. \qquad (5)$$

Algorithm 1 The combined training procedure of $f_{\theta_{AE}}$ and f_{θ_C} with, in addition to the aforementioned terms, I being the number of classes and Y being the set of true labels

1: $\theta_{AE}, \theta_C \leftarrow$ *initial parameters*
2: **while** not converged **do**
3:    **for all** $x \in X, y \in Y$ **do**
4:      // *Generate adversarial examples*
5:      $\hat{x} \leftarrow Attack(\theta_C, x)$
6:      // *Calculate purification loss*
7:      $\mathcal{L}_{AE} \leftarrow \alpha \cdot -SSIM(x, f_{\theta_{AE}}(\hat{x})) + (1-\alpha) \cdot \|x - f_{\theta_{AE}}(\hat{x})\|_1$
8:      // *Calculate and apply gradients*
9:      $\theta_{AE} \leftarrow \theta_{AE} - \nabla_{\theta_{AE}}(\mathcal{L}_{AE})$
10:     **for all** $v \in \{x, f_{\theta_{AE}}(\hat{x})\}$ **do**
11:      // *Calculate classification loss*
12:      $\mathcal{L}_C \leftarrow \sum_{i=1}^{I} y_{.i} \ln(f_{\theta_C}(v)_{.i})$
13:      // *Calculate and apply gradients*
14:      $\theta_C \leftarrow \theta_C - \nabla_{\theta_C}(\mathcal{L}_C)$

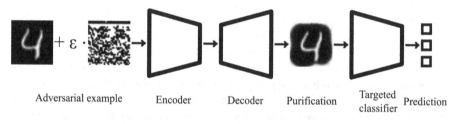

Adversarial example   Encoder   Decoder   Purification   Targeted classifier   Prediction

Fig. 2. Schematic representation of AAP. An adversarial example is reconstructed such that the classifier returns the correct class.

4 Experiments

For our experiments, we use the MNIST dataset [31], which contains gray-scale handwritten digits, the Fashion-MNIST dataset, containing gray-scale clothing images [32], and, finally, the CIFAR-10 dataset [33], consisting of color photographs. As suggested by [34], during training of the algorithms, the images are subjected to random jitter by applying random brightness and random contrast.

We assess the accuracy during inference using the attacks FGSM and iFGSM, and the black-box attack Carlini-Wagner (CW) [35]. Epsilon values of 0.3 and 0.5 are tested for both FGSM and iFGSM on MNIST and Fashion-MNIST, while values 0.06 and 0.3 are used on CIFAR-10. These values enable comparison of our results with those of Adversarial Training [19], Defense-GAN [8], and PuVAE [7]. For completeness, the results of these methods are also reported as shown in Tables 1, 2, 3 and 4. By subjecting the system to these diverse attack methods and epsilon values, we aim to understand the resilience and adaptability of AAP.

4.1 Experimental Setup

In our implementation, the networks use a ResNet-inspired architecture because of its efficiency during training [36]. The encoder is constructed using several repetitions of Dropout [37] with a rate set to 0.2 as proposed in [38] to counteract overfitting, convolutional layers with their numbers of filters set to a multiple of eight for computational efficiency [39], Instance Normalization for decreasing the internal covariate shift during training [40], and Leaky Rectified Linear Unit (Leaky ReLU) activation functions to introduce differentiable non-linearity [41]. A schematic overview of the AE architecture can be seen in Figs. 3 and 4 in the Appendix.

The decoder is similar to the encoder but uses deconvolutions [42] instead of convolutions. f_{θ_C} has an architecture almost identical to the encoder but appends a global max pooling layer and fully connected layers to enable classification. The classifier is trained with cross-entropy loss.

Although the AE and classifier are trained alternatingly, it is expected to be a stable procedure due to its cooperative nature, in contrast to the instability of GANs [23] as employed by Defense-GAN [8]. However, it should be noted that the output produced by $f_{\theta_{AE}}$ is initially semi-random and thus impossible to classify. Therefore, we empirically find it to be paramount to delay the training of f_{θ_C} until $f_{\theta_{AE}}$ has started to converge. Alternatively, the learning rate corresponding to f_{θ_C} could be reduced to avoid converging to a local minimum early on. Nevertheless, we do not choose the latter solution because it would increase the time to convergence. Lastly, the algorithms are trained for 50 000 steps, the equivalent of approximately 51 epochs.

We provide our Python implementation[1], using the TensorFlow 2.11.0 [43] and CleverHans 4.0.0 [44] software libraries.

4.2 Results

We display the classification accuracies under various attacks in Tables 2, 3, and 4 for the MNIST, Fashion-MNIST, and CIFAR-10 datasets respectively. Competitors' results are retrieved from the PuVAE publication [7], due to which some of the desired figures proved to be unavailable. The experiments were performed twice, with the second iteration's results shown due to there not being an observable difference. The gradual fall of overall accuracy on each dataset indicates that the tables are ordered in terms of complexity. However, some consistencies among the results across all datasets are noted as follows.

---

[1] The code is available on this webpage.

When the classifier is tasked with classifying clean images, i.e., those that have not been subjected to an adversarial attack, the accuracies of AAP are slightly lower than those of the classifier used in the PuVAE publication, as exhibited in Table 1. This discrepancy might be caused by the relative complexity of our training procedure, in which the classifier is trained on purified reconstructions of adversarial images produced by the autoencoder. Additionally, for the CIFAR-10 dataset, the authors of PuVAE employ a classification network distinct from the one used for the other datasets, whereas we do not make this distinction. Nevertheless, these accuracies serve as a baseline to compare the remainder of the results to, and getting close to these scores after purification is the goal of this research.

Table 1. Baseline classification accuracies on clean data of AAP's and PuVAE's implementation of the classifier on the various datasets (%).

Method	MNIST	Fashion-MNIST	CIFAR-10
AAP	98.93	91.35	75.19
PuVAE	99.29	93.54	80.30

As anticipated and demonstrated in the remainder of the tables, the accuracy of classification declines significantly when adversarial images are inputted to the classifier and no purification is applied. Nonetheless, our findings suggest a degree of robustness being imparted on the classifier, as the observed degradation suffered is of a lesser extent compared to the prior studies. In fact, the findings suggest that AAP's classifier is robust to such an extent that, even if purification is omitted, it performs better than the other classifiers with incorporation of their respective defense method. This might be the effect of the integration of adversarial examples created in real time in the training procedure. This finding is further discussed in Sect. 5.2.

Furthermore consistent with the desired behavior, purification of the adversarial images seems to significantly restore the accuracy of the classifier and surpasses the performance of competitors. In contrast to the premise that a separation of the parameters of the AE and the classifier will primarily be successful for white-box attacks, the purification is effective against various attacks including the Carlini-Wagner black-box attack. This may further indicate that the model becomes broadly robust against adversarial attacks. Additionally, increases in epsilon for FGSM and iFGSM do not appear to greatly affect the purifier's performance. This might be due to the random jitter and random epsilon incorporated during training.

Table 2. Classification accuracies per defense method on the MNIST dataset (%).

Attack	AAP		Competitors		
	No purification	Purified	Adv. Tr.	Defense-GAN	PuVAE
FGSM ($\epsilon = 0.3$)	97.00	**98.67**	88.49	86.03	88.25
FGSM ($\epsilon = 0.5$)	96.89	**98.74**	-	-	-
iFGSM ($\epsilon = 0.3$)	96.36	**98.29**	93.56	88.55	92.25
iFGSM ($\epsilon = 0.5$)	96.36	**98.29**	-	-	-
CW	98.80	**99.00**	19.20	90.76	92.92

Table 3. Classification accuracies per defense method on the Fashion-MNIST dataset (%).

Attack	AAP		Competitors		
	No purification	Purified	Adv. Tr.	Defense-GAN	PuVAE
FGSM ($\epsilon = 0.3$)	64.74	**89.88**	8.00	56.67	52.46
FGSM ($\epsilon = 0.5$)	63.58	**89.93**	-	-	-
iFGSM ($\epsilon = 0.3$)	58.22	**77.33**	46.97	66.06	71.04
iFGSM ($\epsilon = 0.5$)	58.22	**77.34**	-	-	-
CW	76.70	**88.02**	16.70	67.62	79.42

Table 4. Classification accuracies per defense method on the CIFAR-10 dataset (%).

Attack	AAP		Competitors		
	No purification	Purified	Adv. Tr.	Defense-GAN	PuVAE
FGSM ($\epsilon = 0.06$)	50.42	**51.28**	15.00	30.11	33.20
FGSM ($\epsilon = 0.3$)	47.59	**63.19**	-	-	-
iFGSM ($\epsilon = 0.06$)	46.20	**53.18**	21.51	32.47	34.48
iFGSM ($\epsilon = 0.3$)	46.14	**53.03**	-	-	-
CW	42.14	**60.56**	13.64	28.10	31.70

5 Conclusion

Our experiments investigated the performance of AAP against adversarial attacks on image classification tasks. We have demonstrated the resilience AAP by showing its capacity to maintain high classification accuracy on both clean images and adversarial examples. Moreover, AAP has proven effective in restoring classification accuracy on perturbed data. We have compared our results with cutting-edge techniques and found our approach to be competitive. While challenges remain, our research offers valuable insights and a promising direction for the development of robust defense mechanisms in the field of adversarial machine learning.

5.1 Limitations

While the experimental results demonstrate promising insights into the efficacy of AAP for defending against adversarial attacks, we acknowledge several limitations in our study.

To apply our purification methodology to a more complex dataset than those used in this research, such as ImageNet [45], which comprises over a thousand classes, not only will the classification network have to be larger, but a larger autoencoder is also required for optimal performance. This necessity presents a drawback of the proposed approach, as the required compute power will increase proportional to the increase in required parameters.

Lastly, as elaborated in Sect. 3 AAP relies on the separation of the encoder's parameters from the classifier's parameters, and the assumption that only the classifier is attacked. By feeding the adversarial example into the autoencoder, which is not under attack, we can clean it. However, an attacker may as well attack the encoder as demonstrated in [46], and thereby undermine our method. Minor adjustments to overcome this vulnerability are discussed in Sect. 5.2.

5.2 Future Work

As mentioned in Sect. 4.2, the performance of AAP's classifier on adversarial examples that are not purified exceeds the performance of competing methods. In addition to the aforementioned reasons, this may be caused by the experimental design factors - including the model's architecture, the optimizer and its hyperparameters, the batch size, and the number of epochs - not being identical to the one implemented for the competitors. The capacity of AAP's classifier is further elaborated on in the Appendix. Subsequent studies could replicate this research while maintaining the same classifier across all methodologies.

Further investigation of AAP is also needed to assess whether AE-assisted purification is a byproduct of quality degradation or if it indeed functions as intended. To ascertain this, an alternative test can be conducted by instructing the autoencoder to reconstruct the adversarial examples rather than the original examples. If, as anticipated, these reconstructions are misclassified, it would confirm that the AE operates without suffering from significant quality degradation, thus aligning with its intended purpose.

Moreover, it would be valuable to adapt the training procedure, wherein the autoencoder is trained using the backpropagation signals derived from the classification network instead of relying solely on the reconstruction loss. This adjustment would better simulate real-world scenarios, where obtaining clean images may not always be feasible.

Finally, the possibility of an additional attack on the AE was mentioned in Sect. 5.1. However, excluding the classification head, the architecture of the classifier used in this paper is interchangeable with that of the encoder. Consequently, in further research, two non-weight-sharing pairs consisting of an encoder, a classification head, and a decoder can be used to purify adversarial examples for each other. Thereby, attacks on both the *encoder + decoder*-setup and the *encoder + head*-setup may be mitigated.

Appendix

The architecture of the neural networks as described in Sect. 4.1 is displayed in Fig. 3 and Fig. 4.

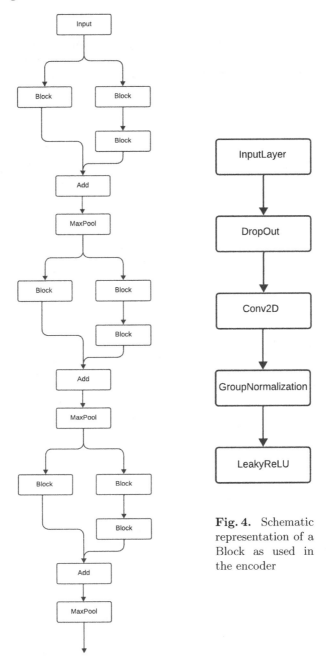

Fig. 3. Schematic representation of the encoder

Fig. 4. Schematic representation of a Block as used in the encoder

In preceding sections, AAP's classifier is stated to possibly be more powerful than the competing methods. Therefore, in Table 5, additional results are displayed. Here, the classifier was trained without the overhead of the purification method. That is, the training samples were neither perturbed nor purified. These accuracies seem comparable to those from Table 1. This does not deviate from expectations. Given the classifier has sufficient capacity, training on a wider variety of data - i.e., both unperturbed and purified images - does not have to decrease its efficacy. Moreover, this finding serves as additional evidence that purified images bear a close resemblance to their unperturbed counterparts.

Table 5. Accuracy of AAP's classifier trained only on clean images (%).

MNIST	Fashion-MNIST	CIFAR-10
97.88	90.45	75.75

References

1. Wang, J., Wang, C., Lin, Q., Luo, C., Wu, C., Li, J.: Adversarial attacks and defenses in deep learning for image recognition: a survey. Neurocomputing **514**, 162–181 (2022)
2. Han, S., Lin, C., Shen, C., Wang, Q., Guan, X.: Interpreting adversarial examples in deep learning: a review. ACM Comput. Surv. (2023)
3. Papernot, N., McDaniel, P., Goodfellow, I., Jha, S., Celik, Z.B., Swami, A.: Practical black-box attacks against machine learning. In: Proceedings of the 2017 ACM on Asia Conference on Computer and Communications Security. ASIA CCS '17, (New York, NY, USA), pp. 506–519. Association for Computing Machinery (2017)
4. Aldahdooh, A., Hamidouche, W., Fezza, S.A., Déforges, O.: Adversarial example detection for DNN models: a review and experimental comparison. Artif. Intell. Rev. **55**, 4403–4462 (2022)
5. Goodfellow, I.J., Shlens, J., Szegedy, C.: Explaining and harnessing adversarial examples (2014)
6. Fidel, G., Bitton, R., Shabtai, A.: When explainability meets adversarial learning: detecting adversarial examples using Shap signatures. In: 2020 International Joint Conference on Neural Networks (IJCNN), pp. 1–8 (2020)
7. Hwang, U., Park, J., Jang, H., Yoon, S., Cho, N.I.: Puvae: a variational autoencoder to purify adversarial examples. IEEE Access **7**, 126582–126593 (2019)
8. Samangouei, P., Kabkab, M., Chellappa, R.: Defense-GAN: protecting classifiers against adversarial attacks using generative models. CoRR, abs/1805.06605 (2018)
9. Y. Song, T. Kim, S. Nowozin, S. Ermon, and N. Kushman, "Pixeldefend: Leveraging generative models to understand and defend against adversarial examples," *CoRR*, vol. abs/1710.10766, 2017
10. Zhang, J., Li, C.: Adversarial examples: opportunities and challenges. IEEE Trans. Neural Netw. Learn. Syst. **31**(7), 2578–2593 (2019)
11. Hinton, G.E., Salakhutdinov, R.R.: Reducing the dimensionality of data with neural networks. Science **313**, 504–507 (2006)

12. Zhang, Y., Li, Y., Li, Y., Guo, Z.: A review of adversarial attacks in computer vision (2023)
13. Qiu, S., Liu, Q., Zhou, S., Wu, C.: Review of artificial intelligence adversarial attack and defense technologies. Appl. Sci. **9**(5), 909 (2019)
14. Dong, Y., et al.: Boosting adversarial attacks with momentum. In: Proceedings of the IEEE Conference on Computer Vision and Pattern Recognition, pp. 9185–9193 (2018)
15. Carlini, N., Wagner, D.: Towards evaluating the robustness of neural networks. In: 2017 IEEE Symposium on Security and Privacy (SP), pp. 39–57. IEEE (2017)
16. Ye, H., Liu, X., Yan, A., Li, L., Li, X.: Detect adversarial examples by using feature autoencoder. In: Sun, X., Zhang, X., Xia, Z., Bertino, E. (eds.) ICAIS 2022. LNCS, vol. 13340, pp. 233–242. Springer, Cham (2022). https://doi.org/10.1007/978-3-031-06791-4_19
17. Tong, L., et al.: Adversarial sample detection framework based on autoencoder. In: 2020 International Conference on Big Data and Artificial Intelligence and Software Engineering (ICBASE), pp. 241–245 (2020)
18. Ye, H., Liu, X.: Feature autoencoder for detecting adversarial examples. Int. J. Intell. Syst. **37**, 05 (2022)
19. Szegedy, C., et al.: Intriguing properties of neural networks (2013)
20. Raghunathan, A., Xie, S.M., Yang, F., Duchi, J.C., Liang, P.: Adversarial training can hurt generalization. arXiv preprint arXiv:1906.06032 (2019)
21. van den Oord, A., Kalchbrenner, N., Kavukcuoglu, K.: Pixel recurrent neural networks (2016)
22. Meng, D., Chen, H.: Magnet: a two-pronged defense against adversarial examples. CoRR, abs/1705.09064 (2017)
23. Goodfellow, I.J., et al.: Generative adversarial networks (2014)
24. Arjovsky, M., Chintala, S., Bottou, L.: Wasserstein GAN (2017)
25. Gulrajani, I., Ahmed, F., Arjovsky, M., Dumoulin, V., Courville, A.: Improved training of Wasserstein GANs (2017)
26. Kingma, D.P., Welling, M.: Auto-encoding variational Bayes (2013)
27. Bond-Taylor, S., Leach, A., Long, Y., Willcocks, C.G.: Deep generative modelling: a comparative review of VAEs, GANs, normalizing flows, energy-based and autoregressive models (2021)
28. Tabacof, P., Tavares, J., Valle, E.: Adversarial images for variational autoencoders (2016)
29. Zhao, H., Gallo, O., Frosio, I., Kautz, J.: Loss functions for neural networks for image processing (2015)
30. Wang, Z., Bovik, A., Sheikh, H., Simoncelli, E.: Image quality assessment: from error visibility to structural similarity. IEEE Trans. Image Process. **13**, 600–612 (2004)
31. Deng, L.: The MNIST database of handwritten digit images for machine learning research. IEEE Sig. Process. Mag. **29**(6), 141–142 (2012)
32. Xiao, H., Rasul, K., Vollgraf, R.: Fashion-MNIST: a novel image dataset for benchmarking machine learning algorithms. arXiv e-prints, arXiv:1708.07747 (2017)
33. Krizhevsky, A., Hinton, G., et al.: Learning multiple layers of features from tiny images (2009)
34. Isola, P., Zhu, J.-Y., Zhou, T., Efros, A.A.: Image-to-image translation with conditional adversarial networks (2016)
35. Tian, S., Yang, G., Cai, Y.: Detecting adversarial examples through image transformation. In: Proceedings of the AAAI Conference on Artificial Intelligence, vol. 32 (2018)

36. He, K., Zhang, X., Ren, S., Sun, J.: Deep residual learning for image recognition (2015)
37. Srivastava, N., Hinton, G., Krizhevsky, A., Sutskever, I., Salakhutdinov, R.: Dropout: a simple way to prevent neural networks from overfitting. J. Mach. Learn. Res. **15**(56), 1929–1958 (2014)
38. Park, S., Kwak, N.: Analysis on the dropout effect in convolutional neural networks, pp. 189–204 (2017)
39. N. Corporation: Convolutional layers user's guide
40. Ulyanov, D., Vedaldi, A., Lempitsky, V.: Instance normalization: the missing ingredient for fast stylization (2016)
41. Maas, A.L.: Rectifier nonlinearities improve neural network acoustic models (2013)
42. Zeiler, M.D., Krishnan, D., Taylor, G.W., Fergus, R.: Deconvolutional networks, pp. 2528–2535 (2010)
43. T. Developers: Tensorflow (2023)
44. Papernot, N., et al.: Technical report on the cleverhans v2.1.0 adversarial examples library. arXiv preprint arXiv:1610.00768 (2018)
45. Deng, J., Dong, W., Socher, R., Li, L.-J., Li, K., Fei-Fei, L.: Imagenet: a large-scale hierarchical image database. In: 2009 IEEE Conference on Computer Vision and Pattern Recognition, pp. 248–255. IEEE (2009)
46. Kos, J., Fischer, I., Song, D.: Adversarial examples for generative models (2018)

Approximate Compression of CNF Concepts

Sieben Bocklandt[1,2](✉)[iD], Vincent Derkinderen[1,2][iD], Angelika Kimmig[1,2][iD], and Luc De Raedt[1,2,3][iD]

[1] Department of Computer Science, KU Leuven, Leuven, Belgium
{sieben.bocklandt,vincent.derkinderen,angelika.kimmig,
luc.raedt}@kuleuven.be
[2] Leuven.AI, KU Leuven Institute for AI, Leuven, Belgium
[3] Center for Applied Autonomous Systems, Örebro University, Örebro, Sweden

Abstract. We consider a novel concept-learning and merging task, motivated by two use-cases. The first is about merging and compressing music playlists, and the second about federated learning with data privacy constraints. Both settings involve multiple learned concepts that must be merged and compressed into a single interpretable and accurate concept description. Our concept descriptions are logical formulae in CNF, for which merging, i.e. disjoining, multiple CNFs may lead to very large concept descriptions. To make the concepts interpretable, we compress them relative to a dataset. We propose a new method named CoWC (Compression Of Weighted Cnf) that approximates a CNF by exploiting techniques of itemset mining and inverse resolution. CoWC compresses the CNF size while also considering the F1-score w.r.t. the dataset. Our empirical evaluation shows that CoWC outperforms alternative compression approaches.

Keywords: Concept learning · Formula compression

1 Introduction

Our work is motivated by two application scenarios. In both of these scenarios we obtain a set of concept descriptions that have been learned from data, and that must be merged and compressed without losing too much accuracy. The first application concerns the automatic generation of music playlists. Here, each learned concept represents a music playlist (e.g., *relaxing music*), as a kind of query to a database of songs [3,9]. The second setting concerns federated learning, where multiple concepts have been learned from distributed, private data in a federated manner [1,20]. The goal in both cases is to obtain a single concept description that is understandable and accurate. We explain our assumptions in more detail.

First, we assume the concept descriptions to be represented as logical formulas in conjunctive normal form (CNF). The key motivation for using CNFs

(style=Jazz ∧ feel=Happy) ∨
(style=Jazz ∧ feel=Exciting) ∨   (style=Jazz ∨ style=Rock) ∧
(style=Rock ∧ feel=Happy) ∨   (feel=Happy ∨ feel=Exciting) ∧
(style=Rock ∧ feel=Exciting) ∨

Fig. 1. A music playlist represented as a logical concept, in DNF (left) and CNF (right). The latter is here preferred.

rather than the more common disjunctive normal form (DNF) is that CNFs are very natural and interpretable when considering an internal disjunction. This is a disjunction over a categorical variable to indicate its possible values, for example 'style=Jazz ∨ style=Rock'. Ryszard Michalski, one of the founders of the field of machine learning, argued for their interpretability from a cognitive perspective, in a concept learning setting [18]. They remain especially relevant today given the prominent use of one-hot encodings. An example CNF (with internal disjunctions) is shown in Fig. 1. Note that it is more readable and interpretable than the shown DNF. CNF is also the desired format for our music playlist application, as formulas like the one in Fig. 1 can be used to retrieve relevant songs from a database. After inspection, the music expert can then also easily tune the formula to obtain the expected behaviour. Finally, another benefit of CNF is that it is the preferred input form for SAT solvers and knowledge compilers [2,12].

Second, we must construct the disjunction of multiple CNFs. Such disjunctions occur naturally when combining learned concepts represented by multiple CNFs, each covering different parts of the data (cf. Fig. 2a). Consider for instance the music playlist use case where multiple playlists can be combined [9]; or the federated learning [1,20] setting with data-privacy constraints where the CNFs, independently learned on distributed private datasets, must be combined. The main challenge of such settings is that disjunction of multiple CNFs requires a cross product of their clauses, which may result in a very large CNF (cf. Fig. 2b).

Third, since we merge and compress descriptions, it is advised to consider the data on which the original CNFs have been learned. For example, a database $\mathcal{D}$ of songs as shown in Fig. 2c, or the federated dataset $\mathcal{D}$ of each individual device. That is why we evaluate not only the degree of compression w.r.t. the disjunction of the CNFs, but also the accuracy on $\mathcal{D}$. To balance these two criteria, we borrow the global criterion principle from the multi-objective learning setting.

This type of compression has - to the best of our knowledge - not been considered before. Compression of large CNF formulas has previously been studied to discover hidden structural knowledge [10] and in the context of SAT solving, as pre-processors to faster determine satisfiability of a given formula [8,16] or the number of models [13,14]. We instead focus on obtaining a CNF of manageable size to facilitate the aforementioned disjunction and obtain more readable CNFs [3]. To achieve this goal, and in deviation of the previously mentioned work, we consider a lossy compression of the CNF formula.

More specifically, we propose a new compression algorithm named CoWC, that is based on four compression rules and uses techniques from itemset mining to find commonly occurring subclauses. CoWC supports both lossy and lossless compression as it is governed by a threshold-parameter that determines the degree of lossiness. We can use the incremental approach in Fig. 2d to compute an approximate disjunction of the set of CNFs (cf. Fig. 2e) by relying on CoWC to maintain CNFs of manageable sizes.

To summarize, the key contributions of our work are: (1) formalizing a novel problem setting for merging and compressing multiple CNF while taking into account a dataset of relevant models; (2) introducing CoWC, a CNF compression algorithm based on inverse resolution and frequent itemset mining as a proposed

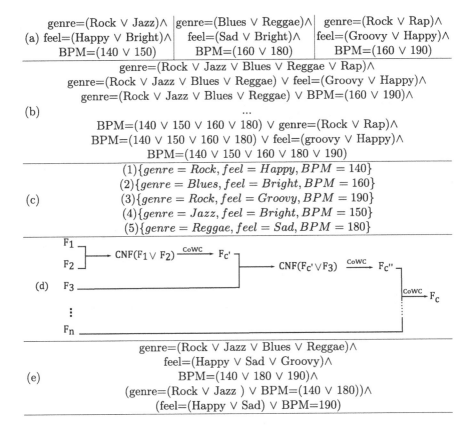

Fig. 2. (a) shows a disjunction of 3 CNF formulas, where we shorten internal disjunctions 'x = A ∨ x = B' as 'x = (A ∨ B)'. (b) contains the disjunction of these formulas in CNF, illustrating the explosive size increase. For larger or more CNFs, this quickly becomes infeasible to compute. (c) shows a dataset of relevant songs $\mathcal{D}$. (d) shows how we can instead use CoWC to create an approximate disjunction of the three CNFs with respect to $\mathcal{D}$, the result of which is in (e). Note that songs 2 and 4 are no longer covered by the result.

solution; (3) evaluating the approach on the two aforementioned application scenarios showing that CoWC achieves significantly more compression with only a minor loss in F1-score.

2 Background

We first review the relevant background on weighted propositional logic and frequent itemset mining, and introduce the necessary notation.

2.1 Propositional Logic

A *literal* l is a Boolean variable or its negation (v or $\neg v$). A *propositional logic formula* F is inductively defined as a literal l, a disjunction of two formulas $F_1 \vee F_2$ (read as 'or'), a conjunction of two formulas $F_1 \wedge F_2$ (read as 'and'), or the negation of a formula $\neg F_1$, with the usual expected semantics. A *clause* C is a disjunction $(l_1 \vee ... \vee l_n)$ of literals l_i, and $|C|$ is the number of literals in the clause. A formula F is said to be in *conjunctive normal form* (CNF) iff it is expressed as a conjunction of clauses $(C_1 \wedge ... \wedge C_k)$. A formula F can be converted into CNF by applying the distributive law, noted as $CNF(F)$. When discussing clauses of a CNF formula, we may also use the notation $C \in F$ to iterate over the clauses of F. Additionally, we may use *subclause* to refer to only part of a clause. The *residue* $\mathfrak{R}(C)$ of a clause C that contains a subclause C_{freq} is the disjunction of literals that are part of the clause, but not of the subclause.

An *interpretation* is a truth assignment over a set of Boolean variables $\mathbf{V}$. We may shorten the representation of an interpretation I, for example, write $I = \{a \mapsto true, b \mapsto false, c \mapsto true\}$ as $\{a, \neg b, c\}$. The literals of I can then be referred to as a $l \in I$. A formula F is *satisfiable* if there is an interpretation I that satisfies all its clauses. An interpretation that satisfies formula F is called a *model* of F. We use $\mathcal{M}(F)$ to denote the set of all models of F. Determining the size of $\mathcal{M}(F)$ is called model counting.

2.2 Weighted Logical Theories

A *weighted logic formula* is a formula F with an associated weight function w, which, for example, can be used to induce a probability distribution over the models [4,6]. In this work, without loss of generality, we assume a weight function w that determines the weight of each literal l, and we define the weight of a model m as a product of the weights of its literals, $w(m) := \prod_{l \in m} w(l)$.

Probabilistic reasoning can then be cast into reasoning over these weighted logic formulas. For example, computing a marginal probability such as $Prob(c = true)$ can be cast as determining the *weighted model count*.

Definition 1 (weighted model count (WMC)). *Given a formula F over variables* $\mathbf{V}$, *and a weight function w that assigns a real to each variable and its* negation, the weighted model count is defined as

$$WMC(F, \mathbf{V}, w) = \sum_{m \in \mathcal{M}(F)} \prod_{l \in m} w(l). \tag{1}$$

When clear from context, we may also write $WMC(F)$ instead of $WMC(F, \mathbf{V}, w)$.

Example 1. Consider formula $F = (\neg a \vee b) \wedge (\neg b \vee \neg c)$ over variables $\mathbf{V} = \{a, b, c\}$, the following two interpretations:

$$I_1 = \{a \mapsto true, b \mapsto false, c \mapsto true\}, \quad I_2 = \{a \mapsto true, b \mapsto true, c \mapsto false\},$$

and a weight function w that maps each variable and its negation to a real:

$$w = \{a \mapsto 0.5, \neg a \mapsto 0.5, b \mapsto 0.25, \neg b \mapsto 0.75, c \mapsto 1, \neg c \mapsto 1\}$$

Formula F has 4 models, $|\mathcal{M}(F)| = 4$, with $I_1 \notin \mathcal{M}(F)$ and $I_2 \in \mathcal{M}(F)$. The weight of I_2 in relation to w is $0.5 \times 0.25 \times 1 = 0.125$. When computing the weight of all models, it is clear w and F induce a probability distribution. The weighted model count of F is $WMC(F) = \underbrace{0.125}_{\{a,b,\neg c\}} + \underbrace{0.125}_{\{\neg a,b,\neg c\}} + \underbrace{0.125}_{\{\neg a,\neg b,c\}} + \underbrace{0.125}_{\{\neg a,\neg b,\neg c\}} = 1$.

2.3 Frequent Itemset Mining

In this work, frequent itemset mining is used to identify subclauses that occur in more than one clause. Once found, the subclauses are subject to operators that compress the formula.

Frequent itemset mining is the task of discovering sets of items that occur together in the transactions of a database. Let $\mathbf{X}$ be a finite set of items x. A transaction $\mathbf{T}$ is a subset of $\mathbf{X}$ and supports an itemset i iff $i \subseteq \mathbf{T}$. The *support* of an itemset i is the amount of transactions in a dataset that include i. The goal of frequent itemset mining is to find $\mathbf{I} = \{i | support(i) \geq minsup\}$, the set of itemsets with a support greater than or equal to a given minimal support value $minsup$. An itemset i is *frequent* iff $i \in \mathbf{I}$ and *closed* iff all supersets have a lower support (i.e., if making the clause more specific would result in a lower support).

3 Problem Setting

We now revisit in more detail the music playlist generation and federated learning tasks, and afterwards formalise the generalised problem.

In the playlist setting, a music playlist is represented as a CNF formula (cf. Fig. 2a) with $\mathcal{D}$ the database of songs (cf. Fig. 2c). As a first problem, we want to compress this CNF to obtain an accurate but interpretable description. A solution to this first problem can then be used to resolve the second problem, merging multiple CNFs (e.g., multiple playlists [9]). As this may result in a very large CNF (cf. Fig. 2b), the goal is to construct an approximate version that is more interpretable and still accurate (cf. Fig. 2e). In the federated learning setting, each CNF represents a model learned on private, distributed data (e.g., constraints in a financial [1] or medical [20] domain). Here, $\mathcal{D}$ is one of the private datasets, as the other distributed data cannot be shared for privacy reasons.

Problem 1: CNF Compression.

Given A dataset $\mathcal{D}$ of relevant models, a CNF formula F that covers relevant models $\mathcal{D}_F = \mathcal{D} \cap \mathcal{M}(F)$, and a function DL that returns the description length for any CNF formula F defined as[1] $DL(F) = \sum_{C \in F} |C|$.

Objective The goal is to find a CNF formula F_c that approximates F, maximising (1) the compression rate $Comp$ whilst ensuring (2) that $\mathcal{D}_{F_c} = \mathcal{D} \cap \mathcal{M}(F_c)$ remains close to $\mathcal{D}_F$.

(1) $\arg\max_{F_c} Comp(F, F_c) = \arg\max_{F_c} \frac{DL(F) - DL(F_c)}{DL(F)}$

(2) $\arg\max_{F_c} F_1\text{-score}(\mathcal{D}_F, \mathcal{D}_{F_c})$ with $F_1\text{-score}(x, y) = \frac{2 \times |x \cap y|}{2 \times |x \cap y| + |x - y| + |y - x|}$

We combine these two objectives using the global criterion principle – a general technique to combine several objectives [17]. More concretely, we optimise the distance of a solution F_c to an ideal (but possibly unreachable) solution whose objective values are denoted as P_{ideal}:

$$\arg\min_{F_c} ||(Comp(F, F_c), F_1\text{-score}(\mathcal{D}_F, \mathcal{D}_{F_c})) - P_{ideal}||_2 \quad (2)$$

with $P_{ideal} = (1, 1)$. The global criterion principle is illustrated in Fig. 3.

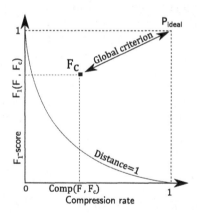

Fig. 3. Visualisation of the global criterion, which is the distance between an ideal point P_{ideal} and the objective values of a solution F_c.

---

[1] The choice for DL is in line with DUCE [19] and the PMC preprocessor [14].

Problem 2: Disjunction of CNFs. The resulting compression algorithm can be used to approximate $CNF(\bigvee_i F_i)$ for a set of CNFs $\{F_1, \ldots, F_n\}$ that would be too large to compute without intermediate compression.

Given A dataset $\mathcal{D}$ of relevant models, and a set of CNF formulas $\{F_1, \ldots, F_n\}$ together forming $F = CNF(\bigvee_i F_i)$ covering relevant models $\mathcal{D}_F = \mathcal{D} \cap \mathcal{M}(F)$.
Objective Compute CNF formula $F_c \approx F$ such that $Comp(F, F_c)$ and F$_1$-score$(\mathcal{D}_F, \mathcal{D}_{F_c})$ are both maximised using the global criterion principle introduced above.

4 Operators

Before introducing the new compression algorithm CoWC, we first elaborate on the four logic compression operators that it relies on. These operators are adapted from DUCE [19], where they are used in the context of feature construction. Apart from the second operator, they are also used in Mistle [11]: a concept learning algorithm that refines a learned formula using the minimum description length principle.

The operators are based on inverse resolution. Resolution is the logical rule used in deductive inference to infer a new clause (bottom) from two other clauses (top): $\frac{x_1 \vee \ldots \vee x_n \vee z, \; y_1 \vee \ldots \vee y_m \vee \neg z}{x_1 \vee \ldots \vee x_n \vee y_1 \vee \ldots \vee y_m}$. Inverse resolution refers to inferring input clauses for resolution based upon the output clauses [21]. Thus, it is an inverse of the process of resolution: $\frac{x_1 \vee \ldots \vee x_n \vee z, \; x_1 \vee \ldots \vee x_n \vee y_1 \vee \ldots \vee y_m}{y_1 \vee \ldots \vee y_m \vee \neg z}$. The first three operators that we will define rely on Theorem 1, that partitions a CNF F into clauses $\mathbf{C}_\in$ that contain a given frequent subclause C_{freq}, and clauses $\mathbf{C}_{\notin}$ that do not contain it. The frequent subclause C_{freq} can be identified using techniques from frequent itemset mining such as LCM [23] or DCI_closed [15]. This theorem easily follows from the definitions of $\mathfrak{R}$ and $\mathcal{C}_\in$.

Theorem 1. *For CNF F and frequent subclause C_{freq}, we have that $F = (C_{freq} \vee \mathfrak{R}) \wedge \mathbf{C}_{\notin}$ where residue $\mathfrak{R} = \bigwedge_{C \in \mathcal{C}_\in} (C)$ is the conjunction of all clause residues.*

Definition 2 (T-operator). *Given a frequent subclause C_{freq}, the truncation operator, or T-operator, removes all residues $\mathfrak{R}$, thus only retaining C_{freq}.*

$$(C_{freq} \vee \mathfrak{R}) \wedge \mathbf{C}_{\notin} \xrightarrow{T\text{-}op} C_{freq} \wedge \mathbf{C}_{\notin} \qquad (3)$$

Example 2 (T-operator). $(a \vee \neg b \vee c) \wedge (a \vee \neg b \vee \neg d) \xrightarrow{T\text{-}op} (a \vee \neg b)$

The T-operator is lossy as it removes all the models that satisfy $\neg C_{freq} \wedge \mathfrak{R} \wedge \mathbf{C}_{\notin}$. It is however lossless in two specific cases. First, when one of the residues is the empty clause, which occurs when the frequent subclause is a clause of the formula. In this case the frequent subclause subsumes all residues.

Example 3 (Subsumption). $(a \vee \neg b) \wedge (a \vee \neg b \vee c) \wedge (a \vee \neg b \vee d \vee e) \xrightarrow{\text{T-op}} (a \vee \neg b)$

Second, when the residue $\mathfrak{R}$ is unsatisfiable, resolution is applied, resulting in lossless compression.

Example 4 (Resolution). $(a \vee \neg b \vee c) \wedge (a \vee \neg b \vee \neg c) \xrightarrow{\text{T-op}} (a \vee \neg b)$

Definition 3 (F-operator). *Given a frequent subclause C_{freq}, the frequent operator, or F-operator, removes C_{freq} and keeps the clause residues $\mathfrak{R}$.*

$$(C_{freq} \vee \mathfrak{R}) \wedge \mathbf{C_{\not\in}} \xrightarrow{\text{F-op}} \mathfrak{R} \wedge \mathbf{C_{\not\in}} \qquad (4)$$

Example 5 (F-operator). $(a \vee \neg b \vee c) \wedge (a \vee \neg b \vee \neg d) \xrightarrow{\text{F-op}} (c) \wedge (\neg d)$

The F-operator is a lossy operator, as it removes all the models that satisfy $C_{freq} \wedge \neg \mathfrak{R} \wedge \mathbf{C_{\not\in}}$.

Definition 4 (V-operator). *Given a frequent subclause C_{freq} and a literal l such that there exists a clause $l \vee C_{freq}$ in F, the V-operator replaces all other occurrences of C_{freq} with $\neg l$.*

$$(l \vee C_{freq}) \wedge (\bigwedge_{\substack{C \in C_\in \\ l \notin C}} C) \wedge \mathbf{C_{\not\in}} \xrightarrow{\text{V-op}} (l \vee C_{freq}) \wedge (\bigwedge_{\substack{C \in C_\in \\ l \notin C}} (\neg l \vee \mathfrak{R}(C))) \wedge \mathbf{C_{\not\in}} \qquad (5)$$

Example 6 (V-operator). $(a \vee \neg b \vee c) \wedge (a \vee \neg b \vee \neg d \vee e) \xrightarrow{\text{V-op}} (a \vee \neg b \vee c) \wedge (\neg c \vee \neg d \vee e)$

The reasoning behind the V-operator is as follows: $l \vee C_{freq}$ is equivalent to $\neg l \Rightarrow C_{freq}$, meaning if l evaluates to false, C_{freq} must be true for the interpretation to be a model of F. Hence, the V-operator enforces C_{freq} in a clause C by using $\neg l$ instead. This is an inverse resolution operator, as $\neg C_{freq}$ can be re-obtained when the output clauses are resolved on l. The V-operator removes all the models that satisfy $l \wedge C_{freq} \wedge \neg \mathfrak{R} \wedge \mathbf{C_{\not\in}}$, thus removing fewer models than the F-operator. If $l \Leftrightarrow C_{freq}$, then the V-operator is lossless.

Definition 5 (W-operator). *Given a frequent subclause C_{freq}, the W-operator introduces a new literal l not yet occurring within F, defines it to be equivalent to C_{freq}, and replaces all occurrences of C_{freq} with l.*

$$\mathbf{C_\in} \wedge \mathbf{C_{\not\in}} \xrightarrow{\text{W-op}} CNF(C_{freq} \Leftrightarrow l) \wedge (\bigwedge_{C \in C_\in} (l \vee \mathfrak{R}(C))) \wedge \mathbf{C_{\not\in}}, \qquad (6)$$

Example 7 (W-operator).

$$(a \vee \neg b \vee c) \wedge (a \vee \neg b \vee \neg d \vee e) \xrightarrow{\text{W-op}}$$
$$\underbrace{(a \vee \neg b \vee \neg w) \wedge (\neg a \vee w) \wedge (b \vee w)}_{a \vee \neg b \Leftrightarrow w} \wedge (w \vee c) \wedge (w \vee \neg d \vee e)$$

The W-operator performs a form of predicate invention and is present in Mining4SAT [10]. It also relates to the Tseitin transformation [22] where predicate invention is used to convert a propositional logical formula to CNF. The W-operator is an inverse resolution operator, as the input can be reconstructed when resolving on the new variable. It does not remove any models, and is hence a lossless operator.

5 CoWC

We introduce a new compression algorithm, named CoWC, that greedily compresses a CNF formula F by repeatedly applying the operators defined in the previous section. We first provide a high level explanation (see Algorithm 1), before delving deeper into its details. In Sect. 5.3 we then discuss how CoWC can be used to solve the second problem setting, that is, how to compute the disjunction of a set of CNFs.

First, we identify frequent subclause candidates using techniques from itemset mining (line 2). This is discussed in more detail in Sect. 5.1. Then, we consider each of the four operators defined in the previous section (line 5), and determine the operator that leads to the best solution (line 6). The details of this step, how CoWC estimates the quality of a solution, how it selects the best operator, and how it determines the literal parameter l of the V-operator, are all discussed in Sect. 5.2. If one of the operators improved the best known solution F_c whilst keeping the F_1-score above the given threshold t_{F_1}, we update F_c and recompute the frequently occurring subclauses (line 9).

Note that this prematurely exits the for-loop in line 4 since a change in F_c means we should reconsider which are the frequently occurring subclauses. If none of the operators improved the best known solution F_c, we consider the next

---

Algorithm 1. CoWC: Compression of Weighted CNF

Input: CNF F; Search threshold t_{search} ; Threshold on F_1-score t_{F_1}
Output: F_c
1: $F_c \leftarrow F$; $nb\_failed\_subclauses \leftarrow 0$
2: $\mathbf{C}_{freqs} \leftarrow$ compute set of frequent subclauses of F_c
3: $\mathbf{C}_{freqs} \leftarrow$ order $\mathbf{C}_{freqs}$ based on estimated compression (high to low)
4: **for** subclause C_{freq} in $\mathbf{C}_{freqs}$ **do**
5:     $F_{c_T}, F_{c_F}, F_{c_V}, F_{c_W} \leftarrow$ apply operators T, F, V and W on F_c
6:     $F'_c \leftarrow$ select best operator result from $\{F_{c_T}, F_{c_F}, F_{c_V}, F_{c_W}\}$
7:     **if** F'_c better than F_c and $F_1(F, F'_c) \geq t_{F_1}$: **then**
8:         $F_c \leftarrow F'_c$
9:         **go to** line 2 to recompute frequent subclauses.
10:     **else**
11:         $nb\_failed\_subclauses \leftarrow nb\_failed\_subclauses + 1$
12:         **if** $nb\_failed\_subclauses \geq t_{search}$ **then**
13:             **return** F_c
14: **return** F_c

identified subclause. We could repeat this process until all subclause candidates are considered. However, there might be many. We instead choose a stopping criterion, which compares the number of subclauses that did not lead to any improvement to a user-specified threshold parameter t_{search} (line 12).

5.1 Frequent Subclause Selection

CoWC minimises the global criterion by applying logical operators to CNF formula F, to obtain compressed F_c. These logical operators operate on a subclause C_{freq} that is decided by techniques from frequent itemset mining. Our implementation of CoWC uses the DCI_Closed algorithm [15] in the LCM frequent itemset miner [23]. The clauses in F can be seen as transactions with each literal being an item in the transaction. A subclause that occurs in more than one clause in F will thus be represented by a frequent itemset in the related mining problem. Recall that a frequent itemset is defined as an itemset with a support of at least $minsup$ (i.e., occurring at least $minsup$ times). In our implementation of CoWC, we unburden the user of this parameter by initially setting $minsup$ to 2, and by dynamically changing it over time. More concretely, the LCM miner is given one minute to return results, otherwise we multiply $minsup$ with 1.5 and try again. This process repeats until LCM finishes in time.

In practice we determine C_{freq} by considering all closed frequent itemsets returned by an itemset miner, so not only the most frequent one. We rank these returned candidates based on their estimated compression value. This is an estimate as it would be too expensive to compute the exact compression value for all frequent itemsets (of which there could be millions).

The estimate is based on the change in the description length. Before applying any operator, a frequent subclause C_{freq} that occurs x times contributes $x \times |C_{freq}|$ to the total description length of F. Recall that the W-operator, that is expected to compress the least, replaces each C_{freq} occurrence with a newly introduced literal l, and adds the definition $C_{freq} \Leftrightarrow l$. The difference in description length is therefore $x \times |C_{freq}| - (DL(C_{freq} \Leftrightarrow l) + x)$. The T-operator, that is expected to compress the most, removes all residues $\mathfrak{R}$ and replaces the x occurrences of C_{freq} by just one occurrence.

As computing the residues is not feasible for all subclauses, CoWC estimates the difference as $x \times |C_{freq}| - |C_{freq}|$. As the applied operator also depends on their effect on the F_1-score, CoWC aggregates the results of these operators into an estimated compression contribution for C_{freq}, using following equation: $x \times |C_{freq}| - \frac{DL(C_{freq} \Leftrightarrow l) + x + |C_{freq}|}{2}$. This estimate is used to produce the ranking of candidate frequent subclauses.

5.2 Operator Selection

In line 6 of Algorithm 1, CoWC selects the result of the best operator. Recall from the problem description in Sect. 3 that the global criterion we minimise is determined by both the Compression rate and the F_1-score, the latter of which compares $\mathcal{D}_F$ to $\mathcal{D}_{F_c}$ (see Eq. 7 and 8). Computing this F_1-score exactly is relatively expensive for a step that would be performed several times. Therefore, CoWC instead compares $\mathcal{M}(F)$ and $\mathcal{M}(F_c)$ as an estimate (see Eq. 9), which is much easier to compute.

$$F_1(\mathcal{D}_F, \mathcal{D}_{F_c}) = \frac{2 \times |\mathcal{D}_F \cap \mathcal{D}_{F_c}|}{2 \times |\mathcal{D}_F \cap \mathcal{D}_{F_c}| + |\mathcal{D}_F - \mathcal{D}_{F_c}| + |\mathcal{D}_{F_c} - \mathcal{D}_F|} \quad (7)$$

$$= \frac{2 \times |\mathcal{D}_{F \wedge F_c}|}{2 \times |\mathcal{D}_{F \wedge F_c}| + |\mathcal{D}_{F \wedge \neg F_c}| + |\mathcal{D}_{F_c \wedge \neg F}|} \quad (8)$$

$$F_1(\mathcal{M}(F), \mathcal{M}(F_c)) = \frac{2 \times |\mathcal{M}(F \wedge F_c)|}{2 \times |\mathcal{M}(F \wedge F_c)| + |\mathcal{M}(F \wedge \neg F_c)| + |\mathcal{M}(F_c \wedge \neg F)|} \quad (9)$$

To compensate for not considering $\mathcal{D}$, CoWC weighs each model of F and F_c to guide the estimation closer to the correct F_1-score. The weight of each literal l is set to the fraction of models in $\mathcal{D}$ where the literal l is true, thus representing the distribution of the literals in $\mathcal{D}$.

Example 8. Consider a dataset $\mathcal{D}$ of three models, $\{a, b, \neg c\}, \{a, \neg b, \neg c\}$ and $\{\neg a, \neg b, \neg c\}$. This would result in the following literal weights:
$w = \{a \mapsto 0.66, \neg a \mapsto 0.33, b \mapsto 0.33, \neg b \mapsto 0.66, c \mapsto 0.0, \neg c \mapsto 1.0\}$

The weighted F_1-score is defined in Eq. 10, where we replaced $|\mathcal{M}(F \wedge F_c)|$ with $WMC(F \wedge F_c)$ (and similarly for the others), and used $WMC(F \wedge \neg F_c) = WMC(F) - WMC(F \wedge F_c)$ and $WMC(F_c \wedge \neg F) = WMC(F_c) - WMC(F \wedge F_c)$.

$$F_1(F, F_c) = \frac{2 \times WMC(F \wedge F_c)}{WMC(F) + WMC(F_c)} \quad (10)$$

This makes it easier to compute with state-of-the-art weighted model counters that rely on the input formula to be a CNF. To select the best operator, all operators are applied on the formula with the selected frequent subclause (Algorithm 1 line 5) from which the one that decreases the global criterion the most is selected. Importantly, we only update F_c with the new solution F'_c if the estimated F_1-score is above the given threshold t_{F_1}. For the V-operator, which also requires a literal l, we consider all valid literals l (i.e., those for which $l \vee C_{freq}$ is in F_c) and only keep the best solution F'_c.

5.3 Approximate Disjunction

We evaluate the performance of CoWC in the context of the second problem: (approximately) computing a disjunction of several CNFs (see Fig. 2). Our proposed approach to solve this problem is based on iterative compression. We start by constructing the disjunction of the first two CNF formulas, and compress the result using CoWC. The resulting compressed CNF is then disjoined with the remainder of the CNF formulas, in a similar fashion until all CNF formulas have been included. Note that this may result in clauses that are not internal disjunctions, as shown in Fig. 2e.

6 Evaluation

Our experiments address the following research questions:

Q1: Do the weights guide CoWC's compression to a better solution F_c?
Q2: Does CoWC outperform other compression approaches?

6.1 Setup

We evaluate CoWC on two example applications: the automatic generation of music playlists using the method from Goyal et al. [9], and the merging of learned logic concepts in a federated learning setting with data privacy constraints. The code of CoWC and the experimental setup are publicly available[2].

Music Playlists. As the dataset of Goyal et al. [9] is not public, we instead generate 20 random music playlists from the Free Music Archive [5] which contains ±9200 annotated songs. Each playlist covers at least 750 songs, from which 10% is then used as input to their concept learning approach that is based on likelihood and item-queries. We repeat this 10 times for each playlist, resulting in 200 sets of CNF formulas.

Federated Learning with Data-Privacy Constraints. We also use categorical benchmark datasets from the UCI repository [7]: cars, tic-tac-toe, congressional voting records, soybean, Hayes-roth, where we encode categorical variables using a one-hot encoding. For each of these datasets, 10 sets of 10 random CNF formulas are generated, each consisting of 5 to 25 random clauses, describing between 10% to 25% of the provided instances.

Comparison. For each set of CNF formulas, we compute the approximate disjunction using the approach described in Sect. 5.3. As a comparison, we replace CoWC with Mining4Sat [10], and with the PMC preprocessor [14]. When obtaining an intermediate result of more than 15 000 clauses, we terminate the instance and label it as infeasible. We additionally compare to KRIMP [24], a concept learner that learns a compressed DNF. However, by using $D_{\neg F}$ as input data, we can negate the resulting DNF to obtain a CNF representing the intended concept. All experiments use search threshold $t_{search} = 10$, which limits the number of frequent subclauses that did not result in compression, thus limiting the search space.

6.2 Results

Q1: Effect of Including Weights. The weighted version of CoWC increases the number of feasible instances compared to the unweighted version, showing that the weighted version leads to better compression (see Table 1). This is especially apparent for the music playlist dataset. For the federated learning dataset, we

---

[2] https://github.com/ML-KULeuven/CoWC.

Table 1. Empirical results. For CoWC we report the weighted (w) and unweighted (u) case, with several F_1-thresholds (0.8, 0.85, ...). To compute the compression rate (Comp), and to compare across methods, we report the global criterion (GC; lower is better), F_1 and Comp averaged across those instances where the disjunction (CNF(F)) can be computed exactly. The ✓-column lists the number of instances for which the approximate disjunction was computed (i.e., no intermediate CNF with > 15 000 clauses).

Method	Music playlists				Federated learning			
	GC	F_1	Comp	✓	GC	F_1	Comp	✓
w CoWC (0.8)	0.62	0.99	0.39	200	0.48	0.53	0.98	49
u CoWC (0.8)	0.62	0.99	0.39	182	0.53	0.51	0.88	48
w CoWC (0.85)	0.62	0.99	0.39	199	0.40	0.6	0.96	50
u CoWC (0.85)	0.62	0.99	0.39	178	0.43	0.64	0.85	45
w CoWC (0.90)	0.62	0.99	0.39	194	0.38	0.66	0.89	47
u CoWC (0.90)	0.62	0.99	0.39	148	0.51	0.69	0.68	40
w CoWC (0.95)	0.62	0.99	0.39	175	0.39	0.78	0.75	44
u CoWC (0.95)	0.62	0.99	0.39	125	0.66	0.85	0.41	38
w CoWC (0.97)	0.62	0.99	0.39	160	0.57	0.89	0.47	40
u CoWC (0.97)	0.62	0.99	0.39	117	0.76	0.91	0.28	34
w CoWC (0.99)	0.62	0.99	0.39	132	0.73	0.97	0.28	36
u CoWC (0.99)	0.62	0.99	0.39	123	0.88	0.99	0.12	33
w CoWC (0.999)	0.62	0.99	0.39	107	0.88	1.0	0.12	33
u CoWC (0.999)	0.62	0.99	0.39	94	0.98	1.0	0.02	32
Krimp	0.98	0.15	0.12	200	0.43	0.69	0.83	50
PMC	1.00	1.0	0.0	66	0.63	1.0	0.37	38
Mining4SAT	1.00	1.0	0.0	66	0.95	1.0	0.07	33
Uncompressed	1.00	1.0	0.0	66	1.0	1.0	0.0	32

also observe a decrease in global criterion, trading a minor loss in F_1-score with a significant increase in compression rate. When considering instances beyond those for which the compression rate is computable, the same observation can be made for the music playlist dataset. For instance, consider the 125 instances that are feasible for both weighted and unweighted CoWC (0.95). The weighted version results in an average description length of 656, versus 1159 for the unweighted version.

Q2: Comparison with Other Approaches. Weighted CoWC clearly outperforms Krimp, the PMC preprocessor, and Mining4SAT on the music playlists dataset. This is shown though the global criterion which we wanted to minimize. Notably, the last two methods do not achieve any compression. On the federated learning dataset, weighted CoWC with $t_{F_1} = \{0.85, 0.90, 0.95\}$ achieves the best global criterion (lower is better), primarily through a significantly higher compression rate. Krimp achieves a very low F_1-score on the music playlist dataset. We suspect this is due to the dataset containing fewer frequent itemsets.

7 Related Work

The problem of compressing a CNF has mostly been studied in the context of preprocessing, for solving satisfiability (SAT) or model counting (#SAT) problems. For example, Sattelite [8] speeds up SAT solving based on three techniques: subsumption, self subsuming resolution and variable elimination by substitution.

The CoPreprocessor [16] extends this by exploiting equivalence-preserving techniques such as hidden tautology elimination, vivication and probing, but is more focused on satisfiability-preserving techniques that eliminate equivalences, blocked clauses and variables. The PMC preprocessor [14] introduces backbone identification, gate detection and replacement techniques and is extended by the B+E-preprocessor [13] that is based on definability of these gates. Finally, Mining4SAT [10] compresses a CNF formula by using frequent itemset mining techniques and predicate invention, similar to what we use in our work.

However, in the context of SAT and #SAT, the compression techniques are lossless with respect to the task result. That is, the satisfiability of a CNF must remain the same, and similar for model counting problems. This is different from the particular setting we study in this work where we mostly consider a lossy compression. Another source of related work exists in the form of compression while learning. Mistle [11], for example, learns a CNF formula from data by relying on predicate invention and inverse resolution to get a concise formula (minimising description length). Similar to our approach, they borrow the logic compression operators from DUCE [19]. However, both Mistle and Duce solve a different problem compared to our approach, namely, they learn a theory from given data instead of merging theories.

Finally, Krimp [24] mines and selects the frequent itemsets that optimally compress a dataset, thus learning a DNF formula.

8 Conclusion

We proposed CoWC, a novel compression algorithm for CNF formulas that balances the compression rate with the F_1-score, the latter of which is defined with respect to a dataset $\mathcal{D}$ of relevant models. The algorithm itself is based on inverse resolution and uses techniques from itemset mining to identify frequently occurring subclauses. A particularly novel addition here is the idea of using weights to guide the compression. We applied this algorithm to approximately compute the disjunction of a set of CNFs, $CNF(\bigvee_i F_i)$ that would otherwise be too large to compute. We evaluated our contribution on two use cases, that of music playlist generation and that of federated learning, observing that CoWC outperforms the other compression alternatives on these tasks. Future work could investigate the efficiency of CoWC. We identify the mining of frequent itemsets in each iteration as the largest computational cost, that could likely be improved to obtain significant performance gains.

Acknowledgements. We thank the music streaming company *Tunify* (https://www.tunify.com/en-gb/) for discussing the use case. VD was supported by the EU H2020 ICT48 project "TAILOR" under contract #952215. This research received funding from the Flemish Government under the "Onderzoeksprogramma Artificiële Intelligentie (AI) Vlaanderen" programme. LDR is also supported by the Wallenberg AI, Autonomous Systems and Software Program (WASP) funded by the Knut and Alice Wallenberg Foundation.

Disclosure of Interests. The authors have no competing interests to declare that are relevant to the content of this article.

References

1. Bharati, S., Mondal, M.R.H., Podder, P., Prasath, V.B.S.: Federated learning: applications, challenges and future directions. Int. J. Hybrid Intell. Syst. **18**(1–2), 19–35 (2022)
2. Biere, A., Heule, M., van Maaren, H., Walsh, T. (eds.): Handbook of Satisfiability, Frontiers in Artificial Intelligence and Applications, vol. 185. IOS Press (2009)
3. Bocklandt, S., Derkinderen, V., Vanderstraeten, K., Pijpops, W., Jaspers, K., Meert, W.: Pruning-based extraction of descriptions from probabilistic circuits (2024)
4. Darwiche, A.: A logical approach to factoring belief networks. In: KR, pp. 409–420. Morgan Kaufmann (2002)
5. Defferrard, M., Benzi, K., Vandergheynst, P., Bresson, X.: FMA: a dataset for music analysis. In: ISMIR, pp. 316–323 (2017)
6. Derkinderen, V., Manhaeve, R., Dos Martires, P.Z., De Raedt, L.: Semirings for probabilistic and neuro-symbolic logic programming. Int. J. Approximate Reasoning 109130 (2024)
7. Dua, D., Graff, C.: UCI machine learning repository (2017). http://archive.ics.uci.edu/ml
8. Eén, N., Biere, A.: Effective preprocessing in SAT through variable and clause elimination. In: Bacchus, F., Walsh, T. (eds.) SAT 2005. LNCS, vol. 3569, pp. 61–75. Springer, Heidelberg (2005). https://doi.org/10.1007/11499107_5
9. Goyal, K., et al.: Automatic generation of product concepts from positive examples, with an application to music streaming. In: BNAIC/BENELEARN. Communications in Computer and Information Science, vol. 1805, pp. 47–64. Springer (2022)
10. Jabbour, S., Sais, L., Salhi, Y.: Mining to compact CNF propositional formulae. CoRR abs/1304.4415 (2013)
11. Jain, A., Gautrais, C., Kimmig, A., Raedt, L.D.: Learning CNF theories using MDL and predicate invention. In: IJCAI, pp. 2599–2605. ijcai.org (2021)
12. Korhonen, T., Järvisalo, M.: SharpSAT-TD in model counting competitions 2021-2023. CoRR abs/2308.15819 (2023)
13. Lagniez, J., Lonca, E., Marquis, P.: Definability for model counting. Artif. Intell. **281**, 103229 (2020)
14. Lagniez, J., Marquis, P.: Preprocessing for propositional model counting. In: AAAI, pp. 2688–2694. AAAI Press (2014)
15. Lucchese, C., Orlando, S., Perego, R.: DCI closed: a fast and memory efficient algorithm to mine frequent closed itemsets. In: FIMI. CEUR Workshop Proceedings, vol. 126. CEUR-WS.org (2004)

16. Manthey, N.: Coprocessor - a standalone SAT preprocessor. CoRR abs/1108.6208 (2011)
17. Masud, C.L.H.A.S.M.: Multiple Objective Decision Making, Methods and Applications: A State-of-the-Art Survey. Springer (1979)
18. Michalski, R.S.: A theory and methodology of inductive learning. Artif. Intell. **20**(2), 111–161 (1983)
19. Muggleton, S.H.: Duce, an oracle-based approach to constructive induction. In: IJCAI, pp. 287–292. Morgan Kaufmann (1987)
20. Rauniyar, A., et al.: Federated learning for medical applications: a taxonomy, current trends, challenges, and future research directions. IEEE Internet Things J. **11**(5), 7374–7398 (2024)
21. Sammut, C., Webb, G.I. (eds.): Inverse Resolution, p. 558. Springer, Boston (2010)
22. Tseitin, G.S.: On the complexity of derivation in propositional calculus, pp. 466–483. Springer, Heidelberg (1983)
23. Uno, T., Kiyomi, M., Arimura, H.: LCM ver. 2: efficient mining algorithms for frequent/closed/maximal itemsets. In: FIMI. CEUR Workshop Proceedings, vol. 126. CEUR-WS.org (2004)
24. Vreeken, J., van Leeuwen, M., Siebes, A.: Krimp: mining itemsets that compress. Data Min. Knowl. Discov. **23**(1), 169–214 (2011)

Computer Vision and Explainable AI

Explainable AI in Time-Sensitive Scenarios: Prefetched Offline Explanation Model

Fabio Michele Russo[1], Carlo Metta[2](✉), Anna Monreale[3], Salvatore Rinzivillo[2], and Fabio Pinelli[1]

[1] IMT School for Advanced Studies Lucca, Lucca, Italy
[2] ISTI-CNR, Pisa, Italy
`carlo.metta@isti.cnr.it`
[3] University of Pisa, Pisa, Italy

Abstract. As predictive machine learning models become increasingly adopted and advanced, their role has evolved from merely predicting outcomes to actively shaping them. This evolution has underscored the importance of Trustworthy AI, highlighting the necessity to extend our focus beyond mere accuracy and toward a comprehensive understanding of these models' behaviors within the specific contexts of their applications. To further progress in explainability, we introduce POEM, Prefetched Offline Explanation Model, a model-agnostic, local explainability algorithm for image data. The algorithm generates exemplars, counterexemplars and saliency maps to provide quick and effective explanations suitable for time-sensitive scenarios. Leveraging an existing local algorithm, POEM infers factual and counterfactual rules from data to create illustrative examples and opposite scenarios with an enhanced stability by design. A novel mechanism then matches incoming test points with an explanation base and produces diverse exemplars, informative saliency maps and believable counterexemplars. Experimental results indicate that POEM outperforms its predecessor ABELE in speed and ability to generate more nuanced and varied exemplars alongside more insightful saliency maps and valuable counterexemplars.

Keywords: Explainability · Trustworthy AI · Machine learning · Artificial intelligence

1 Introduction

Explainable Artificial Intelligence (XAI) within the field of Machine Learning (ML) is gaining increased attention, underlining its critical importance for both the application and advancement of research in this domain. The pervasive deployment of ML across a wide range of automated systems demonstrates its significance. The pace at which ML is evolving in terms of capability and application scope is impressive. However, the adoption of ML models has not been without concerns. There is a growing discourse in both public and academic spheres about the ethical implications and the societal impact of deploying these AI-based technologies [1,14]. Issues such as delegating decision-making processes to machines raise substantial ethical and fairness questions. Moreover, the trustworthiness of automated decision-making systems has been debated. Applications of ML that resulted in unfair or unethical outcomes have underscored the necessity for ML systems to be trustworthy, both in their methodology and application. Introducing regulation into the AI field has provoked a mix of responses. There are concerns that regulations might restrain innovation or fail to achieve their intended goals, thus failing on ethical and technological fronts. The other perspective suggests that thoughtful regulation can prevent the most harmful uses of ML and guide research towards ethical applications [4]. Balancing the vast potential of ML with ethical and legal requirements, such as privacy, security, justice and fairness, is essential. Indeed, the necessity for explanations in ML applications is crucial and particularly challenging in scenarios where *time* is critical. Applications such as autonomous vehicle navigation, real-time medical diagnosis, and high-frequency trading demand accurate and immediate explanations of model decisions.

The ability to provide real-time explanations ensures that autonomous systems are transparent and applicable when decisions need to be made quickly and efficiently. This requirement adds a layer of complexity to the development of explainable models, as they must be designed to operate under stringent time constraints without compromising the quality of explanations. Let's consider the case of a medical diagnosis system that must assist the doctor in the evaluation of a challenging patient's case. The system should provide a suggestion for the diagnosis and explain the reasons behind it timely. In this context, the system's ability to generate timely and accurate explanations is critical for the patient's well-being and the system's reliability. Another example is the deployment of an AI system for monitoring production lines in a factory. In case an atypical condition is found, the system must be able to explain the reasons behind its decisions in real-time to allow the operators to intervene promptly in case of anomalies. A third scenario is the use of AI in the financial sector, where a timely explanation may be crucial to understand the reasons behind a sudden change in the market or to predict future trends.

This paper addresses the challenge of explaining black box image classifications in time-sensitive scenarios. Our work, leveraging an existing XAI method presented in [6], aims to introduce a model-agnostic algorithm for generating timely explanations of image data. Compared to [6], we severely reduce the time needed to generate explanations in the domain of image data and produce higher quality explanations. The implementation of our proposed method, along with the complete code, is publicly available[1], enabling full reproducibility.

The rest of this paper is organized as follows. Section 2 discusses related works. In Sect. 3 we introduce preliminary background on some aspects which are important to understand the details of our approach. Section 4 details POEM while Sect. 5 presents the results of its experimental analysis. Section 6 concludes the paper and identifies some future works.

---

[1] github.com/gatto/poem.

2 Related Work

Research towards the creation of transparent AI systems has led to significant progress in understanding and explaining model decisions [8]. Despite their impressive performance, deep learning models, characterized by their *black box* nature, present considerable challenges in transparency and accountability, motivating the development of XAI methods.

Visual explanations in image classification have drawn attention due to their intuitive appeal, with exemplars and counterexemplars offering straightforward insights into model reasoning by presenting similar or dissimilar training instances [6,13]. Beyond visual methods, local explainability techniques like LIME and SHAP provide insights into model decisions by perturbing input features and assessing the impact on output, facilitating a more detailed understanding of model behavior [7,10,15].

The advancement of model-agnostic methods marks a significant milestone in XAI. This universality enhances the adaptability of XAI solutions, making them valuable tools for developers and researchers working with a diverse array of machine learning models [8]. In parallel, efforts in creating interpretable models from the outset, such as transparent neural networks and decision trees, aim to build systems whose operations can be inherently understood, thereby reducing the dependency on post-hoc explanation methods [16]. Furthermore, the development of synthetic instance generation through approaches like adversarial autoencoders, as seen in ABELE, enriches the landscape of explainability by providing novel ways to explore model decisions in the latent space [12].

Research in XAI also extends to the evaluation of explanations' effectiveness, where metrics and user studies assess how well explanations meet the needs of different stakeholders, including end-users, developers, and regulatory bodies [2,9]. This evaluation is crucial in tailoring explanations to various audiences, ensuring they are technically accurate but also comprehensible and actionable.

Various methods have been developed in the literature to address the need for timely and accurate model explanations, a crucial challenge in time-sensitive scenarios. Approaches like real-time saliency mapping and dynamic feature attribution methods [3] stand as notable competitors, offering insights into model decisions with minimal delay. Techniques such as incremental learning models [17] have also been adapted to provide explanations on-the-fly. Despite the advancements offered by these methods, they often face trade-offs between the speed of explanation generation and the depth or quality of the insights provided. The real-time aspect may come at the cost of oversimplified explanations that might not fully capture the complexity of the decision-making process or require substantial computational resources, limiting their applicability.

3 Background

In this paper, we consider the problem of creating a model that produces on-demand informative explanations for image classification decisions by a black

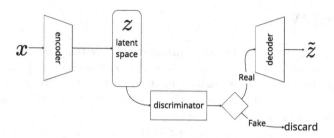

Fig. 1. Components of the Adversarial AutoEncoder. The *encoder* maps an input instance x to a point z into a latent space. Any point in this space is first filtered by the *discriminator* before the *decoder* reconstructs the corresponding image.

box. Given a black box b and an image instance x classified by b with label y, i.e., $b(x) = y$, our goal is to efficiently create an explanation to clarify the internal functioning of the black box that contributed to the specific classification. To this end, we exploit an existing explanation model, ABELE [6], to improve the generated explanations' quality and time efficiency.

ABELE is an explanation algorithm that follows the same strategy of LORE [7]. Starting from an instance x, it generates a neighborhood of synthetic variations of x and uses the black box to assign labels to this set. This annotated version of the neighborhood is used as a training sample for a decision tree. The logical predicates of the tree are used as source for factual and counterfactual rules. For the image domain, this process projects each instance into a latent space by means of an Adversarial Autoencoder (AAE) [11] that encodes an image x into a low-dimensional point z. The AAE is also used to generate synthetic instances decoding one point z from the latent space to a reconstructed image $\tilde{z}$. AAEs ensure synthetic instances retain the original data distribution through a structured framework involving an encoder, decoder, and discriminator, as detailed in Fig. 1. The process aligns the latent space's aggregated posterior distribution with a specified prior, optimizing reconstruction accuracy [11].

Neighborhood Generation. ABELE begins with the input x and its class y to encode the corresponding point z in the latent space Z (Fig. 2(a)). Then, it generates a neighborhood H around point z (Fig. 2(b)). Each element of H is filtered by the discriminator (Fig. 2(c)) and decoded to the corresponding image (Fig. 2(d)). This set of reconstructed images is labeled by the black box b (Fig. 2(f)), and the labels are propagated back to the images' latent representation. ABELE resamples points to guarantee a balanced distribution of the classes within H. Finally, a decision tree is trained on H, replicating the black box model's decisions locally to z (Fig. 2(g)).

Explanation Extraction. The decision tree classifies z, and the corresponding branch represents the factual rule $r = p \rightarrow b(x)$ for z. In other words, the premise p is the conjunction of the splitting conditions in the nodes of the path from the

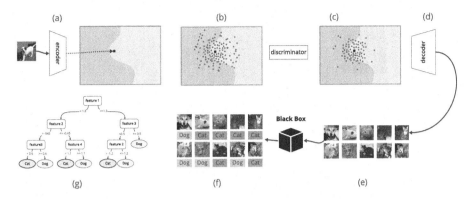

Fig. 2. ABELE workflow, starting from the mapping of the instance x to the extraction of the transparent model: (a) instance x encoded within the latent space; (b) neighborhood generation around x; (c) discriminator filtering the neighborhood; (d) decoder transforming the points into images (e) annotated by the black box; (f) supervised data forming a local training set to (g) learn a decision tree.

root to the leaf that is satisfied by the latent representation z of the instance to explain x (Fig. 3).

The counterfactual rules are determined by following alternative branches with minimal variations. In particular these rules have the form $q \rightarrow \neg b(x)$, where q is the conjunction of splitting conditions for a path from the root to the leaf labeling an instance z' with $\neg b(x)$ and minimizing the number of splitting conditions falsified w.r.t. the premise p of the rule r.

Rules are not directly usable for humans since they refer to latent dimensions that may not directly represent semantic attributes. Thus, to give the user intuition of the rule's meaning, we generate synthetic images as prototypes of the logical rules. ABELE selects points in the latent space that satisfy the factual rule (or counterfactual rule) and pass the discriminator's filter. These points are decoded back and used as exemplars or counterexemplars for explanations.

4 POEM - Prefetched Offline Explanation Model

From the previous section, it is clear that producing an explanation from a given instance requires several steps for generating the neighborhood, decoding these points, classifying them, and learning the corresponding decision tree. This process may imply a large amount of time (see Table 3), which limits the actual use of the explanations, particularly in applications where interactivity or time sensitivity is crucial. Therefore, we propose POEM, an explanation algorithm involving two steps called *offline* and *online*, respectively.

In the *offline* step, POEM leverages ABELE to build an *explanation base* $\tilde{T}$, storing tuples which associate each explained point t to the components crucial for its explanation $e(t)$ produced by ABELE.

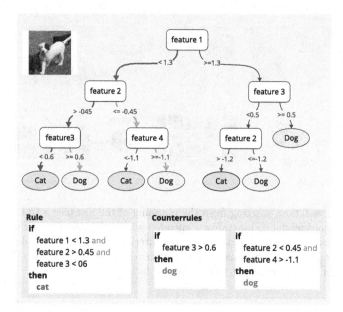

Fig. 3. Extraction of rules and counter-rules from the surrogate model learned by ABELE in the neighborhood H

In the *online* step, given an instance x, POEM selects from the explanation base $\tilde{T}$ the closest point to the latent representation of x and then it uses the corresponding rules and counterfactual rules to generate exemplars and counterexemplars for x. If the explanation base does not contains any tuple corresponding to a point sufficiently close to the point to be explained, the explanation is computed using ABELE, and the result is stored in the explanation base $\tilde{T}$.

During the offline step, we assume to have enough time to dedicate to each explanation in $\tilde{T}$ to guarantee a higher quality of every single explanation.

4.1 Offline Step – Building the *Explanation Base*

During this first phase, we assume the access to a sample of instances D_e belonging to the same distribution of the training set of the black box b. The dataset D_e is sampled in two parts with respect to the class distribution: D_a, used to train the AAE, and D_t used to build the explanation base $\tilde{T}$.

For each point $t \in D_t$, we apply ABELE to determine its corresponding explanation model. Thus we store the latent point z_t, the neighborhood H_t, and the surrogate decision tree DT_t in a lookup index. In Fig. 4, we show a schematic representation of the explanation base: each point t is represented as a black dot, its neighborhood hyperpolyhedron as a dashed polygon and the corresponding decision tree is linked to each point in D_t. Each tuple of objects is stored, and an efficient indexing structure is used to retrieve the closest point to a given query.

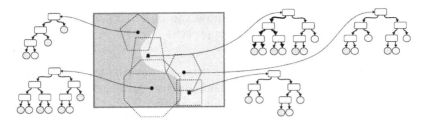

Fig. 4. Representation of the explanation base, where each point $t \in D_t$ is linked to the decision tree and the neighborhood computed during the offline step.

4.2 Online Step – Explanation of New Instances

Given a new instance x to be explained, POEM exploits the explanation base to identify a pre-computed explanation model i.e., a pre-computed neighborhood and decision tree to generate the explanation. The instance x is first mapped into a latent point z_x, then we look for the closest z_t in the explanation base, using the Euclidean distance. We also check if all the following conditions hold:

1. b classifies x identically to the chosen t, i.e., $b(x) = b(t)$;
2. the latent representation z_x is contained within the neighborhood hyperpolyhedron of the candidate point z_t;
3. the branch of the tree DT_t that leads to the classification for z_x is the same branch that holds for the explanation base point z_t.

The first condition guarantees that we are using a pre-computed explanation model generated for explaining the classification of a point as class $b(x)$. This is important because we cannot assume a 100% fidelity of the explanation model generated by ABELE. The second condition ensures that the new point z_x is part of the neighborhood of the previously computed neighborhood of z_t in the explanation base. The third condition restricts the selection only on those surrogate models that share the same predicates on the new instance and the candidate point z_t. This also implies that the classification of DT_t is the same for z_x and z_t, i.e. $DT_t(z_x) = DT_t(z_t)$. We enforce this condition to guarantee that the explanation model is coherent with the new instance x. In future works, we plan to relax this condition to allow for a more flexible explanation model, even accounting for possible inconsistencies in the surrogate model.

If one of the conditions is not satisfied, the search proceeds by checking the second closest point z'_t, and so on until a valid candidate is found or no candidate may be selected. When a candidate $t^* \in \tilde{T}$ is successfully identified, we build the rule and the counter-rules from the corresponding surrogate decision tree, using the same procedure described in Sect. 3 and depicted in Fig. 3. If no candidate is found, the explanation is computed using legacy ABELE approach and the new explanation model is stored in the explanation base.

Once the suitable surrogate decision tree is identified and the rules and counter-factual rules are identified, POEM exploits them for generating exemplars and counter-exemplar images. We propose two novel strategies for generating such images differing from the ABELE approach.

Generation of Exemplars. In this section, we describe the process for generating image exemplars. We assume that the user who needs an explanation for an image classification can request for a specific number k of exemplars. To fulfill such request, POEM applies a procedure involving two phases: *(i) exemplar base generation*, aiming at generating a set of $\nu \gg k$ exemplars; and *(ii) exemplar selection*, which selects the top k most interesting exemplars among the generated ones.

Exemplar base generation: This step generates an *exemplar base*, i.e., a set of ν exemplars where $\nu = k \times \beta$ ($\beta > 1$). In particular, we generate exemplars by generating a new value for each latent feature. Since the latent space learned by the AAE has dimensions that are distributed according to a standard normal distribution [11], we can generate a new value for each latent feature by drawing from a truncated normal distribution constrained to the feature conditions in the premises p of factual rule extracted by the decision tree $r_{t^*} = p \rightarrow b(x)$. For each feature f_i involved in the premises p, we use the corresponding condition for bounding the feature values to be drawn from the standard normal distribution. As an example, if p contains the condition $f_i <$ 0.5 the value for the feature f_i is drawn from the standard normal distribution in the interval $[-\infty, 0.5)$. For features which are not involved in the premises p the value is drawn from the standard normal distribution without any bound. Once for each latent feature we generate a value, the obtained *candidate exemplar* c_e undergoes a discrimination check with the threshold set at some probability α and is classified by the back box b. The candidate exemplar c_e is stored in the exemplar base if its classification matches x's (i.e., $b(c_e) = b(x)$) and it passes the discriminator check; otherwise, it is discarded.

Exemplar selection: This step extracts from the exemplar base the top k most interesting exemplars. To this aim, we select the k most distant exemplars from the point x to be explained. The idea is to show the user exemplars having marked differences with respect to the image to be explained while maintaining the same classification label. In the latent space, this corresponds to selecting the exemplars that are farthest from the point z_x to be explained. To this end, POEM sorts the exemplar base with respect to the pairwise distance from z_x by using the Euclidean distance. Then, the k exemplars furthest from x are presented to the user, where k represents the number of requested exemplars.

An illustrative depiction of the generated and selected exemplars from the *exemplars base* is provided in Fig. 5. We highlight that the figure shows a projection of the features in a two-dimensional space.

Generation of Counterexemplars. In this section, we describe the procedure for generating counterexemplars. It is based on applying to the latent instance

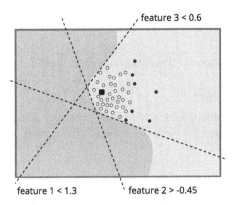

Fig. 5. Generation and selection of exemplars for a point z_x. Dashed lines represent the half-planes determined by the predicates of the rule. A set of points is generated in the latent space and filtered by the rule predicate (gray points). The top k distant points are selected as exemplars (green dots) (Color figure online)

z_x a perturbation guided by the counter-factual rules. In particular, given a candidate t^*, for each counterfactual rule $cr \in \Phi_{t^*}$ extracted from DT_{t^*}, POEM generates a counterexemplar. Given a counterfactual rule $cr = q \rightarrow \neg b(x)$, the premise q is a conjunction of predicates of the form (f op v), where op can be $<, \leq, >,$ or $\geq$ and contains some predicates P falsifying the premise of the factual rule r_{t^*}.

For each feature f in a predicate (f op v) of P, we perturb the value of f assigning a new value by adding (subtracting) to v the value $\epsilon \times i$ in case the comparison operator op of the predicate is $>$ or $\geq$ ($<$ or $\leq$). We highlight that $i = 1, \ldots, m$ while m and ϵ are user parameters. The parameter ϵ allows the user to define the granularity of the search: larger (smaller) values will produce a counter-exemplar more (less) quickly but with a coarser (finer) resolution.

The iteration over i is necessary because adding ϵ to the initial feature value could generate a candidate c that does not pass the discriminator check or is not classified by the black box with the same label as x. Consequently, we iterate on i, increasing step by step the perturbation value until we find the suitable candidate c passing the discriminator check and for which $b(x) \neq b(c)$. In cases where a valid counterexemplar is not produced after m attempts, the counterfactual rule is discarded and does not produce any counterexemplars.

Considering for example an instance x to be explained, mapped to a latent point $z_x = [0.8, -0.34, 0.58, -0.13]$ and the relative counterfactual rule $f_1 < 1.3 \wedge f_2 > -0.45 \wedge f_3 \geq 0.60 \rightarrow$ dog, POEM produces the counterexemplar $[0.8, -0.34, 0.64, -0.13]$ by adding $\epsilon = 0.04$ to the threshold value 0.60 of the predicate and assigning the obtained value to f_3.

As a final note in this part, we remark that these two novel algorithms for exemplar and counterexemplar generation are independent from the rest of the work, i.e. the generation of the explanation base and the distinction in offline and

Table 1. Datasets resolution, train and test dimensions, and AAEs reconstruction error regarding RMSE.

Dataset	Resolution	Size		RMSE	
		Train	Test	Train	Test
MNIST	28 × 28	50k	10k	33.69	40.73
FASHION	28 × 28	50k	10k	28.25	29.81
EMNIST	28 × 28	131k	14k	35.41	41.12

online part. Therefore, the two algorithms could also be implemented directly within ABELE to improve the quality of the generated explanations, without a significant increase in the computational cost, since ABELE spends the most amount of its computational time on neighborhood generation.

Saliency Map Generation. Our method, POEM, generates saliency maps in a very similar way to ABELE. Both approaches aim to clarify why a black box model classifies images in certain ways using saliency maps. ABELE creates these maps using exemplars from the latent feature space, highlighting important areas for classification. In POEM, we also use exemplars to create saliency maps, but our exemplars are produced differently. Although we follow the same steps as ABELE, the novel way we generate our exemplars could lead to more detailed saliency maps. This means our maps might show different insights compared to those from ABELE, even though the underlying construction process is the same.

5 Experiments

This section presents the empirical evaluation of POEM, with the objectives of assessing: *i)* the impact of the *explanation base* size on the model's performance; *ii)* the model's time efficiency compared to the ABELE algorithm; and *iii)* the quality of generated saliency maps in regards to their capacity to identify the most important parts of the image correctly.

5.1 Experimental Setting

Datasets. We conducted our experiments across three widely recognized datasets: MNIST is a dataset of handwritten grayscale digits, FASHION dataset is a collection of Zalando grayscale products (e.g. shirt, shoes, bag, etc.) and EMNIST dataset is a set of handwritten letters. MNIST and FASHION have 10 labels while EMNIST has 26 labels. Details on the datasets are reported in Table 1.

Back Box Classifiers and AAE. For each dataset, we employed two black box classifiers: a Random Forest (*RF*) and a Deep Neural Network (*DNN*), following the original choice made in ABELE [6]. For the *RF* classifier, we used an ensemble

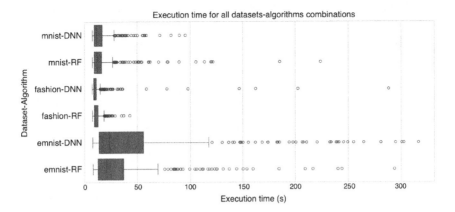

Fig. 6. Time distributions for each dataset and black-box. For MNIST e FASHION, the majority of the instances get a successful hit in the explanation base, and they retrieve an explanation very efficiently. For EMNIST, the large number of class labels makes the search for candidate explanations more complex.

of 100 decision trees, the minimum number of samples in each leaf of 10, and using the Gini coefficient to measure the quality of splits. The *DNN* architecture consisted of three convolutional layers followed by two fully connected layers, with ReLU activations. We used a dropout rate of 0.25 to prevent overfitting and trained the model using the Adam optimizer with a learning rate 0.001. This architecture was selected for its proven effectiveness in image classification tasks.

Overall, classifier performances were consistent with established benchmarks in literature: for both MNIST and FASHION datasets, *RF* achieved accuracies around 95–96%, whereas *DNN* performed better with accuracies ranging in 98–99%. For EMNIST dataset, *RF* accuracy stood at approximately 90%, while *DNN* showed again enhanced performance, reaching around 95%.

Regarding the adversarial autoencoder, we opted for a symmetrical encoder-decoder structure with a tower of 3 convolutional blocks from 64 to 16 channels, a latent space of 4 dimensions for MNIST and 8 for FASHION and EMNIST. The discriminator consists of two fully connected layers of 128 and 64 neurons. Batch Normalization and Dropout were used to improve training and performance. The autoencoders were trained for 10,000 steps with a batch size of 128 images.

We conducted experimental analysis to identify a suitable α value, i.e., the probability threshold for the generated image acceptance by the discriminator. To this end, we used MNIST. We observed that the mean discriminator probability for MNIST is 0.349, with a median of 0.310. We use the highest value between these two to estimate the discriminator's accuracy across the dataset. As these values are extracted from real data, a threshold of $\alpha = 0.35$ is used to decide the validity of generated points for all our experiments.

Table 2. Performance comparison of POEM across different explanation base sizes for the MNIST dataset. The time is the average time in seconds for POEM to generate one explanation.

Black box	Size	POEM Hit %	Time (std dev)
RF	500	86.60	15.94 (14.73)
	1000	92.20	16.58 (16.52)
	2500	96.20	17.12 (19.15)
	5000	98.00	17.07 (19.00)
DNN	500	88.00	12.98 (8.31)
	1000	92.20	13.80 (14.40)
	2500	94.40	13.62 (14.25)
	5000	95.80	15.24 (10.63)

Evaluation Metrics and POEM Parameters. To evaluate the impact of using the explanation base for explaining a classification, we define the measure *hit percentage* as the proportion of explanation requests for which an appropriate pre-computed explanation model $\tilde{t}^*$ is found in the explanation base $\tilde{T}$. The scope of this measure is to assess the effectiveness of the explanation base in providing explanations for new instances, avoiding the need to compute them from scratch.

In order to extract examplars we set the parameter $\beta = 10$, and to extract counterexemplars, we chose the step size $\epsilon = 0.04$ and a maximum number of iterations $m = 40$. The latent space for the counterexemplars generation is explored over a standard normal distribution, and its $95th$ percentile falls between -2 and 2. The step size of 0.04 consists of a 1% of this interval. We chose 40 as the maximum number of iterations.

5.2 Impact of Explanation Base Size

The first experiment consists of an analysis of the size of the explanation base. POEM has a higher efficacy when, for the new instances to be explained, there is a high probability of finding a hit in the explanation base. Thus, the size of the explanation base, i.e., the amount of points used for pre-computing explanations, is crucial to maximize the probability of a hit. This size is strictly related to the dimensionality of the latent space or the number of distinct class labels. In this experiment, we address the question of what is the size of the explanation base $\tilde{T}$ that is suitable in a 10-label dataset to get a *hit percentage* high enough to ensure that the vast majority of explanations are generated through POEM. Also, we want to study the execution time of POEM over different sizes of $\tilde{T}$. We run this experiment with two black boxes, testing an explanation base of increasing sizes of 500, 1000, 2500, and 5000 pre-computed explanations.

Table 3. Average execution times to generate a single explanation using both POEM and ABELE and the hit percentage for POEM. The time measurements for POEM consider the online phase only, since the offline phase is performed once and does not depend on the instance to be explained.

Dataset	Black box	POEM		ABELE
		Hit %	Time (std dev)	Time (std dev)
MNIST	RF	98.00	17.07 (19.00)	301.04 (161.64)
	DNN	95.80	15.24 (10.63)	230.06 (135.52)
FASHION	RF	99.40	11.72 (4.46)	334.75 (159.18)
	DNN	99.40	12.76 (18.85)	281.77 (183.00)
EMNIST	RF	72.89	35.11 (40.23)	333.56 (190.19)
	DNN	72.22	50.75 (63.65)	298.58 (103.25)

Our results on MNIST dataset are reported in Table 2. We can observe that increasing the size of the explanation base improves, as expected, the explanation time and hit percentage. Since explanation bases with 5000 pre-computed explanations report a good level of hit percentage, in the subsequent experiments, we use this value as explanation base size.

5.3 Time Performance Assessment

In this section, we present the results of the time performance of our approach. Table 3 summarises our findings. The results revealed significant enhancements when utilizing POEM compared to ABELE. We observed reductions in execution time ranging from 83.00% to 96.50%, depending on the dataset and black box.

We observe that using the same size of $\tilde{T}$ for a dataset containing 26 class labels instead of 10 leads to reduced performance in terms of both time and hit percentage. This is due to the fact that the *online* step of POEM for explaining a point x only considers pre-computed explanations of points labelled the same as x by the black box. Thus, increasing the number of classes without increasing the explanation base size leads to an under-coverage of the latent space. Lastly, since the standard deviation of completion time is relevant, we analyzed more in detail the distribution of execution time over the instances to be explained. The boxplots in Fig. 6 show that the interquartile range are very small for the three datasets. MNIST and FASHION leverage the coverage of the explanation base, having a very efficient time to find an explanation, since the time variability is very low. EMNIST instead requires more time to explore the explanation base, mainly due to the higher number of class labels.

5.4 Exemplars, Counterexemplars and Saliency Maps

To demonstrate the efficacy of the approach in producing informative explanations, we performed two types of evaluation. The first is a qualitative comparison of the exemplars and counterexemplars produced by POEM. For example, in

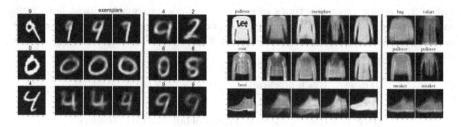

Fig. 7. Three examples of explanations for instances of the MNIST (left) and FASHION (right) datasets. On the left, the image to be classified, with the class assigned by the black box on top. In the middle, a set of exemplars. On the right, two counter exemplars for each instance.

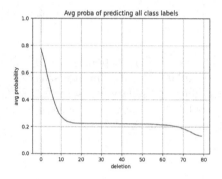

Fig. 8. Deletion experiment over POEM's saliency maps on a sample of 100 instances for each class label in the MNIST dataset. AUC: 19.533.

Fig. 7, we show a selection of images from MNIST and FASHION and their corresponding set of exemplars and counterexemplars. It is evident how the selection of the counterexemplars produces synthetic images that are visually similar to the original instance but with a different label. These counter exemplars push the black box to classify instances that are as close as possible to the decision boundary. This is evident in the second example of MNIST, where the counter exemplars are very similar to the original instance but with a different class label (apparently wrong for the human eye). The first instance of FASHION shows how one counter exemplar is labeled as a t-shirt, whereas at the human vision it resembles a pair of trousers.

To evaluate the efficacy of the saliency maps generated by POEM, we employed the deletion experiment method [5]. This approach assesses the saliency map's capacity to determine each pixel's relevance. A steep decline in classification probability upon pixel removal, in a descending order on the blackbox's correct classification probability, implies a higher map quality. With pronounced drops observed in our deletion experiment curves, our results show that POEM consistently produces saliency maps that accurately identify important pixels for classification. In Fig. 8 it is evident how the deletion impacts the classification

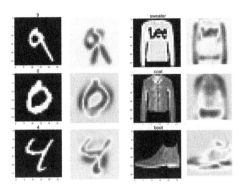

Fig. 9. POEM's saliency maps on MNIST and FASHION datasets. A divergent color scale maps the relevance of each pixel: red, positive contribution; green, negative contribution. The saliency maps highlight the most relevant parts of the image for the classification. For example, the classification of the sweater in the FASHION dataset is influenced by the presence of the sleeves and not by the brand writing. (Color figure online)

probability for a sample of 100 images. Figure 9 shows examples of saliency maps for a sample of images extracted from MNIST and FASHION dataset.

6 Conclusions

In this paper, we show an algorithm called POEM for the model-agnostic explanation of black box decisions on image data that improves on the state-of-the-art of explanations through exemplars, counterexemplars and saliency maps. The proposed approach aims at speeding up the process of providing explanations to the final user while offering comprehensive explanations with good quality. By leveraging deterministic components and consistency checks, POEM ensures the selection of the same pre-computed explanation model for the same instance to be explained. Consequently, identical factual and counterfactual rules are applied across multiple runs, yielding highly stable explanations by design. We tested our approach on benchmarking datasets, which resulted in a significant reduction of execution times. In future work, we intend to explore the possibility of extending this approach to different types of data, such as time series and tabular data. Moreover, it would be interesting to test the approach on real-world data, for example, in the field of medical images and to implement a user study for a qualitative assessment of the POEM's explanations. Another interesting direction would be to investigate the possibility of exploiting the conditions selection of the online phase as an indicator for drifting in the data distribution along the use of the model, for example by triggering a model retraining when a higher number of missing conditions are detected.

Acknowledgements. Research partly funded by PNRR - M4C2 - Investimento 1.3, Partenariato Esteso PE00000013 - "FAIR - Future Artificial Intelligence Research"

- Spoke 1 "Human-centered AI", funded by the European Commission under the NextGeneration EU programme, G.A. 871042 *SoBigData++*, G.A. 101092749 *CREXDATA*, ERC-2018-ADG G.A. 834756 *XAI*, "SoBigData.it - Strengthening the Italian RI for Social Mining and Big Data Analytics" - Prot. IR0000013, G.A. 101120763 *TANGO*.

Disclosure of Interests. The authors have no competing interests to declare that are relevant to the content of this article.

References

1. Barocas, S., Hardt, M., Narayanan, A.: Fairness and abstraction in sociotechnical systems. In: Conference on Fairness, Accountability, and Transparency (2019)
2. Bodria, F., Giannotti, F., Guidotti, R., Naretto, F., Pedreschi, D., Rinzivillo, S.: Benchmarking and survey of explanation methods for black box models. Data Min. Knowl. Disc. **37**, 1719–1778 (2023)
3. Dabkowski, P., Gal, Y.: Real time image saliency for black box classifiers (2017)
4. European Commission, H.L.E.G.o.A.I.: Ethics guidelines for trustworthy AI (2019)
5. Gomez, T., Fréour, T., Mouchère, H.: Metrics for saliency map evaluation of deep learning explanation methods. In: El Yacoubi, M., Granger, E., Yuen, P.C., Pal, U., Vincent, N. (eds.) ICPRAI 2022. LNCS, vol. 13363, pp. 84–95. Springer, Cham (2022). https://doi.org/10.1007/978-3-031-09037-0_8
6. Guidotti, R., Monreale, A., Matwin, S., Pedreschi, D.: Black box explanation by learning image exemplars in the latent feature space. In: ECML/PKDD, vol. 11906, pp. 189–205 (2019)
7. Guidotti, R., Monreale, A., Ruggieri, S., Pedreschi, D., Turini, F., Giannotti, F.: Local rule-based explanations of black box decision systems (2018)
8. Guidotti, R., Monreale, A., Ruggieri, S., Turini, F., Giannotti, F., Pedreschi, D.: A survey of methods for explaining black box models. Comput. Surv. **51**(5) (2019)
9. Hoffman, R.R., Mueller, S.T., Klein, G., Litman, J.: Metrics for explainable AI: challenges and prospects. arXiv preprint arXiv:1812.04608 (2018)
10. Lundberg, S.M., Lee, S.: A unified approach to interpreting model predictions. In: NIPS, pp. 4765–4774 (2017)
11. Makhzani, A., Shlens, J., Jaitly, N., Goodfellow, I.J.: Adversarial autoencoders. CoRR abs/1511.05644 (2015)
12. Metta, C., et al.: Improving trust and confidence in medical skin lesion diagnosis through explainable deep learning. Int. J. Data Sci. Anal. (2023)
13. Metta, C., Guidotti, R., Yin, Y., Gallinari, P., Rinzivillo, S.: Exemplars and counterexemplars explanations for image classifiers, targeting skin lesion labeling. In: IEEE ISCC (2021)
14. O'Neil, C.: Weapons of Math Destruction: How Big Data Increases Inequality and Threatens Democracy. Crown (2016)
15. Ribeiro, M.T., Singh, S., Guestrin, C.: "Why should I trust you?": Explaining the predictions of any classifier. In: KDD, pp. 1135–1144. ACM (2016)
16. Rudin, C.: Stop explaining black box machine learning models for high stakes decisions and use interpretable models instead. Nat. Mach. Intell. **1**(5) (2019)
17. van de Ven, G.M., Tuytelaars, T., Tolias, A.S.: Three types of incremental learning. Nat. Mach. Intell. **4** (2022)

An Attention-Based CNN Approach to Detect Forest Tree Dieback Caused by Insect Outbreak in Sentinel-2 Images

Vito Recchia[1], Giuseppina Andresini[1,2], Annalisa Appice[1,2(✉)], Gianpietro Fontana[1], and Donato Malerba[1,2]

[1] Department of Computer Science, University of Bari "Aldo Moro", Bari, Italy
{vito.recchia,giuseppina.andresini,annalisa.appice,
g.fontana10}@studenti.uniba.it, donato.malerba@uniba.it
[2] CINI - Consorzio Interuniversitario Nazionale per l'Informatica, Bari, Italy

Abstract. Forests play a key role in maintaining the balance of ecosystems, regulating climate, conserving biodiversity, and supporting various ecological processes. However, insect outbreaks, particularly bark beetle outbreaks, pose a significant threat to European spruce forest health by causing an increase in forest tree mortality. Therefore, developing accurate forest disturbance inventory strategies is crucial to quantifying and promptly mitigating outbreak diseases and boosting effective environmental management. In this paper, we propose a deep learning-based approach, named AVALON, that implements a CNN to detect tree dieback events in Sentinel-2 images of forest areas. To this aim, each pixel of a Sentinel-2 image is transformed into an imagery representation that sees the pixel within its surrounding pixel neighbourhood. We incorporate an attention mechanism into the CNN architecture to gain accuracy and achieve useful insights from the explanations of the spatial arrangement of model decisions. We assess the effectiveness of the proposed approach in two case studies regarding forest scenes in the Northeast of France and the Czech Republic, which were monitored using Sentinel-2 satellite in October 2018 and September 2020, respectively. Both case studies host bark beetle outbreaks in the considered periods.

Keywords: CNNs · Attention · XAI · Remote Sensing · Forest tree dieback · Bark beetle outbreak inventory

1 Introduction

Forest tree dieback refers to the phenomenon of a stand of trees losing health and dying due to pathogens, parasites or conditions like acid rain, drought, wind storm or fire. Nowadays, bark beetle outbreaks represent one of the major causes of tree dieback in European spruce forests. These are small insects that reproduce beneath the bark of coniferous trees, depositing their eggs under the bark and swarming when the temperature is higher. Since 2018, Europe has experienced several unprecedented bark beetle outbreaks that are repeatedly ruining vast swathes of conifer forests. The accurate inventory of bark beetle outbreaks has a crucial role in quantifying and promptly mitigating the consequences of bark beetle outbreaks and enables forest managers to perform informed tree sanitation cuts. At present, the bark beetle outbreak inventory is commonly performed in Europe by foresters during laborious, time-consuming and expensive field surveys [7]. However, the increasing amount of Sentinel-2 images acquired within the European Space Agency program of Sentinel-2 missions offers today the opportunity to systematically reduce the amount of forestry fieldwork. Sentinel-2 are high-resolution, multi-spectral satellite images, which are acquired with thirteen spectral bands for the Earth observation. They are recorded every five days with a constellation of two identical satellites in the same orbit. The Copernicus Open Access Hub provides complete, free and open access to Sentinel-2 images. Hence, AI methods used to detect bark beetle outbreaks in Sentinel-2 images of forest areas can allow foresters to scale-up and speed-up the assessment and surveillance of tree dieback caused by bark beetle over large forest areas.

The present study is boosted from amazing results recently achieved with deep learning methods in semantic segmentation of satellite images. The goal of semantic segmentation is to take images and identify imagery patches belonging to specific semantic labels. Bark beetle outbreaks commonly expand into the forests for several tens of meters due to swarming events. Therefore, neighbour information is useful for correctly delineating bark beetle outbreak areas. This makes semantic segmentation approaches appropriate for mapping tree dieback caused by bark beetles. Semantic segmentation is done by processing the image through a deep neural network that performs deep feature extraction and produces a map with a semantic label per pixel. CNNs are among the most widely used deep neural architectures for image analysis due to their ability of leveraging spatial correlation [6] surrounding pixel data during the deep feature extraction. Accordingly, a few studies [17,20] have recently shown that CNN-based semantic segmentation models can achieve a valuable accuracy performance in mapping bark beetle outbreaks in several types of satellite images included Sentinel-2 images. However, CNNs learn opaque models implicitly represented in the form of a huge number of numerical weights, which are difficult to explain due to the complexity of the network structure. Although, the priority of foresters is to count on accurate semantic labels in satellite images of forest areas, easier-to-explain models are also desirable to better develop a symbiotic AI approach that can better gain trust of foresters in semantic segmentation outputs.

The explainability of semantic segmentation models refers partially to how easily humans may comprehend their underlying assumptions and reasoning. For example, the authors of [2,5] use XAI to strengthen the results of bark beetle outbreak mapping by explaining the effect of spectral bands on semantic segmentation decisions. These studies explain post-hoc decisions produced with different machine learning methods (i.e., XGBoost, Random Forest, Support Vector Machine) and draw conclusions that are in line with well known achievements already reported in [1] on which spectral bands are affected from bark beetle outbreaks. Hence, as these explanations show that known spectral patterns underline machine learning reasoning, they allow foresters to gain confidence in approving machine learning decisions. However, they neither contribute to gain accuracy in the semantic segmentation model nor disclose new interpretable knowledge that can facilitate the identification of errors in the semantic segmentation maps.

In this paper, we present a semantic segmentation approach, named AVALON (Attention-based conVolutional neurAL network fOrest tree dieback in seNtinel-2 images), that trains a CNN model for the semantic segmentation of Sentinel-2 images of forest areas. It integrates a pixel attention mechanism [19] that allows the adaptive selection of imagery pixels where the network "sees" the most important spectral information for the local decisions. This mechanism can mitigate local disturbances introduced with useless or noise data (especially along the boundary between patches labelled with opposite semantic labels). On the other hand, the analysis of the results of the attention provides insight into how the CNN model achieves its decisions, by contributing to the enhancement of the explainability of how a pixel neighbourhood contributes to the decision on each semantic label. The experimentation conducted in two case studies, regarding bark beetle outbreaks localised in the Northeast of France in October 2018 and in the Czech Republic in September 2020, show that the attention mechanism allows us to gain semantic segmentation accuracy and provides some overall explanation of the spatial arrangement of decisions. As an additional contribution, we compute the cumulative attention values obtained pixel-wise in every scene to provide a picture of how each pixel contributes to the overall segmentation of the scene. A qualitative analysis of these scene explanation maps shows that these explanations may also help foresters in identifying wrong decisions that may require their fieldwork to be corrected.

The paper is organised as follows. Section 2 introduces relevant related works. Section 3 describes the proposed approach, while Sect. 4 reports the experimental evaluation and discusses the related findings. Finally, Sect. 5 concludes.

2 Related Works

The remote sensing literature has recently explored the performance of several machine learning and deep learning methods for processing Sentinel-2 images and mapping forest tree dieback caused by bark beetle outbreaks. Most of these studies use machine learning algorithms such as Random Forest [2,5,7,9,14], Support Vector Machine [5,9,10] and XGBoost [2,5] combined with spectral vegetation indices extracted with different combination of spectral bands.

On the other hand, deep learning strategies have recently seen a massive rise in popularity in several problems of forest health monitoring. For example, the authors of [5] evaluate the performance of deep learning algorithms such as U-Net and FCN-8 as a contribution of their evaluation study. The authors of both [17,20] analyse the performance of CNN models trained for mapping bark beetle outbreaks in several types of satellite images, including Sentinel-2 images.

However, none of the tested CNN architectures uses the attention mechanism. On the other hand, in [15] the attention mechanism is integrated in a standard U-Net model and uses the defined architecture for detecting deforestation within 3-band and 4-band Sentinel-2 images. In [18], the performance of a U-Net model equipped with the attention mechanism and trained to detect the forest damage induced by bark beetle outbreaks in Sentinel-2 images is evaluated.

Finally, a few recent studies have started the exploration of the explainability of semantic segmentation models trained for mapping bark beetle outbreaks in Sentinel-2 images. In particular, the authors of [5] use SHAP to show that the effect of the spectral input on the decisions yielded by the same model may change according to the classes and, in some cases, according to the machine learning algorithm. In [2], SHAP is used to explain the effect of both temporal spectral data and temporal spectral vegetation indices on decisions yielded from a Random Forest model for bark beetle outbreak mapping. The studies in [15,18] use the attention mechanism with U-Net models trained for mapping bark beetle outbreaks in Sentinel-2 images. However, both these studies neglect the explanation results of attention that is one of the contributions of this study. On the other hand, the work in [16] describes a CNN equipped with attention especially designed for water body extraction from Sentinel-2 images. But, again, the study does not consider explanation information embedded in the attention mechanism. Finally, the authors of [13] have recently explored the use of the attention mechanism in CNNs trained for the scene classification of colour images. Similarly to our work, this study explores achievements of attention in terms of both accuracy and explainability. However, scene classification is a different task where a label is assigned to an entire image, while semantic segmentation is a task where a label is assigned to each imagery pixel.

To the best of our knowledge, this is the first study that couples the evaluation of how an attention mechanism can allow a CNN model to gain accuracy in the inventory of bark beetle outbreaks in Sentinel-2 images (by outperforming several related machine learning and deep learning models) to the analysis of how explanation insights disclosed with attention can help in gaining trust of foresters in obtained maps also thanks to the identification of symptoms of potential errors in the predicted semantic segmentation maps.

3 Proposed Method

AVALON takes S – a collection of labelled Sentinel-2 images – as input. Each Sentinel-2 image $\mathbf{S}$ is a tensor with shape $W_\mathbf{S} \times H_\mathbf{S} \times C$, which covers an Earth scene spanned across $W_\mathbf{S} \times H_\mathbf{S}$ pixels that are observed across C spectral bands. Let $\mathbf{S}(i,j)$ denote the spectral vector with size C recorded at pixel (i,j) in $\mathbf{S}$. In S, every Sentinel-2 image $\mathbf{S}$ is one-to-one associated with a label mask $\mathbf{M}$ with shape $W_\mathbf{S} \times H_\mathbf{S}$ so that every $\mathbf{S}(i,j)$ is assigned to a binary semantic label $\mathbf{M}(i,j)$. In this study, each semantic label assumes the value: "healthy" (0) or "damaged" (1). AVALON learns a semantic segmentation model that predict the unknown label mask of any new Sentinel-2 image. This is done by reformulating and addressing

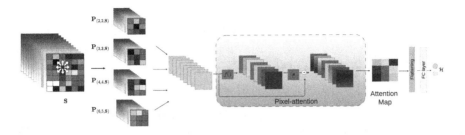

Fig. 1. Schema of AVALON. Abbreviations: FC = Fully Connected

the semantic segmentation problem as an image classification problem. To this aim, we first transform every Sentinel-2 imagery pixel of the input collection into an image that shows the pixel within its pixel neighbourhood. Then, we train a CNN with Attention to image classification. The pipeline of AVALON is shown in Fig. 1.

To achieve a formulation of the semantic segmentation problem as an image classification problem, we transform $\mathcal{S}$ into a new imagery dataset $\mathcal{P}$. Formally, given a target pixel (i, j) recorded into a Sentinel-2 image $\mathbf{S}$, this is mapped into a pixel image $\mathbf{P}_{(i,j,\mathbf{S})}$ with shape $W \times H \times C$, which shows (i, j) within its squared pixel neighbourhood in $\mathbf{S}$. Let us consider $H = W = 2k + 1$ with k as a user-defined, positive, integer parameter to define the neighbourhood size. $\mathbf{P}_{(i,j,\mathbf{S})}$ covers $(2k+1)^2$ pixels of $\mathbf{S}$ spanned across a pixel neighbourhood $N(i, j, k)$ defined as $N(i, j, k) = \{(i', j') | i - k \leq i' \leq i + k, j - k \leq j' \leq j + k\}$. In the case of boundary pixels, we add padding pixels populated with the average spectral values calculated on the remaining neighbour pixels for each spectral band. Every image $\mathbf{P}_{(i,j,\mathbf{S})} \in \mathcal{P}$ is associated with a binary label (0 or 1) that is the label $\mathbf{M}(i, j)$. A CNN model is trained from $\mathcal{P}$ to learn a function to image classification. The used CNN integrates a pixel attention mechanism, as described in [19]. This generates a pixel-wise attention map for all pixels in the input data images. In particular, the pixel attention layer takes as input each tensor $\mathbf{X}^l$ of a given convolutional l-th layer with dimension $W^l \times H^l \times C^l$. It convolves a 1×1 convolution layer followed by a sigmoid function. This is to obtain the attention maps of the level $l + 1$ that are then multiplied with the input pixels at l-th layer. Formally, the pixel attention layer is defined as:

$$\mathbf{X}^{l+1} = \sigma\left(f\left(\mathbf{X}^l\right)\right) \odot \mathbf{X}^l, \tag{1}$$

where $f()$ is a 1×1 convolution layer, σ is the sigmoid function and $\mathbf{X}^{l+1}$ is the resulting output tensor at $l + 1$-th layer. The output of this operation is a new tensor $\mathbf{X}^{l+1}$ with shape $W^{l+1} \times H^{l+1} \times C^{l+1}$ where $W^l = W^{l+1}$, $H^l = H^{l+1}$ and $C^l = C^{l+1}$. To produce a single attention map that explains the importance of pixels in the original image, we apply an average layer, as implemented in [3]. This averages all the channels for each corresponding pixel. More in detail, given the tensor with shape $W^{l+1} \times H^{l+1} \times C^{l+1}$ produced by the pixel attention layer, the average layer produces a single-channel tensor with shape $W^{l+1} \times H^{l+1} \times 1$, for which each attention pixel α_{ij}^{l+1} is equal to:

$$\alpha_{ij}^{l+1} = \frac{1}{C} \sum_{c=1,\ldots,C} \alpha_{ijc}^{l}, \qquad (2)$$

where α_{ijc}^{l} is the attention pixel at position (i,j) in the c-th feature map from the preceding $l-1$ layer. The average layer explains visually which pixels of the input image received more attention from the model for the image classification.

The prediction is done by minimizing a binary cross-entropy loss function:

$$\mathcal{H} = -\left(y \log \hat{y} + (1-y) \log (1-\hat{y})\right), \qquad (3)$$

where $y \in \{0,1\}$ is the ground truth label and $\hat{y}$ is the CNN output for a single pixel image.

4 Evaluation Study

This evaluation study aims at examining the accuracy performance and the explanation insights achieved with AVALON. Section 4.1 describes the case studies considered in the evaluation study and the adopted experimental set-up. Section 4.2 describes the implementation of AVALON. Finally, Sect. 4.3 illustrates the analysis of the accuracy and explainability results achieved in the evaluation.

4.1 Case Studies and Experimental Set-up

We considered two case studies that involve non-overlapping forest scenes spanned across Northeast of France and Czech Republic. The scene collections hosted bark beetle outbreaks. The case study in Northeast of France included 87 forest scenes that hosted bark beetle outbreaks that were likely caused by the hot summer droughts in 2018. The ground truth map of these outbreaks was commissioned by the French Ministry of Agriculture and Food to Sertit (University of Strasbourg), to assess the damage observed in October 2018 in spruce forests of Northeast of France [2,5]. The case study in Czech Republic included 200 forest scenes that hosted bark beetle outbreaks that were likely caused by the hot summer droughts in 2020. The ground truth map of these outbreaks was acquired in September 2020 and recorded in DEFID2 database [12].

Accordingly, cloud-free Sentinel-2 images were downloaded in October 2018 for the scenes of Northeast of France and in September 2020 for the scenes of Czech Republic, respectively. The images were downloaded in the 3857 EPSG system from the ESA Copernicus open access hub using the Google Earth Engine APIs. Sentinel-2 spectral bands with different spatial resolutions from 10 m were all re-sampled at the same pixel size of 10 m resolution. In the dataset of Northeast of France, the size of the Sentinel-2 images varies from 23×39 to 461×425 pixels at $10\,\mathrm{m}^2$ resolution, while the percentage of infested territory per scene

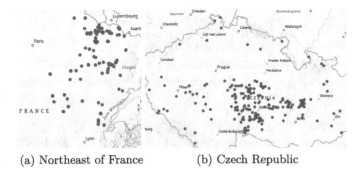

(a) Northeast of France (b) Czech Republic

Fig. 2. Location of the centroids of the study 87 scenes in the Northeast of France area (Fig. 2a), and 200 scenes in the Czech Republic (Fig. 2b). The red circles are the training set scenes, while the blue circles are the testing set scenes. (Color figure online)

varies from 0,33% to 34,8% of the scene surface. The total percentage of damaged territory of the entire scene collection is 2,81%. The semantic segmentation model development and its evaluation were conducted by considering 71 random scenes (covering 1639945 pixels at $10\,m^2$ resolution) as training scene set and 16 left-out scenes (covering 495223 pixels) as testing scene set. In the dataset of Czech Republic, the size of images varies from 33×36 to 260×238 pixels at $10\,m^2$ resolution, while the percentage of infested territory per scene varies from 4,14% to 54,81% of the scene surface. The total percentage of damaged territory of the entire scene collection is 14,59%. The semantic segmentation model development and its evaluation were conducted by considering 160 random scenes (covering 1014708 pixels at $10\,m^2$ resolution) as training scene set and 40 left-out scenes (covering 198253 pixels) as testing scene set. Figures 2a and 2b show the maps of the location of study scenes in Northeast of France and Czech Republic, as well as the partitioning of each scene set in the training scene set and testing scene set.

4.2 Implementation Details

AVALON was developed in Python 3.[1] In particular, the CNN architecture was implemented using the PyTorch library (version 2.0.1). For each dataset, we conducted a Bayesian optimization of the CNN hyperparameters with the tree-structured Parzen estimator algorithm, as implemented in the Optuna library. Specifically, we optimized the set-up of the following hyperparameters: mini-batch size in $\{2^5, 2^6, 2^7, 2^8\}$, learning rate between 0.0001 and 0.001, number of kernel size in $\{2, 3, 4\}$, number of Convolutional layers in $\{1, 2, 3\}$ and dropout between 0 and 1. The hyperparameter optimisation was performed using 20% of the entire training set (selected with stratified sampling) as a validation set. We selected the configuration of hyperparameters that achieved the highest F1

---

[1] Code and parameters of trained model available at https://github.com/s4rgax/AVALON.

computed on the validation set by considering the class "damaged" as the positive class. The CNN architecture was defined with a variable number of Convolutional layers, a Pixel Attention layer placed after the last Convolutional layer in the neural network, and two Fully-Connected layers. A Dropout layer was placed between each pair of Convolutional layers to perform data regularisation and prevent training data overfitting. In all layers, except the final classification layer, the Rectified Linear Unit function (ReLU) was used as the activation function. The Softmax activation function was used in the classification layer. We performed the gradient-based optimisation using the Adam update rule. Finally, we trained the CNN model with the maximum number of epochs set equal to 150, and using an early-stopping approach to retain the best model.

4.3 Result and Discussion

In this Section we illustrated the results achieved in: (1) the study of the sensitivity of the accuracy performance of AVALON to the size of pixel neighbourhoods considered, as well as the ablation study to explore the effect of the attention mechanism on the accuracy performance of the proposed method; (2) the comparative analysis of the accuracy performance of related methods commonly used in the remote sensing literature on bark beetle detection; and (3) the qualitative analysis of the explanations produced with the attention mechanism in AVALON.

Accuracy Performance Analysis. To evaluate the accuracy performance we measured the following metrics: F1 score (F1) computed for the two opposite semantic labels, Macro F1 score (Macro F1) averaged on the two labels and Intersection-over-Union (IoU). The F1 score measures the harmonic mean of Precision and Recall. The Precision=$\frac{TP}{TP+FP}$ is the fraction of pixels correctly classified in a specific class (TP) among pixels of the considered class ($TP+FP$). The Recall=$\frac{TP}{TP+FN}$ is the fraction of pixels correctly classified in a specific class (TP) among pixels classified in the considered class ($TP+FN$). The IoU score is the ratio of the intersected area to the combined area of prediction and ground truth, that is, IoU=$\frac{TP}{TP+FP+FN}$. This is commonly used to evaluate the accuracy of models trained in both semantic segmentation and object detection problems. All metrics are reported in percentages and computed on the Sentinel-2 images collected for the testing scenes. For each metric, the higher the value, the better the performance of the semantic segmentation maps predicted.

Table 1 reports the accuracy performance of AVALON in the two case studies by varying the size of each squared pixel neighbourhood among 3×3, 5×5 and 7×7. This sensitivity analysis of the effect of the neighbourhood size on the accuracy performance of AVALON was performed orthogonal to an ablation study that compared the performance of the AVALON models trained with the attention mechanism to the performance of the AVALON models trained by keeping away the attention mechanism. The results show that each configuration of AVALON run with the attention outperforms the counterpart configuration run without the attention independently of the pixel neighbourhood size. On the other hand, in both case studies, the best accuracy performance is achieved with

Table 1. Accuracy performance of semantic segmentation maps produced with the configurations of AVALON obtained by varying the size of each pixel neighbourhood among 3×3, 5×5 and 7×7, considering the attention mechanism, as well as neglecting it, in the two case studies in Northeast of France and Czech Republic, respectively. The best results are in bold.

AVALON configuration		Northeast of France				Czech Republic			
Neighb.	Attention	F1(d)	F1(h)	Macro F1	IoU	F1(d)	F1(h)	Macro F1	IoU
3 × 3	No	59.60	99.35	79.47	42.45	65.09	92.94	79.02	48.25
5 × 5		66.12	99.39	82.76	49.39	67.76	93.70	80.73	51.24
7 × 7		63.80	99.40	81.60	46.84	66.86	93.09	79.97	50.21
3 × 3	Yes	65.75	99.41	82.58	48.97	69.00	93.73	81.37	52.68
5 × 5		**68.90**	**99.45**	**84.18**	**52.56**	**70.73**	93.64	**82.19**	**54.72**
7 × 7		68.45	99.44	83.94	52.03	69.15	**93.86**	81.51	52.85

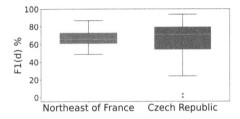

Fig. 3. Box plots of F1(d) scores computed per scene with AVALON in the testing Sentinel-2 images of Northeast of France and Czech Republic, respectively

pixel neighbourhoods with size 5 × 5. By considering that a pixel covers $10\,m^2$, this analysis shows that, independently of the case study, a local area of $50\,m^2$ is a reasonable choice to account for the spatial correlation of spectral bands in bark beetle outbreak mapping. The only exception was observed in the F1(h), that is, the F1 score computed on the majority class "healthy", in the case study in Czech Republic. In this case, the highest F1(h) was achieved in the configuration of AVALON run with the pixel neighbourhood size set equal to 7 × 7, while the highest F1(d), as well as the highest Macro F1 and IoU, were still achieved in the configuration of AVALON with the pixel neighbourhood size set equal to 5 × 5. Foresters are, in general, more interested in better delineating bark beetle outbreak areas than healthy areas. So, the improvement in F1(d) is more desirable than the improvement in F1(h). Hence, we consider the configuration of AVALON using the attention mechanism and processing pixel neighbourhoods with a size set equal to 5 × 5 in the follow-up of this evaluation study.

Figure 3 shows the box plots of F1(d) scores computed with AVALON, scene by scene, in the testing Sentinel-2 images of Northeast of France and Czech Republic. Although the F1(d) score is higher, on average, in Northeast of France than in Czech Republic, we note higher variability of F1(d) in Czech Republic with a few close-to-zero outliers. In the subsequent explanation analysis, we show that explanation maps extracted with AVALON may help foresters to identify signals of potentially inaccurate predictions in these semantic segmentation maps.

Competitor Analysis. We compare the accuracy performance of AVALON to that of machine learning and deep learning methods (i.e., Random Forest, XGBoost, Support Vector Machine, Multi-Layer Perceptron, YOLO8-Seg, U-Net and Attention U-Net) commonly used in literature studies on bark beetle outbreak detection in Sentinel-2 data. As described in [5], a grid search was used to optimise the selection of the label cost in Random Forest, XGBoost and Support Vector Machine, as well as the tree depth in Random Forest and XGBoost. The Multi-Layer Perceptron was composed of a number of fully connected layers varying between 3 and 6. The number of neurons per layer was a power of two, with the final layer having 32 neurons, and the output probabilities were obtained using the softmax activation function. The ReLU activation function was used in all the other hidden layers. The number of layers was selected by maximising the F1 score computed on the class "damaged" for a validation set (20% of the training set). The YOLO8-Seg architecture consists of a revised version of the object detector YOLO (You Only Look Once) for segmentation tasks. In particular, for this experimental evaluation, we fine-tuned the version 8 of YOLO model released in January 2023 by Ultralytics[2] with the RGB images of the two case studies considered in this paper. The U-Net architecture included an encoder repeating blocks with a 3×3 convolutional layer with a spatial Dropout, a Batch Normalization layer and a 2×2 Max Pooling operation layer. At each downsampling step of the encoder, the number of feature channels was increased. In the decoder, the feature map was upsampled through a sequence of up-convolutions that halved the number of feature channels and concatenated the result with the features from the corresponding encoder layer. The classification of each pixel map was obtained by using the Sigmoid activation function. The ReLu was used in all hidden convolution layers. The Tversky loss was used to handle the data imbalance. Finally, the Attention U-Net was implemented by integrating the pixel attention within the skip connections between the downsampling path and the upsampling path of the U-Net.

Table 2 reports the accuracy performance of the compared methods in the two case studies. AVALON achieves the best performance in all the accuracy metrics measured in this evaluation study in both Northeast of France and Czech Republic. The Attention U-Net is the runner-up of this comparative study. Notably, the Attention U-Net uses the same pixel attention module embedded in AVALON, but within the encoder-decoder architecture of a U-Net model, while AVALON integrates pixel attention in a CNN architecture. So, these results show that the attention mechanism can actually help us to better map bark beetle outbreaks in Sentinel-2 images. In addition, the attention-based approach developed in AVALON outperforms the Attention U-Net showing that better accuracy performance can be achieved in the considered task using a CNN architecture.

---

[2] https://docs.ultralytics.com/it/tasks/segment/.

Table 2. Accuracy performance of semantic segmentation maps produced with both AVALON and related methods in Northeast of France and Czech Republic, respectively. The best results are in bold.

Method	Northeast of France				Czech Republic			
	F1(d)	F1(h)	Macro F1	IoU	F1(d)	F1(h)	Macro F1	IoU
Random Forest	61.35	99.32	80.34	44.25	67.43	92.76	80.09	50.86
XGBoost	59.54	99.25	79.39	42.39	67.10	92.20	79.65	50.49
Support Vector Machine	60.38	99.19	79.79	43.25	68.59	92.33	80.46	52.20
Multi-Layer Perceptron	54.89	99.01	76.95	37.82	65.97	92.26	79.12	49.22
YOLO8-Seg	47.07	98.98	73.02	30.78	61.27	90.90	76.09	44.17
U-Net	63.18	99.38	81.28	46.18	67.58	90.78	79.18	51.03
Attention U-Net	65.68	99.29	82.48	48.90	68.65	91.41	80.03	52.27
AVALON	**68.90**	**99.45**	**84.18**	**52.56**	**70.73**	**93.64**	**82.19**	**54.72**

(a) Northeast of France (pvalue=1.04e-11)   (b) Czech Republic(pvalue=3.64e-39)

Fig. 4. Critical difference diagram of Friedman-Nemenyi test performed on F1(d) measured on the semantics segmentation maps of evaluated scenes produced with Random Forest (RF), XGBoost (XGB), Support Vector Machine (SVM), Multi-layer Perceptron (MLP), YOLO8-Seg, U-Net, Attention U-Net, and AVALON in Northeast of France (Fig. 4a) and Czech Republic (Fig. 4b)

Figures 4a and 4b report the Friedman-Nemenyi test of the F1(d) scores measured for both AVALON and the related methods on the testing scenes in both Northeast of France and Czech Republic. The results of this comparative test support the conclusion that AVALON generally achieves higher F1(d), measured on the testing scenes, than the related methods.

Explainability Analysis. Finally, we perform a qualitative analysis of explanations produced with the pixel attention mechanism integrated in AVALON. We recall that AVALON addresses a semantic segmentation task as an image classification problem by mapping each Sentinel-2 image **S** of a forest scene into a collection of pixel neighbourhood-based images. It assigns a semantic label to each imagery scene pixel by assigning a class label to each associated image of the pixel neighbourhood. Accordingly, AVALON produces the attention maps of the neighbourhood-based images of the scene pixels to explain the decisions produced for the scene pixels accounting for the information enclosed in their neighbour pixels. Specifically, for each pixel (i, j) in **S**, the pixel attention map

Algorithm 1: Attention Scene Map of a Sentinel-2 image **S**

Data: A Sentinel-2 image **S** with shape $W_S \times H_S \times C$; the Attention-CNN model; the neighbourhood size parameter k
Result: The attention map **A** computed for **S** with shape $W_S \times H_S$

1 begin
2   **A**= zeros(W_S, H_S)
3   **Counter**= zeros(W_S, H_S)
4   for $(i = 1; i \leq W_S; i++)$ do
5   for $(j = 1; j \leq H_S; j++)$ do
6           $\hat{y}_{(i,j,S)}$ = predict(CNN, $\mathbf{P}_{(i,j,S)}$)
7           $\mathbf{a}_{(i,j,S)}$ = attentionMap(CNN, $\mathbf{P}_{(i,j,S)}$)
8   if $\hat{y}_{(i,j,S)}$ == 1 then
9             $\mathbf{A}[i-k:i+k, j-k:j+k]+ = a_{(i,j,S)}$
10   else
11             $\mathbf{A}[i-k:i+k, j-k:j+k]+ = -1 \cdot a_{(i,j,S)}$
12   **Counter**=$[i-k:i+k, j-k:j+k]+ = 1$
13   **A** =minmaxScaler(**A**/**Counter**)
14   return **A**

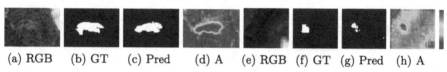

(a) RGB   (b) GT   (c) Pred   (d) A   (e) RGB   (f) GT   (g) Pred   (h) A

Fig. 5. RGB images, Ground truth (GT) label masks, Predicted semantic segmentation masks (Pred) and Scene Attention maps (A) of two scenes in Northeast of France. Figures 5a–d refer to a scene for which AVALON achieved F(d) = 0.87. Figures 5e–h refer to a scene for which AVALON achieved F(d) = 0.49. (Color figure online)

of the pixel neighbourhood-based image $\mathbf{P}_{(i,j,S)}$ sees which pixels in the considered neighbourhood received more attention from the CNN model to predict the semantic label of (i,j). Based on these premises, we combined all pixel attention maps produced in a scene, as described in Algorithm 1, to yield the overall attention map of a Sentinel-2 imagery scene. In particular, for each pixel in **S**, we sum up the attention values that are computed for the pixel in every neighbourhood-based image extracted from **S** where the pixel appears as a target pixel or neighbour pixel. Notice that each attention value is a positive, real value. To distinguish between pixels that are seen to predict "healthy" labels and pixels that are seen to predict "damaged" labels, attention values of pixels are multiplied by -1 when they are part of a pixel-neighbourhood image predicted in the class "healthy", while they are considered in their original positive format when they are part of a pixel-based neighbourhood image predicted in the class "damaged". Finally, every accumulated pixel attention value is averaged with respect to the number of single attention values accumulated for the pixel in the scene and scaled between 0 and 1 to be visualised in a heatmap.

Figures 5 and 6 show the RGB bands, ground truth label masks, predicted semantic segmentation masks and scene attention maps of two Sentinel-2 images selected from the testing sets of Northeast of France and Czech Republic,

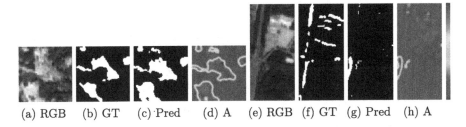

(a) RGB (b) GT (c) Pred (d) A (e) RGB (f) GT (g) Pred (h) A

Fig. 6. RGB images, Ground truth label masks (GT), Predicted semantic segmentation masks (Pred) and Scene Attention maps (A) of two scenes in Czech Republic. Figures 6a–d refer to a scene for which AVALON achieved $F(d) = 0.71$. Figures 6e–h refer a scene where AVALON achieved $F(d) = 0.04$. (Color figure online)

respectively. In both case studies, the two scenes were selected as scenes where AVALON achieved high $F1(d)$ (i.e., $F1(d) = = 0.87$ and $F1(d) = 0.71$ in Figs. 5c and 6c) and low $F1(d)$ (i.e., $F1(d) = 0.49$ and $F1(d) = 0.04$ in Figs. 5g and 6g), respectively. The scene attention maps plotted in Figs. 5d and 6d show that the zones that received higher attention delimit well the patches where AVALON predicted the presence of a bark beetle outbreaks in Figs. 5c and 6c, respectively. We recall that AVALON achieved high $F1(d)$ in both these scenes, so this type of explanation outcomes can be used as a visual explanation for foresters of the certainty of the predictions produced for these scenes. On the other hand, the scene attention maps plotted in Figs. 5h and 6h show signals of low certainty for predictions produced from AVALON in these two scenes and plotted in Figs. 5g and 6g. We recall that AVALON achieved low accuracy performance in both these scenes, so this scene explanation outcome can be seen as an alert for foresters who may decide to integrate inventory results of AVALON in critical zones with the inventory produced with traditional fieldwork.

In particular, the predicted mask plotted in Fig. 5g shows a small patch wrongly detected as "damaged" in the bottom-right of the scene. However, the attention map of this scene reported in Fig. 5h shows that AVALON devoted little attention to seeing a bark beetle outbreak in this patch. Differently, the patch correctly predicted in the same scene is highlighted with high attention values in the corresponding scene attention map. Similarly, the predicted mask plotted in Fig. 6g shows that AVALON fails in delimiting outbreak zones in the centre-up of the scene. However, the attention map of the scene reported in Fig. 6h shows some uncertainty in this area. In fact, attention values take on a light shade of blue in the missed outbreak patches (wrongly detected as "healthy"), which is in contrast to the dark shade of blue we observe in the attention value of patches correctly labelled as "healthy". Further conclusions can be drawn from the scene attention map shown in Fig. 6d in correspondence with the predicted mask shown in Fig. 6c. In this case, AVALON wrongly labels a patch in the centre-up of the scene as "damaged", and the attention values of the patch show that AVALON actually pays attention to the zone. However, the red extension of the attention (where, anyway, we note a prevalence of yellow and green) is limited

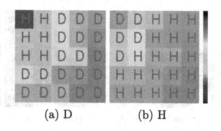

Fig. 7. Pixel attention map for a "damaged" (D) pixel (Fig. 7a) and a "healthy" (H) pixel (Fig. 7b) selected from the Sentinel-2 image shown in Fig. 6a.

compared to other patches correctly predicted as "damaged" in the same scene. In general, we note that this analysis is coherent with our symbiotic view of AI, where an AI semantic segmentation model explains its decisions so that foresters may have an active role in revising the AI-based decision process to achieve the best inventory outcome in problems of bark beetle outbreak detection.

Finally, Fig. 7 shows the pixel attention maps produced with AVALON for two pixels of opposite classes in the scene of Fig. 6a. Both pixels were selected along the boundary of opposite classes. The plots show that the CNN is able to give more attention to the inner pixels that have the same class as the target one, while it gives less attention to pixels of the opposite class. This outcome supports our idea that the attention mechanism allows the CNN model of AVALON to recognise neighbour pixels of a fixed-size neighbourhood, which presumably belong to the same class of the target pixel and which plausibly provide the more relevant spectral information of the neighbourhood for the decision.

5 Discussions and Conclusions

In this paper, we presented an Attention-CNN architecture trained to perform the inventory of bark beetle outbreaks in Sentinel-2 images of forest scenes. The accuracy of the proposed method AVALON was evaluated in two case studies, which refer to forest scenes infested by bark beetles in Northeast of France and Czech Republic, respectively. The obtained results show that AVALON outperforms related models trained with machine learning and deep learning methods commonly considered in the literature studies addressing the same task. Furthermore, the qualitative evaluation highlights that attention-based explanations may disclose useful information on the effect of the spatial arrangement of imagery data on predictions. Notably, they may also provide a visual guideline that can help foresters identify potentially incorrect predictions.

Beyond the valuable results achieved by AVALON in terms of accuracy and explanation, several limitations still require further investigations. With this regard, we intend to investigate how the proposed approach can be effectively used to map tree dieback areas caused by different families of insect and fungal pathogens. In addition, we plan to explore how the semantic segmentation model trained with the proposed approach can be used to detect tree dieback areas in geographic areas different from the geographic areas considered for the model training. With regard to the XAI contribution, we plan to explore how explanation information disclosed with the pixel attention module integrated with AVALON can be confirmed or integrated using an XAI post-hoc explanation approach such as Grad-CAM [8]. In addition, we plan to leverage recent results achieved in the cybersecurity domain [11] to distil attention information for gaining accuracy in problems of bark beetle outbreak detection. Finally, we intend to extend the current research to consider Attention-CNNs trained from time-series of Sentinel-2 images. Based on literature [2,7], feature extraction to identify temporal spectral changes may facilitate the early detection of bark beetle outbreaks. Notably, this direction is coherent with an emerging remote sensing research trend that is assigning a crucial role to change detection analysis [4] in Sentinel-2 images.

Acknowledgments. Annalisa Appice and Donato Malerba acknowledge support from the SWIFTT project, funded by the European Union under Grant Agreement 101082732. Giuseppina Andresini and Vito Recchia are supported by the project FAIR - Future AI Research (PE00000013), Spoke 6 - Symbiotic AI, under the NRRP MUR program funded by the European Union- NextGenerationEU.

Authorship Contribution Statement

Vito Recchia. Conceptualization, Methodology, Data curation, Software, Validation, Visualization, Investigation, Writing - original draft **Giuseppina Andresini**: Conceptualization, Methodology, Validation, Investigation, Supervision, Writing - original draft, Writing - review & editing **Annalisa Appice**: Conceptualization, Methodology, Validation, Investigation, Writing - original draft, Writing - review & editing, Supervision, Project administration. **Gianpietro Fontana**: Software. **Donato Malerba**: Conceptualization, Methodology, Investigation, Writing - original draft, Writing - review & editing.

References

1. Abdullah, H., Skidmore, A.K., Darvishzadeh, R., Heurich, M.: Sentinel-2 accurately maps green-attack stage of European spruce bark beetle (ips typographus, l.) compared with Landsat-8. Remote Sens. Ecol. Conserv. **5**(1), 87–106 (2019)
2. Andresini, G., Appice, A., Malerba, D.: Leveraging sentinel-2 time series for bark beetle-induced forest dieback inventory. In: The 39th ACM/SIGAPP Symposium on Applied Computing. SAC 2024, pp. 875–882. ACM (2024). https://doi.org/10.1145/3605098.3635908
3. Andresini, G., Appice, A., Caforio, F.P., Malerba, D., Vessio, G.: ROULETTE: a neural attention multi-output model for explainable network intrusion detection. Expert Syst. Appl. 117144 (2022). https://doi.org/10.1016/j.eswa.2022.117144
4. Andresini, G., Appice, A., Ienco, D., Malerba, D.: Seneca: change detection in optical imagery using Siamese networks with active-transfer learning. Expert Syst. Appl. **214**, 119123 (2023). https://doi.org/10.1016/j.eswa.2022.119123

5. Andresini, G., Appice, A., Malerba, D.: SILVIA: an explainable framework to map bark beetle infestation in sentinel-2 images. IEEE J. Sel. Top. Appl. Earth Observ. Remote Sens. **16**, 10050–10066 (2023). https://doi.org/10.1109/JSTARS.2023.3312521
6. Appice, A., Malerba, D.: Segmentation-aided classification of hyperspectral data using spatial dependency of spectral bands. ISPRS J. Photogramm. Remote. Sens. **147**, 215–231 (2019). https://doi.org/10.1016/j.isprsjprs.2018.11.023
7. Bárta, V., Lukeš, P., Homolová, L.: Early detection of bark beetle infestation in Norway spruce forests of central Europe using Sentinel-2. Int. J. Appl. Earth Obs. Geoinf. **100**, 102335 (2021). https://doi.org/10.1016/j.jag.2021.102335
8. Caforio, F.P., Andresini, G., Vessio, G., Appice, A., Malerba, D.: Leveraging grad-CAM to improve the accuracy of network intrusion detection systems. In: Soares, C., Torgo, L. (eds.) DS 2021. LNCS (LNAI), vol. 12986, pp. 385–400. Springer, Cham (2021). https://doi.org/10.1007/978-3-030-88942-5_30
9. Candotti, A., De Giglio, M., Dubbini, M., Tomelleri, E.: A Sentinel-2 based multi-temporal monitoring framework for wind and bark beetle detection and damage mapping. Remote Sens. **14**(23) (2022). https://doi.org/10.3390/rs14236105
10. Dalponte, M., Solano-Correa, Y.T., Frizzera, L., Gianelle, D.: Mapping a European spruce bark beetle outbreak using Sentinel-2 remote sensing data. Remote Sens. **14**(13) (2022). https://doi.org/10.3390/rs14133135
11. De Rose, L., Andresini, G., Appice, A., Malerba, D.: Vincent: cyber-threat detection through vision transformers and knowledge distillation. Comput. Secur. 103926 (2024). https://doi.org/10.1016/j.cose.2024.103926
12. Forzieri, G., Dutrieux, L., et al.: The database of European forest insect and disease disturbances: DEFID2. Glob. Change Biol. **29**(21), 6040–6065 (2023). https://doi.org/10.1111/gcb.16912
13. Hou, Y.E., Yang, K., Dang, L., Liu, Y.: Contextual spatial-channel attention network for remote sensing scene classification. IEEE Geosci. Remote Sens. Lett. **20**, 1–5 (2023). https://doi.org/10.1109/LGRS.2023.3304645
14. Huo, L., Persson, H.J., Lindberg, E.: Early detection of forest stress from European spruce bark beetle attack, and a new vegetation index: normalized distance red & SWIR (NDRS). Remote Sens. Environ. **255**, 112240 (2021). https://doi.org/10.1016/j.rse.2020.112240
15. John, D., Zhang, C.: An attention-based U-Net for detecting deforestation within satellite sensor imagery. Int. J. Appl. Earth Obs. Geoinf. **107**, 102685 (2022). https://doi.org/10.1016/j.jag.2022.102685
16. Parajuli, J., Fernandez-Beltran, R., Kang, J., Pla, F.: Attentional dense convolutional neural network for water body extraction from Sentinel-2 images. IEEE J. Sel. Top. Appl. Earth Observ. Remote Sens. **15**, 6804–6816 (2022). https://doi.org/10.1109/JSTARS.2022.3198497
17. Turkulainen, E., Honkavaara, E., Näsi, R., et al.: Comparison of deep neural networks in the classification of bark beetle-induced spruce damage using UAS images. Remote Sens. **15**(20) (2023). https://doi.org/10.3390/rs15204928
18. Zhang, J., Cong, S., Zhang, G., Ma, Y., Zhang, Y., Huang, J.: Detecting pest-infested forest damage through multispectral satellite imagery and improved UNet++. Sensors **22**(19) (2022). https://doi.org/10.3390/s22197440
19. Zhao, H., Kong, X., He, J., Qiao, Yu., Dong, C.: Efficient image super-resolution using pixel attention. In: Bartoli, A., Fusiello, A. (eds.) ECCV 2020. LNCS, vol. 12537, pp. 56–72. Springer, Cham (2020). https://doi.org/10.1007/978-3-030-67070-2_3

20. Zwieback, S., Young-Robertson, J., et al.: Low-severity spruce beetle infestation mapped from high-resolution satellite imagery with a convolutional network. ISPRS J. Photogramm. Remote. Sens. **212**, 412–421 (2024). https://doi.org/10.1016/j.isprsjprs.2024.05.013

Open Access This chapter is licensed under the terms of the Creative Commons Attribution 4.0 International License (http://creativecommons.org/licenses/by/4.0/), which permits use, sharing, adaptation, distribution and reproduction in any medium or format, as long as you give appropriate credit to the original author(s) and the source, provide a link to the Creative Commons license and indicate if changes were made.

The images or other third party material in this chapter are included in the chapter's Creative Commons license, unless indicated otherwise in a credit line to the material. If material is not included in the chapter's Creative Commons license and your intended use is not permitted by statutory regulation or exceeds the permitted use, you will need to obtain permission directly from the copyright holder.

Explaining Image Classifiers with Visual Debates

Avinash Kori[✉], Ben Glocker, and Francesca Toni

Imperial College London, London, UK
{a.kori21,b.glocker,f.toni}@icl.ac.uk

Abstract. Current deep learning-based models for image classification are effective at making decisions, but their lack of transparency can be a significant concern in high-stakes settings. To address this issue, many state-of-the-art methods define (post-hoc) explanations as visual heatmaps or image segments deemed responsible for the classifiers' outputs. However, the static nature of these explanations often fails to align with human explanatory practices. To obtain human-oriented explanations, we propose an alternative, novel form of dialogue-based interactive explanations for image classifiers: *visual debates* between two fictional players who interact to argue for and against the classifiers' outputs. Specifically, in our method, the players propose *arguments*, which are (abstract) features drawn from classifiers' latent representations and these arguments are countered by the opposing player. We present a realization of visual debates based on quantization for extracting arguments, recurrent networks for modelling player behaviour, and network dissection for argument visualization. Experimentally, we show that our visual debates satisfy the desiderata of *dialecticity*, *convergence*, and *faithfulness*.

Keywords: Explainability · Debates · Image Classification

1 Introduction

Current deep learning models are effective at classifying images but their lack of transparency can be a serious concern in high-stakes settings. Transparency can be enforced by design [9] or, as we do in this paper, by deploying *post-hoc* explanation methods generating reasons why models are making particular decisions.

Fig. 1. *Visual debates* with two steps.

Conventionally, post-hoc explanations take various forms, including feature attributions [23,29,33], attention maps [32], counterfactual explanations [14], or concept-based explanations [13,21]. These tend to be *static* captures of input-output relations and mostly fail to align with human explanatory practices, which favor, in particular, *contrastive, selective* and *social* explanations [24]. For illustration, imagine a scenario where you are trying to explain to some interlocutor that a given image (e.g. as in Fig. 1) belongs to the class 'Cat': a contrastive explanation would focus on why the image is of a 'Cat' rather than, say, of a 'Dog'; a selective explanation would focus on some selected reasons, rather than a complete portfolio thereof; and a social explanation would amount to a conversation or interaction. A natural way to combine these three characteristics of explanations is via dialogues [24]. In our scenario, the dialogue may unfold with you highlighting cat-specific features and your interlocutor attempting to contrast these features, progressively, with others applicable to dogs. Inspired by this view of explanation, we aim to define/realize a novel notion of *visual debates* as explanations for image classifiers. As in Fig. 1, the debates are between two fictional players ($\mathcal{P}^1$, $\mathcal{P}^2$) putting forward arguments for and against (respectively). In the figure, the classifier predicts the image to be a 'Cat' and the players claim it to be a 'Cat' and a 'Dog', respectively. The players' claims are supported by arguments amounting to (abstract) features (z_i) drawn from the classifier's latent representations and visualised as regions in the image (in the figure, these are also equipped with human-oriented descriptions). Our definition/realization of visual debates is driven by three desiderata: *dialecticity*, amounting to players offering different perspectives for contrastiveness, *convergence*, amounting to players reaching consensus, and *faithfulness*, amounting to the visual debates reflecting the behaviour of the classifiers (as standardly required for explanations).

Related work. While different explanation methods have different pros and cons, it is generally acknowledged that explanations have an important role to play towards improving the transparency and understanding of the model's reasoning process [10], uncovering model biases [19], identifying biases in the data-generating process, and fulfilling legal obligations for model deployment.

Debates [17] or dialogues [7,22], have already been advocated as an interesting perspective for extracting classifier reasoning. Specifically, in [17], debates are advocated as an intrinsically transparent and human aligned model, demonstrated on a toy setting with the MNIST dataset, while in [7], a method is proposed to learn natural language dialogues cooperatively for reasoning about the considered environment. Unlike these works, visual debates provide contrastive, selective and social explanations for image classifiers. We give a method for realizing them by using (abstract) features to distill the knowledge of the classifier into a discrete surrogate model obtained by quantization as in [36] and we visualize the features by means of a deterministic process adapted from [1]. We train the two players in debates by adapting [17], where the players learn to pick the relevant discrete features for a given image as arguments. Finally, all the players' arguments are used to estimate utilities that progressively help the players learn.

Alignment of explanations with human *concepts* is rare, with exceptions [13,21] focusing on generating disentangled concepts and traces between them to explain models' reasoning. We draw inspiration from recent approaches [20,30] striving toward faithfulness to the models by generating explanations using the model's latent knowledge rather than the original data. At a high-level of detail, debates can be interpreted as simple instances of argumentation frameworks (AFs, e.g. of the kinds advocated in [11] or in [2]), and thus our approach can be seen as a form of argumentative XAI [6]. To the best of our knowledge, our proposed debates are the first to use models' latent knowledge to extract arguments and counterarguments for explaining image classifiers via debates as interactive game-playing among learning players. Indeed, other approaches using AFs for explainable image classification mirror the mechanics of the model, e.g. as in [35].

Several metrics for the evaluation of XAI methods have been proposed and studied in the literature (e.g. see [26] for some overviews). Amongst these, *faithfulness* (also referred to as correctness, fidelity or descriptive accuracy [26]) is widely regarded as crucial, as it amounts to explanations being truthful to the model they aim to explain. In addition, *completeness* is also widely advocated, relating to the extent to which explanations actually explain models [26]. For both metrics, high values are preferable. We will adopt these measures, customizing them to our specific setting of explanation as debates. Furthermore, we will define and assess, in the experiments, additional, novel metrics relating to the specific nature of our explanations as debates.

Contributions. Overall, our contributions include: (i) a debate framework (§ 2) formalizing visual debates; (ii) a realization of the debate framework (§ 3), based on first distilling the knowledge of a classifier into a discrete surrogate model (by quantization) and then training the players involved in the debates to learn to pick, from the discrete model, arguments as relevant features to the input image; (iii) visual debates' experimental evaluation (§ 4) along metrics for assessing their satisfaction of the three desiderata of dialecticity, convergence and faithfulness. Our approach is overviewed in Fig. 2.

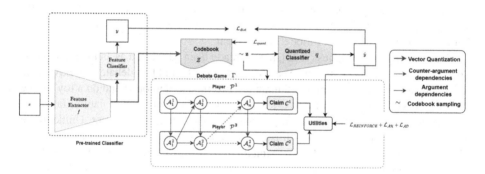

Fig. 2. Overview of the realization (§ 3) of the debate framework(§ 2). The classifier to be explained is $C = g \circ f$, whereas q is the quantized classifier using codebook $\mathcal{Z}$ for distilling the continuous latent knowledge of C in a discrete form. Γ is a debate game between players $\mathcal{P}^1$ and $\mathcal{P}^2$: for $i \in \{1, 2\}$, $\mathcal{A}_k^i$ is the argument put forward by $\mathcal{P}^i$ at iteration k and C^i is the claim made by $\mathcal{P}^i$ at the end of the debate, when each player gets a utility, resulting from argument and claim rewards.

2 Formalisation of Visual Debates

Preliminaries. Let $\mathcal{D} \subseteq \mathcal{X} \times \mathcal{Y}$ be a dataset, such that $\mathcal{X} \in \mathbb{R}^{H \times W \times C}$ and $\mathcal{Y} = \{1, \ldots, N\}$, where $H \times W \times C$ corresponds to images resolution, and $N \geq 2$ is the number of classes. Let $C : \mathcal{X} \to \mathcal{Y}$ be a *classifier*, pre-trained on $\mathcal{D}$: given an *input image* sampled from $\mathcal{X}$ ($x \sim \mathcal{X}$), $\mathcal{C}(x) \in \mathcal{Y}$ is the predicted class. As conventional in image classification with deep learning, we assume that $\mathcal{C}$ can be decomposed as $g \circ f$, where g is a *feature classifier* and f is a *feature extractor*.

We aim to explain individual predictions by $\mathcal{C}$ in terms of debates whereby players exchange arguments, amounting to features supporting classes. To achieve this, we assume a model q corresponding to a *quantized classifier* which behaves like a proxy to g, where $q : \mathcal{Z} \to \mathbb{P}(\mathcal{Y})$ for $\mathcal{Z}$ a discrete set of *quantized features* [36] and $\mathbb{P}(S)$, for any set S, is a probability distribution over S (e.g. determined by a softmax operation) which may be interpreted as providing confidence scores on the elements of S. The quantized classifier operates on discrete features rather than continuous features as the pre-trained classifier. The discrete features $z \sim \mathcal{Z}$ are obtained from the input image x as the result of distilling the continuous latent space of $\mathcal{C}$, when applied to x (see Fig. 2).

For $z \sim \mathcal{Z}, y \in \mathcal{Y}$, we use $q(z)_y$ to denote the confidence score for class y as estimated by the quantized classifier q on z, and, with an abuse of notation, we reserve $q(z)$ to indicate simply the class in $\mathcal{Y}$ with the highest confidence score. Finally, we use the following notation for $z \subseteq \mathcal{Z}, z' \subseteq z, y \in \mathcal{Y}$: $q(z; do(z'))_y$ stands for the confidence score for y estimated by q with just the quantized features z' and all the other features masked, i.e. $\hat{z} = 0, \forall \hat{z} \in z \setminus z'$; and $q(z; do(z'))$ stands for the class with the highest confidence score after masking.

Debate Framework. The debate framework Γ consists of two *players* $\mathcal{P}^1, \mathcal{P}^2$ who sequentially argue about a common *question* $Q \in \mathcal{Q}$ for a fixed number of n iterations to make *final claims* $\mathcal{C}^1, \mathcal{C}^2$, respectively.[1] In the context of image classification, the question Q relates to the classifier's prediction $y = \mathcal{C}(x)$ for the input image $x \in \mathcal{X}$, and may be something like *"Why did the classifier predict the image x as a cat?"*, and the claims may be $\mathcal{C}^1 = Cat$ and $\mathcal{C}^2 = Dog$, as illustrated in Fig. 1. Player $\mathcal{P}^1$'s objective is to provide relevant *arguments* ($\mathcal{A}^1 = \{\mathcal{A}^1_1, \ldots, \mathcal{A}^1_n\}$) supporting the decision made by the classifier on x, while player $\mathcal{P}^2$'s objective is to provide relevant counterarguments ($\mathcal{A}^2 = \{\mathcal{A}^2_1, \ldots, \mathcal{A}^2_n\}$) that oppose the classifier's decision, where all arguments are drawn from $\mathcal{Z}$. Finally, the players are equipped with *utilities* $\mathcal{U}^1, \mathcal{U}^2$, sanctioning how effective their choices of arguments are towards answering the question with their respective claims. Formally:

$$\Gamma = \langle \{\mathcal{Q}, \mathcal{Z}\}, \{\mathcal{P}^1, \mathcal{P}^2\}, \{\mathcal{A}^1, \mathcal{A}^2\}, \{\mathcal{C}^1, \mathcal{C}^2\}, \{\mathcal{U}^1, \mathcal{U}^2\} \rangle.$$

Intuitively, players argue using discrete features from $\mathcal{Z}$ for supporting ($\mathcal{P}^1$) or for opposing ($\mathcal{P}^2$) the classifier's decision. We will show how players can be obtained in § 3. Here, we define arguments, claims and utilities.

Definition 1. *An argument $\mathcal{A}^i_k$, for $i \in \{1, 2\}$ and $k \in \{1, \ldots, n\}$, is a triple (z^i_k, c^i_k, s^i_k), where $z^i_k \in z$ is a particular quantized feature for $x \in \mathcal{X}$, $c^i_k =$ argmax $q(z; do(\{z^i_k\}))$ is the argument claim, and $s^i_k \in \{-1, 1\}$ is the argument strength, where, for $\Delta = |q(z)_y - q(z; do(\{z^1_k, z^2_k\}))_y|$, with $\tau \in (0, 1)$ a given threshold:*

---

[1] The framework is applicable to any number of players, but we focus for simplicity on two. We assume that n is fixed up-front depending on the cognitive needs of users using visual debates as explanations (we experiment with various choices of n in § 4).

$$s_k^1 = \begin{cases} 1, & \Delta \leq \tau \\ -1, & \text{otherwise} \end{cases} \qquad s_k^2 = \begin{cases} 1, & \Delta > \tau \\ -1, & \text{otherwise} \end{cases}$$

Here, Δ measures the effective contribution of a particular pair of argument (by $\mathcal{P}^1$) and counterargument (by $\mathcal{P}^2$) towards the quantized classifier's decision. Note that a low-value of Δ indicates that the majority of the latent information is encoded in the latent features pair (z_k^1, z_k^2). Also, the notion of argument strength differs between players: player $\mathcal{P}^1$, supporting the classifier's decision, considers a higher value of Δ to be a 'weak' (negative) argument, while player $\mathcal{P}^2$ considers that to be a 'strong' argument, in the spirit of zero-sum games. Further, note that the claim of an argument depends only on the chosen quantized feature in that argument. Also, given τ, the strength of an argument depends only on the quantized feature: thus, we will often equate arguments $\mathcal{A}_k^i$ with their quantized features z_k^i. Also, for $i \in \{1, 2\}$, $-i$ will stand for $j \in \{1, 2\} \setminus \{i\}$.

Definition 2. Claim $\mathcal{C}^i$ is given by $agg^i(\mathcal{A}^1, \mathcal{A}^2)$ for some aggregation function agg^i.

The notion of player's claim is thus distinguished from that of argument claim: the former is cumulative and results from the full set of features in arguments (for $\mathcal{P}^1$) and counterarguments (for $\mathcal{P}^2$) at the end of the debate, while the latter only depends on the feature put forward at a particular step. In practice, in § 3, to obtain agg^i we will use hidden state vectors of recurrent neural networks as aggregated arguments, followed by a linear layer to map arguments to classes. Note that different players may perceive the effectiveness of the arguments and counterarguments differently, so agg^i are different for the two players, leading to different players' claims at the end of the debate (step n).

Definition 3 (Utilities). *The utility of player* $\mathcal{P}^i$ *is* $\mathcal{U}^i = r_\Gamma^i + \sum_{k=1}^{k=n} s_k^i$ *for* $y' = q(z, do(\mathcal{A}^1 \cup \mathcal{A}^2))$, *and* $r_\Gamma^i = \begin{cases} 1 & \text{if } y' = \mathcal{C}^i, y' \neq \mathcal{C}^{-i}; \\ -1 & \text{if } y' \neq \mathcal{C}^i, y' = \mathcal{C}^{-i}; \\ 0 & \text{if } y' \neq \mathcal{C}^i, y' \neq \mathcal{C}^{-i} \text{ or } y' = \mathcal{C}^i, y' = \mathcal{C}^{-i}. \end{cases}$

Here, we treat the argument strength s_k^i as *argument reward*. The utility is a function of the argument rewards and of an overall debate *claim reward* r_Γ^i, obtained by masking all the quantized features not present in the arguments and comparing the prediction by the quantized classifier after masking (y') with the claims by the players. Note that, when y' matches neither or both players' claims, the claim reward is 0, as basically there is no debate between the players in those cases. Note also that the utilities, by design, have a zero-sum nature (i.e. $\mathcal{U}^1 = -\mathcal{U}^2$), reflecting that players should focus on different concepts (quantized features) for contrastiveness. For debate frameworks to provide faithful explanations, we need to guarantee that the classifier's reasoning is encoded in the selected arguments: in § 4, this faithfulness will be measured by computing the accuracy of the debate framework with respect to the classifier's prediction [3].

Note that, given the complexity of the latent space when classifying images, the discretization by the quantized classifier helps both in limiting the argument space for the players and, alongside the choice of n, in generating *cognitively tractable* explanations [5]. Finally, note that our debates could be interpreted from the perspective of computational argumentation, e.g. players' arguments may be seen as forming *rebuttal attacks* against the other player's arguments, leading to argumentation frameworks in the spirit of [11] and the aggregation function could be viewed as a form of *gradual semantics*, in the spirit of [2]. A re-interpretation of our debates in formal computational argumentation terms and generalizations thereof to accommodate more complex argumentation are outside the scope of our paper and are left for future work.

3 Realization of Visual Debates

In this section we describe the implementation details (discretization by quantization, players, model training and visualization to obtain our *visual debates*).

Discretization by Quantization. The output of the feature extractor (f, see Fig. 2) is continuous, posing multiple challenges for extracting meaningful explanations about the classifier's latent reasoning. To address this, we first distill the classifier's latent representations into a *codebook* $\mathcal{Z}$ with $\tilde{n}$ discrete features, each of dimension d, using the process of vector quantization, similarly to [36]. However, whereas [36] sample across pixels, we sample along channels, using the Hessian penalty of [27] to disentangle these channels and obtain varied features in $\mathcal{Z}$. The quantization is achieved by deterministically mapping $\tilde{z} = f(x)$ for $x \in \mathcal{X}$ to the nearest codebook vector $z \in \mathcal{Z}$, formally described in equation $\boxed{z = \{\mathrm{argmin}_{z_j \in \mathcal{Z}} \tilde{d}(\tilde{z}, z_j) | z_j \in \tilde{z}\}}$, Where $\tilde{d}$ is some convex distance function (this is enforced by $\mathcal{L}_{quant}$ in Fig. 2, see the accompanying codebase for details). In § 4, we consider two distance functions for sampling the codebook: (i) Euclidean sampling, where $\tilde{d}(a, b) = \|a - b\|_2^2$ and (ii) cosine sampling, where $\tilde{d}(a, b) = -\langle a, b \rangle$ (negative cosine distance). To distill the knowledge from the continuous to the discrete space, we introduce a *quantized classifier* q, which maps the average pooled sampled vector z to the classifier's output $\mathcal{C}(x)$ for any input $x \in \mathcal{X}$. The parameters in the quantized classifier are trained using knowledge distillation loss $\mathcal{L}_{dist}$ (striving towards the faithfulness desideratum), amounting to cross-entropy loss between the classifier's output $y = \mathcal{C}(x)$ and the quantized classifier's output $\tilde{y} = q(z)$ (see Fig. 2).

Players. To capture the sequential nature of debates, we formalise the players as recurrent models using gated recurrent units (GRU) as the *backbone network* ζ^i, with context vector h_k^i capturing the debate information at iteration k. In addition, each player includes a *policy network* $\Pi_{\theta^i}^i$ which considers the context vector h_k^i along with previous step argument made by both the players to define the player's strategy for selecting discrete features $z \sim \mathcal{Z}$ as a next argument (here θ^i represents all the learnable parameters for player $\mathcal{P}^i$), a *modulator network* $\mathcal{M}^i$ converting arguments to the hidden state dimension in ζ^i (referred to as *modulated arguments*), and a *claim network* estimating the player's final claim $\mathcal{C}^i$ (with an abuse of notation, we call this network also $\mathcal{C}^i$). We also equip players with a *baseline network* $\mathcal{B}^i$ for value estimation, in the spirit of [25], so as to minimize the variance of reward, as discussed later.
At every step $0 < k < n$ (with step 0 some random initialisation), $\mathcal{P}^i$ receives an argument $\mathcal{A}_{k-1}^{-i}$ and decides an argument $\boxed{\mathcal{A}_k^i \sim \Pi_{\theta^i}^i(h_k^i \mid \hat{z}_k, \hat{y})}$, where $\hat{y}$ is the output

of quantized classifier q, and $\hat{z}_k = (z; do(\{\mathcal{A}_k^1, \mathcal{A}_k^2\}))$ is the masked environmental state (updating state information with previous step arguments). The argument $\mathcal{A}_k^i$ is then projected onto the modulated argument $\boxed{e_k^i = \mathcal{M}^i(\mathcal{A}_k^i)}$ using the modulator network. The modulated argument is used to update the player's context $\boxed{h_{k+1}^i = \zeta^i(h_k^i, e_k^i)}$. Finally, the baseline value is estimated using the context vector: $\boxed{b_k^i = \mathcal{B}^i(h_k^i)}$, which is later used in learning objective. Player $\mathcal{P}^1$ estimates $\mathcal{A}_1^1$ using a sampled feature and randomly initialised context (hidden state) vector, followed by the other player. Once estimated, arguments are considered common knowledge, and both players can use this information to estimate their subsequent arguments.

When training the players, we adopt two novel loss components, striving towards the dialecticity desideratum. They are defined in terms of argument entropy and diversity, as follows.

Definition 4 (Argument Entropy ($\mathcal{AH}$)). *For $z \subseteq \mathcal{Z}$, $\mathcal{AH}(z) = -\mathbb{E} \log p$, where p is the probability of considering a particular feature as an argument.*

This notion ensures that the probability distribution over features is focused on selected features, also leading to manageable, cognitively tractable explanations.

Definition 5 (Argument Diversity ($\mathcal{AD}$)). $\mathcal{AD}(\mathcal{A}^1, \mathcal{A}^2) = (\mathbb{E}((\tilde{\mathcal{A}}^1 - \mathcal{A}^2)^2) + \mathbb{E}((\mathcal{A}^1 - \tilde{\mathcal{A}}^2)^2)) - \lambda \sum_{i \in \{1,2\}} \mathbb{E}((\mathcal{A}^i - \tilde{\mathcal{A}}^i)^2)$, *where* $\tilde{\mathcal{A}}^i = \frac{1}{n} \sum_{k=1}^{k=n} \mathcal{A}_k^i$.

Intuitively, to preserve diversity and encourage coherence between arguments, we maximize the variance between inter-player arguments (first two terms in the definition of $\mathcal{AD}$) while minimizing the variance between intra-player arguments (last term in the definition of $\mathcal{AD}$) with hyperparameter λ.

Similarly to [25], we use baseline values, b_k^i, to reduce the variance in the estimated reward and average the estimated policy over M Monte Carlo samples, resulting in the reinforce learning rule described as:

$$\frac{1}{M} \sum_m \sum_k \nabla_{\theta^i} \log \Pi_{\theta^i}^i (h_k^i \mid \hat{z}_k, \hat{y})(\mathcal{U}^i - b_k^i),$$

where $\mathcal{U}^i$ is the player's utility (see Def. 3) and b_k^i corresponds to the baseline value estimated, for which the parameters of the baseline network are learned by reducing the squared error between $\mathcal{U}^i$ and b_k^i. We consider a loss function supporting the above REINFORCE learning rule described as: $\mathcal{L}_R^i = - \sum_k \log \Pi_{\theta^i}^i (h_k^i \mid \hat{z}_k, \hat{y})(\mathcal{U}^i - b_k^i)$. To train the policy networks $\Pi_{\theta^i}^i$ to learn θ_i, this loss is combined with argument entropy and argument diversity regularisation terms as per Defs. 4 and 5, represented respectively by $\mathcal{L}_{\mathcal{AH}}^i = -\mathcal{AH}(\mathcal{A}_k^i, z')$ and $\mathcal{L}_{\mathcal{AD}} = -\mathcal{AD}(\mathcal{A}^1, \mathcal{A}^2)$. These regularisation terms ensure that players make a unique arguments at each step. Our training setup consists of two steps: (i) *supportive* training, where both players are trained to support the classifier's decisions distilling the classifier knowledge into a quantized classifier and learning codebook embeddings with the combined loss as described as:

$$\mathcal{L}_{sup.} = \sum_i \mathcal{L}_R^1 + \lambda_2 \mathcal{L}_{\mathcal{AH}}^i + \lambda_3 \mathcal{L}_{quant} + \lambda_4 \mathcal{L}_{dist},$$

where $\mathcal{L}_R^1$ is the reinforce loss for the supporting player $\mathcal{P}^1$, providing positive reinforcement for correct predictions for both players, $\mathcal{L}_{\mathcal{AH}}^i$ encourages the varieties in the arguments, $\mathcal{L}_{quant}$ correspond to quantization loss, and $\mathcal{L}_{dist}$ correspond to a distillation loss corresponding to cross entropy between the predicted quantized classifier prediction and pre-trained classifier; and (ii) *contrastive* training, where the players are trained for debating, generating arguments and counterarguments with the combined loss as described below

$$\mathcal{L}_{con.} = \sum_i \mathcal{L}_R^i + \lambda_2 \mathcal{L}_{\mathcal{AH}}^i + \lambda_3 \mathcal{L}_{\mathcal{AD}},$$

where the players are rewarded with respect to utilities defined in Def. 3, encouraging players to debate for and against the classifier. In a contrastive setting, we use additional regularisation to maximise the diversity between arguments made by both players ($\mathcal{L}_{\mathcal{AD}}$). Supportive training is a pre-debate step to provide a common knowledge base for the players to generate debate.

Visualization. For our debate frameworks to serve as explanations as *visual debates*, e.g. as illustrated in Fig. 1, we need to visualize arguments so that they can be comprehensible to humans. To do so, we follow a deterministic approach based on [1]. We consider, from any specific argument $\mathcal{A}_k^i$, a low-resolution attention map $\mathcal{F}$ drawn from the quantized feature z in $\mathcal{A}_k^i$. We then compare $\mathcal{F}$ with the original input image by resizing it to the input dimension using bilinear interpolation, anchoring the interpolation with respect to the feature's receptive field. Specifically, we first normalize the resized attention map between 0 and 1 using min-max normalization and overlay it on the original image.

4 Experiments

In this section, we describe several experiments with the concrete realization of the debate framework from § 3, empowering a quantitative and qualitative evaluation against standard feature attribution explanation methods, namely LIME [29], DeepSHAP [23], deepLIFT [33], and gradCAM [32].[2] We conducted experiments with three image classifiers (a vanilla model and two deeper models: DenseNet-121 [16] and ResNet18 [15]) trained on the SHAPE [18], MNIST [8] and AFHQ [4] datasets (we use images of dimension 32×32 with 1 channel for SHAPE and MNIST, and 128×128 with 3 channels for AHFQ). We report accuracy for the DenseNet-121, Vanilla[3], and ResNet-18 models in the second column of Tables 1, 2, and 3.

---

[2] Additional details and further experiments are described in https://arxiv.org/abs/2210.09015.
[3] We use a tiny architecture with 7 convolutional layers with 3×3 kernel with batch-norm and ReLU activation layer. Finally, we project the global average pooled vector onto a class probability space using a linear layer followed by softmax activation.

Quantitative Evaluation Measures. Our definition and realization of visual debates is driven by three desiderata: *dialecticity*, amounting to players offering different perspectives in visual debates for contrastiveness, *convergence*, amounting to players in visual debates reaching consensus towards the output of the underlying classifier to reflect the reasoning thereof, and *correctness*, amounting to the visual debates reflecting the behaviour of the underlying classifier and data. The first two are bespoke requirements for visual debates, whereas the latter is a standard requirement for post-hoc explanations. We define (and measure) these metrics in the context of a test set $\mathcal{T} \subseteq \mathcal{X} \times \mathcal{Y}$. We use $\Gamma(x) = q(z, do(\mathcal{A}^1, \mathcal{A}^2))$ to indicate the outcome of debate Γ (as in § 2) on input x.

To assess dialecticity, we need to assess how players are polarised around the two different viewpoints. We measure how well diversity across players is achieved in terms of the notion of *partition ratio* Z_R below, where $\mathcal{Z}_1 \subseteq \mathcal{Z}$ and $\mathcal{Z}_2 \subseteq \mathcal{Z}$ are the sets of

Table 1. Ablation results with DenseNet121 (whose accuracy is given in the second column, in brackets) wrt debate length (4, 6 or 10) with mean and variance over 3 runs.

PPTY → DATASET ↓	FEATURE EXT. ↓	CODEBOOK SAMPLING ↓	COMPLETENESS 4	6	10	CORRECTNESS 4	6	10	CONSENSUS 4	6	10	PARTITION RATIO (Z_R) 4	6	10
SHAPE (0.96±0.02)	DENSENET	EUCLIDIAN	0.78±0.13	0.89±0.04	0.99±0.02	0.81±0.07	0.93±0.02	0.99±0.01	0.63±0.07	0.68±0.05	0.92±0.03	0.36	0.42	0.60
		COSINE	0.80±0.14	0.91±0.03	0.98±0.02	0.83±0.13	0.94±0.06	0.98±0.01	0.57±0.08	0.65±0.06	0.90±0.03	0.38	0.44	0.58
MNIST (0.99±0.00)	DENSENET	EUCLIDIAN	0.38±0.09	0.54±0.05	0.83±0.06	0.41±0.15	0.55±0.06	0.86±0.01	0.31±0.21	0.76±0.04	0.87±0.02	0.39	0.49	0.51
		COSINE	0.42±0.18	0.56±0.04	0.77±0.05	0.41±0.19	0.56±0.09	0.84±0.03	0.28±0.23	0.74±0.06	0.85±0.01	0.44	0.49	0.51
AFHQ (0.65±0.08)	DENSENET	EUCLIDIAN	0.49±0.12	0.54±0.11	0.58±0.08	0.54±0.12	0.60±0.06	0.62±0.02	0.31±0.11	0.40±0.09	0.56±0.04	0.60	0.63	0.68
		COSINE	0.41±0.23	0.40±0.08	0.65±0.02	0.43±0.13	0.43±0.04	0.65±0.03	0.33±0.16	0.48±0.02	0.60±0.00	0.72	0.72	0.77

Table 2. Ablation results with the Vanilla model (accuracy in the second column, in brackets) wrt debate length (4, 6 or 10) with mean and variance over three random runs.

PPTY → DATASET ↓	FEATURE EXT. ↓	CODEBOOK SAMPLING ↓	COMPLETENESS 4	6	10	CORRECTNESS 4	6	10	CONSENSUS 4	6	10	PARTITION RATIO (Z_R) 4	6	10
SHAPE (0.94±0.01)	VANILLA	EUCLIDIAN	0.55±0.12	0.80±0.14	0.85±0.02	0.55±0.14	0.79±0.07	0.89±0.03	0.33±0.15	0.64±0.06	0.78±0.01	0.58	0.60	0.60
		COSINE	0.58±0.23	0.84±0.06	0.82±0.01	0.58±0.19	0.81±0.03	0.88±0.02	0.34±0.16	0.76±0.07	0.82±0.01	0.48	0.56	0.56
MNIST (0.98±0.00)	VANILLA	EUCLIDIAN	0.36±0.22	0.45±0.03	0.73±0.05	0.38±0.18	0.49±0.01	0.83±0.00	0.24±0.17	0.62±0.06	0.74±0.04	0.44	0.45	0.56
		COSINE	0.46±0.11	0.58±0.03	0.79±0.02	0.53±0.08	0.61±0.05	0.78±0.00	0.22±0.11	0.58±0.02	0.82±0.01	0.42	0.45	0.55
AFHQ (0.96±0.02)	VANILLA	EUCLIDIAN	0.81±0.09	0.86±0.06	0.89±0.05	0.89±0.02	0.91±0.04	0.94±0.03	0.35±0.11	0.43±0.08	0.82±0.04	0.39	0.44	0.50
		COSINE	0.31±0.10	0.68±0.02	0.77±0.02	0.33±0.09	0.71±0.04	0.82±0.01	0.38±0.11	0.63±0.05	0.76±0.02	0.44	0.48	0.49

Table 3. Ablation results with ResNet-18 (accuracy in the second column, in brackets) wrt debate length (4, 6 or 10) with mean and variance over three random runs.

PPTY → DATASET ↓	FEATURE EXT. ↓	CODEBOOK SAMPLING ↓	COMPLETENESS 4	6	10	CORRECTNESS 4	6	10	CONSENSUS 4	6	10	PARTITION RATIO (Z_R) 4	6	10
SHAPE (0.98±0.01)	RESNET	EUCLIDIAN	0.74±0.23	0.83±0.14	0.96±0.03	0.76±0.12	0.81±0.06	0.98±0.02	0.41±0.12	0.58±0.05	0.89±0.02	0.38	0.54	0.58
		COSINE	0.73±0.20	0.81±0.08	0.97±0.02	0.76±0.21	0.83±0.01	0.97±0.02	0.44±0.09	0.62±0.05	0.91±0.01	0.44	0.58	0.58
MNIST (0.99±0.00)	RESNET	EUCLIDIAN	0.42±0.19	0.46±0.08	0.81±0.02	0.45±0.13	0.51±0.02	0.85±0.01	0.18±0.21	0.57±0.09	0.76±0.03	0.54	0.58	0.61
		COSINE	0.38±0.22	0.42±0.03	0.78±0.00	0.42±0.12	0.49±0.05	0.81±0.00	0.21±0.09	0.58±0.02	0.78±0.00	0.43	0.60	0.60
AFHQ (0.98±0.01)	RESNET	EUCLIDIAN	0.76±0.06	0.84±0.01	0.91±0.01	0.81±0.07	0.90±0.02	0.97±0.00	0.30±0.08	0.69±0.03	0.93±0.02	0.38	0.41	0.43
		COSINE	0.58±0.08	0.72±0.03	0.88±0.00	0.64±0.06	0.75±0.02	0.92±0.01	0.40±0.18	0.63±0.08	0.72±0.07	0.40	0.42	0.41

Table 4. Correctness and completeness for the baselines on the DenseNet121 model.

METHODS →	COMPLETENESS				CORRECTNESS			
DATASET ↓	GRADCAM	DEEPLIFT	GRADIENTSHAP	LIME	GRADCAM	DEEPLIFT	GRADIENTSHAP	LIME
SHAPE	0.78±0.08	0.73±0.06	0.74±0.07	0.62±0.11	0.84±0.07	0.80±0.08	0.80±0.08	0.66±0.14
MNIST	0.74±0.08	0.74±0.11	0.68±0.12	0.59±0.19	0.77±0.09	0.77±0.11	0.73±0.13	0.58±0.17
AFHQ	0.70±0.10	0.66±0.15	0.67±0.17	0.54±0.21	0.75±0.10	0.71±0.11	0.71±0.14	0.52±0.19

quantized features from which players $\mathcal{P}^1$ and $\mathcal{P}^2$ draw their respective arguments in the debates $\Gamma(x)$ for all $(x,y) \in \mathcal{T}$:[4]

$$Z_R = \frac{|\mathcal{Z}_1|}{|\mathcal{Z}_1| + |\mathcal{Z}_2|} \quad (1)$$

Remark 1. For dialecticity to hold, the codebook needs to be partitioned into two sets, one per player, and thus Z_R needs to be significantly greater than 0 and lower than 1.

To assess convergence, we define a *consensus* metric as follows, where, for input x, $\mathcal{A}^i(x)$ refers to the arguments of player $\mathcal{P}^i$ and $\mathcal{C}^i(\mathcal{A}^i(x))$ refers to the claim by player $\mathcal{P}^i$ based on these arguments:

$$|\{\mathcal{C}^1(\mathcal{A}^1(x)) \mid \mathcal{C}^1(\mathcal{A}^1(x)) = \mathcal{C}^2(\mathcal{A}^2(x)), (x,y) \in \mathcal{T}\}|/|\mathcal{T}| \quad (2)$$

Basically, consensus captures the idea of persuasion, *i.e.*, it captures how often player $\mathcal{P}^1$ persuades the other player $\mathcal{P}^2$.

Remark 2. For convergence to hold, we expect that, with sufficiently long debates, if the explained classifier is accurate then player $\mathcal{P}^1$ should be able to persuade player $\mathcal{P}^2$ and thus the longer the debates the higher we expect the consensus to be.

The need for suitable metrics for the evaluation of correctness explanation methods is well understood in the XAI literature, and several such metrics have been proposed and studied (e.g. see [26,34] for some overviews). Here, we use two such metrics (*correctness* and *completeness*, with respect to the given classifier and the data, respectively), framed in our setting as follows. Correctness measures the proportion of matches between debates' outcomes and classifier's predictions for all inputs in $\mathcal{T}$:

$$|\{\Gamma(x)|\Gamma(x) = (g \circ f)(x), (x,y) \in \mathcal{T}\}|/|\mathcal{T}| \quad (3)$$

Remark 3. Higher values of correctness are desirable, as they correspond to a more extensive distillation of the classifier's latent knowledge in the players arguments.

Completeness measures data-specific knowledge encoded within the players' arguments; we compute the accuracy of debates with respect to ground truth labels:

$$|\{\Gamma(x)|\Gamma(x) = y, (x,y) \in \mathcal{T}\}|/|\mathcal{T}| \quad (4)$$

Remark 4. Higher completeness values are desirable, as they signify that the acquired arguments possess significance and are closely tied to the provided dataset.

---

[4] Here and below $|S|$ represents the cardinality of set S.

Quantitative Evaluation Results. We measured correctness, completeness, consensus, and partition ratio by varying debate length and codebook sampling criterion.

Tables 1, 2, and 3 give results for DenseNet121, Vanilla and ResNet18, respectively. For comparison, we also provide an evaluation of the properties of completeness and correctness for LIME, DeepSHAP, deepLIFT, and gradCAM in Table 4 for DenseNet121 (note that the novel metrics of consensus and partition ratio are only applicable to our visual debates and thus cannot be assessed for these existing baselines for explaining image classifiers, or any other existing baseline, to the best of our knowledge).

The results show that debate length has implications on both completeness and correctness. It is important to note that, since the debate's objective is for players to select the most *relevant* features, in some cases, completeness and correctness scores are greater than the classifier's accuracy (given in the second column in the table). Another observation from Tables 1, 2, and 3, for classifiers with higher accuracy, the completeness score is similar to the correctness score (differing by $\sim 5\%$). Also, when compared against baseline methods as illustrated in Table 4, we can see that visual debates as explanations result in higher correctness and completeness measures.

Table 1 also shows desired behaviour for consensus: in the case of SHAPE and MNIST, when the classifier has high accuracy, consensus increases with debate length as expected, while in the case of AFHQ, when the classifier has low accuracy, the opposite behaviour can be observed. Note also that higher values of correctness indicate a greater extent of distillation of the classifier's latent knowledge, giving lower possibilities of debate, and thus higher consensus. The last three columns in Tables 1, 2, and 3 (partition ratio) show that the players' arguments differ about 40–60% of the time, as expected. This is reflected in the illustrations in Fig. 3, where the arguments are largely different across players (note that difference in arguments does not mean difference in final claims, as the latter result from the combined effect of all arguments).

Tables 6 and 5 describes the ablation results for the DenseNet121 model, wrt completeness and partition ratio respectively, while varying the codebook size for generating visual debates. Based on this we can see that, the debate length helps in achieving better debate accuracy irrespective of codebook size. However, the improvement in performance plateaus after certain length, depending upon a selected dataset. However in the case of partition ratio codebook size after some threshold makes it easier for players to partition features, resulting in higher partition ratio in the case of larger codebooks.

Qualitative Evaluation. Figure 3(a)-(c) illustrate visual debates obtained on all three datasets, for the DenseNet121 model. Figure 3(d) focuses on explanations with LIME, DeepSHAP, deepLIFT and gradCAM with AFHQ, for comparison with Fig. 3(c), highlighting the dynamic, fine-grained nature of visual debates as opposed to existing static, coarse-grained explanations (highlighting whole regions in a block rather than individual sub-regions as arguments as in visual debates). For all visual debates, the first image is the input, while the first and second rows correspond respectively to $\mathcal{P}^1$'s and $\mathcal{P}^2$'s visual arguments. By comparing Fig. 3(c) and (d), it can be seen that the collective arguments in visual debates result in similar explanations as existing methods, *i.e.*, the combination of all the arguments result in similar regions being highlighted compared to other methods, while the debates provide additional information by partitioning

Table 5. Debate correctness by varying codebook size (total number of discrete features) on SHAPE, MNIST, and AFHQ datasets, for the DenseNet121 model.

scDatasets (→)	SHAPE				MNIST				AFHQ			
$\mathcal{Z}$ scSize (↓)	4	6	10	20	4	6	10	20	4	6	10	20
1024	-	-	-	-	-	-	-	-	0.61	0.74	**0.83**	**0.79**
512	-	-	-	-	-	-	-	-	0.63	0.79	**0.83**	**0.83**
256	-	-	-	-	-	-	-	-	0.61	**0.78**	**0.78**	**0.82**
128	0.59	0.82	**0.93**	**0.94**	0.52	0.64	0.73	**0.74**	0.61	**0.81**	**0.79**	**0.81**
64	0.58	0.80	0.93	**0.94**	0.52	0.64	0.88	**0.94**	-	-	-	-

Table 6. Partition ratio by varying codebook size (total number of discrete features) on SHAPE, MNIST, and AFHQ datasets, for the DenseNet121 model.

scDatasets (→sc)	SHAPE				MNIST				AFHQ			
$\mathcal{Z}$ scSize (↓)	4	6	10	20	4	6	10	20	4	6	10	20
1024	-	-	-	-	-	-	-	-	0.53	0.52	**0.60**	**0.60**
512	-	-	-	-	-	-	-	-	0.53	**0.58**	0.56	0.56
256	-	-	-	-	-	-	-	-	0.50	**0.58**	**0.58**	**0.58**
128	0.59	**0.60**	**0.58**	**0.59**	0.40	0.47	**0.58**	**0.57**	0.53	**0.59**	**0.59**	**0.59**
64	0.58	0.58	**0.60**	**0.58**	0.38	0.46	**0.56**	**0.56**	-	-	-	-

the regions of interest into multiple sub-regions as contrasting arguments. The existing methods (Fig. 3(d)) basically only provide input-output explanations, disregarding the inner mechanism of the classifier and thus limiting the use of explanations for debugging or reasoning about the classifier. Instead, the arguments can help in better understanding the model. We demonstrate additional visual debates in the accompanying codebase.

5 Conclusion

Our main focus in this paper was on generating highly expressive post-hoc explanations for image classification, pointing towards classifiers' inner reasoning as captured by a surrogate (quantized) model. To achieve this goal, we re-thought explanations as visual debates between (fictional) players, whose contrastive nature forces them to unearth the role of (quantized) features towards and against the classifiers' outputs. Whereas standard feature attribution methods for explaining the outputs of image classifiers amount to block regions in images, our visual debates provide fine-grained compositions of sub-regions, with the (fictional) player playing devil's advocate against the classifiers' outputs pointing to classifiers' uncertainties. We implemented our general methodology onto a novel, practical debate framework for generating post-hoc explanations for image classification. Experimentally, with three datasets and different models, we showed that our visual debates satisfy desiderata of dialecticity and convergence, measured in terms

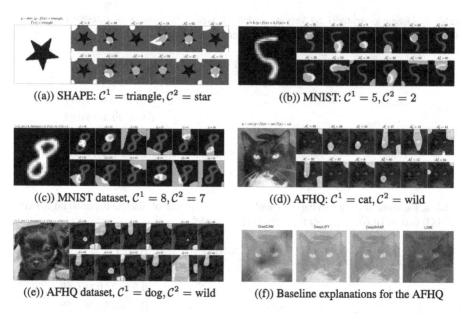

Fig. 3. Explanations obtained for the DenseNet121 model trained on the respective datasets, with (a), (b), (c), (d) and (e) illustrating visual debates of length 6 and (f) illustrating explanations obtained from GradCAM, DeepLIFT, DeepSHAP, and LIME from left to right, respectively. In (a), (b), (c), (d) and (e), $y, (g \circ f)(x)$ and $\Gamma(x)$ on top of the input image correspond to, respectively, the ground-truth label, the classifier's prediction and the debate outcome, whereas C^1 and C^2 amount to the players' respective final claims.

of new metrics of partition rati an consensus tailored to explanations as debates, and faithfulness (competitively with standard feature attribution methods).

Our work opens several avenues for future work. Our qualitative evaluations show that, even when the quantized features selected by the players are unique, arguments may overlap, indicating the need for better discretization. It would also be interesting to explore whether our method can help unearth shortcuts in the classifiers better than standard methods by virtue of their fine-grained nature. In addition, it would be useful to use our method with domain experts to assign semantic meaning to quantized features/arguments (as we have done in Fig. 1), possibly to support alignment among humans and models' latent knowledge. Further, given the tractability of our debate framework (finite action, finite player setting), visual debates can be effectively scaled (as demonstrated with the experiments on the more complex AFHQ dataset) – thus it would be interesting to apply our approach to settings where quantized representation learning is already explored, from natural images to medical data [12,31,36]. Finally, it would be interesting to explore the use of our visual debates to develop standalone, transparent, human-aligned neuro-symbolic models, in the spirit of [28], whereby debates are learnt for classification, rather than to explain as we do.

Acknowledgements. This research was partially supported by ERC under the EU's Horizon 2020 research and innovation programme (grant agreement No. 101020934), by J.P. Morgan and the Royal Academy of Engineering under the Research Chairs and Senior Research Fellowships scheme and by UKRI through the CDT in Safe and Trusted Artificial Intelligence.

References

1. Bau, D., Zhou, B., Khosla, A., Oliva, A., Torralba, A.: Network dissection: quantifying interpretability of deep visual representations. In: CVPR (2017)
2. Cayrol, C., Lagasquie-Schiex, M.: Graduality in argumentation. J. Artif. Intell. Res. **23**, 245–297 (2005). https://doi.org/10.1613/jair.1411
3. Chen, Z., Subhash, V., Havasi, M., Pan, W., Doshi-Velez, F.: What makes a good explanation?: A harmonized view of properties of explanations. In: Workshop on Trustworthy and Socially Responsible Machine Learning, NeurIPS 2022 (2022)
4. Choi, Y., Uh, Y., Yoo, J., Ha, J.W.: Stargan v2: diverse image synthesis for multiple domains. In: CVPR (2020)
5. Čyras, K., Letsios, D., Misener, R., Toni, F.: Argumentation for explainable scheduling. In: AAAI (2019)
6. Čyras, K., Rago, A., Albini, E., Baroni, P., Toni, F.: Argumentative XAI: a survey. In: IJCAI (2021)
7. Das, A., Kottur, S., Moura, J.M., Lee, S., Batra, D.: Learning cooperative visual dialog agents with deep reinforcement learning. In: ICCV (2017)
8. Deng, L.: The MNIST database of handwritten digit images for machine learning research [best of the web]. IEEE Signal Process. Mag. **29**(6), 141–142 (2012)
9. Donnelly, J., Barnett, A.J., Chen, C.: Deformable protopnet: an interpretable image classifier using deformable prototypes. In: IEEE/CVF Conference on Computer Vision and Pattern Recognition (CVPR 2022), pp. 10255–10265. IEEE (2022)
10. Doshi-Velez, F., Budish, R., Kortz, M.: The role of explanation in algorithmic trust. In: Technical Report, Artificial Intelligence and Interpretability Working Group (2017)
11. Dung, P.M.: On the acceptability of arguments and its fundamental role in nonmonotonic reasoning, logic programming and n-person games. Artif. Intell. **77**(2), 321–357 (1995)
12. Esser, P., Rombach, R., Ommer, B.: Taming transformers for high-resolution image synthesis. In: Proceedings of the IEEE/CVF Conference on Computer Vision And Pattern Recognition, pp. 12873–12883 (2021)
13. Ghandeharioun, A., Kim, B., Li, C.L., Jou, B., Eoff, B., Picard, R.W.: Dissect: Disentangled simultaneous explanations via concept traversals (2021). arXiv:2105.15164
14. Goyal, Y., Wu, Z., Ernst, J., Batra, D., Parikh, D., Lee, S.: Counterfactual visual explanations. In: ICML. PMLR (2019)
15. He, K., Zhang, X., Ren, S., Sun, J.: Deep residual learning for image recognition. In: CVPR, pp. 770–778 (2016)
16. Huang, G., Liu, Z., Van Der Maaten, L., Weinberger, K.Q.: Densely connected convolutional networks. In: CVPR, pp. 4700–4708 (2017)
17. Irving, G., Christiano, P., Amodei, D.: AI safety via debate (2018). arXiv:1805.00899
18. Kaggle: Shapes dataset (2016). https://www.kaggle.com/datasets/smeschke/four-shapes?resource=download
19. Kim, B., et al.: Interpretability beyond feature attribution: quantitative testing with concept activation vectors. In: ICML (2018)
20. Kori, A., Glocker, B., Toni, F.: GLANCE: Global to local architecture-neutral concept-based explanations (2022). arXiv:2207.01917

21. Kori, A., Natekar, P., Srinivasan, B., Krishnamurthi, G.: Interpreting deep neural networks for medical imaging using concept graphs. In: International Workshop on Health Intelligence (2021)
22. Lakkaraju, H., Slack, D., Chen, Y., Tan, C., Singh, S.: Rethinking explainability as a dialogue: A practitioner's perspective (2022). arXiv:2202.01875
23. Lundberg, S.M., Lee, S.I.: A unified approach to interpreting model predictions. In: NeurIPS (2017)
24. Miller, T.: Explanation in artificial intelligence: insights from the social sciences. Artif. Intell. **267**, 1–38 (2019)
25. Mnih, V., et al.: Recurrent models of visual attention. NeurIPS **27** (2014)
26. Nauta, M., et al.: From anecdotal evidence to quantitative evaluation methods: a systematic review on evaluating explainable AI. ACM Comput. Surv. **55**(13s), 1–42 (2023)
27. Peebles, W., Peebles, J., Zhu, J.Y., Efros, A., Torralba, A.: The hessian penalty: a weak prior for unsupervised disentanglement. In: ECCV (2020)
28. Proietti, M., Toni, F.: A roadmap for neuro-argumentative learning (2023)
29. Ribeiro, M.T., Singh, S., Guestrin, C.: Why should I trust you? Explaining the predictions of any classifier. In: 22nd ACM SIGKDD Int. Conf. on Knowledge Discovery and Data Mining (2016)
30. Santhirasekaram, A., et al.: Hierarchical symbolic reasoning in hyperbolic space for deep discriminative models. arXiv:2207.01916 (2022)
31. Santhirasekaram, A., Kori, A., Winkler, M., Rockall, A., Glocker, B.: Vector quantisation for robust segmentation. In: International Conference on Medical Image Computing and Computer-Assisted Intervention, pp. 663–672. Springer (2022)
32. Selvaraju, R.R., Cogswell, M., Das, A., Vedantam, R., Parikh, D., Batra, D.: Grad-cam: visual explanations from deep networks via gradient-based localization. In: ICCV (2017)
33. Shrikumar, A., Greenside, P., Kundaje, A.: Learning important features through propagating activation differences. In: ICML (2017)
34. Sokol, K., Flach, P.: Explainability fact sheets: a framework for systematic assessment of explainable approaches. In: Proceedings of the 2020 Conference on Fairness, Accountability, and Transparency, pp. 56–67 (2020)
35. Sukpanichnant, P., Rago, A., Lertvittayakumjorn, P., Toni, F.: Neural QBAFs: explaining neural networks under LRP-based argumentation frameworks. In: AIxIA 2021 - Advances in AI - 20th International Conference of the Italian Association for AI (2021)
36. Van Den Oord, A., et al.: Neural discrete representation learning. Adv. in NeurIPS **30** (2017)

Resource-Constrained Binary Image Classification

Sean Park(✉), Jörg Wicker, and Katharina Dost

University of Auckland, Auckland, New Zealand
spar610@aucklanduni.ac.nz, {j.wicker,katharina.dost}@auckland.ac.nz

Abstract. Deep convolutional neural networks (CNNs) have achieved state-of-the-art performance in image classification tasks by automatically learning discriminative features from raw pixel data. However, their success often relies on large labeled training datasets and substantial computational resources, which can be limiting in resource-constrained scenarios. This study explores alternative, lightweight approaches. In particular, we compare a lightweight CNN with a combination of randomly initialized convolutional layers with an ensemble of weak learners in a stacking framework for binary image classification. This method aims to leverage the feature extraction capabilities of convolutional layers while mitigating the need for large datasets and intensive computations. Extensive experiments on seven datasets show that under resource constraints, the decision as to which model to use is not straightforward and depends on a practitioner's prioritization of predictive performance vs. training and prediction time vs. memory requirements.

Keywords: Binary image classification · Resource-constraints

1 Introduction

Deep neural networks have become the dominant and most common approach for image classification tasks in recent years due to their ability to automatically learn feature representations from raw pixel data [10]. This means that they perform preprocessing and feature extraction automatically, a significant advantage over traditional methods that require manual feature engineering. Deep learning models, particularly those based on convolutional neural networks (CNNs) [10], have consistently outperformed traditional machine learning algorithms in various benchmarks and real-world applications. CNNs excel in capturing spatial hierarchies in images through layers of convolutions, pooling, and non-linear activations, leading to remarkable improvements in tasks such as object detection, facial recognition, and medical image analysis.

However, despite their impressive performance, deep learning models come with several drawbacks [1]. They usually require large labeled training datasets to achieve high accuracy. Additionally, these models are computationally expensive to train, often necessitating specialized hardware like GPUs to handle the

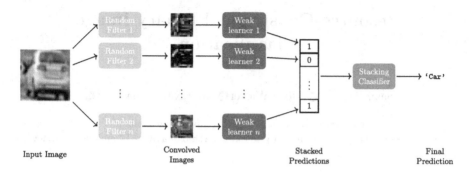

Fig. 1. Overview of `FilterStack`.

large number of parameters and the extensive computations involved in backpropagation. Both can be a significant limitation in domains where labeled data is scarce and computational resources are limited, such as in drones used for agricultural monitoring that classify crop health and identify weeds or pests; autonomous vehicles that classify road signs, pedestrians, and other vehicles in real-time; environmental monitoring devices used in wildlife conservation to count population or ensure health; or in factories to identify defective products.

In certain domains where training data is limited, deep learning may not be the optimal solution, despite recent efforts to reduce its computational burden [3,11]. Traditional machine learning models, such as support vector machines (SVMs), decision trees, and nearest neighbor methods, can potentially offer simpler and more practical solutions. These models are frequently less computationally intensive, typically requiring only a CPU for training and inference, and can be effective with smaller datasets. Traditional machine learning methods, however, do not perform as well on raw images, as our experiments confirm. To perform competitively, they require intensive feature engineering, where relevant features must be manually extracted from the raw image.

Given these considerations, the aim of this study is to explore non-deep learning approaches that could provide viable alternatives to deep neural networks for image classification tasks, particularly in resource-constrained scenarios. We propose `FilterStack`, a new approach combining convolutional filters used to train individual weak traditional learners with an ensemble based on their stacked predictions (see Fig. 1), that achieves similar performance to a lightweight CNN at lower training times. Our experiments on binary image classification show that not only `FilterStack` but also simple SVMs, random forests, and hybrid versions of both are strong competitors to CNNs and that the image classification landscape does not need to be as narrow as recent trends suggest.

Section 2 reviews related work. Section 3 introduces the models used in this study and proposes `FilterStack`. Section 4 describes our experimental setup whereas Sect. 5 discusses results. Finally, Sect. 6 concludes the paper.

2 Previous Work

Image classification is a fundamental problem in computer vision with numerous applications. Over the years, researchers have developed various techniques and algorithms to tackle this challenging task. In this section, we review some of the previous influential work in image classification.

Early approaches to image classification relied heavily on hand-crafted features and traditional machine learning models. A notable example is the Bag-of-Visual-Words (BoVW) model [4], which represents images as histograms of quantized local features. This approach involves several steps: detecting key points in an image, extracting local features around these key points (using methods such as SIFT or SURF), quantizing these features into a vocabulary of visual words, and finally representing the image as a histogram of these visual words.

Another popular feature descriptor is the Histogram of Oriented Gradients (HOG) [5]. HOG descriptors divide the image into small connected regions called cells and compute a histogram of gradient directions or edge orientations for each cell. The descriptor is then a concatenation of these histograms.

Convolutional Neural Networks (CNNs) [10] revolutionized the field of image classification. CNNs employ learnable kernels that act as automatic feature extractors, hierarchically capturing spatial hierarchies in images through layers of convolutions, pooling, and non-linear activations. CNNs leverage deep layers and large datasets to learn complex representations directly from raw pixel data, eliminating the need for manual feature engineering. Research has been dedicated to limiting CNN's resource usage [8,14] but not as vigorously as we do in our experimental setup. Pruning approaches limit the memory requirements, but not the training complexity [13].

Transfer learning is another powerful technique that has been widely adopted to boost the performance of image classification models, especially when dealing with limited data. There are three main approaches to transfer learning in the context of image classification:

Fine-tuning Pre-trained Models: This approach [12] involves using a pretrained neural network (trained on a large dataset like ImageNet) and retraining it on the target dataset by adjusting the final layers. Typically, the earlier layers of the network, which capture more general features, are kept frozen, while the later layers, which capture more specific features, are fine-tuned to the new task. This method leverages the feature extraction capabilities learned from the large dataset, thereby improving performance on the target task even with limited training data.

Feature Extraction with Traditional Machine Learning: In this approach, a pre-trained neural network is used solely as a feature extractor. The image is passed through the pre-trained network, and the output of one of the intermediate layers is taken as the input for a traditional machine learning model, e.g., a Random Forest [2].

Transfer Learning with Untrained Networks: Interestingly, research has shown that even randomly initialized neural networks can serve as effective feature extractors. Jarrett et al. [7] demonstrate that a randomly initialized CNN, without any prior training, can still produce useful feature representations that, when combined with traditional classifiers, can perform competitively in certain tasks. We employ this technique in our experiments to construct "hybrid" models. Further, it serves as an inspiration for the design of FilterStack.

Despite the remarkable progress, image classification remains an active area of research, with ongoing efforts to improve the accuracy, efficiency, robustness, and interpretability of the models, as well as to address domain-specific challenges and real-world applications.

3 Methodology

We are interested in exploring lightweight methods for binary image classification. In addition to CNNs that have demonstrated remarkable performance in the past, we also explore traditional machine learning algorithms that could potentially achieve comparable classification performance to CNNs, especially when dealing with limited training data. We compare (i) CNN with (ii) *Random Forest (RF)* and *Support Vector Machine (SVM)* directly applied on the images, (iii) RF and SVM applied on a CNN-based embedding (we call this a "hybrid" approach), and (iv) we propose a new neural network-free model, FilterStack. Inspired by CNN's convolutional layers, FilterStack randomly initializes filters, convolves the image, trains a weak learner for each of the filters, combines them in a stacking fashion, and trains an ensemble model for the final prediction. We subsequently discuss each of the four image classification methods.

3.1 CNN

We designed a simple CNN using the Keras framework, chosen for its lightweight architecture suitable for binary image classification with limited computational resources. The CNN accepts 32×32 pixel grayscale images. It starts with a Conv2D layer with 32 filters (3×3) using ReLU activation, followed by a 2×2 MaxPooling layer to reduce spatial dimensions and retain significant features. The next Conv2D layer has 64 filters (3×3), again followed by a 2×2 MaxPooling layer. An additional Conv2D layer with 64 filters (3×3) allows the network to learn more complex features. The output is then flattened into a 1D vector. A Dense layer with 64 neurons and ReLU activation processes the data, followed by an Output layer with 2 neurons for classification. Refer to Fig. 2 (left) for the CNN architecture diagram.

3.2 RF and SVM on the Raw Image

As a first competitor and baseline to our experiments, we include Random Forest and Support Vector Machines, trained on the raw images. We expect this to be problematic due to the high dimensionality of the data and since neither considers spatial relationships among pixels. Although they might derive higher-level features like edges or shapes, these methods are disadvantaged since we need to flatten the images pixel-by-pixel before training the models.

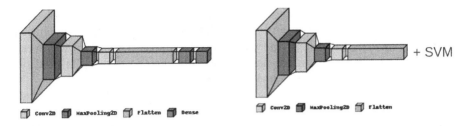

Fig. 2. Visualization of the simple CNN architecture (left) and the hybrid approach (right)

3.3 Hybrid Approach: RF and SVM on the Embedded Image

To overcome the challenges of training RF and SVM on the raw images, we include the hybrid approach suggested by Jarrett *et al.* [7]: To provide the models with higher-level features, we leverage feature maps produced by random filters. Specifically, we generate random filters, convolve the image, capturing small-scale shapes, and flatten the results of all convolutions into a vector. This high-dimensional vector then serves as the input to RF and SVM. Refer to Fig. 2 (right) for a diagram of the transfer learning approach.

3.4 FilterStack

To strengthen the hybrid approach and reduce the dimensionality (and thereby the training time), we propose FilterStack. We create random filters to convolve the image, similar to CNNs, to provide a learner with more context than individual pixels. Each convolved version of the image is then fed into an individual weak learner. The outputs of these learners are stacked, yielding the input for a second model predicting the overall image class. Figure 1 provides an overview of the approach.

In particular, the algorithm follows three stages: convolving the image with random filters, training the weak learners, and training the ensemble model on the stacked results.

Convolution With Random Filters. We begin by generating multiple filters randomly. The intuition behind using randomly initialized filters is to avoid optimizing the filters to reduce computational costs. The downstream stacked model will be able to select those filters that are meaningful. The random initialization of the kernels in the convolutional layer is based on the Kaiming He initialization [6], which is an effective technique that helps neural networks learn efficiently. For each filter, He initialization calculates a standard deviation based on the number of input and output channels in a convolution layer, then randomly sets the filter's weights using a normal distribution with a mean of 0 and the calculated standard deviation.

Each filter is a small matrix (e.g., 3×3 or 5×5) that slides across the input image, performing element-wise multiplication and summation at each position.

The filter acts as a feature detector, capturing specific patterns or features in the input image, such as edges, textures, or shapes. As in CNNs, the convolution operation produces a feature map, which is a 2D representation of the filtered image. Multiple filters can be applied to the input image, resulting in multiple feature maps that capture different types of features.

We apply the Rectified Linear Unit (ReLU) activation function to introduce non-linearity, which is crucial for learning complex patterns and representations in the data. Without a non-linear activation function, the convolutional layers would essentially be performing a series of linear operations, limiting their ability to model non-linear relationships in the data.

Max pooling is a downsampling operation we apply to the feature maps after the convolution and ReLU activation. It partitions the feature map into small rectangular regions (e.g., 2×2 or 3×3) and selects the maximum value from each region. This operation reduces the spatial dimensions of the feature map while retaining the most significant features. As a result, max pooling helps in making the model more robust to small shifts and distortions in the input image. It also reduces the computational complexity by decreasing the number of features for the subsequent models.

Weak Learner Training. The convolved feature maps obtained from the convolution operation serve as inputs to multiple weak learners. Each weak learner is trained independently on the output of a specific feature map. These weak learners, which can be simple classifiers such as tree-based algorithms or logistic regression, aim to capture different aspects of the image features. For the binary classification task, each weak learner outputs predictions as binary values (0 s and 1 s).

Stacking Ensemble. The predictions from all weak learners are aggregated to form a stacking dataset. In the context of binary classification, the stacking dataset consists of binary values (0 s and 1 s), representing the predictions of each weak learner for each instance. The stacking dataset encapsulates diverse feature representations and serves as the input for the stacking model. A classifier, such as SVM, is then trained on the stacking dataset to combine the output of the weak learners into a final prediction. This ensemble method leverages the strengths of each weak learner and enhances the overall classification performance.

4 Experimental Setup

To evaluate the proposed framework, we compare the performance of the models described in Sect. 3 on several image classification datasets. We evaluate the models in a 5-fold stratified cross-validation in terms of accuracy, precision, and recall. For the sake of comparability, all experiments are run on AMD Ryzen

5600u CPU with 16GB RAM on Linux 6.6.34-1-its kernel. For the sake of running time measurements, we restrict all methods to one single thread. Code and additional results can be found in our repository[1].

4.1 Model Setup

The CNN model is compiled with Adam as the optimizer for its efficiency and adaptive learning rate properties, which help in faster convergence. As a loss function, we use Sparse Categorical Cross Entropy because it is well-suited for multi-class classification problems. Additionally, it computes the loss between the true class labels and the predicted class probabilities, enabling the model to learn to assign higher probabilities to the correct class during training. We train the model for 20 epochs, ensuring adequate learning without overfitting. A batch size of 64 is selected to balance between memory usage and training efficiency.

The RF trained on the raw image data is set up with 100 trees. The corresponding SVM uses an RBF kernel and $C = 1$. We use the same RF and SVM configuration in the hybrid approach for the sake of comparability. As an embedding, we leverage the feature maps produced by the third convolutional layer of our CNN.

For FilterStack, 100 randomly initialized filters were generated. As weak learners, we use random forests with 10 trees and a maximum depth of 5. SVM is used as the stacking classifier. Additional details regarding these choices are provided in our repository.

4.2 Data

Since we are interested in lightweight models that are capable of learning patterns from small datasets, we include a number of diverse, and mostly small, image classification datasets, i.e.:

- Car vs Bike[2]: a collection of 4,000 color images in 2 classes, cars and bikes, with 2,000 images per class.
- CIFAR-10 [9]: 60,000 32 × 32 color images in 10 classes, with 6,000 images per class. The dataset is obtained from Keras' datasets module. From this, the first two classes were extracted (airplane and automobile), with 500 random samples of images per class.
- Pizza or Not Pizza?[3]: a collection of 1,966 color images in 2 classes, pizza and not_pizza, with 983 images per class.
- Healthy and Bleached Corals Image Classification[4]: A binary image classification dataset that contains 438 images of healthy corals and 485 images of bleached corals.

---

[1] github.com/NovemberDays/FilterStack.
[2] kaggle.com/datasets/utkarshsaxenadn/car-vs-bike-classification-dataset.
[3] kaggle.com/datasets/carlosrunner/pizza-not-pizza.
[4] kaggle.com/datasets/vencerlanz09/healthy-and-bleached-corals-image-classification.

- Egg Image Dataset[5]: A dataset consisting of images capturing various eggs in real-world environments (kitchen, farms, markets, and more) to detect whether an egg is damaged (Damaged) or not (Not Damaged), with 632 and 162 images, respectively.
- Chest X-Ray Images (Pneumonia)[6]: A chest X-ray images dataset that contains two classes, 1.583 images of NORMAL and 4,273 images PNEUMONIA classes.
- Covid-19 Image Dataset[7]: A dataset that consists of three classes from chest X-ray images, Covid, Normal, and Viral Pneumonia. Covid and Normal classes are extracted from the dataset for binary classification, which consists of 137 and 90 images respectively.

To reduce the memory requirements, all images are converted to greyscale, with pixel values ranging from 0 to 1. Furthermore, all images are resized to a uniform resolution of 32×32 pixels as in Yuan *et al.* [14].

5 Results and Discussion

In this section, we present the results of our experiments, including detailed performance metrics and comparisons across various models and datasets. We investigate several questions, ranging from a performance comparison between models, including their running time and performance on small datasets, to investigating characteristics of `FilterStack`.

5.1 Predictive Performance

First, we compare the predictive performance of all models in terms of test set accuracy, precision, and recall, averaged over the 5 folds. Table 1 displays the results. We observe that although the simple CNN shows the strongest performance overall, `FilterStack` is a strong competitor. Although slightly weaker, the two hybrid methods and the methods operating on the raw images are not far behind, achieving statistically equivalent accuracy on 4 out of 7 datasets.

Apart from performance comparisons among models, we also observe that the datasets we selected pose diverse levels of difficulty, particularly due to the greyscale pre-processing.

5.2 Training and Prediction Times

Table 2 provides insights into the computational efficiency of the competitors, averaged over all datasets. The times are measured per training/test set, not per individual example.

---

[5] kaggle.com/datasets/abdullahkhanuet22/eggs-images-classification-damaged-or-not.
[6] kaggle.com/datasets/paultimothymooney/chest-xray-pneumonia/data.
[7] kaggle.com/datasets/pranavraikokte/covid19-image-dataset.

Table 1. Accuracy, Precision, and Recall (average over cross-validation runs ± standard deviation) evaluated on the test sets. Bold is the best result and all others that are not significantly worse (based on pairwise t-test with 95% significance level)

Dataset		SimpleCNN	SVM	RF	Hybrid SVM	Hybrid RF	FilterStack
Car vs Bike		**.90±.03**	.82±.01	.82±.01	.87±.01	.86±.00	.86±.02
CIFAR-10		**.86±.03**	.81±.03	.82±.03	.81±.03	**.86±.02**	**.86±.02**
Pizza		**.73±.01**	.67±.02	.66±.01	.71±.02	.71±.02	.72±.01
Corals	Accuracy	.61±.03	**.66±.04**	.65±.05	.65±.02	.65±.03	.64±.02
Eggs		.82±.04	**.83±.01**	.83±.04	.80±.00	.82±.02	.81±.03
X-Ray		**.95±.01**	**.95±.01**	.94±.01	**.95±.00**	.94±.01	.94±.00
Covid-19		**1.00±.01**	.96±.03	.97±.02	.97±.03	.97±.02	.98±.02
Car vs Bike		**.90±.05**	.82±.01	.84±.01	.87±.01	**.88±.00**	.86±.01
CIFAR-10		**.87±.07**	.79±.04	.79±.03	.79±.03	**.84±.03**	.82±.04
Pizza		**.75±.04**	.70±.03	.66±.01	**.76±.02**	.73±.02	.71±.01
Corals	Precision	.62±.03	.67±.04	.66±.05	**.68±.01**	.67±.03	.65±.02
Eggs		**.89±.04**	.83±.01	.84±.02	.80±.00	.84±.01	**.89±.01**
X-Ray		.95±.01	**.96±.01**	.95±.01	.95±.00	.95±.01	.95±.01
Covid-19		**.99±.02**	.93±.07	.97±.04	.96±.05	.96±.01	.97±.03
Car vs Bike		**.92±.04**	.82±.02	.79±.02	.86±.02	.85±.01	.86±.02
CIFAR-10		.87±.10	.85±.04	.87±.04	.84±.04	.89±.03	**.91±.03**
Pizza		.70±.06	.61±.03	.64±.03	.63±.02	.66±.03	**.74±.02**
Corals	Recall	.69±.05	.69±.04	.69±.05	.61±.03	.68±.05	**.70±.04**
Eggs		.89±.06	.99±.01	.96±.02	**1.00±.00**	.97±.02	.87±.03
X-Ray		**.99±.01**	.97±.00	.97±.00	**.98±.00**	**.98±.00**	.96±.00
Covid-19		**1.00±.00**	.98±.03	.97±.03	.98±.03	.98±.03	.99±.02

The training times varied substantially among the models, with Hybrid SVM training the longest, followed by the simple CNN and Hybrid RF. This is due to two factors: First, these three methods convolve the image, requiring repeated passes over the entire image. Second, they subsequently train on all convolved images, which slows down the training due to the high dimensionality. Hybrid RF is less affected than Hybrid SVM due to its intrinsic feature sampling. In contrast, SVM and RF are comparatively fast. These models train directly on the pictures of the input image, circumventing both the convolution step and the high dimensionality. FilterStack lies in between both. It convolves the image but then reduces the dimensionality due to its filter-based models. These findings underscore the trade-offs between model complexity and computational efficiency, highlighting the importance of selecting appropriate models based on application-specific requirements.

In terms of test times, Hybrid SVM stands out with comparatively high prediction times. Random forest predicts the fastest. All other methods show similar results.

5.3 Model Space Requirements

In addition to training and test times, we provide memory requirements of the different models when serialized (using Python's `pickle` function) and written to file. The model sizes are listed in Table 3. For `FilterStack`, file sizes vary greatly, depending on the configuration of its weak and stacked learners. See Sect. 5.7 for a discussion. We add the possible range in brackets. We observe that when tailored towards low memory requirements, `FilterStack` outperforms its competitors at a small cost to its predictive performance. Although CNN requires comparatively little space when trained on a large dataset, its memory requirements are not competitive on small datasets.

Table 2. Training and prediction times (in s), averaged over all datasets and folds

	SimpleCNN	SVM	RF	Hybrid SVM	Hybrid RF	FilterStack
Training	20.22±17.38	0.70±0.93	2.53±2.23	43.72±50.65	11.82±14.04	7.15±6.20
Test	0.16±0.06	0.22±0.28	0.01±0.00	7.84±9.04	0.13±0.09	0.23±0.15

Table 3. Model size (in kB) when trained on a small and a large dataset. Sizes for `FilterStack` vary greatly depending on the configuration; see Sect. 5.7.

Dataset	SimpleCNN	SVM	RF	Hybrid SVM	Hybrid RF	FilterStack
Covid-19 (Small)	1,459	691	171	15,311	158	1,981 (189–1,981)
Car vs Bike (Large)	1,459	21,077	5,752	397,853	4,526	5,268 (481–6,242)

5.4 Performance on Small Datasets

To further analyze the performance of the two best-performing methods, simple CNN and `FilterStack`, we evaluate them across different train-test split ratios, focusing on how both models would perform when there is limited training data. Figure 3 illustrates the classification performance with varying test set proportions ranging from 5% to 95% (inclusive) in increments of 5%.

We observe a downward trend in test accuracies for both approaches as the test set size increases on most datasets. This is expected since the test set size increase implies a shrinking training set. Furthermore, we expect the simple CNN to be superior for lower test ratios (= larger training sets) and inferior for high

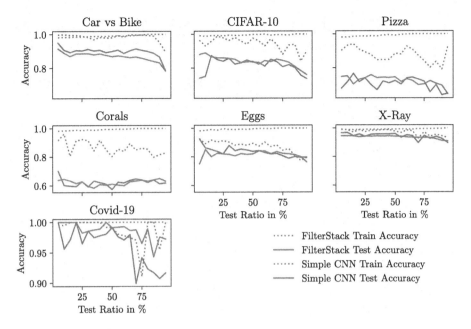

Fig. 3. Classification performance for one run of CNN and `FilterStack` on different training/test splits to simulate the behavior of the methods when trained on small datasets

test rations (= smaller training sets) due to its deeper architecture. However, this cannot be confirmed based on our results, except for the Eggs dataset. The results indicate that the simple CNN performs competitively even with fewer training samples, highlighting its ability to learn robust representations from limited data.

Interestingly, `FilterStack`'s training accuracy is higher than the simple CNN's for most test ratios, and it shows a slightly increasing trend for larger test ratios, whereas the simple CNN's train accuracy displays a downward trend. This observation suggests a tendency for `FilterStack` to overfit the training data, exacerbated by small training sizes. This is potentially due to the limited capacity of the weak learners to grasp the complex patterns of the convolved image or the stacking ensemble's inability to effectively generalize to unseen data.

5.5 Best-Performing Filters

To gain an understanding of `FilterStack`'s behavior and to find potential reasons for the previously observed findings, we investigate the top 10 best-performing filters that provided the highest test accuracy for each weak learner. Best-performing filters were identified using each weak learner's accuracy to its corresponding filter's convolved images. This allows us to assess how well-performing filters convolve images, specifically, what feature maps they output to the weak learners. For instance, Fig. 4 (top) shows that for the Car vs Bike

Fig. 4. A sample from the Car vs Bike dataset (top left) and the convolved images obtained through the top-10 best-performing filters (top right). Similarly, the bottom row shows one sample from the Pizza dataset (bottom left) and the convolved images obtained through the top-10 best-performing filters (bottom right).

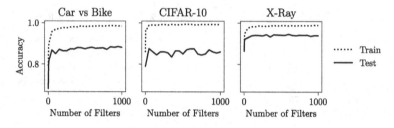

Fig. 5. Sensitivity of FilterStack to the number of filters

dataset, the filters effectively highlighted the distinguishing features of cars and bikes, which might have led to better classification performance. These filters, when applied to the images, primarily detected edges of objects or segmented objects against the background, leaving non-informative pixels as black. This behavior illustrates why convolutional filters are effective for image classification: they capture essential features, such as edges and shapes, which are critical for distinguishing between different classes.

In contrast, Fig. 4 (bottom) shows that for the Pizza dataset, the filters do not perform well in extracting relevant features. This poor performance could be due to the complexity and variability within the pizza images or the lack of color, making it challenging to find a good filter. The simple CNN, on the other hand, tunes the filters rather than guessing randomly, which creates a meaningful advantage over random filters.

5.6 Sensitivity to the Number of Filters

As discussed previously, the quality of filters differs. To investigate how many filters are required to capture the essential dataset patterns, we train FilterStack with varying numbers of filters and evaluate the performance of the resulting models. Figure 5 shows the results for three datasets; we omit the others since they show similar trends. We observe that the test performance gets saturated with less than 100 filters for all datasets. Further, the training curve, although on a higher level, shows a similar saturation trend, which implies that an elbow criterion on the training curve could be used to decide when to stop adding filters in practice. Future research will explore this.

5.7 Ablation Study

Overall, based on the previously presented results, `FilterStack` emerges as a strong contestant for low-resource applications. Here, we validate the design decisions of `FilterStack`.

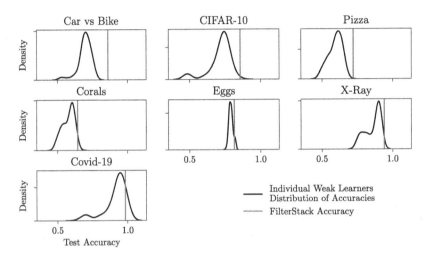

Fig. 6. Distribution of individual accuracies of the weak learners (KDE plot) compared with the accuracy of a model trained on all of the weak learners (`FilterStack`)

First, RF or SVM trained directly on the raw image leads to inferior performance, so capturing higher-level features of the image seems essential.

Second, we have investigated different image pre-processors than convolution in preliminary experiments (including BoVW using SIFT and k-means, HOG, Superpixels, and image compression), but convolution has proven to be substantially stronger.

Third, we have trained models directly on the convolved images (Hybrid RF/SVM), but their training times are higher at a lower performance overall.

Fourth, we can train individual models per filter and combine their predictions using, for example, a majority vote. Figure 6 shows the accuracy of the individual weak learners taking the convolved image from one of 1000 random filters as an input, compared to `FilterStack`'s accuracy. We observe that the majority of these individual models perform substantially worse than `FilterStack`, which justifies a more sophisticated ensemble technique than a majority vote.

Fifth, instead of using random forests as weak learners and SVM on top, we could replace both choices with alternative traditional machine learning methods, such as decision trees (DT) or logistic regression (LR). As can be seen from

Table 4, using RF as a weak learner generally leads to improved predictive performance. However, in some use-cases, trading off a small performance drop versus substantially smaller model sizes (e.g., for DT+LR) might be worthwhile. Note that the training time with random forest was significantly shorter compared to using a single decision tree. This efficiency is attributed to the fact that random forest employed only ten trees with feature selection and bootstrapping.

Table 4. Predictive performance of alternative versions of `FilterStack`, averaged over all datasets and folds, and size (in kB) of the trained model for a small (Covid-10) and a large (Car vs Bike) dataset

Weak Learner	Stacked Model	Accuracy	Precision	Recall	Model Size (S)	Model Size (L)	Training Time	Test Time
DT	LR	.82±.12	.83±.13	.86±.10	189	481	13.51±13.1	.14±.09
	RF	.82±.12	.83±.12	.85±.11	205	2435	13.36±13.2	.14±.10
	SVM	.82±.12	.83±.13	.85±.12	209	1338	13.40±13.2	.15±.11
RF	LR	.83±.12	.83±.12	.87±.11	1958	4471	7.03±6.1	.21±.12
	RF	.83±.11	.84±.12	.86±.11	1973	6242	7.09±6.2	.21±.13
	SVM	.83±.11	.84±.11	.86±.10	1981	5268	7.07±6.2	.23±.14

5.8 Limitations

While our analysis and `FilterStack` in particular demonstrate promising results, several limitations remain. `FilterStack` currently suffers from overfitting, as evident from the experimental results. Implementing regularization techniques and guidance on hyperparameter choices might be able to mitigate overfitting and enhance generalization. Developing a mechanism to identify and discard the worst-performing filters during training. This approach aims to maintain a lower computational cost compared to traditional neural networks while improving the overall performance by retaining only the most informative filters and potentially improving overfitting. A potential method could include a cosine similarity between flattened kernels could be employed to discard similar kernels. Exploring alternative kernel initialization methods beyond the He [6] initialization currently used could generate well-performing filters that help with classification.

6 Conclusion

CNNs have shown impressive performance in image classification tasks; however, they are known for high data requirements and resource-intensive training. In resource-constrained scenarios, this may not be an option. In this study, we explored lightweight alternatives and proposed a novel approach, `FilterStack`, that employs a stacking ensemble model consisting of randomly initialized convolutional filters and weak learners.

Our results show that while CNNs generally demonstrate reliably high predictive performance, there are slightly weaker alternatives that provide benefits in terms of training time, prediction time, or memory. `FilterStack` emerges as a strong option, but depending on the concrete memory, data, training and test time constraints of an application, other alternatives may be better suited.

Future research will be targeted at extending this study beyond binary grayscale image classification to both colored images and multi-class problems. Additionally, we will address `FilterStack`'s limitations mentioned above. In doing so, we will further refine the proposed method and enhance its applicability to various image classification tasks, both binary and multi-class.

Acknowledgements. KD is funded by the Ministry for the Environment, New Zealand.

References

1. Alzubaidi, L., et al.: Review of deep learning: concepts, CNN architectures, challenges, applications, future directions. J. Big Data **8**(1), 1–74 (2021). https://doi.org/10.1186/s40537-021-00444-8
2. Bansal, M., Kumar, M., Sachdeva, M., Mittal, A.: Transfer learning for image classification using vgg19: Caltech-101 image data set. J. Ambient Intell. Humanized Comput. 1–12 (2023)
3. Blok, M.P.M.: Resource constrained neural network training. Sci. Reports **14**(2421) (2024). https://doi.org/10.1038/s41598-024-52356-1
4. Csurka, G., Dance, C., Fan, L., Willamowski, J., Bray, C.: Visual categorization with bags of keypoints. In: Workshop on statistical learning in computer vision, ECCV. vol. 1, pp. 1–2. Prague (2004)
5. Dalal, N., Triggs, B.: Histograms of oriented gradients for human detection. In: 2005 IEEE Computer Society Conference on Computer Vision and Pattern Recognition (CVPR'05). vol. 1, pp. 886–893 (2005). https://doi.org/10.1109/CVPR.2005.177
6. He, K., Zhang, X., Ren, S., Sun, J.: Delving deep into rectifiers: surpassing human-level performance on imagenet classification. In: 2015 IEEE International Conference on Computer Vision (ICCV), pp. 1026–1034. IEEE Computer Society, Los Alamitos, CA, USA (2015). https://doi.org/10.1109/ICCV.2015.123
7. Jarrett, K., Kavukcuoglu, K., Ranzato, M., LeCun, Y.: What is the best multi-stage architecture for object recognition? In: 2009 IEEE 12th International Conference on Computer Vision, pp. 2146–2153 (2009). https://doi.org/10.1109/ICCV.2009.5459469
8. Khan, T.M., Robles-Kelly, A., Naqvi, S.S.: T-net: a resource-constrained tiny convolutional neural network for medical image segmentation. In: Proceedings of the IEEE/CVF Winter Conference on Applications of Computer Vision (WACV), pp. 644–653 (2022)
9. Krizhevsky, A.: Learning multiple layers of features from tiny images (2009). https://www.cs.toronto.edu/~kriz/learning-features-2009-TR.pdf
10. LeCun, Y., Bengio, Y., Hinton, G.: Deep learning. Nature **521**, 436–444 (2015). https://doi.org/10.1038/nature14539
11. Liu, H.I., et al.: Lightweight deep learning for resource-constrained environments: a survey. ACM Comput. Surv. (2024). https://doi.org/10.1145/3657282

12. Plested, J., Gedeon, T.: Deep transfer learning for image classification: a survey. ArXiv **abs/2205.09904** (2022)
13. Xu, S., Huang, A., Chen, L., Zhang, B.: Convolutional neural network pruning: a survey. In: 2020 39th Chinese Control Conference (CCC), pp. 7458–7463 (2020). htttps://api.semanticscholar.org/CorpusID:221591820
14. Yuan, H., Cheng, J., Wu, Y., Zeng, Z.: Low-res mobilenet: an efficient lightweight network for low-resolution image classification in resource-constrained scenarios. Multimedia Tools Appl. (2022). https://doi.org/10.1007/s11042-022-13157-8

Towards a Multimodal Framework for Remote Sensing Image Change Retrieval and Captioning

Roger Ferrod[1](✉)[iD], Luigi Di Caro[1][iD], and Dino Ienco[2,3][iD]

[1] University of Turin, Turin, Italy
{roger.ferrod,luigi.dicaro}@unito.it
[2] INRAE, UMR TETIS, University of Montpellier, Montpellier, France
dino.ienco@inrae.fr
[3] INRIA, University of Montpellier, Montpellier, France

Abstract. Recently, there has been increasing interest in multimodal applications that integrate text with other modalities, such as images, audio and video, to facilitate natural language interactions with multimodal AI systems. While applications involving standard modalities have been extensively explored, there is still a lack of investigation into specific data modalities such as remote sensing (RS) data. Despite the numerous potential applications of RS data, including environmental protection, disaster monitoring and land planning, available solutions are predominantly focused on specific tasks like classification, captioning and retrieval. These solutions often overlook the unique characteristics of RS data, such as its capability to systematically provide information on the same geographical areas over time. This ability enables continuous monitoring of changes in the underlying landscape.

To address this gap, we propose a novel foundation model for bi-temporal RS image pairs, in the context of change detection analysis, leveraging Contrastive Learning and the LEVIR-CC dataset for both captioning and text-image retrieval. By jointly training a contrastive encoder and captioning decoder, our model add text-image retrieval capabilities, in the context of bi-temporal change detection, while maintaining captioning performances that are comparable to the state of the art. We release the source code and pretrained weights at: https://github.com/rogerferrod/RSICRC.

Keywords: Remote Sensing · bi-temporal change detection · image captioning · text-image retrieval · contrastive learning

1 Introduction

Modern Earth observation systems allow acquiring systematic information, under the shape of satellite imagery, to monitor and characterize the evolution of the underlying Earth surface. Among all the applications, the possibility to detect and characterized particular changes in the land surfaces is of paramount importance in a variety of applications such as environmental protection, disaster monitoring and land planning [16].

More precisely, the value of such information comes from the detailed comparison between bi-temporal remote sensing imagery and all the related features. The majority of current change detection approaches (e.g., [2,13]) exhibit a limited level of interaction from a user perspective. Given a pair of images, they mainly highlight the spatial areas affected by some change phenomena, without any additional information or any strategies to guide the process via possible user query. To express its full potential and allow an appropriate interaction with users, features related to changed areas must be effectively described and searchable, making it possible to interact with the system in natural language. For these reasons, there is a strong interest in the community to develop models that go beyond simple Change Detection (CD) strategies and try instead to accurately describe bi-temporal changes occurring or retrieve a pair of images associated to a given textual prompt.

Those aims can be achieved through multimodal foundation models (e.g., [14,20,24,29]) that, once pretrained on a large-scale dataset, can be used in many down-stream tasks with remarkable performance. Foundation vision-language models, led by CLIP [17], have already demonstrated excellent capabilities in remote sensing applications, too ([7,14,30]). However, despite great success in other domains, Vision-Language models for Remote Sensing applications still suffer from the scarcity of large-scale datasets. In fact, while there are plenty of works focused on specific tasks – such as classification [1], captioning [2] or retrieval [31] – foundational models are still at their infancy stage in the RS domain.

Motivated by the lack of Vision-Language models for Remote Sensing applications that manage bi-temporal change detection information, here, we propose a foundation model specifically designed for pairs of bi-temporal remote sensing images. To the best of our knowledge, it is the first attempt to fill this gap in the literature. Given the shortage of resources focused on changes in RS imagery, we propose to adopt a remote sensing image change captioning dataset (LEVIR-CC) to assess the potential of our framework considering both bi-temporal captioning and bi-temporal text-image retrieval tasks. By jointly training a contrastive encoder and captioning decoder, we provide a single model that, simultaneously, allows bi-temporal text-image retrieval capabilities, preserving captioning abilities comparable to the state of the art.

2 Related Works

Our work follows a line of research started recently by [13] who introduced the task and the dataset we used. The dataset, described more in details in Sect. 3.1, is accompanied by a model named RSICCformer. Another model (Chg2Cap) was then proposed by [2] the following year, obtaining state-of-the-art results on the same benchmark. Both models are limited to the captioning task. Therefore, their efforts are focused on building a solid encoder attached to a simple decoder and train them with a captioning loss. Similar solutions, not limited to the RS domain, include [4–6,21,22,27].

With foundation models, instead, a single pretrained model can be used for many downstream tasks, with only minor fine-tuning and supervision efforts. Such techniques are now getting more and more attention also in the (RS) domain, with recent advances such as RemoteCLIP [14], EarthGPT [30] and RSGPT [7]. However, such initiatives are still focused on the analysis of static information and do not take into account the temporal dimension, i.e., the evolution or changes that might occur between two or more remote sensing image acquisitions.

With the aim to combine two different paradigms (retrieval and captioning) into a single model, though, some architectural modifications are required. For this purposed, we based our framework on CoCa [28]. In CoCa the authors propose a strategy to train a multimodal foundation model that can perform various tasks, such as captioning, text-image and image-text retrieval and visual recognition. Such work, however, is limited to static, natural images, and it cannot be easily extended to cope with bi-temporal satellite imagery.

With regard to the retrieval task, Vision-Language contrastive models like CLIP [17] and derivatives, such as ALBEF [11] and ALIGN [9] can be adopted. These multimodal AI models have already contributed to ameliorate state-of-the-art results in many downstream task. Recently, extensions of these models for the RS domain have been proposed, like [18] and RemoteCLIP [14], achieving state-of-the-art performance in RS image retrieval task but, unfortunately, such models are still focused on the analysis of static remote sensing imagery, and they cannot be employed for the analysis of bi-temporal remote sensing imagery for downstream change detection applications.

3 Method

Building on top of state-of-the-art approaches, we propose a new model to address the challenges posed by retrieving a pair of images related to temporal changes in remote sensing images via natural language prompts. More precisely, our goal is to provide a model to cope with both captioning and text-image retrieval tasks at once, for the case of remote sensing data. To achieve this objective, we integrate a SOTA image-encoder specifically designed for image change captioning with a new decoder that enables a joint training of the two objectives: an autoregressive captioning loss and a contrastive loss.

Inspired by CoCa [28] we split the decoder in two parts: an unimodal module that encodes the textual input only and a multimodal module with cross attention that combines textual and visual embeddings. In both cases, the decoder prohibits tokens from attending to future tokens in the sequence. More details on the decoder are provided in Sect. 3.3.

Differently from CoCa, thought, our model needs to deal with a pair of images, not a single one. Having experimented with RSICCformer [13] and

Chg2Cap [2] we have decided to adopt Chg2Cap's encoder, with only minor revisions, given its excellent behavior exhibited on the captioning task. The encoder is responsible for encoding the images individually, through a pretrained model, and then combining the features into a single representation. The details related to the encoder are supplied in Sect. 3.2. The overall architecture is depicted in Fig. 1.

To test our model, we relied on the LEVIR-CC dataset introduced in [13], which was the first research work to address the Remote Sensing Image Change Captioning (RSICC) task introducing the RSICCformer architecture. Since no datasets are available, today, for remote sensing image retrieval with pairs of images (before change/after change), we exploited the LEVIR-CC dataset both for captioning and retrieval, although this choice required some special precautions discussed in Sect. 3.1.

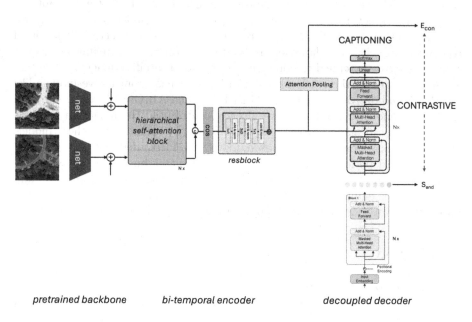

Fig. 1. The overall architecture of our model; once the pair of images is encoded through two siamese pretrained models, the information is processed by a bi-temporal encoder that merges the two representations, then a single embedding can be retrieved through attentive pooling and contrastively compared with the corresponding textual embedding or used directly as input for the cross-modal decoder; the decoder is splited in two parts: unimodal layers that only encode the textual representation and multimodal layers that generate the captions.

3.1 Dataset

The LEVIR-CC dataset consists of 10,077 image pairs and 5 captions each describing the changes happening between the two images cover the same area

(Fig. 2). Half of the pairs do not have any significant change, while the other half portrays changes such as new building or road being built. Often the description includes detailed spatial references (e.g., "on the top-left" or "at the bottom of the scene") that identify the changes. Although the dataset is vast and has over 50k sentences, the possible changes that could occur are limited by the fact that the acquired images are taken almost in the same geographical zone (Texas, USA), within a similar urban context. Therefore, while it is naturally possible to exploit the data for the captioning task, adapting the dataset for image retrieval is more challenging. Given a textual query, indeed, more images that match the description could be retrieved. Within a contrastive learning framework, those examples are considered false negatives and could make the training stage challenging.

Fig. 2. Examples of items taken from the LEVIR-CC dataset, where each image pair (before/after) is accompanied by 5 human annotated captions (only one is shown here).

3.2 Encoder

The encoder follows the same architecture proposed by [2], with a pretrained siamese network to encode the image pair, a hierarchical attention mechanism to combine the features and a residual block with cosine mask. Whereas the original Chg2Cap model was proposed with a resnet-101 backbone – and tested on other models pretrained on ImageNet – we included in our experiments networks especially pretrained on remote sensing images. If our model is set to fine-tune the encoder's backbone, only the weights from the last two convolutional layers of the ResNet architecture, or the transformer layers of ViT, are updated.

After a linear projection head used to bring the feature dimension to the desired size, we continue as in the original Chg2Cap model, in particular once we have an embedding representation F_i for each image X_i:

$$F_1 = net(X_1)$$
$$F_2 = net(X_2) \qquad (1)$$

with $F_i \in \mathbb{R}^{h \times w \times D}$ and D the feature dimension, we add learnable positional embeddings:

$$F_i = F_i + F_{pos} \qquad (2)$$

with $F_{pos} \in \mathbb{R}^{h \times w \times D}$, and pass the extracted features to a hierarchical self-attention block that will apply the attention mechanism across the two images:

$$I_i, I_2 = HSA(F_1, F_2) \qquad (3)$$

where $I_i \in \mathbb{R}^{h \times w \times D}$. When fusing the two representations, through a concatenation over the hidden dimension, a cosine mask is applied as follows:

$$F_{fus} = [I_1; I_2] + Cos(I_i, I_2) \qquad (4)$$

where $F_{fus} \in \mathbb{R}^{h \times w \times 2D}$, $[;]$ is the concatenation operation and $Cos(\cdot, \cdot)$ the cosine similarity between the two tensors. After applying a 2D convolution layer with 1×1 kernel size and dimensionality reduction, F_{fus} is processed by a residual block (1×1 conv, 3×3 conv, 1×1 conv) obtaining a final encoding (E_{cap}) for the image pair:

$$C = conv_{1 \times 1}(F_{fus})$$
$$E_{cap} = ReLU(ResBlock(C) + C) \qquad (5)$$

where $C \in \mathbb{R}^{h \times w \times D}$ and $E_{cap} \in \mathbb{R}^{h \times w \times D}$.

E_{cap} can then be used directly as input for the decoder cross-attention layer to generate a caption. For the contrastive loss used for the text-image retrieval task, instead, a single embedding vector representing the pair of images is derived. To this aim, we added a single query multi-head attention layer as an attention pooling operation:

$$E_{con} = MHA(E_{cap}) \qquad (6)$$

with $E_{con} \in \mathbb{R}^D$

3.3 Decoder

The transformer-based decoder requires some modifications to both captioning and contrastive learning. Following CoCa [28], we split the decoder in two parts: the first one will encode the text and can be used directly by the contrastive learning loss, while the second part will combine the textual encoding with visual embeddings and produce the caption. Both apply casually-masked attention to prevent the current token to attend to future tokens, but only the second part will apply cross-attention, combining the textual and visual representations. More formally: input tokens are mapped into a word embedding using an embedding layer and added to positional embeddings:

$$E = emb + pos \tag{7}$$

where $emb : \mathbb{R}^n \to \mathbb{R}^{n \times D}$, n the number of input tokens, D the embedding dimension and pos the sinusoidal positional encoding. Then the unimodal decoder layers follow, which will produce two results: the self-attention outputs with shape $n \times D$ and a single representation of the sequence taken from the last token:

$$\begin{aligned} S_{seq} &= MHA(E) \\ S_{end} &= S_{seq}[n] \end{aligned} \tag{8}$$

where $S_{seq} \in \mathbb{R}^{n \times D}$ and $S_{end} \in \mathbb{R}^D$. While S_{end} is used directly by the contrastive loss, S_{seq} will pass through other decoder layers, this time with cross-attention:

$$E_{cross} = MHA(S_{seq}, E_{cap}) \tag{9}$$

where $E_{cross} in \mathbb{R}^{n \times D}$, S_{seq} the textual sequence embeddings with shape $n \times D$ and E_{cap} the image embeddings with shape $h \times w \times D$. Finally, a linear layer (LN) is used for tokens prediction:

$$C = LN(E_{cross}) \tag{10}$$

with $LN \in \mathbb{R}^{n \times |V|}$ where V is the vocabulary.

Differently from the encoder, which in part exploits pretrained models, the decoder in trained from scratch. The particularity lies in the vocabulary used, that is derived from the dataset itself. Both [2,13], indeed, compose the vocabulary with the words appearing at least 5 times in the dataset, resulting in 463 tokens. Any attempt to go beyond these limits adopting a pretrained decoder with thousands of tokens will drop the performance, making the results not comparable. Therefore, we have decided to follow the same strategy adopted by previous works. We also add the possibility to tie embedding weights with the decoder, resulting in shared parameters.

3.4 Training

As already mentioned, this work aims to combine captioning capabilities with text-image retrieval using only a single model. We achieve this with a modified architecture which allows the two paradigms to coexist.

Possible Issues Related to False Negatives. However, moving from a single image to a pair of images (before and after a change occurred) is not straightforward and requires a mechanism to tackle false negatives, i.e., examples identified as negative by the contrastive loss, which in fact should be considered as positive. This problem arises from using a dataset originally intended for captioning only.

The phenomenon is common in many other tasks in which contrastive learning is employed, especially in a self-supervised setting, and solutions exist to cope with this particular issue [3,8,26]. Following the common literature, we implemented False Negative Elimination (FNE) and False Negative Attraction (FNA) strategies. With FNE, the detected false negative is simply removed from the loss computation, while the FNA strategy considers the false negative as a positive example and changes its label accordingly. In our work, the false negative detection phase is done by comparing the captions similarities. If two captions belonging to different image pairs (therefore considered negative examples) have a cosine similarity higher than a given threshold (θ), the FNE or FNA mechanisms are activated. As shown in Fig. 3, two different captions with similar semantics can then be considered as the same caption (FNA) or excluded from the same batch (FNE).

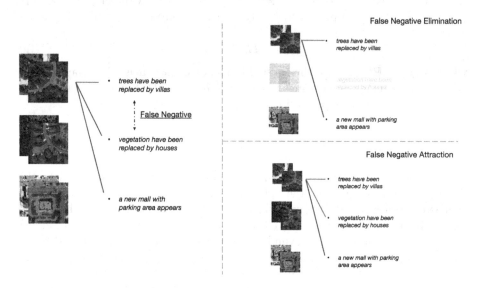

Fig. 3. By performing contrastive learning, an anchor (image pair) is compared with the captions inside the batch, the corresponding textual description is considered a positive example, the others negative. If a False Negative is detected (caption similarities higher than θ), one can exclude it from the loss computation (False Negative Elimination) or consider it as positive (False Negative Attraction).

Since the textual decoder is trained from scratch, for the similarity computation we rely on a third-party model specifically designed for sentence comparison. We pre-compute the embeddings with Sentence-BERT [19] and then the cosine similarity before applying the contrastive loss.

Contrastive Learning. To combine the visual representation and textual embeddings, we rely on the InfoNCE loss, a popular contrastive learning loss.

Visual and textual encoders are jointly optimized by contrasting pairs of images and captions inside the batch:

$$\mathcal{L}_{con} = -\frac{1}{N}(\sum_i^N log\frac{exp(e_i \cdot s_i/\tau)}{\sum_j^N exp(e_i \cdot s_j/\tau)} + \sum_i^N log\frac{exp(s_i \cdot e_i/\tau)}{\sum_j^N exp(s_i \cdot e_j/\tau)}) \quad (11)$$

where N is the batch size, τ the temperature and e_i and s_j the (normalized) visual and textual embeddings, respectively.

More in detail, the batch is composed by N triples $<e_i, s_i, l_i>$, where l_i is the label of the $(e,s)_i$ pair. We then compute all the possible combinations of contrastive pairs and assign a positive or negative label accordingly to l_i, i.e., when s_i is the corresponding caption for the e_i image pair. At this point, we have to correct the labels by detecting and removing false negatives. We take the pre-computed normalized Sentence-BERT embeddings (t_i) and compute a $N \times N$ similarity matrix in which we identify the combination of captions – within the batch – that have a relative similarity higher than a given threshold θ. Formally, positive and negative examples are defined as:

$$(e_i, s_j) = \begin{cases} (e_i, s_j)_+ & \text{if } l_i = l_j \\ (e_i, s_j)_- & \text{if } l_i \neq l_j \wedge t_i \cdot t_j < \theta \end{cases} \quad (12)$$

when using False Negative Elimination (FNE), or:

$$(e_i, s_j) = \begin{cases} (e_i, s_j)_+ & \text{if } l_i = l_j \vee t_i \cdot t_j \geq \theta \\ (e_i, s_j)_- & \text{if } l_i \neq l_j \wedge t_i \cdot t_j < \theta \end{cases} \quad (13)$$

when using False Negative Attraction (FNA).

Captioning. While the contrastive loss compares the whole representation of the image pair with an analogous representation of the caption, the captioning task acts at a more fine-grained scale. Indeed, the cross-attention of the decoder considers the entire sequence of tokens and the spatial features from the images. Given this information, the decoder will predict the sequence of tokens autoregressively:

$$\mathcal{L}_{cap} = -\sum_{i=1}^n log P(y_i|y_{<i}, e) \quad (14)$$

where n in the sequence length and P the probability of generating the y_i token given the previous words and the image encoding (e).

Training Objective. Finally, the model is jointly trained by combining both $\mathcal{L}_{con}$ and $\mathcal{L}_{cap}$ and weighting the relative contribution in the main loss:

$$\mathcal{L} = \mathcal{L}_{cap} + \lambda \cdot \mathcal{L}_{con} \quad (15)$$

4 Experiments and Results

We trained our model on the LEVIR-CC dataset and compared its results with other remote sensing models. Of those, only [2,13] deal with a pair of images as input, yet they are focus only on the captioning task. Other remote sensing models, on the other hand, are trained for text-image retrieval and/or captioning, but only on single images.

4.1 Setup

Training and evaluation are performed on a single NVIDIA A40 GPU with 48 GB of memory. We evaluate the model after each epoch, and we select the model that minimizes the loss. The maximum number of epochs is set to 50 and the batch size to 32. The AdamW optimizer is used, with $\epsilon = 10^{-8}$ and linear warm up reaching a target learning rate of 10^{-4}. The InfoNCE temperature τ is set to 0.01.

4.2 Experiments

We compared our results with the state of the art and other pretrained remote sensing models. For those models that were only trained for single-image ranking ([9,11,14]) we encoded the image pair as the difference between RGB values, specifically: $E = |X_{after} - X_{before}|$.

We also add the possibility of including hard negatives examples in the batches. A hard negative pair is an example in which the image and textual encodings have a high similarity despite being negative examples, i.e., the text is not the intended caption for that specific image. Therefore, at each epoch we compute and update an index, sampling the negatives examples globally (i.e., across the entire dataset). For this purpose, we rely on the FAISS library [10] and cosine similarity.

We evaluated the captions with standard text quality metrics like BLEU score [15] and ROUGE-L [12] as well as CIDEr [23] for image description evaluation. The same metrics were used in SOTA models like [2,13]. For the text-image retrieval task, instead, we relied on Precision (P), Recall (R) and Mean Reciprocal Rank (MRR) over *top-k* results. Specifically, the metrics are defined as follows:

$$P@k = \frac{relevant\ items\ in\ top\text{-}k\ results}{k} \quad (16)$$

$$R@k = \frac{relevant\ items\ in\ top\text{-}k\ results}{relevant\ items} \quad (17)$$

$$MRR@k = \frac{1}{Q} \sum_{i=1}^{Q} \frac{1}{rank_i} \quad (18)$$

where Q is the number of queries, $rank_i$ the position of the first relevant item in the *top-k* results for the *i-th* query. For each metric, the relevant items are computed by taking into account the θ threshold used for the false negative detection phase during the training. Therefore, if two examples are merged during the training because their similarity is higher than θ, the same will happen during the evaluation phase.

4.3 Results

When evaluating the results, it is important to consider the uneven distribution of the dataset. Since half of the dataset consists of examples without changes (approximately 5,000 items), a *top-5* search of an unchanged scene would yield a low Recall score (0.1%), even if all five matches are relevant.

At the same time, since for each caption we can have only one corresponding image in the original dataset – or slightly more if we consider the aggregation results induced by the use of the θ threshold – for most of the queries the Precision will be near $1/k$, or 20% if $k = 5$.

In Table 1 we compare the baselines with our framework, without contrastive learning, with False Negative Elimination and with False Negative Attraction. As highlighted by the obtained results, the contrastive learning strategy allows our model to outperform the baseline approaches. At the same time, we maintained comparable performances on the captioning task, despite having a smaller backbone than the one originally used in Chg2Cap (ResNet-50 instead of ResNet-101). It is also interesting to note that the contrastive learning loss had a positive influence also on the captioning task, with small advantages compared to the baseline without contrastive learning. Among the two strategies, False Negative Attraction came out as the best and confirms the finding already reported in the literature [8].

Table 1. Text-image retrieval and captioning results of SOTA captioning system (Chg2Cap), Retrieval baselines (ALIGN, Remote-CLIP and ALBERF) and our models (without contrastive learning, with False Negative Elimination or False Negative Attraction); the best results are highlighted in bold, the second best are underlined.

P@5	R@5	MRR@5	BLEU-1	BLEU-4	ROUGE-L	CIDEr	model
-	-	-	**82.84**	**60.92**	**72.72**	**130.97**	Chg2Cap
21.33	0.61	27.59	-	-	-	-	ALIGN
41.95	0.71	<u>52.32</u>	-	-	-	-	Remote-CLIP RN50
40.74	0.28	51.85	-	-	-	-	ALBEF
30.1	0.52	16.87	79.55	55.76	62.15	117.14	w/o contrastive learning
<u>41.6</u>	<u>1.2</u>	52.15	69.38	37.92	55.21	106.61	FNE
52.32	**2.85**	**53.51**	<u>82.34</u>	<u>59.04</u>	<u>62.99</u>	<u>120.14</u>	FNA

Before choosing the proper threshold value for the false negative detection task, we manually annotated 150 captions randomly sampled from the test set, by merging textual description with the same meaning. Based on these annotations, we can then evaluate the quality of different threshold values, as reported in Table 2 (a). In particular, with θ above 0.96 (in terms of cosine similarity) the false negative detection is perfectly aligned with human judgment, before dropping sharply with lower similarity value. From the Table 2 (b), instead, we can appreciate the difference between False Negative Elimination and False Negative Attraction compared with a simpler training method that does not apply a false negative detection strategy. We confirm that the False Negative Attraction strategy exhibits the best behavior, when the similarity threshold is set to 1.0. With this setting, we can achieve better ranking scores and comparable captioning performances compared to the baseline approaches.

Table 2. a) evaluation of the threshold over 150 manually annotated samples and b) impact on the model.

ACC	F1	P	R	θ
99.90	54.55	37.50	100	0.92
99.92	60.00	42.86	100	0.94
99.98	85.71	75.00	100	0.95
100	100	100	100	0.96
100	100	100	100	1.0

(a)

R@5	BLEU-1	BLEU-4	model
2.23	**82.64**	**59.76**	w/o FN detection
1.93	77.96	54.27	FNE $\theta = 0.96$
2.58	80.95	57.73	FNA $\theta = 0.96$
1.2	69.38	37.92	FNE $\theta = 1.0$
2.85	82.34	59.04	FNA $\theta = 1.0$

(b)

Although the main difference of our framework compared to previous approaches is the decoder component, we also propose to replace the original backbone used in Chg2Cap (a ResNet-101 pretrained on ImageNet) with a Remote Sensing pretrained model. Given the good results obtained by Remote-CLIP (Table 1), we have chosen to adopt its encoder as backbone for our framework. Despite the fact that it has fewer parameters than ResNet-101, the pre-training phase on remote sensing images proved to be beneficial, and the model was able to outperform the original encoder pre-trained on ImageNet. In Table 3 we compare the fine-tuned ResNet-50 backbone we adopted, with a non fine-tuned version (i.e., with frozen weights) and the same architecture but trained on ImageNet. Albeit a frozen version of the Remote-CLIP backbone already permits to achieve more than reasonable results on the ranking task, we have observed that fine-tuning even ameliorate the results, as briefly mentioned in Sect. 3.2. To this end, we only fine-tune the last two layers.

Concerning the importance of the (global) hard negatives sampling mechanism, no clear evidence is available. Although this solution represents the best practice in Contrastive Learning [25], it proved to be counterproductive in our case, as shown in Table 4. Since we rely on self-supervised labels for the ranking task rather than high-quality labeled data, the hard negative mechanism might nullify the effect of false negative detection, thereby hindering the final results.

Table 3. Comparison between different backbones: RN50 from RemoteCLIP finetuned/non finetuned and the same architecture pretrained on ImageNet.

R@5	BLEU-1	BLEU-4	model
1.83	73.95	45.38	ImageNet RN50 finetune
2.9	78.47	52.47	RemoteCLIP RN50 no finetune
2.85	**82.34**	**59.04**	RemoteCLIP RN50 finetune

Table 4. Impact of the hard negative mechanism on performance.

R@5	BLEU-1	BLEU-4	model
2.85	**82.34**	**59.04**	w/o hard negatives
0.98	72.69	43.57	w/ hard negatives

5 Conclusion and Future Works

In this work, we present a new multimodal foundation model designed for bi-temporal remote sensing images, with captioning and text-image retrieval capabilities. Building on top of state-of-the-art solutions, we adapt pretrained visual encoders and a novel multitask decoder with the aim to combining two paradigms into a single model. Given the absence of benchmarks for RS retrieval of bi-temporal images, we propose to exploit a change detection captioning dataset (LEVIR-CC) for both tasks, with some precautions that aim to mitigate the problem of false negatives under contrastive learning training. Although there is still room for improvement in the absolute metric results for the retrieval task, our model maintains captioning performance comparable to the state of the art, with the added benefit of also providing a solution for the text-image retrieval task.

We believe that natural language prompting is crucial to facilitate the exploration of Remote Sensing image archives by non-expert end-users. Our work is a step towards this direction, but there is still work to be done in order to achieve fully reliable solutions. For instance, efforts must be done to create and curate a benchmark for bi-temporal image retrieval, which can be shared with the community to stimulate research activities in this important area. Another possible future research can be devoted to the design and development of generic foundational visual language remote sensing models capable of handling multiple tasks simultaneously, rather than relying on single-task models that require systematic adaptation for each downstream task.

Acknowledgements. This work was partially supported by the French National Research Agency under the grant ANR-23-IAS1-0002 (ANR GEO ReSeT).

References

1. Adegun, A.A., Viriri, S., Tapamo, J.R.: Review of deep learning methods for remote sensing satellite images classification: experimental survey and comparative analysis. J. Big Data **10**, 1–24 (2023)
2. Chang, S., Ghamisi, P.: Changes to captions: an attentive network for remote sensing change captioning. IEEE Trans. Image Process **32**, 6047–6060 (2023)
3. Chen, T.S., Hung, W.C., Tseng, H.Y., Chien, S.Y., Yang, M.H.: Incremental false negative detection for contrastive learning. In: International Conference on Learning Representations (2022)
4. Chouaf, S., Hoxha, G., Smara, Y., Melgani, F.: Captioning changes in bi-temporal remote sensing images. In: 2021 IEEE International Geoscience and Remote Sensing Symposium IGARSS, pp. 2891–2894 (2021)
5. Guo, Z., Wang, T., Laaksonen, J.T.: CLIP4IDC: clip for image difference captioning. ArXiv abs/2206.00629 (2022)
6. Hoxha, G., Chouaf, S., Melgani, F., Smara, Y.: Change captioning: a new paradigm for multitemporal remote sensing image analysis. IEEE Trans. Geosci. Remote Sens. **60**, 1–14 (2022)
7. Hu, Y., Yuan, J., Wen, C., Lu, X., Li, X.: RSGPT: a remote sensing vision language model and benchmark (2023)
8. Huynh, T., Kornblith, S., Walter, M.R., Maire, M., Khademi, M.: Boosting contrastive self-supervised learning with false negative cancellation. In: 2022 IEEE/CVF Winter Conference on Applications of Computer Vision (WACV), pp. 986–996 (2022)
9. Jia, C., et al.: Scaling up visual and vision-language representation learning with noisy text supervision. In: Meila, M., Zhang, T. (eds.) Proceedings of the 38th International Conference on Machine Learning. Proceedings of Machine Learning Research, vol. 139, pp. 4904–4916. PMLR (18–24 Jul 2021) (2021)
10. Johnson, J., Douze, M., Jégou, H.: Billion-scale similarity search with GPUs. IEEE Trans. Big Data **7**(3), 535–547 (2019)
11. Li, J., Selvaraju, R.R., Gotmare, A.D., Joty, S.R., Xiong, C., Hoi, S.C.H.: Align before fuse: vision and language representation learning with momentum distillation. In: Neural Information Processing Systems (2021)
12. Lin, C.Y.: ROUGE: a package for automatic evaluation of summaries. In: Text Summarization Branches Out, pp. 74–81. Association for Computational Linguistics, Barcelona, Spain (2004)
13. Liu, C., Zhao, R., Chen, H., Zou, Z., Shi, Z.: Remote sensing image change captioning with dual-branch transformers: a new method and a large scale dataset. IEEE Trans. Geosci. Remote Sens. **60**, 1–20 (2022)
14. Liu, F., et al.: RemoteCLIP: a vision language foundation model for remote sensing (2024)
15. Papineni, K., Roukos, S., Ward, T. and Zhu, W.J.: BLEU: a method for automatic evaluation of machine translation, pp. 311–318 (2002)
16. Zhu, Q., Guo, X., Li, Z.L., Li, D.: A review of multi-class change detection for satellite remote sensing imagery. Geo-spatial Inf. Sci. **27**(1), 1–15 (2024)
17. Radford, A., et al.: Learning transferable visual models from natural language supervision. In: International Conference on Machine Learning (2021)
18. Rahhal, M.M.A., Bazi, Y., Alsharif, N.A., Bashmal, L., Alajlan, N.A., Melgani, F.: Multilanguage transformer for improved text to remote sensing image retrieval. IEEE J. Sel. Top. Appl. Earth Observ. Remote Sens. **15**, 9115–9126 (2022)

19. Reimers, N., Gurevych, I.: Sentence-BERT: sentence embeddings using Siamese BERT-networks. In: Proceedings of the 2019 Conference on Empirical Methods in Natural Language Processing. Association for Computational Linguistics (2019)
20. Sun, X., et al.: RingMo: a remote sensing foundation model with masked image modeling. IEEE Trans. Geosci. Remote Sens. **61**, 1–22 (2023)
21. Tu, Y., Li, L., Su, L., Du, J., Lu, K., Huang, Q.: Viewpoint-adaptive representation disentanglement network for change captioning. IEEE Trans. Image Process. **32**, 2620–2635 (2023)
22. Tu, Y., Li, L., Su, L., Lu, K., Huang, Q.: Neighborhood contrastive transformer for change captioning. IEEE Trans. Multimedia **25**, 9518–9529 (2023)
23. Vedantam, R., Zitnick, C.L., Parikh, D.: Cider: Consensus-based image description evaluation. In: 2015 IEEE Conference on Computer Vision and Pattern Recognition (CVPR), pp. 4566–4575 (2015)
24. Wang, D., et al.: Advancing plain vision transformer toward remote sensing foundation model. IEEE Trans. Geosci. Remote Sens. **61**, 1–15 (2022)
25. Wang, F., Liu, H.: Understanding the behaviour of contrastive loss. In: 2021 IEEE/CVF Conference on Computer Vision and Pattern Recognition (CVPR), pp. 2495–2504 (2020)
26. Xu, L., Xie, H., Wang, F.L., Tao, X., Wang, W., Li, Q.: Contrastive sentence representation learning with adaptive false negative cancellation. Inf. Fusion **102**(C), 102065 (2024)
27. Yao, L., Wang, W., Jin, Q.: Image difference captioning with pre-training and contrastive learning. In: AAAI Conference on Artificial Intelligence (2022)
28. Yu, J., Wang, Z., Vasudevan, V., Yeung, L., Seyedhosseini, M., Wu, Y.: CoCa: contrastive captioners are image-text foundation models. Trans. Mach. Learn. Res. **36**(3), 3108–3116 (2022)
29. Zhang, J., Zhou, Z., Mai, G., Mu, L., Hu, M., Li, S.: Text2Seg: remote sensing image semantic segmentation via text-guided visual foundation models. ArXiv abs/2304.10597 (2023)
30. Zhang, W., Cai, M., Zhang, T., Zhuang, Y., Mao, X.: EarthGPT: a universal multi-modal large language model for multi-sensor image comprehension in remote sensing domain (2024)
31. Zhou, W., Guan, H., Li, Z., Shao, Z., Delavar, M.R.: Remote sensing image retrieval in the past decade: achievements, challenges, and future directions. IEEE J. Sel. Top. Appl. Earth Observ. Remote Sens. **16**, 1447–1473 (2023)

Classification Models

Improving the Performance of Already Trained Classifiers Through an Automatic Explanation-Based Learning Approach

Andrea Apicella, Salvatore Giugliano(✉), Francesco Isgrò, and Roberto Prevete

Dipartimento di Ingegneria Elettrica e delle Tecnologie dell'Informazione, Università degli Studi di Napoli Federico II, Naples, Italy
salvatore.giugliano2@unina.it

Abstract. While much of the existing XAI literature focuses on explaining AI systems, there has recently been a growing interest in using XAI techniques to improve the performance of AI systems without human involvement. In this context, we propose a novel explanation-based learning approach that aims to improve the performance of an already trained Deep-Learning (DL) classifier M without the need for extensive retraining. Our approach involves augmenting the responses of M with specific and relevant features obtained from a predictor P of explanations, which is trained to highlight relevant information in terms of input encoded features. These encoded features, together with the responses provided by M, are then fed into an additional simple classifier to produce a new classification. Importantly, P is constructed so that its training is less computationally expensive than training M from scratch, or equivalent to fine-tuning M. This approach avoids the computational cost associated with training a complex DL model from scratch. To evaluate our proposal, we used 1) three different well-known DL models as M, specifically EfficientNet-B2, MobileNet, LeNet-5, and 2) three standard image datasets, specifically CIFAR-10, CIFAR-100 and STL-10. The results show that our approach uniformly improves the performance of all already trained DL models for all the inspected datasets.

Keywords: XAI · performance improvement · deep learning · explanations

1 Introduction

Explainable Artificial Intelligence (XAI) aims to provide insights into the inner workings of AI models and the rationale behind their decisions, enabling users to understand the results produced by Machine Learning (ML) models, and Deep Learning (DL) models in particular. XAI approaches are becoming crucial for explaining several ML models applied to different kinds of inputs, including, but not limited to, images [2,18], natural language processing [19], clinical decision support systems [21], and others.

Nevertheless, it is important to highlight that while a significant portion of the existing XAI literature focuses on providing explanations for AI systems [2,5,18,20], one of the main motivations of XAI studies is to improve ML systems, especially DL, with respect properties such as robustness and performance, however relatively little attention has been paid to this aspect in the literature [24]. Furthermore, the literature on this topic often presents approaches in which the role of the human is predominant.

In this paper, which builds up on what described in [7], we propose a system aimed at improving the performance of an already trained (i.e. pre-trained) classifier. Unlike similar approaches, our approach considers a scenario where a pre-trained model, whose performance we want to improve, is available, without the need to retrain the model from scratch or possibly just fine-tune it. More specifically, in order to improve the performance of a pre-trained classifier M, our proposal relies on two external modules, namely i) a predictor (P) and ii) a simple classifier (C), the former being able to infer accurate explanation encodings tailored to the specific model under consideration, and the latter being able to provide a classification based on both the model M and the predictor outputs. Consequently, the final classifier, armed with the corrected explanation encoding alongside the M output, can move towards producing more accurate predictions than M alone.

An important aspect of our approach is how to obtain P. As a first key constraint, we consider that P should be designed so that its training requires less computational resources than training M from scratch, or is comparable to fine-tuning M. To this end, we propose a two-step learning process, as described in Sect. 3 and Fig. 1, which exploits the explanations of the output of M over the training set and their encoding by an auto-encoder network. In other words, we propose a method to merge together the knowledge given by a pre-trained classifier with the knowledge given by an XAI method, avoiding the computational efforts in training a complex DL model from scratch. To achieve this, we will assume that P and the auto-encoder are simpler models than M, or a version of M that requires fine-tuning (see Sect. 4.3). To evaluate our proposal, in this work we report the results on three different well-known DL models adopted as M, which are EfficientNet-B2, MobileNet, LeNet-5, and three standard image datasets, which are CIFAR-10, CIFAR-100 and STL-10.

The paper is organized as follows: in Sect. 2, we provide an overview of the main related works that have employed XAI to enhance ML systems. Sections 3 detail our proposed method, while Sect. 4 outlines the experimental assessment conducted to validate our system. The results of the experimental assessment are presented in Sect. 5, and in Sect. 6, these results are discussed comprehensively, concluding with final remarks.

2 Related Works

From a general point of view, XAI methods can also be interpreted and used as a way to recover external knowledge about the dataset in order to improve an ML system, see for example [3,13]. In particular, combining DL models with the domain knowledge that humans use to understand the world can avoid that purely data driven learning strategies go toward counter-intuitive results [13]. For example, in [22], a mechanism to interactively querying the user (or some other information source) about visual explanations during the training stage to obtain the desired outputs of the data points is proposed (eXplanatory interactive Learning , XiL). The effectiveness of two types of user feedback used in XiL on model performance and explanation accuracy is inspected in [11].

More recently, there has been an increasing focus on the use of XAI techniques to autonomously improve DL systems, especially in terms of performance, without human intervention [24]. The basic underlying idea is that explanations about model outputs can contribute to better parameter tuning of ML systems. An example of method relying on XAI methods as attention mechanism is provided in [10]. However, most of the literature that has exploited XAI to improve automatically classification performances relied on changing the training procedure or the loss of the model to focus on relevant information. Therefore, in these cases DL models have to be trained from scratch using appropriate learning strategies. Thus, this process can be computationally expensive due to the large number of DL model parameters involved. In addition, as mentioned above, current strategies often include human feedback in the training process. Conversely, our work does not need any human feedback or knowledge for integration with the model output, relying solely on insights provided by XAI in an automatic way.

In [24] a survey of this type of approaches is reported. In particular, [24] divides the current strategies into four main categories: i) augmenting the data, where explanations are leveraged to generate artificial samples able to provide information against undesired behavior (e.g., [22]), ii) augmenting the intermediate features, consisting in adopting the feature-wise information provided by explanations to determine the importance of intermediate features (e.g., [1,4,6]), iii) augmenting the loss, consisting in augmenting the loss training function with additional regularization terms relying on explanations (e.g., [16]), and iv) augmenting the model, used to modify a model on an estimation of parameters' importance (e.g., [26]).

However, among the mentioned categories, the prevailing trend is to leverage XAI to train ML models from scratch, without necessarily focusing on scenarios where the goal is to enhance an already trained (pre-trained) ML model. By contrast, our proposal is aimed at this last specific scenario, where an already trained ML model is to be improved using XAI methods without direct human interaction.

3 Method Description

The proposed approach builds upon some ideas introduced in [7], where a general framework for merging knowledge from XAI methods with classifier outputs is outlined. In this section, we provide an overview of these ideas and then propose a novel approach for effectively integrating attribution-based explanations to improve the classification process in an automated manner.

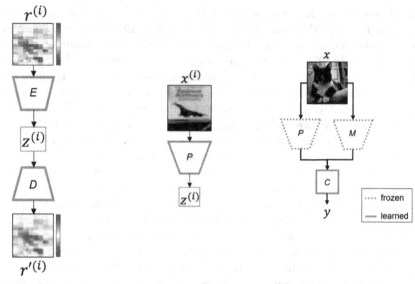

(a) Dataset S^P generation. **(b)** Training the Predictor P. **(c)** The final architecture for the classification task will be obtained by training solely the classifier C.

Fig. 1. A general description of the proposed framework. (a) The Auto-Encoder ($E-D$) architecture takes as input the attribution map $r^{(i)}$ corresponding to the image $x^{(i)}$. The $z^{(i)}$ encodings are then stored. (b) The dataset $S^P = \{x^{(i)}, z^{(i)}\}$ is used to train the Predictor P. (c) C is trained using the output from M and the encoding z generated by P. M and P are represented by dotted lines since their weights are frozen during the training phase of C. Information from M and the encoded representation z produced by P is utilized by C for the final classification task.

Assuming that we have a set of n labeled data $S^T = \{(x^{(i)}, y^{(i)})\}_{i=1}^n$ and a set of m unlabeled data $S^E = \{x^{(j)}\}_{j=1}^m$, we want to know the correct labels of S^E relaying on a neural network M trained on S^T.

In our proposal, we consider a further DL model P, which we will refer as *Predictor*. We want that P provides information in terms of attribution of the input $x^{(j)}$ regarding model M's behavior if it is fed with $x^{(j)}$ and output the correct class $y^{(j)}$. In particular, we introduce a 'attribution encoding', denoted as $z^{(j)} = P(x^{(j)})$ able to capture information to what M should pay attention to in order to output $y^{(j)}$ when given $x^{(j)}$ as input. Our hypothesis is that $z^{(j)}$ can be exploited together with the M output to correctly classify the input $x^{(j)}$ with the support of an additional classifier C, which should be properly trained.

In Fig. 1c the proposed approach is described, while Table 1 reports the inputs and outputs for the different architecture modules. Taken an input $x^{(j)}$ and fed to both M and P, the outputs of M and P are concatenated and fed to the classifier C, returning the predicted class relying on both the knowledge given by M, trained to perform a classification task, and P, trained to extract relevant information. Consequently, on an unlabeled data point $\boldsymbol{x}^{(j)}$, the newly estimated output $\hat{y}_C^{(j)}$ is defined as $\hat{y}_C^{(j)} = \arg\max softmax\big(C\big(M(\boldsymbol{x}^{(j)}), P(\boldsymbol{x}^{(j)})\big)\big)$.

Since we assume that the model M can be any already trained neural network, the proposed framework can be obtained training P and the classifier C without using unlabeled data S^E. The following of this section describes a possible strategy to train both P and C.

Table 1. Overview of the architecture modules used in the model. Each module is defined by its specific input and output: the Encoder (E) processes the attribution map to generate attribution encodings, the Decoder (D) reconstructs the attribution map from encodings, the Predictor (P) produces attribution encodings from the input image, the Pre-trained Model (M) predicts scores based on the input image, and the Classifier (C) utilizes concatenated encodings and score predictions to output the final classification.

Architecture module	Input	Output
E (Encoder)	Attribution map r corresponding to the image x	z attribution encodings
D (Decoder)	Attribution encodings z	Reconstructed attribution map r'
P (Predictor)	Image x	Attribution encodings z
M (Pre-trained Model)	Image x	Score prediction $M(x)$
C (Classifier)	Attribution encodings z and score prediction $M(x)$	Classification output y

To train P, it is necessary to have suitable attribution encodings $\boldsymbol{z}^{(i)}$ for a given set of inputs $\boldsymbol{x}^{(i)}$. These encodings should be related to the aspects that M needs to focus on to produce the desired output $y^{(i)}$ when given $\boldsymbol{x}^{(i)}$ as inputs. Therefore, we needs a new labeled dataset, S^P, which maps each input $\boldsymbol{x}$ to an encoding $\boldsymbol{z}$ containing useful information about $\boldsymbol{x}$ and $M(\boldsymbol{x})$. Since we assume to have the dataset S^T which is used during the training stage, we extract $\boldsymbol{z}^{(i)}$ from the data $\boldsymbol{x}^{(i)}$ in S^T, obtaining $S^P = \{(\boldsymbol{x}^{(i)}, \boldsymbol{z}^{(i)})\}_{i=1}^n$ (see Fig. 1a). Therefore, the synthesis of the model P requires a two step process: firstly, the creation of the training dataset S^P, secondly the training of the model P using S^P. Being P a classical DNN, it can be trained in a classical supervised way. The following of this section describe how to generate the dataset S^P.

3.1 Generating the Training Dataset S^P

$S^P = \{(\boldsymbol{x}^{(i)}, \boldsymbol{z}^{(i)})\}_{i=1}^n$ can be seen as a dataset which maps, for each data $\boldsymbol{x}^{(i)}$, an attribution encoding $\boldsymbol{z}^{(i)}$ containing meaningful information about $\boldsymbol{x}^{(i)}$. In

Fig. 1a the $z^{(i)}$ generation process is summarized. In particular, the S^P generation architecture consists in an Auto-Encoder AE. Indeed, similarly to [6], we assume that an appropriately trained encoder E can distill essential information from a given attribution map $r^{(i)}$. We emphasize that differently from approaches such as [10], our aim is to improve the performance of an already trained model M, without the necessity of training it from the beginning. More in depth, to generate $z^{(i)}$, the following steps are made:

i) Let us assume the existence of an attribution method R able to generate the attribution map $r^{(i)}$ about $x^{(i)}$ corresponding to the true class label $y^{(i)}$ relying on $M(x^{(i)})$. To obtain the attribution map with respect to the true class label, we adopt the SHAP XAI method [17]. The attribution map is computed with respect to the true class label $y^{(i)}$ provided in the training data, that is $r^{(i)} = r_{y^{(i)}} \in SHAP(M, x^{(i)})$.

ii) We train an encoder E to extract useful information from $r^{(i)}$, i.e. $z^{(i)} = E(r^{(i)})$. Achieving this requirement involves employing a Decoder $D(z^{(i)})$ proficient in reconstructing the original attribution map $r^{(i)}$. When considering E and D together, they act as an Auto-Encoder AE, where the output $z = E(x)$ from E feeds into D. To train AE, Mean Squared Error (MSE) between $r^{(i)}$ and $AE(r^{(i)})$ can be used as loss function. The schematic representation of the overall encoding generation architecture is illustrated in Fig. 1a.

3.2 Training the Predictor P

Once that a training set $S^P = \{(x^{(i)}, z^{(i)})\}_{i=1}^n$ has been generated, it can be used to train a Predictor model P, using $x^{(i)}$ and $z^{(i)}$ as model input and target, respectively (see Fig. 1b). To train P, Mean Squared Error (MSE) between $z^{(i)}$ and $P(x^{(i)})$ can be used as loss function. Notice that $x^{(i)}$ in S^P does not overlap with the test set where M will be evaluated.

3.3 Training the Classifier C

A final classifier C is trained using the output from M and the encoding z generated by P (see Fig. 1c). In this final step, the weights of M and P are frozen. Information from M and the encoded representation z produced by P is utilized by C for the final classification task. For each new input, P can predict an attribution encoding that guides the C model to the correct class, improving the classification of the M model.

4 Experimental Assessment

To validate our approach, the experimental pipeline involved several key steps:
1) *Baseline synthesis.* Training the model M on a given dataset $S^T = \{x^{(i)}, y^{(i)}\}_{i=1}^n$ and evaluating its accuracy on a test set $S^E = \{x^{(j)}, y^{(j)}\}_{j=1}^m$ where the labels $y^{(j)}$ will be used only for evaluation purposes.
2) *Attribution computation.* Computing attribution maps $\{r^{(i)}\}_{i=1}^n$ using the

already trained model M and method R with each $x^{(i)}$ instance in the training set S^T, with respect to their true class $y^{(i)}$, as described in Sect. 3.1.

3) S^P *generation*. Learning the parameters of the modules E and D (see Fig. 1a and Sect. 3.1) using training data S^T. Storing the encoding $z^{(i)}$ for each $x^{(i)}$ to create the dataset S^P.

4) *P training*. Training the P model using the dataset S^P. The Predictor P takes the $x^{(i)}$ input and reconstructs / predicts the encoding associated with the original attribution map r^i (see Fig. 1b and Sect. 3.2).

5) *C training*. Learning the parameters of the module C. The parameters of M and P models remain frozen during this step (see Fig. 1c and Sect. 3.3).

6) *Proposal evaluation*. Utilizing the outputs of the Predictor P and model M concatenated into module C to predict labels for the unseen data S^E (see Fig. 1c) and comparing their performance with that of baseline M on the same data.

In the following of this section, details about the adopted datasets, attribution generator R, and models adopted are given.

4.1 Datasets

CIFAR-10, CIFAR-100 and STL-10 datasets were used as benchmark datasets. CIFAR-10 [14] is a collection of 60,000 color images grouped into ten categories . The dataset offers 50,000 training images and 10,000 test images, all of size 32×32 pixels.

CIFAR-100 [14] is a collection of 60,000 color images grouped into hundred categories. The dataset offers 50,000 training images and 10,000 test images, all of size 32×32 pixels.

The STL-10 [8] dataset consists of images belonging to ten different classes. Each image has a size of 96×96 pixels. The dataset offers 5,000 training images and 8,000 test images.

4.2 Adopted Attribution Generator Algorithm

SHAP, a well-established XAI method, offers insights into how various features contribute to a model's predictions. In this work, we adopted the *Partition explainer algorithm* and to ensure a sufficient level of granularity for superpixels explanations were built with 2000 evaluations [17].

4.3 Adopted Models and Training

We evaluate our proposed method on three different models M: i) EfficientNet-B2 [23] pre-trained on the Noisy Student weights [25], ii) MobileNet [12] pre-trained on the ImageNet weights [9], and iii) LeNet-5 [15].

Baseline performance are computed after a fine-tuning stage using the training set provided with each investigated dataset. SHAP attribution maps are then computed on the M baselines for each input belonging to the training data considering their true class labels.

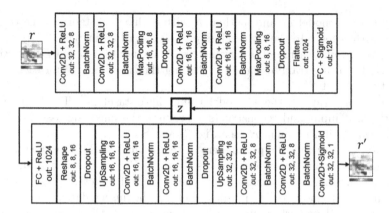

Fig. 2. The *Auto-Encoder* architecture, designed for the CIFAR-10 and CIFAR-100 datasets. The architecture is composed of convolutional (Conv2D), fully-connected (FC), and Batch Normalization (BatchNorm) layers. The kernel size is 3×3 for all the convolutional layers, while the number of filters is given by the third dimension of the output shape.

The AE architecture adopted for CIFAR-10/100 datasets is shown in Fig. 2. Due to the different input dimension of STL-10 dataset respect to CIFAR-10/100 datasets, a further $Conv - BatchNorm - Conv - BatchNorm - MaxPooling - Dropout$ was added to the encoder and decoder of the AE architecture when used with this dataset.

The construction of the Predictor P model involves replacing the last fully-connected layers of the M model architecture with a fully-connected layer that has a dimensionality of 128 and a *sigmoid* as activation function. The training of P utilizes the generated dataset S^P.

The classifier C is a shallow fully-connected neural network with a number of outputs equal to the number of classes. While maintaining the internal parameters of M and P fixed, the parameters of C are learned during the training phase (see Fig. 1c). Furthermore, to assess whether the encoded explanation provided by P effectively focus on relevant parts of the input, the performance is reassessed using z obtained by feeding the trained P with random images. In this way, we want to assess the specificity of the encoded explanations provided

Table 2. Variation ranges for the hyperparameters, considered during the grid search optimization strategy for the M baseline models and E, D, C modules.

Hyperparameter	Range
Batch Size	{16, 32, 64}
Learning Rate	[0.0001, 0.01] with step of 0.0005
Validation Fraction	{0.05, 0.1, 0.2}

by P for the given input and their importance for the classifier's revaluation of the responses.

Optimal hyperparameters for the M baselines, Encoder E, Decoder D and classifier C were determined using a grid-search strategy with ranges detailed in Table 2.

5 Results

Table 3 compares the performance of different models M (LeNet-5, MobileNet and EfficientNet-B2) on CIFAR-10, CIFAR-100, and STL-10 datasets with and without our approach. The 'Proposal' column represents the accuracy of the proposed method, which involves the utilization of the Predictor P capable of extracting useful information from the input. Moreover, the performance is compared to the results obtained when random images are used as input for the Predictor P. This evaluation is conducted to assess the contribution and specificity of the encoded explanations provided by P, as discussed in Sect. 4.3.

Table 3. Accuracy (%) scores on test sets on CIFAR-10, CIFAR-100 and STL-10 test sets. The performance is compared with the results obtained when random images are provided as input to the predictor P.

	Model	Baseline	Proposal	
			Real	Random
CIFAR-10	LeNet-5	67.96	**70.47**	67.84
	MobileNet	94.63	**94.95**	94.66
	EfficientNet-B2	98.06	**98.17**	98.02
CIFAR-100	LeNet-5	36.23	**37.18**	36.26
	MobileNet	76.72	**76.97**	75.74
	EfficientNet-B2	87.88	**88.42**	87.98
STL-10	LeNet-5	52.59	**55.77**	52.71
	MobileNet	91.07	**92.51**	90.95
	EfficientNet-B2	98.76	**98.86**	98.79

Results on CIFAR-10. Adopting LeNet-5 as baseline model M, the accuracy obtained by the proposal approach increased from 67.96% (baseline) to 70.47%. The proposal demonstrates an improvement, with an increase in accuracy compared to the baseline. Instead, if we consider MobileNet and EfficientNet-B2 as model M, the proposal maintains similar but higher performance with respect to the baselines (94.63% vs. 94.95%, 98.06% vs. 98.17%).

Results on CIFAR-100. Adopting LeNet-5 as baseline model M, the accuracy increased from 36.23% (Baseline) to 37.18% (proposal). Instead, considering MobileNet as baseline model M, the accuracy goes from the 76.72% (baseline) to 76.97% (proposal). A greater improvement resulted with EfficientNet-B2, going from an accuracy of 87.88% (baseline) to 88.42% (proposal).

Results on STL-10. Adopting LeNet-5 as baseline model M, the accuracy on STL-10 improved from 52.59% (Baseline) to 55.77% (proposal). Our proposed method demonstrates a substantial improvement in accuracy compared to the baseline. Similarly, MobileNet has an improvement going from 91.07% to 92.51%. Instead, considering EfficientNet-B2 as baseline model M, the proposal maintains a high accuracy similar to the baseline, but with a slight improvement (98.76% vs 98.86%).

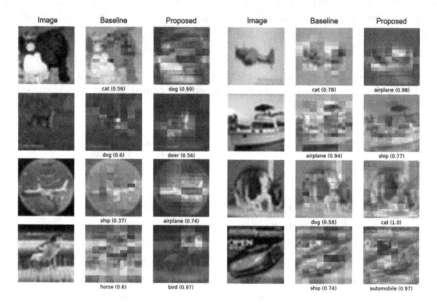

Fig. 3. Some images from the CIFAR-10 test set, where the proposed method corrects the classification of the MobileNet model. The *Baseline* column shows the explanation $r_{M(x)}$ on the class predicted by M, i.e. $r_{M(x)} \in SHAP(M, x)$. The *Proposed* column presents the explanation reconstructed by $D(P(x))$. For each image, the class and score given by M and our proposal are provided. The images have been filtered for better visualisation.

Reconstructed Explanations. In Fig. 3, 4 and 5 images from the CIFAR-10, CIFAR-100 and STL-10 test sets where the proposed method corrects the classification of the MobileNet model are shown. Images are provided together with the explanation (Baseline column) on the class predicted by M, i.e. $r_{M(x)} \in$

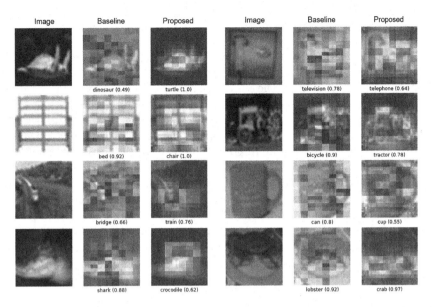

Fig. 4. Some images from the CIFAR-100 test set, where the proposed method corrects the classification of the MobileNet model. The *Baseline* column shows the explanation $r_{M(x)}$ on the class predicted by M, i.e. $r_{M(x)} \in SHAP(M, x)$. The *Proposed* column presents the explanation reconstructed by $D(P(x))$. For each image, the class and score given by M and our proposal are provided. The images have been filtered for better visualisation.

$SHAP(M, x)$ and the reconstruction (Proposed column) $D(P(x))$ using P as encoder and D as decoder. For each image, the class and score given by M and our proposal are provided. It is possible to show that, in the majority of cases, the reconstructed explanations concentrate on input areas that are intuitively more related to the real output compared to the explanation. This highlights the P's capability to generate an encoding of the explanation that effectively focuses on the pertinent part of the input. The reconstructed explanations tend to highlight the areas or features of the input that a human expert might consider important for reaching a similar conclusion as the model. This correlation is evaluated using qualitative criteria, focusing on visual elements that a human observer would likely identify as key to correctly classifying the image. For example, in Fig. 3 the reconstructed explanations appear to emphasize essential parts to determine the true class. This is particularly evident in the deer (back and belly), aeroplanes (fuselage and wings), bird (head) and ship (bow and hull) images. Figure 4 shows similar patterns for images such as the crocodile (head and eye), telephone (handset and dial), tractor (wheel and cab) and crab (claws) images. In Fig. 5 this property is visible in all the images shown, in particular: monkeys (head), dog (ears) and car (wheels).

6 Discussions

The comparative performance evaluation of various models (LeNet-5, MobileNet, and EfficientNet-B2) on CIFAR-10, CIFAR-100, and STL-10 datasets, as depicted in Table 3, highlights the efficacy of our proposed approach in extracting relevant information from the input data. The observed improvements resulting from the proposed framework exhibit variability across datasets and models. Particularly significant is the performance enhancement achieved by the proposed method based on LeNet-5 for CIFAR-10 and STL-10 datasets, surpassing the baseline models significantly. In other cases, the impact of the proposed approach remains consistent, achieving higher performance against the respective baselines, indicating a slight but discernible improvement. However, although the improvements might be considered marginal, they are consistently present across all examined scenarios. Furthermore, the magnitude of improvement may be contingent upon the baseline performance. In other words, when the baseline models have low accuracy, the proposed method offers a more substantial improvement, as less effective models gain greater benefits from the features identified by the Predictor P.

To better understand whether the improvements achieved by our approach were specific to a few classes or distributed across all classes, we also computed a confusion matrices for the STL-10 test set, using MobileNet as model M, with

Fig. 5. Some images from the STL-10 test set, where the proposed method corrects the classification of the MobileNet model. The *Baseline* column shows the explanation $r_{M(x)}$ on the class predicted by M, i.e. $r_{M(x)} \in SHAP(M, x)$. The *Proposed* column presents the explanation reconstructed by $D(P(x))$. For each image, the class and score given by M and our proposal are provided.

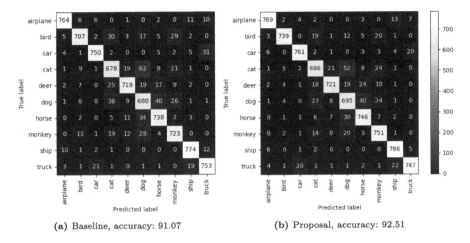

Fig. 6. Confusion matrices for the STL-10 test set, using MobileNet as model M, (a) without the proposed method and (b) with the proposed method. The main diagonal in (b) illustrates that the enhancements introduced by our approach are evenly distributed across all classes, highlighting the positive impact of the proposed method on the model's performance across a diverse range of categories.

and without the proposed approach (see fig. 6). From the confusion matrices, one can note that the improvements introduced by our approach are uniformly distributed across all classes of the inspected dataset. This observation underscores the positive and complete impact of the proposed method on enhancing the model's performance across a diverse spectrum of categories.

As illustrated in Fig. 3, 4 and 5, the reconstructed explanations exhibit a tendency to focus on the pertinent areas of the image, and they frequently exhibit qualitative superiority to the explanations generated by the SHAP method. The reconstructions proposed by our method provide a more clear and focused representation of the critical features for correct classification. This qualitative superiority of the reconstructed explanations demonstrates the effectiveness of our approach in capturing salient and relevant information, thereby improving not only performance but also the interpretability and comprehensibility of the model's decisions.

7 Conclusions

The proposed method demonstrates enhancements over the baseline models, indicating the efficacy of the introduced Predictor P in extracting crucial information for achieving higher accuracy in several machine learning tasks. It is important to note that P is constructed on the basis of the explanations with respect to the true classes extracted by a specific XAI method on the training set where the model was trained. Furthermore, the method's ability to reconstruct explanations through the decoder D effectively clarifies the reasoning behind the model results, improving both interpretability and performance.

It would be beneficial for future work to examine the impact of this approach by utilising different XAI methods and datasets that differ from those used to train the model M. Additionally, we intend to assess the potential of our approach by retraining the entire architecture, in conjunction with M, from the outset, gradually integrating the contributions of the Predictor P during the learning phase.

Acknowledgement. This work was partially supported by the European Union - FSE-REACT-EU, PON Research and Innovation 2014–2020 DM1062/2021 contract number 18-I-15350-2, by the Ministry of University and Research, PRIN research project "BRIO – BIAS, RISK, OPACITY in AI: design, verification and development of Trustworthy AI.", Project no. 2020SSKZ7R, and by the PNRR MUR project PE0000013-FAIR (CUP: E63C22002150007).

Reviews

Reviewer 1
Questions

2. Summary of the paper

The paper proposes a framework for improving the accuracy of a pre-trained neural network (M) using XAI-based features. More precisely, the framework consists of three functions, the pre-trained network M, a XAI-based feature generator P and the final classifier C using outputs of M and P as inputs. To train P, the attention map r_i (generated by the SHAP given instance x_i) is used to construct the encoding z_i with an auto-encoder. P is trained with the sample S^P consisting of pairs x_i and z_i. The experimental results show that the proposed framework consistently improve the accuracy of the pre-trained network.

3. Indicate three strong points of the paper:

The framework is simple but interesting enough. The idea is clearly stated and the paper itself is well-written in general. Experimental results supports the claim of the paper.

4. Indicate three weak points of the paper:

So far, the proposed framework is shown to be effective for image classification tasks only. It is not fully clear how generic the proposed framework is for other tasks. In particular, how the explanation improves the accuracy of the pre-trained network is not clear yet.

5. Please enter your detailed comments regarding the paper:

As mentioned above, the framework is interesting but pre-matured as a generic method. It might be interesting to investigate more generic framework by generalizing the proposed methods.

6. Please rate the replicability and reproducibility of the paper:

Medium (some code, data or documentation is missing)

7. Please give an overall rating to the paper into the following categories.
Weak Accept: Borderline, tending towards accept

9. Should this paper be considered for the best paper award?
No

10. If you think there are ethical issues (i.e., you answered yes to the previous question), please briefly describe them.
N/A

11. If the paper is not accepted, would you consider it for the late-breaking work/poster session?
Yes

Reviewer 2

Questions

2. Summary of the paper

This paper presents a method to improve classifiers by training a separate predictor model using SHAP attribution maps, and augmenting the original model with the predictions from this classifier.

The method is evaluated on an image classification task, using common datasets and models. Efforts are made to analyse the suitability of the method to different tasks and across classes.

3. Indicate three strong points of the paper:

- The ambition of the paper to eliminate the need for human feedback aligns well with the aims of the conference.
- The method is outlined reasonably well, and the figures included are helpful.
- The method was evaluated on a range of datasets and models, and effort was made to inspect whether the effects of the method were class-specific.

4. Indicate three weak points of the paper:

- The detail of input and output for the various models is a little hard to follow.
- The authors tend to use sensational language. Prefer plain language.
- Some claims are not sufficiently substantiated (see details below).

5. Please enter your detailed comments regarding the paper:

The paper is in general well-written, with some helpful figures. Some of the presentation of the method lacks clarity, but I think this could be addressed with small changes to the structure of the text or the inclusion of a table (see below).

Areas for improvement:

- As mentioned, the detail of input and output for the various models is a little hard to follow. This could be alleviated by small modifications to Fig. 1 or §3, maybe with a list or table of the models, their inputs and outputs.
- In"§5 Results - Reconstructed explanations", it is stated that: "It is possible to show that, in the majority of cases, the reconstructed explanations concentrate on input areas that are intuitively more related to the real output compared to

the explanation." However this is not shown systematically. We get some visual examples in Figs. 3-5, but it is not demonstrated for a majority, nor is "intuitively more related to the real output" defined in sufficient detail.

- In the discussion it is stated that "the magnitude of improvement may be contingent upon the baseline performance." It seems like there is enough data to actually perform a test of this, at the moment this claim is left hanging.

- Also in the discussion it is claimed that "The reconstructions proposed by our method provide a more clear and focused representation of the critical features for correct classification". How is the assessment of "critical features" made? There is no systematic analysis of this, as mentioned above, so the support for this claim is weakened.

- Also in the discussion, it is claimed that there is a "comprehensive impact of the proposed method". What is meant here by comprehensive?

6. Please rate the replicability and reproducibility of the paper:
   Medium (some code, data or documentation is missing)

7. Please give an overall rating to the paper into the following categories.
   Accept: Good paper

9. Should this paper be considered for the best paper award?
   Yes

10. If you think there are ethical issues (i.e., you answered yes to the previous question), please briefly describe them.
    None

11. If the paper is not accepted, would you consider it for the late-breaking work/poster session?
    Yes

References

1. Anders, C.J., Weber, L., Neumann, D., Samek, W., Müller, K.R., Lapuschkin, S.: Finding and removing clever hans: using explanation methods to debug and improve deep models. Inf. Fusion **77**, 261–295 (2022)
2. Apicella, A., Isgrò, F., Prevete, R., Sorrentino, A., Tamburrini, G.: Explaining classification systems using sparse dictionaries. In: ESANN 2019 - Proceedings, 27th European Symposium on Artificial Neural Networks, Computational Intelligence and Machine Learning, pp. 495–500 (2019)
3. Apicella, A., Corazza, A., Isgrò, F., Vettigli, G.: Integration of context information through probabilistic ontological knowledge into image classification. Information **9**(10), 252 (2018)
4. Apicella, A., Di Lorenzo, L., Isgrò, F., Pollastro, A., Prevete, R.: Strategies to exploit XAI to improve classification systems. Commun. Comput. Inf. Sci. **1901 CCIS**, 147–159 (2023). https://doi.org/10.1007/978-3-031-44064-9_9

5. Apicella, A., Giugliano, S., Isgrò, F., Prevete, R.: Exploiting auto-encoders and segmentation methods for middle-level explanations of image classification systems. Knowl.-Based Syst. **255**, 109725 (2022)
6. Apicella, A., Giugliano, S., Isgrò, F., Prevete, R.: SHAP-based explanations to improve classification systems. In: Proceedings of the 4th Italian Workshop on Explainable Artificial Intelligence co-located with 22nd International Conference of the Italian Association for Artificial Intelligence(AIxIA 2023), Roma, Italy, November 8, 2023. CEUR Workshop Proceedings, vol. 3518, pp. 76–86. CEUR-WS.org (2023)
7. Apicella, A., Giugliano, S., Isgrò, F., Prevete, R.: Towards a general framework for improving the performance of classifiers using XAI methods. arXiv preprint arXiv:2403.10373 (2024)
8. Coates, A., Ng, A., Lee, H.: An analysis of single-layer networks in unsupervised feature learning. In: Proceedings of the Fourteenth International Conference on Artificial Intelligence and Statistics, pp. 215–223. JMLR Workshop and Conference Proceedings (2011)
9. Deng, J., Dong, W., Socher, R., Li, L.J., Li, K., Fei-Fei, L.: ImageNet: a large-scale hierarchical image database. In: 2009 IEEE Conference on Computer Vision and Pattern Recognition, pp. 248–255. IEEE (2009)
10. Fukui, H., Hirakawa, T., Yamashita, T., Fujiyoshi, H.: Attention branch network: learning of attention mechanism for visual explanation. In: Proceedings of the IEEE/CVF Conference on Computer Vision and Pattern Recognition, pp. 10705–10714 (2019)
11. Hagos, M.T., Curran, K.M., Mac Namee, B.: Impact of feedback type on explanatory interactive learning. In: International Symposium on Methodologies for Intelligent Systems, pp. 127–137. Springer (2022)
12. Howard, A.G., et al.: MobileNets: efficient convolutional neural networks for mobile vision applications (2017)
13. Hu, Z., Ma, X., Liu, Z., Hovy, E., Xing, E.: Harnessing deep neural networks with logic rules. arXiv preprint arXiv:1603.06318 (2016)
14. Krizhevsky, A., Hinton, G., et al.: Learning multiple layers of features from tiny images (2009)
15. LeCun, Y., Bottou, L., Bengio, Y., Haffner, P.: Gradient-based learning applied to document recognition. Proc. IEEE **86**(11), 2278–2324 (1998)
16. Liu, F., Avci, B.: Incorporating priors with feature attribution on text classification. arXiv preprint arXiv:1906.08286 (2019)
17. Lundberg, S.M., Lee, S.I.: A unified approach to interpreting model predictions. In: Guyon, I., Luxburg, U.V., Bengio, S., Wallach, H., Fergus, R., Vishwanathan, S., Garnett, R. (eds.) Advances in Neural Information Processing Systems 30, pp. 4765–4774. Curran Associates, Inc. (2017)
18. Montavon, G., Binder, A., Lapuschkin, S., Samek, W., Müller, K.R.: Layer-wise relevance propagation: an overview. Explainable AI: Interpreting, Explaining Visualizing Deep Learn 193–209 (2019)
19. Qian, K., Danilevsky, M., Katsis, Y., Kawas, B., Oduor, E., Popa, L., Li, Y.: XNLP: a living survey for XAI research in natural language processing. In: 26th International Conference on Intelligent User Interfaces-Companion, pp. 78–80 (2021)
20. Ribeiro, M.T., Singh, S., Guestrin, C.: " why should i trust you?" explaining the predictions of any classifier. In: Proceedings of the 22nd ACM SIGKDD International Conference on Knowledge Discovery and Data Mining, pp. 1135–1144 (2016)

21. Schoonderwoerd, T.A., Jorritsma, W., Neerincx, M.A., Van Den Bosch, K.: Human-centered XAI: developing design patterns for explanations of clinical decision support systems. Int. J. Hum Comput Stud. **154**, 102684 (2021)
22. Schramowski, P., et al.: Making deep neural networks right for the right scientific reasons by interacting with their explanations. Nat. Mach. Intell. **2**(8), 476–486 (2020)
23. Tan, M., Le, Q.: EfficientNet: rethinking model scaling for convolutional neural networks. In: International conference on machine learning, pp. 6105–6114. PMLR (2019)
24. Weber, L., Lapuschkin, S., Binder, A., Samek, W.: Beyond explaining: opportunities and challenges of XAI-based model improvement. Inf. Fusion (2022)
25. Xie, Q., Luong, M.T., Hovy, E., Le, Q.V.: Self-training with noisy student improves imagenet classification. In: Proceedings of the IEEE/CVF Conference on Computer Vision and Pattern Recognition, pp. 10687–10698 (2020)
26. Yeom, S.K., et al.: Pruning by explaining: a novel criterion for deep neural network pruning. Pattern Recogn. **115**, 107899 (2021)

A Simple Method for Classifier Accuracy Prediction Under Prior Probability Shift

Lorenzo Volpi[iD], Alejandro Moreo[✉][iD], and Fabrizio Sebastiani[iD]

Istituto di Scienza e Tecnologie dell'Informazione, Consiglio Nazionale delle Ricerche,
56124, Pisa, Italy
{lorenzo.volpi,alejandro.moreo,fabrizio.sebastiani}@isti.cnr.it

Abstract. The standard technique for predicting the accuracy that a classifier will have on unseen data (*classifier accuracy prediction* – CAP) is cross-validation (CV). However, CV relies on the assumption that the training data and the test data are sampled from the same distribution, an assumption that is often violated in many real-world scenarios. When such violations occur (i.e., in the presence of *dataset shift*), the estimates returned by CV are unreliable. In this paper we propose a CAP method specifically designed to address *prior probability shift* (PPS), an instance of dataset shift in which the training and test distributions are characterized by different class priors. By solving a system of n^2 independent linear equations, with n the number of classes, our method estimates the n^2 entries of the contingency table of the test data, and thus allows estimating any specific evaluation measure. Since a key step in this method involves predicting the class priors of the test data, we further observe a connection between our method and the field of "learning to quantify". Our experiments show that, when combined with state-of-the-art quantification techniques, under PPS our method tends to outperform existing CAP methods.

Keywords: Classifier accuracy prediction · Prior probability shift · Label shift · Quantification

1 Introduction

In machine learning, estimating the accuracy that a classifier will have on unseen data (*classifier accuracy prediction* – CAP) is a fundamental step towards ensuring that the classifiers we deploy are effective. The accuracy of a classifier (where "accuracy" is broadly understood as any effectiveness metric – e.g., vanilla accuracy, or F_1) is typically estimated by means of cross-validation (CV) on the training data. However, CV relies on the so-called IID assumption, according to which the training data and the test data are expected to follow the same distribution. Such an assumption is often violated in many real scenarios. When this happens, the use of CV often leads to unreliable estimates of classifier accuracy, ultimately compromising the choice of the classifier to be deployed, or misleading its optimization process. CAP is thus an open problem when the IID assumption is not verified, i.e., when *dataset shift* [21] is present.

One specific type of dataset shift is *prior probability shift* (PPS), which is defined as the type of dataset shift in which (a) the class priors can change between the training distribution and the test distribution and (b) the classconditional distribution of the covariates is stationary.

In this paper we propose a simple yet effective method for CAP which is specifically tailored to address PPS; we dub this method LEAP (for *Linear Equation -based Accuracy Prediction*). Given that the vast majority of evaluation measures are computed on a contingency table that relates the true labels to the predicted labels, we approach CAP by treating the entries of this table as unknowns in a system of linear equations. In a classification problem with n classes, there are n^2 such unknowns; our method involves a set of n^2 independent linear equations, whose solution results in the prediction of the n^2 contingency table entries. From these, the accuracy of the classifier on the test data can be estimated, according to one or multiple accuracy measures at the same time. This is in contrast to previously proposed methods, that estimate accuracy according to a single measure and need to be retrained if the estimate of a different measure is needed.

We further observe that a key step in our method involves predicting the prevalence values (i.e., the priors, or relative frequencies) of the classes in the test data. This suggests a connection with the field of *learning to quantify* [5,8,11], the supervised learning task of predicting the distribution of the class labels in the test data. Through several experiments we show that, when equipped with a state-of-the-art quantification technique, our CAP method often outperforms other competing methods in scenarios characterized by PPS.

The rest of the paper is structured as follows. Section 2 reviews the related literature on classifier accuracy prediction. In Sect. 3 we define the notation and the concepts we use in the rest of the paper. Section 4 is devoted to explaining our novel method, the set of equations on which it is based, and its connections with quantification learning. We present our experimental results in Sect. 5, while Sect. 6 wraps up and discusses potential ideas for future work.

2 Related Work

A few methods for CAP under dataset shift have emerged in recent years. *Average Thresholded Confidence* (ATC) [10] estimates model accuracy by evaluating the expected proportion of test items for which the confidence score of the model exceeds a threshold learned on the training set. The authors propose two variants for computing such scores, ATC-MC (that relies on maximum confidence) and ATC-NE (that instead relies on negative entropy). *Difference of Confidence* (DoC) [12] also leverages the confidence a model shows in its predictions for training a regressor that estimates the accuracy of the original model on the test set. *Mandoline* [4] belongs instead to the family of "importance-weighting methods" [22], and relies on user-defined so-called "slicing" functions to create common representations and compare training data with test data. Chen et al. [3] exploit model agreement in an iterative framework, training an ensemble of auxiliary models for each iteration and using their agreement ratio to estimate the accuracy of the original model. *Generalisation Disagreement Equality* (GDE) [13] is instead based on the observation that the degree of disagreement on in-distribution (i.e., IID) data between identical neural architectures that have been initialised differently correlates strongly with their accuracy on (non-IID) test data.

A common trait of all these methods, and an important difference between them and our proposed method, is that the former are trained to estimate one specific evaluation measure (typically: vanilla accuracy). In cases where more than one evaluation measure is needed, these methods should be retrained from scratch.

3 Background

3.1 Notation

Throughout this paper we use the following notation. By $\mathcal{X} = \mathbb{R}^d$ and $\mathcal{Y} = \{1,\ldots,n\}$ we denote the input space (vectors of covariates) and the output space (classes), respectively; in binary problems we write instead $\mathcal{Y} = \{0,1\}$. By $D = \{(\mathbf{x}_i, y_i)\}_{i=1}^m$ we denote a generic labelled set consisting of pairs $\mathbf{x}_i \in \mathcal{X}$ and $y_i \in \mathcal{Y}$, that we use for training or testing our models.

A (crisp) classifier $h : \mathcal{X} \to \mathcal{Y}$ is a function mapping vectors of covariates into classes. A probabilistic classifier has instead the form $h : \mathcal{X} \to \Delta^{n-1}$, i.e., it maps vectors of covariates into vectors of posterior probabilities lying in the probability simplex $\Delta^{n-1} = \{(p_1,\ldots,p_n) : p_i \geq 0, \sum_{i=1}^n p_i = 1\}$. From a probabilistic classifier one can easily obtain a crisp classifier by returning the class corresponding to the largest posterior probability.

3.2 Prior Probability Shift

Dataset shift is defined as the situation in which the training set is drawn from a distribution P and the test set is drawn from another distribution Q such that

$$P(X,Y) \neq Q(X,Y) \qquad (1)$$

Prior probability shift is a type of dataset shift (to be found in *anti-causal learning* [19], i.e., the class of learning problems in which $\mathcal{Y}$ represents the phenomenon to be predicted and $\mathcal{X}$ represents *symptoms* of this phenomenon) defined as the case in which

$$\begin{aligned} P(Y) &\neq Q(Y) \\ P(X|Y) &= Q(X|Y) \end{aligned} \qquad (2)$$

i.e., where the prevalence of the classes can change between the training distribution and the test distribution while the class-conditional distribution of the covariates does not change. PPS is sometimes called *target shift* [24] or *label shift* [15]. For all X-measurable functions $f : \mathcal{X} \to \mathbb{R}$, from $P(X|Y) = Q(X|Y)$ it also follows that $P(Z|Y) = Q(Z|Y)$ with $Z = f(X)$ [23]. In particular, if we take $f = h$, it follows that the distribution of the class-conditional predictions $\hat{Y}$ issued by the classifier is stationary between the training distribution P and the test distribution Q, i.e., $P(\hat{Y}|Y) = Q(\hat{Y}|Y)$.

3.3 Problem Setting

We assume a classifier h trained on a set L of labelled items which may no longer be available. We also assume a validation set V of labelled items on which we train a classifier accuracy predictor. CAP consists of predicting the accuracy that our classifier h will exhibit on a given test set U (for which we assume the labels are not available) in terms of an accuracy measure A. We assume L and V are drawn from the same distribution P, while U is instead drawn from a different distribution Q, such that P and Q are related by PPS. Note that h is assumed already trained, and our goal is not to improve its performance on U but just to estimate how well it will fare on U.

4 Method

Most popular measures used for evaluating the performance of a classifier can be computed in terms of a *contingency table* (also known as a *confusion matrix*) that relates the outcomes of the random variable Y (which denotes the true class) with those of the random variable $\hat{Y}$ (which denotes the predicted labels). A contingency table is a matrix where entry (i, j) denotes the number of items whose true class is i and whose predicted class is j. However, without loss of generality, from now on we will take the value of each entry to be normalized by the total number of items, i.e., we will consider entry (i, j) of a contingency table as the *fraction* of items whose true class is i and whose predicted class is j; in other words, we will see a contingency table as an empirical *probability distribution*.

Let us consider the case in which the contingency table is obtained by applying a classifier h to a validation set $V = \{(\mathbf{x}_i, y_i)\}_{i=1}^m$ drawn from the training distribution P. Applying h to V brings about a transformation $V' = \{(\hat{y}_i, y_i)\}_{i=1}^m$, with $\hat{y}_i = h(\mathbf{x}_i)$, from which we can obtain a contingency table $\mathbf{V}$ with entries

$$v_{ij} = \frac{|\{(\hat{y}_k, y_k) \in V' : y_k = i \wedge \hat{y}_k = j\}|}{|V'|} \tag{3}$$

where symbol v reminds us that the classified items are taken from the validation set V. Note that $v_{ij} \approx P(\hat{Y} = j, Y = i)$ when $|V|$ is large enough.

As noted in the introduction, the contingency table $\mathbf{V}$ cannot be used to reliably estimate the accuracy of h on the test data U if V and U are drawn from different distributions (P and Q) related by PPS. Ideally, to estimate the accuracy of h on U we would like to directly estimate the entries u_{ij} of the contingency table **U** that derives from U, but this is not trivial, since on U we can observe the predicted labels (by simply applying h to the items of U) but not the true labels. If we had access to these true labels, any evaluation measure could be computed; for example, vanilla accuracy for a test set U can be expressed as

$$\text{Acc}(h, U) = \frac{\sum_{i=1}^{n} u_{ii}}{\sum_{i=1}^{n} \sum_{j=1}^{n} u_{ij}} = \sum_{i=1}^{n} u_{ii} \qquad (4)$$

In the binary case we will indicate fractions v_{11} and u_{11} as TP_V and TP_U, respectively, where TP stands for "true positives" and the V and U subscripts indicate from which set of items (the validation set V or the test set U) these fractions originate;[1] analogously, FP, FN, TN, will stand for "false positives", "false negatives", "true negatives", respectively. We will also use the shorthands

$$\text{tpr} = \frac{\text{TP}}{\text{TP} + \text{FN}} \qquad \text{fpr} = \frac{\text{FP}}{\text{FP} + \text{TN}} \qquad (5)$$

to refer to the "true positive rate" and the "false positive rate" of classifier h, and we will write tpr_V, tpr_U, fpr_V, fpr_U to indicate whether these ratios are computed on the contingency tables for sets V or U.

4.1 LEAP: A System of n^2 Linear Equations for CAP

We approach CAP by treating the entries of contingency table **U** as unknowns in a system of linear equations. In a classification problem with n classes, there are n^2 such unknowns; LEAP involves a set of n^2 independent linear equations, whose solution results in the prediction of the n^2 entries of the contingency table. From these, the accuracy of classifier h on the test data U can be estimated, according to one or more accuracy measures. In the following we derive LEAP for the general multiclass case $(n > 2)$, and discuss how it looks like in the binary case $(n = 2)$, in which our 4 unknowns are TP_U, FP_U, FN_U, and TN_U.

The first equation is trivial: the entries of the contingency table are probabilities of disjoint events that cover the entire event space, and they thus must sum to one, i.e.,

$$\sum_{i=1}^{n} \sum_{j=1}^{n} u_{ij} = 1 \qquad (6)$$

In the binary case, this corresponds to equation

$$\text{TP}_U + \text{FP}_U + \text{FN}_U + \text{TN}_U = 1 \qquad (7)$$

---

[1] The term "true positives" is typically used to refer to the *number* (and not the *prevalence*) of items which have correctly been predicted to belong to class 1; however, in this paper it is useful for us to always think in terms of prevalence values.

The second batch of equations relates the sum of the counts by columns to the fraction of predicted positives for each class, that is $\sum_{i=1}^{n} Q(\hat{Y}, Y = i) = Q(\hat{Y})$. Note that $Q(\hat{Y})$ is observed and can be obtained by simply classifying and counting the fraction of instances attributed to each class by h. Let us define a vector of observed normalized counts (fractions) $\mathbf{c} = (c_1, \ldots, c_n) \in \Delta^{n-1}$ where

$$c_i = \frac{1}{|U|} \sum_{\mathbf{x} \in U} \mathbb{1}[h(\mathbf{x}) = i] \tag{8}$$

This adds $(n-1)$ independent equations (note that the n-th one is constrained) which correspond to adding

$$\sum_{i=1}^{n} u_{ij} = c_j, \quad \text{for } j \in \{1, \ldots, n-1\} \tag{9}$$

In the binary case, this amounts to adding one equation imposing that the fraction of the instances classified as positive (c_1) must coincide with the sum of FP_U and TP_U, that is,

$$\text{FP}_U + \text{TP}_U = c_1 \tag{10}$$

The third batch of equations imposes that the class-conditional rates should remain stationary across the training and test distributions. This follows from the PPS assumptions (Eq. 2) and from the fact (noted at the end of Sect. 3.2) that $P(X|Y) = Q(X|Y)$ implies $P(\hat{Y}|Y) = Q(\hat{Y}|Y)$. Let us define a matrix of class-conditional rates $\mathbf{R}$ with entries $r_{ij} = \frac{v_{ij}}{\sum_{k=1}^{n} v_{ik}}$ that we compute on the validation set V. This allows us to add the $(n-1)^2$ independent equations

$$\frac{u_{ij}}{\sum_{k=1}^{n} u_{ik}} = r_{ij}, \quad \text{for } i, j \in \{1, \ldots, n-1\} \tag{11}$$

which can be more conveniently written as

$$u_{i1} r_{ij} + \cdots + u_{ij}(r_{ij} - 1) + \cdots + u_{in} r_{ij} = 0, \quad \text{for } i, j \in \{1, \ldots, n-1\}$$

In the binary case, this comes down to taking $r_{11} \equiv \text{tpr}_V$ (as computed on the validation set V) as an estimate of tpr_U. That is, we add the equations

$$\frac{\text{TP}_U}{\text{TP}_U + \text{FN}_U} = \text{tpr}_V \tag{12}$$

which can be more conveniently written as

$$\text{FN}_U \cdot \text{tpr}_V + \text{TP}_U \cdot (\text{tpr}_V - 1) = 0 \tag{13}$$

The fourth batch of equations is more complicated since it relates the sum of the entry values by rows to the class prevalence values in the test set, which we do not know; that is, $\sum_{j=1}^{n} Q(\hat{Y} = j, Y) = Q(Y)$. However, it turns out that the class prevalence values can be estimated from the training counts with the aid of the PPS assumptions (Eq. 2). This is better shown in the binary case by noting that

$$Q(\hat{Y} = 1) = Q(\hat{Y} = 1|Y = 1)Q(Y = 1) + Q(\hat{Y} = 1|Y = 0)Q(Y = 0)$$
$$= P(\hat{Y} = 1|Y = 1)Q(Y = 1) + P(\hat{Y} = 1|Y = 0)Q(Y = 0) \quad (14)$$

where the first step derives by the law of total probability and the second step derives by the already observed fact that, by the PPS assumptions, $P(\hat{Y}|Y) = Q(\hat{Y}|Y)$. By taking into account the fact that $Q(Y = 0) = 1 - Q(Y = 1)$, and $P(\hat{Y} = 1|Y = 1) \approx \text{tpr}_V$ and $P(\hat{Y} = 1|Y = 0) \approx \text{fpr}_V$, and the fact that $Q(\hat{Y} = 1)$ is observed (we called its estimate c_1), it then follows that

$$Q(Y = 1) \approx \frac{c_1 - \text{fpr}_V}{\text{tpr}_V - \text{fpr}_V} \quad (15)$$

The above solution is well-known in the literature, but we will come back to this in the next section. Since $c_1 = \text{FP}_U + \text{TP}_U$ and $Q(Y = 1) \approx \text{FN}_U + \text{TP}_U$, Eq. 15 leads to the following equation in the binary case

$$\text{FP}_U + \text{FN}_U \cdot (\text{fpr}_V - \text{tpr}_V) + \text{TP}_U \cdot (1 + \text{fpr}_V - \text{tpr}_V) = \text{fpr}_V \quad (16)$$

In the multiclass case, there are $(n-1)$ such independent equations. The derivation is slightly more complicated, since it entails writing Eq. 14 as a system of linear equations, one per class, that involves the class-conditional rates $P(\hat{Y} = j|Y = i)$. This problem can be written in matrix form as $\mathbf{c} = \mathbf{R}^\top \mathbf{q}$, where $\mathbf{c} \in \Delta^{n-1}$ is our vector of (normalized) counts, $\mathbf{R}$ is our matrix of class-conditional rates, as before, and $\mathbf{q} = (q_1, \ldots, q_n) \in \Delta^{n-1}$ is the sought class distribution on the test data. The solution thus comes down to solving the system as $\mathbf{q} = (\mathbf{R}^\top)^{-1}\mathbf{c}$, then taking $(n-1)$ test prevalence values from $\mathbf{q}$ and adding the corresponding equations

$$q_i = \sum_{j=1}^{n} u_{ij}, \quad \text{for } i \in \{1, \ldots, n-1\} \quad (17)$$

This batch of equations adds $(n-1)$ new independent equations. Adding the first equation (Eq. 6), the $(n-1)$ equations from the second batch (Eq. 9), the $(n-1)^2$ equations from the third batch (Eq. 11), and the $(n-1)$ equations from the second batch (Eq. 17), we obtain a system of exactly $(1+(n-1)+(n-1)^2+(n-1)) = n^2$ independent equations with n^2 unknowns.

Finally, note that the system can be written in matrix form as $\mathbf{AX} = \mathbf{b}$, with $\mathbf{X}$ our unknowns. In degenerate cases in which the PPS assumptions do not hold, or in which the estimated class prevalence values deviate much from the true value, the system of equations might not be solvable via matrix inversion ($\mathbf{X} = \mathbf{A}^{-1}\mathbf{b}$), or the solution might be unfeasible (e.g., some of the $\hat{u}_{ij}$ might not be in $[0,1]$). In cases like this, we instead resort to solving the constrained problem $\mathbf{X}^* = \arg\min_{\mathbf{X} \in \Delta^{n-1}} ||\mathbf{AX} - \mathbf{b}||_2$. Either way, the solution to the system is something modern software packages solve very quickly.

Recap for the binary case: If we write $q_1 = \frac{c_1 - \text{fpr}_V}{\text{tpr}_V - \text{fpr}_V}$ (Eq. 15), in the binary case the system of four equations is:

$$\begin{cases} \text{TN}_U + \text{FN}_U + \text{FP}_U + \text{TP}_U = 1 \\ \text{FP}_U + \text{TP}_U = c_1 \\ \text{TP}_U \cdot (\text{tpr}_V - 1) + \text{FN}_U \cdot \text{tpr}_V = 0 \\ \text{FN}_U + \text{TP}_U = q_1 \end{cases}$$

which has the solution

$$\begin{aligned} \text{TN}_U &= 1 - c_1 + q_1(\text{tpr}_V - 1) \\ \text{FN}_U &= q_1(1 - \text{tpr}_V) \\ \text{FP}_U &= c_1 - q_1 \cdot \text{tpr}_V \\ \text{TP}_U &= q_1 \cdot \text{tpr}_V \end{aligned} \tag{18}$$

We can now derive the closed form of many well-known evaluation measures under PPS. For example, vanilla accuracy and F_1 as computed on the test contingency table $\mathbf{U}$ (Eqs. 19 and 22) can be expressed in compact form with the aid of $\mathbf{c}$ and $\mathbf{q}$ (Eqs. 20 and 23), or expanded as functions that use only $\mathbf{c}$ along with counts from the validation contingency table $\mathbf{V}$ (Eqs. 21 and 24), as

$$\text{Acc}(h, U) = \frac{\text{TP}_U + \text{TN}_U}{\text{TP}_U + \text{TN}_U + \text{FN}_U + \text{FP}_U} \tag{19}$$

$$= 1 + q_1(2\text{tpr}_V - 1) - c_1 \tag{20}$$

$$= \frac{((\text{TN}_V + \text{FP}_V) \cdot c_1 - \text{FP}_V)(\text{TP}_V - \text{FN}_V)}{\text{TN}_V \text{TP}_V - \text{FP}_V \text{FN}_V} + 1 - c_1 \tag{21}$$

$$F_1(h, U) = \frac{2 \cdot \text{TP}_U}{2 \cdot \text{TP}_U + \text{FP}_U + \text{FN}_U} \tag{22}$$

$$= \frac{2 q_1 \cdot \text{tpr}_V}{c_1 + q_1} \tag{23}$$

$$= \frac{2 \cdot \text{TP}_V(c_1 \cdot \text{TN}_V - c_0 \cdot \text{FP}_V)}{c_1 \cdot \text{TN}_V(2 \cdot \text{TP}_V + \text{FN}_V) - \text{FP}_V(\text{FN}_V + \text{TP}_V \cdot c_0)} \tag{24}$$

Note that the compact formulas above (Eqs. 20 and 23) are generic, and can be used also in cases in which the class prevalence $\mathbf{q}$ is obtained by means of other techniques. We discuss this topic in the following section.

4.2 Enhancing LEAP via Quantification

The fourth batch of equations from the previous section requires estimating the class prevalence values q_i of the test set. The solution we have shown in Eqs. 14 and 15 are actually borrowed from a well-known method in the quantification literature called Adjusted Classify & Count (ACC) [1,8,9]; the method is also dubbed Black-Box Shift Estimator (BBSE) in [15]. This method is of particular interest for our derivation since it only involves components (as the tpr and fpr) which can be derived from the validation contingency table. ACC was originally devised for binary quantification but it was later extended to the multiclass case by Firat [7] and later improved by Fernandes Vaz et al. [6] and Bunse [2]. The implementation we use in this paper does indeed adhere to the improvements brought about by [2], which comes down to better handling the cases in which the matrix of the class-conditional rates $\mathbf{R}^\top$ is not invertible.

ACC is by no means the only method available for estimating class prevalence, nor the most sophisticated one. There is an entire body of literature devoted to devising better ways for predicting class prevalence [5,11]. In this section, we propose an enhanced version of our LEAP method that defers the fourth batch of equations to more sophisticated quantification algorithms.

More formally, a quantifier is a function $\lambda : \mathbb{N}^\mathcal{X} \to \Delta^{n-1}$ mapping bags (or multi-sets) of instances from the input space $\mathcal{X}$ to the probability simplex. We focus our attention to the so-called distribution matching approaches, that take a permutation-invariant function Φ to represent samples, and solve for $\mathbf{q}$ (the sought test prevalence vector) the equation

$$\mathbf{t} = \mathbf{z}^\top \mathbf{q} \tag{25}$$

where $\mathbf{t} = \Phi(U)$ is the representation of test sample U and $\mathbf{z} = [\Phi(V_1), \ldots, \Phi(V_n)]$ contains the class-wise representations of validation sets $V_i = \{\mathbf{x}_k : (\mathbf{x}_k, y_k) \in V, y_k = i\}$. Specifically, we adopt the KDEy-ML variant proposed in [17] (hereafter simply called KDEy) which relies on kernel density estimation for representing the samples as density functions in the probability simplex. KDEy then solves Eq. 25 via maximum likelihood, as the minimization problem

$$\mathbf{q}^* = \arg\min_{\mathbf{q} \in \Delta^{n-1}} \mathcal{D}_{\mathrm{KL}}(\mathbf{t} || \mathbf{z}^\top \mathbf{q}) \tag{26}$$

where $\mathcal{D}_{\mathrm{KL}}$ is the well-known Kullback-Leibler divergence between the density model of the test set and the density model of a mixture parameterized by $\mathbf{q}$. We call this extension LEAP$_{\mathrm{KDEy}}$.

5 Experiments

In this section we turn to describing the experiments we have carried out in order to assess the effectiveness of our LEAP variants.

5.1 Experimental Setup

The **effectiveness measure** we use in order to assess the quality of the predictions of CAP methods, is absolute error (AE). For a generic classifier accuracy measure A, AE is defined as $|A(h,U) - \hat{A}(h,U)|$, where $A(h,U)$ is the true accuracy of h on U (which in lab experiments we can determine since we know the actual labels of U), and where $\hat{A}(h,U)$ is the accuracy of h on U as predicted by the CAP method.

As our **classifier accuracy measure** A we here employ vanilla accuracy (Acc – see Eq. 4). Aside from the fact that it is one of the most important measures of classifier effectiveness, the reason why we choose Acc is that some of our baselines only work for this evaluation measure. Therefore, for the sake of fair experimental comparison, we adopt Acc as our only target measure.

The **experimental protocol** we adopt is described as follows. Given a labelled collection $D = \{(\mathbf{x}_i, y_i)\}_{i=1}^m$, we split it into a training set (70%) and a test set U (30%) via stratified sampling. We further split the training set into a set L (50%) that we use for training a classifier h, and a validation set V (50%) that we use for training a CAP method, using stratification. We train and test all competing methods on the same partitions. We then extract, from U, 1000 test samples $U_1, U_2, \ldots, U_{1000}$ of $|U_i| = 100$ elements each. For each such extracted test sample, we first draw, uniformly at random, a vector of prevalence values from the unit simplex Δ^{n-1} (using the Kraemer sampling algorithm [20]), and then draw bags of instances with replacement from the corresponding classes in U so as to satisfy the required prevalence distribution. For each test sample U_i we ask the CAP method to predict the accuracy (Acc) of h. We then compare (via AE) the predicted score with the true one and report averaged values across all 1000 tests. Note that L, V (as well as the original U) come from the same distribution P, while the test samples U_i extracted from U instead come from a different distribution Q which is related to P via PPS (since the extraction protocol indeed simulates PPS).

Although our methods are natively multiclass, in this paper we mainly concentrate on **binary classification**, and defer a more in-depth experimentation on multiclass problems to future work. As our **datasets**, we use the 29 datasets from the UCI machine learning repository [14] used in [17]. The number of instances vary from a minimum of 150 instances (iris.2, iris.3) to a maximum of 5,473 (pageblocks.5), while the number of features vary from a minimum of 3 (haberman) to a maximum of 256 (semeion). The training sets display varying degrees of balance, ranging from datasets with only 2.1% positive instances (pageblocks.5), to almost perfectly balanced datasets (mammographic), to datasets with 77.8% positive instances (ctg.1).

For generating the **classifiers** we consider four learning algorithms. In particular, we report results for the case in which h is learned via logistic regression (LR), kNN with $k = 10$, support vector machines (SVM) with the radial basis function kernel, and multilayer perceptron (MLP) with hidden sizes of 100 and 15 neurons and the ReLU activation function. In all cases, we rely on the implementations provided in scikit-learn [18], leaving the rest of the hyperparameters at their default values.

As the **baseline methods** against which we test the performance of our LEAP methods, we use DoC [12] and ATC-MC (hereafter ATC) [10] (see Sect. 2).

We do not report results for Mandoline [4] since the results we have obtained for it were not competitive with the rest of the methods.[2] We also report the results of Naive, a method that assumes that the training and test data are IID, and report the vanilla accuracy score obtained in the contingency table generated in the validation set V by h. For ATC we borrow the implementation provided by the authors, while the rest of the methods are implemented by us.

In order to assess the extent to which the quality of the surrogate quantifier impacts on the estimated accuracy, we include two additional baselines. The first one acts as a lower-bound baseline, in which the quantifier is taken to be the simplest possible method: the Classify and Count (CC); we denote this variant LEAP_{CC}. The second one, instead, acts as an upper-bound baseline, in which the quantifier is replaced by an oracle that always returns, as the prevalence values of the test set, the true class proportions; we denote this variant LEAP_Φ. For the sake of consistency, we will hereafter denote LEAP_{ACC} to our "vanilla" LEAP method. As recalled from Sect. 4.2, we denote $\text{LEAP}_{\text{KDEy}}$ to our enhanced variant in which we employ KDEy as our quantifier.

For the implementation of the quantification algorithms (ACC and KDEy) we rely on the QuaPy[3] package [16]. The code to reproduce all our experiments is available on GitHub.[4]

5.2 Results

Table 1 shows the results we have obtained in our experiments in terms of AE (lower is better). Overall, our results show that $\text{LEAP}_{\text{KDEy}}$ obtains the best results (averaged across all datasets) for all four classifiers. $\text{LEAP}_{\text{KDEy}}$ obtains 14 best results for LR, 7 best results for kNN, 14 best results for SVM, and 7 best results for MLP, out of 29 datasets. Conversely, the vanilla variant LEAP fares better than ATC on average for LR and kNN, and better than DoC on average for SVM, but it only manages to beat both methods simultaneously for MLP. The superiority of $\text{LEAP}_{\text{KDEy}}$ over LEAP_{ACC} speaks in favor of replacing the basic quantifier ACC with a more sophisticated quantification method like KDEy for predicting the test class prevalence values.

However, these results also show that all methods (including our proposed variants) sometimes fail loudly in certain cases where other methods would instead fare reasonably well. This is the case of, e.g., (sonar, LR) in which DoC yields the smallest error (0.070) while $\text{LEAP}_{\text{KDEy}}$ obtains the worst score (0.113). Figure 1 shows a selection of prototypical diagonal plots for LR, including sonar (the full set of plots for all datasets and all classifiers can be consulted online[5]).

---

[2] We ran experiments using the code provided by the authors. The likely reason why the performance of Mandoline is not competitive with other baselines in our experiments is that Mandoline heavily relies on user-defined transformations, called *slicing functions*, which require manual intervention, for each dataset, on the part of the designer.
[3] https://github.com/HLT-ISTI/QuaPy.
[4] https://github.com/lorenzovolpi/LEAP.
[5] https://github.com/lorenzovolpi/LEAP/tree/main/plots.

Table 1. Values of AE obtained in our experiments for different CAP methods. **Boldface** indicates the best method for a given (dataset, classifier) pair. Superscripts † and ‡ denote the methods (if any) whose scores are not statistically significantly different from the best one according to a Wilcoxon signed-rank test at different confidence values: symbol † indicates $0.001 < p\text{-value} < 0.01$ while symbol ‡ indicates $0.01 \leq p\text{-value}$. Cells are colour-coded so as to facilitate readability and allow for quick comparisons across results. Intense green highlights the best result while intense red highlights the worst one; milder colours are used in the obvious way.

	LR					k-NN					SVM					MLP				
	Naive	ATC	DoC	LEAP$_{ACC}$	LEAP$_{KDEy}$	Naive	ATC	DoC	LEAP$_{ACC}$	LEAP$_{KDEy}$	Naive	ATC	DoC	LEAP$_{ACC}$	LEAP$_{KDEy}$	Naive	ATC	DoC	LEAP$_{ACC}$	LEAP$_{KDEy}$
balance.1	.020	.034	.026	.023	.025	.033	.032	.056	**.018**	.020	**.015**	.023	.016	.017	.026	.021	**.012**	.016	.020	.021
balance.3	.024	.048	.024	.016‡	**.015**	.019	.076	.019	**.015**	.029	.037	.039	.036‡	**.035**	.050	.033	.019†	.029	.031	**.018**
breast-cancer	.030	.026	.016	.017	**.013**	.019	.035	**.012**	.013	.014	.014	.012†	.015	**.016**	.011	.030	.034	.020	**.016**	.019
cmc.1	.072‡	.080	.106	.069	**.067**	.130	.236	.106	.071	**.061**	.160	.128	**.181**	.105	.086	.077	.094	.081	**.036**	.036‡
cmc.2	.267	.183	**.102**	.124	.074	.267	.336	**.072**	.188	.074†	.320	.235	**.142**	.252	.157	.226	.134	**.043**	.078	.080
cmc.3	.184	.167	**.098**	.174	.086	.157	.260	**.073**	.098	.078‡	.270	.192	**.172**	.252	.205	.104	.061	.059	.065	**.051**
ctg.1	.045	.041	.030	.033	**.019**	.081	.039	.035	.032	**.026**	.155	.070	.085	.050	**.026**	.041	.038	.024	**.016**	.019
ctg.2	.087	.037	**.021**	.039	.061	.165	.194	**.063**	.132	.116	.376	.372	.383	.252	**.219**	.074	.022†	.021	.022	.026
ctg.3	.091	.036	.072	**.033**	.049	.200	.192	**.027**	.057	.038	.315	.239	.166	.144	**.050**	.078	.032	**.024**	.026	.041
german	.160	.124	**.050**	.128	.039	.226	.368	**.122**	.240	.171	.279	.211	.248	.200	**.113**	.099	.060	**.038**	.042	.053
haberman	.224	.162	**.154**	.183	.120	.253	.175	.363	.386	**.101**	.298	.158	**.090**	.252	.236	.174	**.089**	.379	.271	.210
ionosphere	**.063**	.064‡	.109	.084	.096	.118	**.037**	.074	.095	.092	.046	**.022**	.028	.076	.094	**.045**	.059	.198	.099	.119
iris.2	.203	.065	.102	.135	**.076**	.030	**.018**	.024	.037	.036	.056	.051	**.084**	.058	.041	.022	.018	**.016**	.026	.037
iris.3	.030	.051	.065	.075	**.118**	.113	.049	**.034**	.082	.046	**.083**	.021	.027	.032	.035	.035	.035	.035	.035	**.034**
mammographic	.064	.087	.074	.062	**.046**	.068	.148	.052	.070	**.043**	.060	.035†	**.062**	.059	.034	.054	.069	.082	**.040**	.057
pageblocks.5	.368	.293	.114	.102	**.068**	.477	.378	**.101**	.496	.177	.449	.223	**.137**	.170	.160	.367	.266	**.089**	.097‡	.092‡
semeion	.125	.067	.037	.036	**.031**	.170	.060	**.024**	.073	.037	.160	.113	.073	.045	**.029**	.102	**.023**	.043	.040	.030
sonar	.113	.109	**.070**	.108	.113	.172	**.112**	.298	.266	.234	.155	.108	**.082**	.210	.219	.083	.112	.104	**.060**	.073
spambase	.022	.026	**.020**	.021†	.023	.083	.054	**.038**	.043	.042†	.139	.086	.110	.062	**.054**	.048	.040	.034	.030‡	**.029**
spectf	.157	**.064**	.139	.145	.102	.142‡	**.135**	.174	.236	.157	.340	.280	.219	.496	**.107**	.124	.175	.084	**.055**	.046
tictactoe	.018	.018	.013	.008	**.008†**	.079	.074	.061	**.021**	.024	.022	.023	**.024**	.010	.011‡	.026	.016	.024	.016	**.013**
transfusion	.269	.194	**.081**	.122	.072	.286	.357	**.091**	.201	.099‡	.289	.241†	**.232**	.286	.296	.290	.176	**.128**	.197	.087
wdbc	.021	.040	.037	.024‡	**.027**	.045	.021	.024	.021	**.019**	.058	.048	.055	.023	**.021**	.135	.088	.101	**.040**	.065
wine.1	.039	.056	**.022**	.061	.084	.041	.069	**.036**	.078	.115	.030	.069	.025‡	**.042**	.027	.032†	.034	.052	**.059**	—
wine.2	.023	**.020**	.023	.027	.033	.105	.311	.234	**.051**	.056	.082	.093	.160	.065	**.056**	.046	.035‡	.035	.039	.044
wine.3	.021	.028	.013	.012	**.005**	.093	.095	.083	.064‡	**.058**	.304	.320	**.116**	.252	.251	.013	.013	.013	**.013**	.013
wine-q-red	.049	.048	.045‡	.049	**.043**	.045	.068	.046	**.034**	.043	.135	.111	.116	.078	**.098**	.033	.040	.034	.030	.033
wine-q-white	.130	.084	**.051**	.058	.053‡	.100	.269	.055	.057	**.050**	.282	.221	**.156**	.426	.367	.035	.059	.037	**.032**	.035
yeast	.220	.131	**.085**	.209	.123	.123	.095	.062	**.123**	.097	.168	.107	.111	.106	**.084**	.089	.059	**.036**	.039†	.040†
Average	.108	.082	.062	.075	**.058**	.132	.148	**.085**	.114	.074	.174	.135	.115	.140	**.103**	.087	.066	.064	**.054**	.051

These plots display the estimated accuracy as a function of the true accuracy; a perfect CAP method would only contain values lying on the diagonal from (0,0) to (1,1). For the sake of clarity, we have omitted the Naive method, which always predicts accuracy values lying on a horizontal line and is thus uninteresting. Note that ATC often overestimates accuracy (haberman, cmc.2), while DoC tends to be less biased in this respect. DoC slightly overestimates accuracy in cmc.2 and underestimates it in iris.2, but generally produces values close to the diagonal (e.g., german). LEAP$_{KDEy}$ is generally closer to the diagonal but also produces high errors at the extremes in some datasets (see haberman). Notably, LEAP$_{ACC}$ often shows a marked pattern in which the predicted values appear to lie on parallel horizontal and equidistant lines (haberman, cmc.2). The explanation for this pattern is that the estimation for the class prevalence in LEAP$_{ACC}$ relies on the crisp counts of the classifier predictions (Eq. 15), which are naturally discrete. This effect is corrected in the enhanced variant LEAP$_{KDEy}$, which instead allows for continuous predictions via KDEy. Turning back to the results of Table 1, it is also noteworthy that there are some cases (few though) in which the Naive method beats all other competitors. All in all, this suggests that further research is still needed to better determine which method to apply in which scenario.

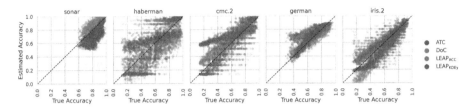

Fig. 1. Diagonal plots illustrating the correlation between true accuracy (x-axis) and estimated accuracy (y-axis) of a classifier trained via LR, for different CAP methods.

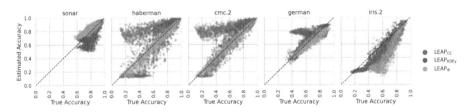

Fig. 2. Diagonal plots showing the same experimental setup as 1, but confronting LEAP$_{KDEy}$ (lower-bound) with LEAP$_{CC}$ and LEAP$_{\Phi}$ (upper-bound).

Concerning the baselines ATC and DoC, our experiments seem to indicate that ATC is weaker than DoC when facing PPS. On average, DoC beats ATC consistently in all cases. ATC shows erratic behavior for kNN, obtaining 4 best results and 14 worst results, thus performing, on average, even worse than the Naive baseline for this classifier. DoC obtains 6 best results for LR, 13 for kNN, 10 for SVM, and 8 for MLP, out of 29 datasets. Still, LEAP$_{KDEy}$ shows a relative error reduction, with respect to DoC, of 5.96% in LR, 12.7% in kNN, 11.1% in SVM, and 20.3% in MLP.

To better understand the performance leeway that depends on the choice of the quantifier, we confront LEAP$_{KDEy}$ against LEAP$_{CC}$ (the worst quantifier) and against LEAP$_{\Phi}$ (the perfect quantifier). As shown in Fig. 2, LEAP$_{CC}$ performs markedly worse than LEAP$_{KDEy}$, thus confirming that KDEy plays a

Table 2. Pearson correlation between the random variables "classifier accuracy" (in terms of Acc) and "CAP performance" (in terms of AE).

	Naive	ATC	DoC	LEAP	LEAP$_{KDEy}$
LR	−0.8340	−0.7241	−0.5050	−0.5560	**−0.3392**
k-NN	−0.7724	−0.8221	**−0.3130**	−0.6519	−0.3987
SVM	−0.8640	−0.8228	−0.5975	−0.5338	**−0.3785**
MLP	−0.7222	−0.6807	−0.3955	**−0.3686**	−0.3825
Average	−0.7981	−0.7624	−0.4527	−0.5276	**−0.3747**

key role in the effectiveness of the method. On the other side, LEAP$_\Phi$, represents an ideal upper bound for LEAP. Nonetheless, the results show that the gap between LEAP$_\Phi$ and LEAP$_{KDEy}$ is relatively narrow, which indicates our proposed variant is a good method in practice.

Regarding the classifiers under study, the accuracy of MLP seems to be the easiest one to determine, as witnessed by the fact that the averaged CAP errors are the smallest across the four classifiers. While CAP accuracy should not be confused with classifier accuracy, we observed a strong negative Pearson correlation between these two random variables, approximately $r = -0.6$ for all classifiers and CAP methods (p-value $\ll 0.0001$). This indicates a general trend where better classifier performance is associated with lower CAP error. This observation is consistent with the average CAP results obtained for our classifiers, which achieved the following average accuracy scores across all datasets: $\text{Acc}(\text{LR}, U_i) = 0.7843 \pm 0.2052$, $\text{Acc}(k\text{NN}, U_i) = 0.7275 \pm 0.1963$, $\text{Acc}(\text{SVM}, U_i) = 0.7089 \pm 0.2622$, and $\text{Acc}(\text{MLP}, U_i) = 0.8148 \pm 0.1739$. This strong correlation is undesirable because we want our CAP methods to perform well regardless of classifier accuracy. Table 2 shows the correlation values for each pair of (CAP method, classifier). Although still biased, LEAP$_{KDEy}$ exhibits the smallest correlation on average. This, along with the fact that LEAP$_{KDEy}$ achieves the lowest CAP errors on average across all classifiers, suggests that LEAP$_{KDEy}$ is a noteworthy new contender in the CAP arena.

In preliminary experiments we have carried out (here omitted for reasons of space), we did not find any method (including ours) that stands out in terms of performance in the multiclass case. For this experiment, we considered 21 multiclass datasets from the UCI machine learning repository [14]; our results seem to indicate there is no statistically significant difference among the methods tested (including Naive). The number of classes surely makes the problem so difficult that it ends up hindering the relative merits of the different systems. We plan to investigate this issue more thoroughly in the near future.

Concerning running times, all CAP methods exhibit fast performance, never exceeding 3.5 s during training and 23 milliseconds during test. DoC tends to be the slowest method during training, with LEAP$_{ACC}$ and LEAP$_{KDEy}$ requiring times up to two orders of magnitude lower. However, LEAP$_{KDEy}$ appears to be the slowest method of the lot at test time, performing up to ten times slower than the baselines. Nevertheless, test times range from milliseconds to tens of milliseconds for both LEAP$_{\text{ACC}}$ and LEAP$_{\text{KDEy}}$, indicating that both variants are well-suited for real-world applications.

6 Conclusions

In this paper we have proposed a new method for classifier accuracy prediction (CAP) specifically designed to address situations affected by prior probability shift (PPS). Our method consists of estimating the entries of the contingency table that results from classifying the test set, from which any evaluation measure can be computed. As a result, and in contrast to previously existing methods from the literature, LEAP is not limited to any specific evaluation measure. We have also drawn a connection with the field of quantification, and through many experiments we have shown that, when our method is endowed with state-of-the-art quantifiers, it tends to outperform existing CAP methods under PPS.

Possible directions for future work include considering multi-objective optimization problems, in which our method could be especially useful due to its ability to estimate more than one evaluation measure simultaneously. We also plan to conduct a systematic examination of the multiclass setting, aiming to enhance the accuracy estimation of our methods in such scenario. We intend to extend our method to other types of dataset shift, thus accommodating different assumptions on the data distributions, such as covariate shift. Additionally, we plan to investigate the ability of our method to estimate the accuracy of classifiers based on deep learning.

Acknowledgments. Funded by the QuaDaSh project (P2022TB5JF) "Finanziato dall'Unione europea—Next Generation EU, Missione 4 Componente 2 CUP B53D23026250001".

References

1. Bella, A., Ferri, C., Hernández-Orallo, J., Ramírez-Quintana, M.J.: Quantification via probability estimators. In: Proceedings of the 11th IEEE International Conference on Data Mining (ICDM 2010), pp. 737–742. Sydney, AU (2010)
2. Bunse, M.: On multi-class extensions of Adjusted Classify and Count. In: Proceedings of the 2nd International Workshop on Learning to Quantify (LQ 2022), pp. 43–50. Grenoble, IT (2022)
3. Chen, J., Liu, F., Avci, B., Wu, X., Liang, Y., Jha, S.: Detecting errors and estimating accuracy on unlabeled data with self-training ensembles. In: Proceedings of the 35th Conference on Neural Information Processing Systems (NeurIPS 2021), pp. 14980–14992. Virtual Event (2021)
4. Chen, M.F., Goel, K., Sohoni, N.S., Poms, F., Fatahalian, K., Ré, C.: MANDOLINE: model evaluation under distribution shift. In: Proceedings of the 38th International Conference on Machine Learning (ICML 2021), pp. 1617–1629. Virtual Event (2021)
5. Esuli, A., Fabris, A., Moreo, A., Sebastiani, F.: Learning to quantify. Springer Nature, Cham, CH (2023). https://doi.org/10.1007/978-3-031-20467-8

6. Fernandes Vaz, A., Izbicki, R., Bassi Stern, R.: Prior shift using the ratio estimator. In: Proc. of the International Workshop on Bayesian Inference and Maximum Entropy Methods in Science and Engineering, pp. 25–35. Jarinu, BR (2017)
7. Firat, A.: Unified framework for quantification. arXiv:1606.00868v1 [cs.LG] (2016)
8. Forman, G.: Counting positives accurately despite inaccurate classification. In: Gama, J., Camacho, R., Brazdil, P.B., Jorge, A.M., Torgo, L. (eds.) ECML 2005. LNCS (LNAI), vol. 3720, pp. 564–575. Springer, Heidelberg (2005). https://doi.org/10.1007/11564096_55
9. Forman, G.: Quantifying counts and costs via classification. Data Min. Knowl. Disc. **17**(2), 164–206 (2008)
10. Garg, S., Balakrishnan, S., Lipton, Z.C., Neyshabur, B., Sedghi, H.: Leveraging unlabeled data to predict out-of-distribution performance. In: Proceedings of the 10th International Conference on Learning Representations (ICLR 2022). Virtual Event (2022)
11. González, P., Castaño, A., Chawla, N.V., del Coz, J.J.: A review on quantification learning. ACM Comput. Surv. **50**(5), 74:1–74:40 (2017)
12. Guillory, D., Shankar, V., Ebrahimi, S., Darrell, T., Schmidt, L.: Predicting with confidence on unseen distributions. In: Proceedings of the IEEE/CVF International Conference on Computer Vision (ICCV 2021), pp. 1134–1144. Montreal, CA (2021)
13. Jiang, Y., Nagarajan, V., Baek, C., Kolter, J.Z.: Assessing generalization of SGD via disagreement. In: Proceedings of the International Conference on Learning Representations (ICLR 2022). Virtual Event (2022)
14. Kelly, M., Longjohn, R., Nottingham, K.: The UCI machine learning repository. https://archive.ics.uci.edu
15. Lipton, Z.C., Wang, Y., Smola, A.J.: Detecting and correcting for label shift with black-box predictors. In: Proceedings of the 35th International Conference on Machine Learning (ICML 2018), pp. 3128–3136. Stockholm, SE (2018)
16. Moreo, A., Esuli, A., Sebastiani, F.: QuaPy: a Python-based framework for quantification. In: Proceedings of the 30th ACM International Conference on Knowledge Management (CIKM 2021), pp. 4534–4543. Gold Coast, AU (2021)
17. Moreo, A., González, P., del Coz, J.J.: Kernel density estimation for multiclass quantification. arXiv:2401.00490 [cs.LG] (2024)
18. Pedregosa, F., et al.: Scikit-learn: machine learning in Python. J. Mach. Learn. Res. **12**, 2825–2830 (2011)
19. Schölkopf, B., Janzing, D., Peters, J., Sgouritsa, E., Zhang, K., Mooij, J.M.: On causal and anticausal learning. In: Proceedings of the 29th International Conference on Machine Learning (ICML 2012). Edinburgh, UK (2012)
20. Smith, N.A., Tromble, R.W.: Sampling Uniformly From the Unit Simplex. Johns Hopkins University, Tech. rep. (2004)
21. Storkey, A.: When training and test sets are different: characterizing learning transfer. In: Quiñonero-Candela, J., Sugiyama, M., Schwaighofer, A., Lawrence, N.D. (eds.) Dataset shift in machine learning, pp. 3–28. The MIT Press, Cambridge, US (2009)

22. Sugiyama, M., Nakajima, S., Kashima, H., Buenau, P., Kawanabe, M.: Direct importance estimation with model selection and its application to covariate shift adaptation. In: Proceedings of the 21st Conference on Advances in Neural Information Processing Systems (NIPS 2007), pp. 1433–1440. Vancouver, CA (2007)
23. Tasche, D.: Comments on Friedman's method for class distribution estimation. arXiv:2405.16666 [cs.LG] (2024)
24. Zhang, K., Schölkopf, B., Muandet, K., Wang, Z.: Domain adaptation under target and conditional shift. In: Proceedings of the 30th International Conference on Machine Learning (ICML 2013), pp. 819–827. Atlanta, US (2013)

Pairwise Difference Learning for Classification

Mohamed Karim Belaid[1,2(✉)], Maximilian Rabus[2], and Eyke Hüllermeier[3]

[1] IDIADA Fahrzeugtechnik GmbH, Munich, Germany
karim.belaid@idiada.com
[2] Dr. Ing. h.c. F. Porsche AG Stuttgart, Stuttgart, Germany
{extern.karim.belaid, maximilian.rabus2}@porsche.de
[3] Ludwig-Maximilians-Universität Munich, Munich, Germany
eyke@if.lmu.de

Abstract. Pairwise difference learning (PDL) has recently been introduced as a new meta-learning technique for regression. Instead of learning a mapping from instances to outcomes in the standard way, the key idea is to learn a function that takes two instances as input and predicts the difference between the respective outcomes. Given a function of this kind, predictions for a query instance are derived from every training example and then averaged. This paper extends PDL toward the task of classification and proposes a meta-learning technique for inducing a PDL classifier by solving a suitably defined (binary) classification problem on a paired version of the original training data. We analyze the performance of the PDL classifier in a large-scale empirical study and find that it outperforms state-of-the-art methods in terms of prediction performance. Last but not least, we provide an easy-to-use and publicly available implementation of PDL in a Python package.

Keywords: Supervised learning · Multiclass classification · Meta-learning

1 Introduction

Pairwise difference learning (PDL) has recently been introduced independently by Tynes et al. [19] and Wetzel et al. [22] as a meta-learning technique for regression, which transforms the original task of learning to predict outcomes for individual inputs into the task of learning to predict *differences* between the outcomes of input *pairs*: Noting that the value of a function f at a point x can be written "from the perspective" of any other point x' as $f(x) = f(x') + \Delta(x, x')$ with $\Delta(x, x') = f(x) - f(x')$, the simple idea of PDL is to train an approximation $\tilde{\Delta}$ of the difference function Δ and obtain predictions of new outcomes $y = f(x)$ by averaging over the predicted differences to the outcomes in the training data:

$$y \approx \frac{1}{N} \sum_{i=1}^{N} y_i + \tilde{\Delta}(x, x_i) \qquad (1)$$

One of the main motivations of PDL is the quadratic increase of the training data: If the original training data contains N data points $(x_1, y_1), \ldots, (x_N, y_N)$, the difference function can be trained on potentially $\mathcal{O}(N^2)$ training examples of the form $((x_i, x_j), y_i - y_j)$. This increase might be specifically useful in the "small data" regime (even if the transformed examples are of course no longer independent of each other). Moreover, note that the prediction (1) benefits from a statistically useful averaging effect.

Building on the basic idea of PDL, we make the following contributions. We extend the idea of PDL toward the task of classification and propose the PDL classifier, a meta-learning approach that transforms any multiclass classification problem into a single binary problem. This innovative method leverages the concept of learning inter-class differences, leading to demonstrably improved average prediction accuracy (Sect. 3). We introduce the "pairwise difference learning library" (pdll) on PyPI, which incorporates our implementation of the PDL classifier and ensures compatibility with any Sklearn ML model (Sect. 3.5). We conduct a large-scale experimental analysis of PDL and compare the results to state-of-the-art ML estimators (Sect. 4). We discuss the architecture of PDL and how it can lead to an improvement of the accuracy (Sect. 5).

2 Related Work

Tynes et al. introduced pairwise difference regressor [19], a novel meta-learner for chemical tasks that enhances prediction performance, compared to random forest and provides robust uncertainty quantification. In computational chemistry, estimating differences between data points helps mitigate systematic errors [19]. In parallel, Wetzel et al. used twin neural network architectures for semi-supervised regression tasks (TNNR), focusing on predicting differences between target values of distinct data points [22]. The approach of Wetzel et al. enabled training on unlabelled data points when paired with labeled anchor data points. By ensembling predicted differences between target values, the method achieved high prediction performance for regression problems. While conceptually similar to the pairwise difference regressor in emphasizing differences between data points, it is specialized to neural network architectures for semi-supervised regression tasks [23]. TNNR draws inspiration from Siamese networks, which use identical neural networks to learn similarities between input pairs. Unlike Siamese networks, which focus on the similarity task, TNNR adapts this architecture to predict target value differences for regression tasks.

The pairwise difference learning (PDL) literature has since then, evolved into diverse methodologies and applications. Spiers et al. measured sample similarity in chemistry, emphasizing spectral shape differences using metrics like Euclidean and Mahalanobis distances. They extended the approach by calculating a Z-score which offers insights into prediction accuracy, facilitating outlier detection and model adaptation [18]. PDL was developed mainly for regression tasks. It can also be adapted to targets that might be known or only bounded. Example of target annotations could be $y = 5.3$, $y < 2.1$, or $y > 6.5$. Predicting an increase/decrease between a pair is a possible solution [8].

PDL regressor with its variants has demonstrated efficacy across various applications, including regression with image input [11], learning chemical properties [7], quantum mechanical reactions [5], and drug activity ranking [21]. PDL has demonstrated scalability, successfully training on large datasets like Fe-DFTB, with 11,656 features and 12,726 data points [19]. This underscores PDL's capability to handle extensive datasets, reinforcing its utility in large-scale problems.

3 PDL Classification

Consider a standard setting of supervised (classification) learning: Given a set of training data

$$\mathcal{D} = \{(x_i, y_i)\}_{i=1}^N \subset \mathbb{R}^d \times \mathcal{Y},$$

comprised of training instances in the form of feature vectors $x_i \in \mathbb{R}^d$ together with observed discrete labels $y \in \mathcal{Y} = \{1, \ldots, K\}$, and assumed to be generated i.i.d. according to an underlying (unknown) joint probability measure P, the task is to learn a predictor PDC : $\mathbb{R}^d \to \mathcal{Y}$ with low risk (expected loss). The PDL classifier transforms the original training data $\mathcal{D}$ into the new data

$$\mathcal{D}_{pair} = \{(z_{i,j}, y_{i,j}) \mid 1 \leq i, j \leq N\}, \tag{2}$$

where $z_{i,j} = \phi(x_i, x_j)$ is a *joint* feature representation of the instance pair (x_i, x_j) and

$$y_{i,j} = \begin{cases} 0 & \text{for } y_i \neq y_j, \\ 1 & \text{for } y_i = y_j \end{cases}. \tag{3}$$

Thus, we seek a binary classifier $\gamma : \mathbb{R}^d \times \mathbb{R}^d \to [0, 1]$ that, given two instances x and x' as input, predicts whether or not the respective classes y and y' are the same. More specifically, we assume γ to be a probabilistic classifier, so that $\gamma(x, x') \in [0, 1]$ is the probability that $y = y'$. Deterministic classifiers that return a binary label as a prediction are treated as degenerate $\{0, 1\}$-valued probabilistic classifiers. Leveraging the joint feature representation, γ is of the form $\gamma(x, x') = h(\phi(x, x'))$, where h is trained on the transformed data (2). To this end, any binary classification method can be used. Note, however, that the binary problem might be quite imbalanced, as the transformation (3) will produce much more negative (unequal) than positive (equal) examples. One can solve this issue by introducing class weights [13] to equalize the loss function of the classifier γ. As for the joint feature representation, the original proposal was to define $z_{i,j}$ as a concatenation of x_i and x_j. It turned out, however, that expanding this vector by the difference $x_i - x_j$ has a positive influence on performance [19], wherefore we also adopted this representation in our work.

Since (class) equality is a symmetric relation, γ is naturally expected to be symmetric in the sense that $\gamma(x_i, x_j) = \gamma(x_j, x_i)$. By adding both $(\phi(x_i, x_j), y_{i,j})$

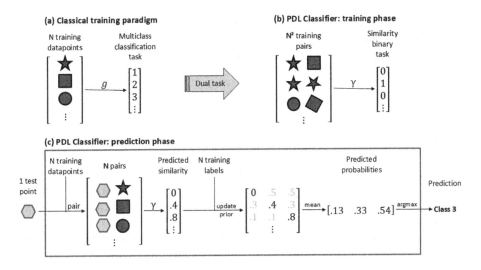

Fig. 1. Illustration of the PDL classifier.

and $(\phi(x_j, x_i), y_{j,i})$ to $\mathcal{D}_{pair}$, this symmetry can also be reflected in the training data. But even then, however, γ is not necessarily guaranteed to preserve symmetry. Therefore, we additionally "symmetrize" the predictor as follows:

$$\gamma_{sym}(x_i, x_j) = \frac{\gamma(x_i, x_j) + \gamma(x_j, x_i)}{2} \quad (4)$$

Given a query x_q, we finally estimate the probability of class labels $y \in \mathcal{Y}$ as follows: Considering each training example (x_i, y_i) as a piece of evidence for the unknown class y_q, the semantics of the above prediction suggests that the probability of the event $y_q = y_i$ is given by (4). More formally, $P(E) = \gamma_{sym}(x_q, x_i)$, where E denotes the event $y_q = y_i$ (and hence $P(\neg E) = 1 - \gamma_{sym}(x_q, x_i)$). Let p denote the prior distribution on the class labels $\mathcal{Y}$ (which can easily be estimated by relative frequencies on the training data). This distribution is then updated by conditioning it on the (uncertain) event E, which yields the following posterior suggested by (x_i, y_i):

$$p_{post,i}(y) = \begin{cases} \gamma_{sym}(x_q, x_i) & \text{if } y = y_i \\ \dfrac{p(y) \cdot (1 - \gamma_{sym}(x_q, x_i))}{1 - p(y_i)} & \text{otherwise} \end{cases} \quad (5)$$

Thus, the (posterior) probability of y_i is fixed to $\gamma_{sym}(x_q, x_i)$, and all other probabilities are rescaled in a proportional way, to guarantee that the sum of posterior probabilities adds to 1. Finally, we average over the evidences from all training examples to obtain

$$p_{post}(y) = \frac{1}{N} \sum_{i=1}^{N} p_{post,i}(y). \quad (6)$$

In case a deterministic prediction is sought, the class with the highest (estimated) probability is chosen:

$$\hat{y}_q = \arg\max_{y \in \mathcal{Y}} p_{post}(y) \tag{7}$$

3.1 Uncertainty Quantification

Interestingly, the PDL approach also offers a natural approach to uncertainty quantification, a topic that has received increasing attention in the recent machine learning literature. In particular, recent research has focused on the distinction between so-called *aleatoric* uncertainty (caused by inherent randomness in the data) and *epistemic* uncertainty (caused by the learner's incomplete knowledge of the true data-generating process) — we refer to [12] for a detailed exposition of this topic.

Within the Bayesian approach, these two types of uncertainty can be captured by properties of the posterior predictive distribution, which in turn can be approximated through ensemble learning [15]. In a sense, PDL parallels this approach, with each anchor playing the role of an ensemble member, and (6) mimicking Bayesian model averaging. This suggests the following quantification of aleatoric (AU), epistemic (EU), and total uncertainty (TU) of a prediction, with H denoting Shannon entropy.:

$$\text{TU} = H(p_{post}(y)) = H\left(\sum_{i=1}^{N} p_{post,i}(y)\right)$$

$$\text{AU} = \frac{1}{N}\sum_{i=1}^{N} H(p_{post,i}(y))$$

$$\text{EU} = \text{TU} - \text{AU}$$

Theoretically, these measures are justified based on a well-known result from information theory, according to which entropy additively decomposes into conditional entropy and mutual information [6]. Broadly speaking, the more uniform the (averaged) distribution p_{post}, the higher the total uncertainty, and the more diverse the individual predictions $p_{post,i}$, the higher the epistemic uncertainty.

3.2 Illustration

Thanks to its novel structure, the PDL classifier can solve a multiclass classification task by training exactly one instance of a base learner on a binary task. Figure 1 illustrates the PDL classifier algorithm, showcasing both the training and prediction phases on a simple multiclass task. Figure 1a shows a traditional multiclass classifier g that maps each of the N training data points to their assigned unique class label (star, square, or circle). In Fig. 1b, PDL classifier transforms the data by creating N^2 pairs of data points. During training, a binary classifier γ learns to distinguish between pairs that belong to the same class (positive label) from pairs of different classes (negative label). In Fig. 1c, given one query input, the PDL classifier pairs it with each of the N training data points. For each pair, the classifier predicts a probability of similarity (belonging to the same class). Predicted probabilities are mapped to the column corresponding to the initial label of each training point. Missing posterior probabilities, in grey, are estimated by updating the prior probabilities, assuming a uniform distribution in this example. Finally, averaging across all training points yields the predicted probabilities for each class. The class with the highest predicted probability is chosen as the final class label for the query point (e.g., Class 3).

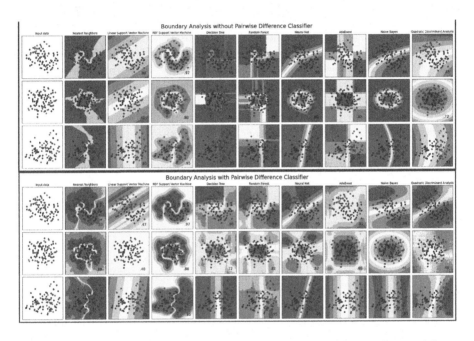

Fig. 2. Comparing learned patterns using PDL classifiers and baseline models.

Figure 2 illustrates the patterns learned by nine baseline models across three 2D datasets. The baseline 3-Nearest-Neighbor (3-NN) classifier can only predict four probabilities: $0, \frac{1}{3}, \frac{2}{3}$, and 1. This is evident in the figure, where each dataset shows only four discrete regions. In contrast, when using PDL on top of 3-NN, the predicted probability is derived from the averaging over N discrete predictions. This results in more refined and precise probability estimates. Despite the simplicity of some estimators, PDL leverages more complex patterns. The contrast between DecisionTree with and without PDL clearly illustrates PDL's capability to learn non-linear patterns. The underfitting observed when incompatible base models learn corrupted patterns underscores the critical role of the choice of base learners.

3.3 Choice of Base Learners

As already said, PDL can theoretically be implemented with any (probabilistic) binary classifier as a base learner — or, stated differently, it can be used as a wrapper for any (binary or multinomial) classifier. Practically, however, some classifiers might be more suitable as base learners and others less.

One thing one should keep in mind is that even if the original data $\mathcal{D}$ is i.i.d., independence will be lost for $\mathcal{D}_{pair}$ as soon as the same instance x_i is paired with various other instances. This is very similar to the setting of metric learning, where models are also trained on pairs of data points [2]. In practice, although many machine learning algorithms turn out to be quite robust against violations of the i.i.d. assumption [14], some methods may be concerned more than others.

Another important aspect is the joint feature representation $z = \phi(x, x')$. For example, by defining z as a concatenation of x, x', and the difference $x - x'$, one obviously introduces (perfect) multicollinearity. Again, while this is problematic for some machine learning methods, notably linear models [19, p.8], others can deal with this property more easily.

While an in-depth analysis of the suitability of different base learners is beyond the scope of this paper, we generally found that non-parametric methods are more robust and tend to show better performance than parametric ones. In our experimental evaluation, we will therefore mainly use tree-based methods, which have the additional advantage of being fast to train.

3.4 Complexity

Looking at the complexity of PDL, suppose the complexity of a base learner to be $\mathcal{O}(p(N, M, F, K))$, where $p(\cdot)$ is polynomial in the number of training points (N), the number of test points (M), the number of input features (F), and the number of output classes (K). The complexity of PDL is then $\mathcal{O}(p(N^2, 2MA, 3F, 2))$: The training points are scaled to N^2 pairs; the features are scaled to $3F$ (F features of point x_i, F features of point x_j, and F features of the difference $x_i - x_j$. This feature construction technique for PDL has demonstrated previously improved results [19]); Each test point is paired with the A anchor points. Pairs are duplicated twice to obtain their symmetry. Thus, M test predictions of PDL require $2MA$ predictions using the base learner. The number of output classes K shrinks to 2 since the model is asked to predict whether the pair of points has a similar class.

3.5 PDL Library

Our library[1] includes a Python implementation of the PDL classifier, adhering to the Scikit-learn standards. Consequently, integrating the PDL classifier into existing codebases is straightforward, requiring minimal modifications. As demonstrated in the example below, only two additional lines of code are needed:

---

[1] Link: https://github.com/Karim-53/pdll.

```
1  !pip install pdll
2  from pdll import PairwiseDifferenceClassifier    # Added
3  X, y = load_data()
4  model = RandomForestClassifier()
5  model = PairwiseDifferenceClassifier(model)      # Added
6  model.fit(X, y)
7  ...
```

4 Evaluation

In this section, we test PDC on various public datasets from OpenML [20] and compare it to 7 Scikit-learn state-of-the-art learners.

4.1 Data

OpenML provides a diverse range of datasets, many of which are small, with 37% having less than 600 data points. This study focuses on small datasets, for which the pairwise learning approach is presumably most effective. We applied grid search CV for parameter tuning, leveraging the search space from TPOT [16]. To accommodate our grid search setup, we subsampled the search space to 1,000 parameter combinations per estimator. Following dataset selection constraints similar to the OpenML-CC18 benchmark [3], we randomly selected 99 datasets (see summary statistics in Fig. 3). Although these datasets are relatively small, the effective data size for PDC is quadrupled due to the pairing, reaching 360000 data points. We also monitored class imbalance using the "minority class" metadata, which represents the percentage of the minority class relative to the total size of each dataset. Considering the 7 baseline models, we performed 5 times 5-fold CV with an inner 3-fold grid search CV, totaling 66 528 000 train-test runs and 3 weeks wall-time on an HPC.

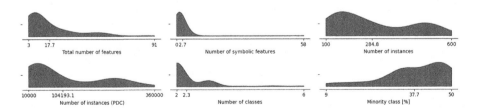

Fig. 3. Distribution of key characteristics of the 99 OpenML classification datasets (minimum, mean, maximum).

4.2 Data Processing Pipeline

Using scikit-learn [17], we implemented a common data processing pipeline for all runs, with standardization for numeric features, one-hot encoding for nominal features, and ordinal encoding for ordinal features. Since PDL needs the pair difference $x_i - x_j$ as additional inputs, processed features are all treated as numeric when applying the difference.

4.3 Performance Measures

We measure performance in terms of the (macro) F1 score, which is arguably more meaningful than the standard misclassification rate in the case of imbalanced data. In binary classification, the F1 score is defined as the harmonic mean of precision and recall. For multinomial problems, the macro version of this score is the (unweighted) mean of the F1 scores for the individual class:

$$\mathrm{Macro} F1 = \frac{1}{K} \sum_{i=1}^{K} F1_i,$$

where $F1_i$ is the F1 score on the ith class (treating test examples of this class as positive and all others as negative). We also report the improvement of PDL over the base learner in terms of the difference $\Delta F1 = \mathrm{Macro} F1_{PDC} - \mathrm{Macro} F1_{base}$. We aggregate the results using the mean ± standard error.

To aggregate the results of all datasets, we count the number of wins/losses by comparing the average performance of models over 25 runs (5 times 5-fold CV) per dataset. A win is counted when PDC's average score is higher than the baseline; a loss is counted otherwise. To determine the number of significant wins/losses, a Student's t-test is conducted for each dataset to assess the statistical significance of the difference in performance. A significant win/loss is recorded when the p-value of the t-test is below a predetermined threshold $\alpha = 0.05$. In some cases, there may be a tie in the average scores, leading to instances where the number of wins and losses does not sum to 99, which is the total number of datasets benchmarked.

As an alternative to counting wins and losses, and despite being aware of the questionable nature of this statistic, we also average performance over datasets. While average performance may offer a preliminary overall impression, it should always be interpreted with caution. Additionally, to provide a more comprehensive assessment, we report further metrics and results within the repository of the PDL library. This allows for a more nuanced evaluation of the method's performance across different scenarios, enabling users to explore detailed outcomes beyond average performance.

4.4 Results

First, the PDL classifier, on top of ExtraTrees, obtained the best average Macro F1 score over the 99 datasets, outperforming all baselines, see Fig. 4. In Table 1, the ratio of significant wins demonstrates an advantage for the PDL classifier, suggesting that, in a one-to-one comparison, PDL is more likely to outperform its equivalent baseline.

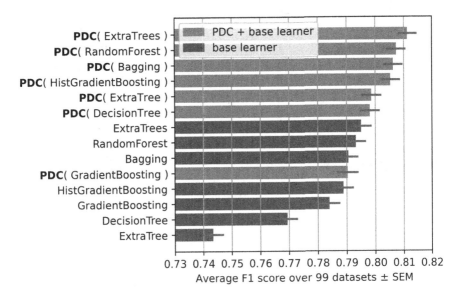

Fig. 4. Comparing average Macro F1 score of optimized baseline classifiers and PDL classifiers.

The PDL classifier can be viewed as a method to simplify the trained model. As shown in Fig. 4, the test performance of PDC(DecisionTree) is equivalent to or better than that of the seven benchmarked state-of-the-art estimators. This indicates that, with the help of PDL, training a single tree can compete with ensemble methods that typically train around 100 trees. In this context, explaining a single tree may provide a more straightforward solution.

Analyzing the Unique Contribution. While PDL classifiers have high probabilities of outperforming baseline models in a one-to-one comparison, the ultimate goal of a data scientist is to obtain the best performance on each dataset. Before introducing PDC, the maximum achievable Macro F1 score was 0.8112 ± 0.0035 averaged over the 99 datasets. With the help of PDC, we achieve higher scores in 75 datasets, and the new record becomes 0.8243 ± 0.0031. This advance showcases the unique contribution of PDC to the field of ML compared to existing algorithms. Moreover, PDC offers not only an important unique contribution to the record but also the highest contribution. Indeed, PDC's leave-one-out contribution to this record is $0.8243 - 0.8112 = 0.0131$ while popular estimators like HistGradientBoosting get no unique contribution, i.e., they are not able to outperform all other estimators on any of the 99 datasets, see Table 2. PDC's contribution is even 32 times more important than the best baseline.

Table 1. Comparing baseline classifiers to PDC using 99 datasets.

Classifier	Significant wins		Wins		Average Test Macro F1			
	base	PDC	base	PDC	base	± sem	PDC	± sem
Bagging	3	26	27	70	0.7906	0.0035	**0.8062**	0.0034
DecisionTree	2	50	22	76	0.7694	0.0037	**0.7982**	0.0034
ExtraTree	1	61	9	90	0.7434	0.0037	**0.7987**	0.0035
ExtraTrees	6	24	21	77	0.7951	0.0036	**0.8113**	0.0035
GradientBoosting	9	23	25	72	0.7839	0.0037	**0.7903**	0.0039
HistGradientBoosting	2	32	15	82	0.7888	0.0037	**0.8053**	0.0035
RandomForest	5	27	22	73	0.7933	0.0035	**0.8073**	0.0034

Table 2. Unique contribution of each estimator to the average Macro F1 score using the best optimized model on each dataset.

Estimator	Unique contribution	Wins
ExtraTree	0	0
HistGradientBoosting	0	0
RandomForest	0.00002	1
Bagging	0.00004	2
GradientBoosting	0.00006	2
DecisionTree	0.00020	10
ExtraTrees	0.00041	9
PDC	**0.01312**	**75**

Analyzing the Overfitting. PDL classifiers demonstrate a clear advantage in reducing overfitting. Indeed, looking at the 199 cross-validation (CV) runs in which both the baseline and PDL classifier obtain non-significant differences in train Macro F1 scores, we notice that PDL classifiers consistently exhibited a smaller train-test gap. A lower overfitting is observed when grouping by base classifier, see Table 3. This even remains true without conditioning on non-significantly different train scores. To further substantiate this, we analyzed the impact of dataset size on overfitting. As shown in Fig. 5, coloring by the number of instances reveals that our PDL method improves PDL's test scores and reduces overfitting, regardless of dataset size. This underscores the robustness of the PDL approach, particularly in small datasets, reinforcing its utility across various data scales.

Table 3. Comparing test Macro F1 on the subset of runs where train scores are not significantly different.

Estimator	# CV runs	Baseline Macro F1 Train	Baseline Macro F1 Test	PDC Macro F1 Train	PDC Macro F1 Test	Test $\Delta F1$	Test p-value
Bagging	20	0.998	0.835	0.999	**0.859**	0.024	10^{-15}
DecisionTree	14	0.950	0.884	0.955	**0.895**	0.011	10^{-05}
ExtraTree	11	0.915	0.844	0.924	**0.861**	0.017	10^{-04}
ExtraTrees	26	0.985	0.828	0.991	**0.853**	0.025	10^{-16}
GradientBoosting	58	0.930	0.822	0.926	**0.840**	0.018	10^{-17}
HistGradientBoosting	52	0.961	0.820	0.963	**0.839**	0.019	10^{-19}
RandomForest	18	0.992	0.855	0.997	**0.881**	0.026	10^{-11}
Total	199	0.958	0.832	0.960	**0.852**	0.020	10^{-74}

5 Why Does PDL Yield Improved Performance?

The empirical results reveal that the PDL classifier significantly improves over the baseline methods. In this section, we elaborate on possible reasons for this improvement.

5.1 Combining Instance-Based and Model-Based Learning

A distinguishing feature of PDL is a unique combination of (local) *instance-based* learning and (global) *model-based* learning. Like the well-known nearest-neighbor principle, a prediction for a new query is produced by other instances from the

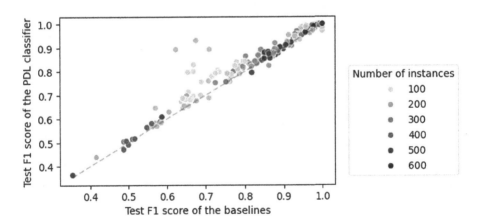

Fig. 5. Overfitting analysis: Comparison of test F1 scores between baselines and PDCs on datasets with non-significantly different training scores (Sizes ranging from 100 to 600 instances).

training set, namely the anchor points; yet, as opposed to NN, these instances are not restricted to nearby cases but can be located anywhere in the instance space. This becomes possible through the model-based component of PDL, namely the classifier γ, which is a global model that generalizes over the entire instance space. Broadly speaking, by constructing γ, the classifier learns how to transfer class information from one data point to another.

Of course, there are other learning methods with similar characteristics. For example, instead of using a predefined distance function, the nearest neighbor method can be instantiated with a distance function δ that is learned on the training data. Metric learning typically proceeds from sets of similar instances (belonging to the same class) and dissimilar instances (belonging to different classes), and seeks to learn a function δ that keeps the distance low for the former while making it high for the latter [2,10]. In a sense, this is indeed quite comparable to PDL, especially because both δ and γ are two-place functions taking pairs of instances as input. Moreover, γ could indeed also be seen as a kind of distance measure, if "distance" is defined in terms of "probability of belonging to the same class". Yet, PDL is arguably more flexible, because γ is not required to satisfy properties of a distance or metric.

5.2 Simplification Through Binary Reduction

Another advantage of PDL is *simplicity*: The original classification task is effectively reduced to a *binary* problem, namely, to decide whether or not two instances share the same class label. This is comparable to binary decomposition techniques such as one-vs-rest and all-pairs [4, p.202], which reduce a single multinomial classification problem to several binary problems. Instead, PDL constructs a *single* binary problem, although the total number of training examples produced essentially coincides for all methods (it is roughly quadratic in the size of the original data). In any case, binary problems are normally easier to solve, which explains the improved classification accuracy commonly reported for reduction techniques. In this regard, a decomposition can even be useful for methods that are able to handle multinomial problems right away (such as decision trees).

5.3 Error Reduction Through Averaging

Last but not least, by instantiating the global model for every anchor and collecting predictions from all of them, PDL benefits from a kind of ensemble effect and reduces error through *averaging*. In particular, since prediction errors of individual anchors can be compensated by other anchors, PDL is able to reduce the variance of the prediction error. Again, this is somewhat comparable to the nearest-neighbor method. Given the model γ, the anchor predictions can even be considered as independent[2], which, under the simplified assumption of

---

[2] Of course, this independence is lost if the anchor points are also part of the data used to train γ.

homoscedasticity, means that the prediction error is reduced by a factor of $1/\sqrt{A}$, with A the number of anchors [23, p.4].

Even if these assumptions may not be completely satisfied, an expected improvement through averaging can clearly be observed in empirical studies. Figure 6 represents four cases encountered with four different datasets and DecisionTree as a baseline. We compare the loss of the baseline (baseline loss) with the actual PDL loss, i.e., the loss given all available anchors. The empirical approximation curve is meant to show how the loss depends on the number of anchor points. Its value at A is produced by averaging the performance over randomly selected anchor subsets of size A. The curve goes from the average loss when only one anchor is used (γ loss) until reaching the actual PDL loss. The theoretical approximation curve is an optimal fit of a theoretical model to the empirical approximation, namely, the decrease of the error under the ideal assumption of independent prediction errors distributed normally with mean μ and standard deviation σ. As can be seen, even if this assumption may not fully hold, the two curves deviate but slightly.

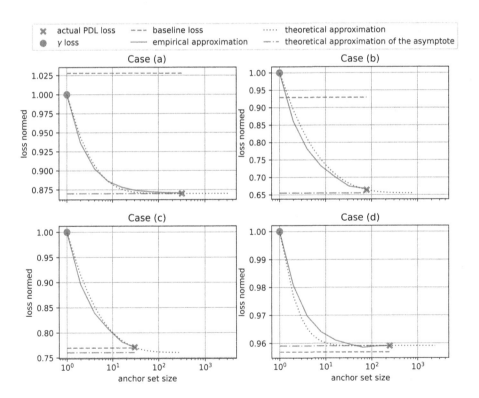

Fig. 6. Effect of the anchor set size on PDC's loss relative to the baseline.

In case (a), the loss of the PDC's γ estimator is better than the loss of the baseline model. As expected in this case, PDC is better than the baseline with any number of anchors. In case (b), the baseline loss is between γ loss and PDC's loss. With the theoretical approximation, we estimate how many anchors are enough to outperform the baseline. In case (c), the baseline model is better than PDL. Nevertheless, the theoretical approximation allows us to estimate the additional anchors needed to outperform the baseline and the best reachable loss. It becomes less and less efficient to improve the score by adding more anchors. It might become more interesting, starting from a certain size, to work more on the base learner or the data quality. In case (d), the baseline model is even better than the approximated asymptote because learning the dual problem is more difficult. Adding more anchors is less likely to help.

6 Conclusion

Building on the concept of pairwise difference learning (PDL), we proposed the PDL classifier (PDC), a meta-learner able to reduce a multiclass classification problem into a binary problem. Our extensive empirical evaluation across 99 diverse datasets demonstrates that PDL consistently outperforms state-of-the-art machine learning models, resulting in improved F1 scores in a majority of cases. This highlights PDL's effectiveness in enhancing performance over baseline methods, facilitated through its straightforward integration via our Python package. To explain its strong performance, we also elaborated on several properties and features of PDC.

Future research directions include the exploration of instance (anchor) weighting through regularization or Shapley data importance [9] and interaction [1]. Moreover, we plan to elaborate more closely on PDC's potential to quantify predictive uncertainty (cf. Sect. 3.1).

In conclusion, PDL emerges as a practical solution for improving ML models, offering versatility and performance improvements across diverse applications. Its adaptability and robust performance make it a valuable addition to the ML toolkit, promising more accurate and reliable predictions in various domains.

References

1. Belaid, M.K., El Mekki, D., Rabus, M., Hüllermeier, E.: Optimizing Data Shapley Interaction calculation from $\mathcal{O}(2^N)$ to $\mathcal{O}(TN^2)$ for KNN models. arXiv preprint arXiv:2304.01224 (2023)
2. Bian, W., Tao, D.: Learning a distance metric by empirical loss minimization. In: Proc. IJCAI, International Joint Conference on Artificial Intelligence (2013)
3. Bischl, B., et al.: OpenML benchmarking suites. arXiv preprint arXiv:1708.03731 (2017)
4. Bishop, C.: Pattern recognition and ML. Springer **2**, 183 (2006)
5. Chen, Y., Ou, Y., Zheng, P., Huang, Y., Ge, F., Dral, P.O.: Benchmark of general-purpose ML-based quantum mechanical method AIQM1 on reaction barrier heights. J. Chem. Phys. **158**(7) (2023)

6. Depeweg, S., Hernandez-Lobato, J., Doshi-Velez, F., Udluft, S.: Decomposition of uncertainty in Bayesian deep learning for efficient and risk-sensitive learning. In: Proc. ICML, 35th International Conference on Machine Learning. Stockholm, Sweden (2018)
7. Fralish, Z., Chen, A., Skaluba, P., Reker, D.: DeepDelta: predicting ADMET improvements of molecular derivatives with deep learning. J. Cheminformatics **15**(1), 101 (2023)
8. Fralish, Z., Skaluba, P., Reker, D.: Leveraging bounded datapoints to classify molecular potency improvements. RSC Med. Chem. (2024)
9. Ghorbani, A., Zou, J.: Data Shapley: equitable valuation of data for ML. In: International Conference on ML, pp. 2242–2251. PMLR (2019)
10. Globerson, A., Roweis, S.: Metric learning by collapsing classes. Adv. Neural Inf. Process. Syst. **18** (2005)
11. Hu, J., et al.: Exploring a general convolutional neural network-based prediction model for critical casting diameter of metallic glasses. J. Alloy. Compd. **947**, 169479 (2023)
12. Hüllermeier, E., Waegeman, W.: Aleatoric and epistemic uncertainty in machine learning: an introduction to concepts and methods. Mach. Learn. **110**(3), 457–506 (2021). https://doi.org/10.1007/s10994-021-05946-3
13. King, G., Zeng, L.: Logistic regression in rare events data. Polit. Anal. **9**(2), 137–163 (2001)
14. Kutner, M.H., Nachtsheim, C.J., Neter, J., Li, W.: Applied linear statistical models. McGraw-hill (2005)
15. Lakshminarayanan, B., Pritzel, A., Blundell, C.: Simple and scalable predictive uncertainty estimation using deep ensembles. In: Proceedings of NeurIPS, 31st Conference on Neural Information Processing Systems. Long Beach, California, USA (2017)
16. Olson, R.S., Bartley, N., Urbanowicz, R.J., Moore, J.H.: Evaluation of a tree-based pipeline optimization tool for automating data science. In: Proceedings of the Genetic and Evolutionary Computation Conference 2016, pp. 485–492 (2016)
17. Pedregosa, F., et al.: Scikit-learn: ML in Python. J. ML Res. **12**, 2825–2830 (2011)
18. Spiers, R.C., Norby, C., Kalivas, J.H.: Physicochemical responsive integrated similarity measure (PRISM) for a comprehensive quantitative perspective of sample similarity dynamically assessed with NIR spectra. Anal. Chem. (2023)
19. Tynes, M., et al.: Pairwise difference regression: a ML meta-algorithm for improved prediction and uncertainty quantification in chemical search. J. Chem. Inf. Model. **61**(8), 3846–3857 (2021)
20. Vanschoren, J., van Rijn, J.N., Bischl, B., Torgo, L.: OpenML: networked science in ML. SIGKDD Explor. **15**(2), 49–60 (2013). https://doi.org/10.1145/2641190.2641198, http://doi.acm.org/10.1145/2641190.264119
21. Wang, Y., King, R.D.: Extrapolation is not the same as interpolation. In: International Conference on Discovery Science, pp. 277–292. Springer (2023)
22. Wetzel, S.J., Melko, R.G., Tamblyn, I.: Twin neural network regression is a semi-supervised regression algorithm. ML: Sci. Technol. **3**(4), 045007 (2022)
23. Wetzel, S.J., Ryczko, K., Melko, R.G., Tamblyn, I.: Twin neural network regression. Appl. AI Lett. **3**(4), e78 (2022)

SoBigData++: City for Citizens and Explainable AI

Explaining Urban Vehicle Emissions in Rome

Matteo Bohm[1]((✉)), Patricio Reyes[2], Mirco Nanni[3], and Luca Pappalardo[3,4]

[1] Sapienza University of Rome, Rome, Italy
bohm@diag.uniroma1.it
[2] Barcelona Supercomputing Center (BSC), Barcelona, Spain
patricio.reyes@bsc.es
[3] ISTI-CNR, Pisa, Italy
{mirco.nanni,luca.pappalardo}@isti.cnr.it
[4] Scuola Normale Superiore, Pisa, Italy

Abstract. Urban emissions are a significant challenge for city livability. Our work focuses on studying vehicle emissions in cities, using spatial and non-spatial models to understand their relationships with various urban features. We find that the spatial model demonstrates better performance and provides powerful insights into the influence of different predictors in various city areas. Our findings reveal that CO2 emissions in Rome are primarily linked to the presence of main arterial roads, population density, and road network density. However, the importance of these factors varies across different areas of the city. We also performed a what-if analysis to show that limiting the circulation of highly polluting vehicles may help reduce emissions, especially in city centres. Our research contributes to a better understanding of the complex relationships between the urban environment and the spatial variability of vehicle emissions in Rome.

Keywords: spatial analysis · GHG emissions · transport · GWR

1 Introduction

Scientists have been studying the relationship between traffic and the urban environment since the 1990s when it was first suggested that multiple urban factors must work together to create transportation benefits [8]. Achieving this synergy is challenging because the relationships between these dimensions are complex and can result in multifaceted outcomes [24,25,35].

One of the most extensively examined factors influencing urban emissions is population density. Nevertheless, the scientific literature yields conflicting outcomes. Some studies suggest that as cities grow, changes in population density exert a greater impact on emissions [11,13,22,30]. Other studies argue that larger or denser cities are not more emissions-efficient than smaller ones [10]. There is evidence indicating that higher population density may lead to reduced emissions per capita by improving accessibility and promoting efficient public transport [7]. At the same time, there are other works showing that it can also increase emissions due to road congestion [35]. The conflicting outcomes on the relationship between population density and urban emissions can be attributed to differences in definitions, datasets, and methodological approaches, even within the same geographical context.

Beyond population density, other factors such as socioeconomic characteristics, urban morphology, and the structure of city networks also have a noticeable impact on urban emissions [14,20]. For example, household size and personal wealth have a significant effect on emissions [2] and total road length has a super-linear relationship with CO2 emissions [20]. In China, CO2 emissions increase with economic growth, road density, and freight turnover, while urbanization level, population, and the number of private and public vehicles have the opposite effect [28]. At the same time, another study suggests that urbanization, economic growth, and industrialization drive emissions upwards, while service and technology levels act as mitigating factors [32]. Urban development patterns also have mixed effects: for example, $PM_{2.5}$ and NO_2 concentrations are correlated negatively with the city's compactness and positively with its fragmentation [15]. At the same time, polycentric networks can reduce emissions [17,31] but can also increase them by causing cross-commuting and traffic congestion [29,33].

The relationships between urban factors and vehicle usage or transport emissions can vary greatly depending on location. It is thus important to consider the specific characteristics of each city when studying urban emissions.

In this paper, we focus on the urban area of the Metropolitan City of Rome, Italy, which is participating in the EU Mission for climate-neutral and smart cities by 2030.[1] We test two hypotheses: (1) that nearby areas of Rome suffer similar levels of pollution and (2) that specific factors contribute to the vehicle emissions in Rome's neighbourhoods. To achieve these goals, we use both spatial and non-spatial models to analyze the emissions-city relationship. The models include simple linear regression, random forest regressor, and Geographically Weighted Regression (GWR) [5], a spatially-varying coefficient (SVC) model that captures spatial non-stationarity in the data. We compare the results of these models and show that a spatial regression model performs better in modelling the relationship between urban features and vehicle emissions. It also provides valuable insights into what influences the level of vehicle emissions and their variability in each city neighbourhood. In addition, we perform a what-if analysis in which we block the circulation of the most polluting vehicles and observe changes in regression coefficients. We find that while the importance of certain features remains unchanged, the coefficients of other features change significantly, leading to a shift in the relationship between features and emissions.

---

[1] https://ec.europa.eu/commission/presscorner/detail/en/IP_22_2591.

Our work goes into the direction of defining urban interventions tailored to city's needs [1,9,24], as a support for local emissions mitigation and urban sustainability plans.

2 Methodology

Rome's road network is quite sparse and varied, with considerable differences between neighbourhoods and a clear distinction between the centre and the periphery [18]. These spatial disparities significantly impact private transportation: Rome ranks first in Italy for the number of cars per inhabitant (629 for every 1000 inhabitants [4]) and 13th globally for the highest hours lost in traffic congestion (107 h per year [16]).

We consider the spatial subdivision of Rome in 155 urban areas (UAs) defined by the Italian Statistics Bureau (see Fig. 1). This choice represents a trade-off between the different official subdivisions within Rome: the Municipi (15 in total) are too broad, while the census sections (13,656 in total) are overly detailed. UAs have been widely recognized for their usefulness in research and have previously provided a valuable framework for studying spatial inequality in Rome [18]. Furthermore, a similar level of spatial granularity has proven effective in exploring accessibility patterns within Barcelona [12].

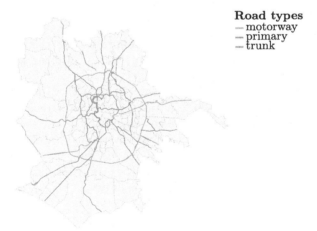

Fig. 1. The subdivision of Rome into the 154 urban areas (UAs) considered (in grey). Also, we show the roads that, in OpenStreetMap, are tagged as motorway (green), trunk (blue), or primary (orange) roads. Note that we exclude the area of Martignano as it is completely isolated from the others – i.e. it has no neighbours – and it is almost non-urban, with very sparse data. (Color figure online)

We have collected several open data for each of the urban areas (UAs). **Road Network.** We collect the road network of Rome from OpenStreetMap (OSM) [23] and use it to compute the following features for each UA:

- Road density: it is the total length of all roads within a UA divided by its area (in km^2), capturing the concentration of road infrastructure within a UA;
- High-speed road density: it is the total length of "motorway," "trunk," or "primary" roads within a UA (the three highest-priority road types according to OSM) divided by its area (in km^2), capturing the prevalence of major transportation arteries within a UA;
- Road betweenness centrality: it evaluates the road network's betweenness centrality for each UA, highlighting the significance of roads within the road network. We consider only the roads whose betweenness centrality falls within the 95th percentile of the distribution of betweenness centrality values across the entire city.

We also collect, for each UA, the density of traffic lights and the number of car accidents from the open data portal of the Municipality of Rome[2]. This dataset offers more detailed and granular information than the corresponding data available from OSM.

Points of Interest (POIs). The presence of amenities in UAs represents a source of human activity and, thus, possible causes of congestion and emissions. We collect the number and density of shops, education, services, and food amenities (130,244 POIs in total) from the open data portal of the Municipality of Rome [26]. We then compute, for each UA, the density (number of amenities per km^2) of each type of amenity and the density of all POIs in each UAs, regardless of their type.

Population. We collect the number of inhabitants per km^2 (population density) and the number of inhabitants living within 10 min of walking from the closest metro/tram station from MappaRoma [18][3].

Vehicles' Emissions. We use a dataset of road-level CO2 emissions generated by thousands of private vehicles in Rome [6]. The dataset is publicly available on the SoBigData platform.[4] Using these emission estimates at the road level, we compute the average CO2 emissions per road within each UA. Figure 2 shows the distribution of these emissions across the UAs and reveals a noticeable pattern: while some UAs exhibit notably high average CO2 emissions per road, most areas have low emissions levels along their roadways.

2.1 Feature Selection

To avoid multicollinearity, we select only features with a Variance Inflation Factor (VIF) < 5 [19]. We observe high VIF values for features related to shop density (10.2), food amenities (7.1), and service amenities (6.0) (see Table 1 for details). As a result, we replace the four features measuring the density of POIs with an overall measure of the density of all the types of POIs. After this adjustment, we recompute the VIFs and end up with ten features with VIF < 5 (see

---

[2] https://dati.comune.roma.it/.
[3] https://www.mapparoma.info.
[4] https://bit.ly/SBD_emissions.

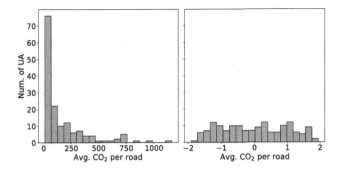

Fig. 2. Histograms showing the distribution of the average CO_2 emissions per road in each UA before (left panel) and after (right panel) Yeo-Johnson transformation.

Table 2). These ten features are used as predictors to estimate the average emissions per road, our dependent variable, in each UA. We then standardise the ten features and apply a Yeo-Johnson transformation [34] to the dependent variable to make its distribution more Gaussian-like (see Fig. 2).

Table 1. The initial set of features and the associated Variance Inflation Factor (VIF).

feature	VIF
street_density	3.6
density_traffic_lights	3.2
motorway_roads_density	1.2
trunk_roads_density	1.1
primary_roads_density	1.2
perc_roads_bc	1.2
car_accidents	1.6
density_shops	10.2
density_food_amenities	7.1
density_education_amenities	1.2
density_service_amenities	6.0
perc_people_close_to_public_transport	1.8
pop_density	3.7

2.2 Geographically Weighted Regression

In our analysis, we employ a Geographically Weighted Regression (GWR) as a spatial model [5]. One of the key attributes of GWR is its ability to produce regression coefficients that may vary across different locations. This disparity

Table 2. The ten features used as predictors of the average CO_2 emissions per road in our experiments. We also show each feature's Variance Inflation Factor (VIF).

category	feature	description	VIF
Road network	street_density	overall streets' length per km^2	3.5
Road network	density_traffic_lights	number of traffic lights per km^2	3.0
Road network	motorway_roads_density	overall length of motorway roads per km^2	1.2
Road network	trunk_roads_density	overall length of trunk roads per km^2	1.1
Road network	primary_roads_density	overall length of primary roads per km^2	1.2
Road network	perc_roads_bc	share of roads with a betweenness centrality above the 95th percentile of its distribution across the roads of the whole network	1.2
Road network	car_accidents	number of car accidents registered	1.4
POIs	density_all_amenities	total number of amenities (education, service, food) per km^2	2.2
Population	perc_people_close_to_public_transport	share of inhabitants living within 10 min walking from the closest metro/tram station	1.8
Population	pop_density	number of inhabitants per km^2	2.5

occurs because GWR constructs distinct Ordinary Least Squares (OLS) regressions, each corresponding to a specific UA. In each of these regression equations, the dependent and predictor variables are considered for the areas falling within a certain predefined bandwidth from the focal location.

The motivation for using GWR is twofold: first, non-spatial models cannot capture the spatial dimension of the phenomenon; second, GWR offers local interpretability with spatially varying coefficients. This allows us to understand how different areas of the city are affected by vehicle emissions, unlike non-spatial models that only provide insights into overall city-wide interactions.

We compare GWR with linear regression (LR) and random forest regression (RF) as baselines and assess the model performance using the coefficient of determination R^2 and the Median Absolute Error (MAE). Moreover, we examine the spatial autocorrelation of its standardized residuals by computing Moran's I, which measures the intensity of relationships among close data units [3,21]. The expected value of Moran's I under the null hypothesis of no spatial autocorrelation is $E(I) = \frac{1}{1-N}$, where N is the number of UA in our case. Therefore, the closer the Moran's I of the model's residuals to $E(I)$, the more randomly the residuals are distributed across the space.

3 Results

GWR achieves $R^2 = 0.84$ and MAE $= 0.29$ (Fig. 3c). This represents a 16% in R^2 and a 24% improvement in MAE over the best baseline (Table 3). GWR can account for the spatial heterogeneity of the phenomenon: the spatial autocorrelation of GWR's errors is lower than that of the baselines, with a Moran's I equal to 0.080 (Table 3 and Fig. 6c). This value is 52% lower than the best result obtained with non-spatial models (0.168), indicating that the errors are closer to being randomly distributed across the space.

Table 3. Metrics obtained for the three models (LR, RF, GWR).

metric	LR	RF	GWR
R^2	0.65	0.72	**0.84**
MAE	0.42	0.36	**0.29**
Moran's I	0.168	0.249	**0.080**

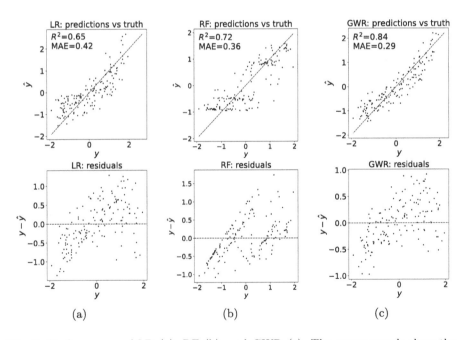

Fig. 3. Performances of LR (a), RF (b), and GWR (c). The upper panels show the predicted values (y-axis) against the true values (x-axis) of the normalized average CO_2 emissions per road in each UA. The lower panels show the residual plots.

Figure 4 shows the coefficients estimated by GWR for each UA and feature. The coefficients related to motorway density are significant for almost all the UAs (Fig. 4a), while those related to primary roads and POIs density are non-significant almost everywhere (Figs. 4c,g). The trunk road density is critical in the southern part of the city (Fig. 4b), where street density also shows significance but with opposite effects (Fig. 4d). This might be due to a sparse road network in these UAs, with a few important roads (e.g., Via Cristoforo Colombo, Via Pontina, and Via Del Mare) serving as main sources of traffic and emissions. The northeastern part of the city exhibits negative coefficients associated with population density (Fig. 4h). A few UAs in this area also show significant coefficients associated with the roads with high betweenness centrality (Fig. 4e). Finally, many UAs in the southern part of the city centre have significant coefficients linked to car accidents (Fig. 4f), indicating high traffic volume in these neighbourhoods. These results can be summarized by identifying the most important feature for each UA, as shown in Fig. 5. Motorway density has a significant impact from the south-east to the north-west. In the south-western part of the city, emissions are influenced by street density, while in the northeast, they are affected by population density, or in some UAs, by roads with high centrality in the network.

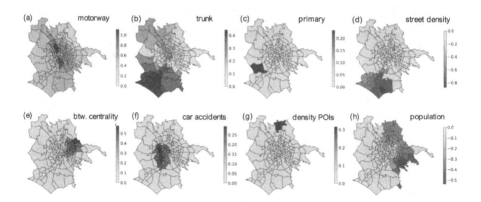

Fig. 4. Coefficients of GWR in each UA. Predictors `density_traffic_lights` and `perc_people_close_to_public_transport` are excluded because they have non-significant coefficients. Red and blue areas indicate positive and negative coefficients, respectively, with the intensity of the colour according to its (absolute) value. Grey areas indicate non-significant coefficients (with 95% confidence level).

What-if Analysis

Recent studies show that a few vehicles, known as gross polluters, are responsible for the greatest share of emissions in the city [6]. Given this finding, we aim to investigate the impact of blocking the circulation of these gross polluters. Furthermore, we seek to understand how this intervention would alter the relationships between various factors and emissions.

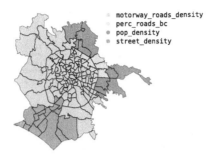

Fig. 5. The most important features from GWR in each UA.

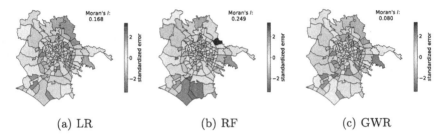

Fig. 6. The standardized errors obtained for LR (a), RF (b), and GWR (c), in each UA. Red areas correspond to positive errors (i.e. overestimation), while blue areas to negative ones (i.e. underestimation). We also show Moran's I computed on the errors. (Color figure online)

To answer these questions, we recompute the total emissions in the city by subtracting those generated by gross polluters. Then, we conduct a linear regression analysis to observe how the coefficients change compared to the scenario where emissions from gross polluters are included. Figure 7 shows the median coefficient of each feature across five repetitions of 5-fold cross-validation. We examine four scenarios: when the average level of emissions on the roads comes from all the vehicles (including gross polluters), and when it stems from all the vehicles except the top 1%, 5% or 10% most polluting ones.

As we gradually remove the most polluting vehicles, we observe that the coefficients of certain features do not change significantly. For instance, the density of motorways, trunk roads, and primary roads still strongly impact emissions (they contribute to generating more emissions). However, the coefficients related to population density, car accidents, and percentage of roads with high (betweenness) centrality become closer to zero, while the coefficients for traffic light density and all amenities density change from positive to negative. When fewer highly polluting vehicles circulate, the impact of population density and highly congested roads decreases, while the impact of traffic lights and amenity density increases. Since city centres have more traffic lights and amenities per square kilometre, restricting the circulation of highly polluting vehicles may help lower emissions in metropolitan centres. These results are supported by the

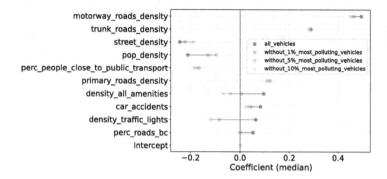

Fig. 7. Sensitivity analysis of LR coefficients within different scenarios of emissions. For each feature (y-axis), we show the median of the 25 coefficients obtained with repeated 5-fold cross-validation of LR when the emissions stem from all the vehicles (all_vehicles), and from all the vehicles but the top 1%, 5% and 10% most polluting ones (without_x%_most_polluting_vehicles), respectively.

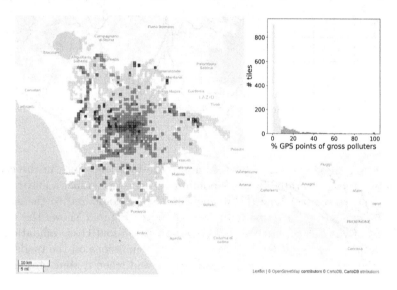

Fig. 8. Mobility of the 1% most polluting vehicles. We show a tessellation (1×1 km) of Rome. Each tile's colour represents the percentage of GPS points belonging to the 1% most polluting vehicles in it. The colouring of the tiles reflects the groups resulting from a k-means clustering with $k = 4$. The histogram shows the distribution of the percentage of GPS points describing the trips made by the 1% most polluting vehicles during the year across the tiles, with the bars coloured coherently with the four clusters.

observation of the movements of gross polluters, which exhibit slightly higher activity in the city centre compared to other parts of the city (see Fig. 8).

4 Conclusion

This work investigated the relationship between city characteristics and CO2 emissions in Rome. We showed that a spatial regression model outperforms non-spatial regression models and used the potentiality of spatially varying coefficients to understand the roles of different factors in each neighbourhood. We discovered that main arterial roads, population density, and road network density heavily influence the average CO2 emissions in neighbourhoods. In particular, the sparsity of Rome's road network, combined with the city having only a few arterial roads, negatively impacts the southwestern part of Rome. In the northeastern neighbourhoods, the scarcity of population and the presence of highly central roads are significant factors. In the rest of the city, motorways play a prominent role. Finally, blocking the circulation of the most polluting vehicles leads to lower emissions in areas with higher traffic lights and amenities densities. This measure would be particularly beneficial for the city centre.

The methodology we employed is flexible, scalable and reproducible. It can effectively identify spatial effects among various urban indicators. A potential extension of this work would be to test a framework proposed in a recent paper [27], the A-GWR, which combines machine learning models with spatial ones. Also, future works may involve more cities to validate the model in different contexts, as well as temporal variations in traffic dynamics, and to study how these factors influence the results obtained. In the meantime, our work contributes to understanding the complex relationships between the city and vehicle emissions on the roads while also providing valuable insights to urban policymakers to develop resilient and sustainable urban mobility plans.

Data and Code Availability.

The data and Python code to reproduce our analyses are collected in a public repository and are accessible at https://github.com/matteoboh/urbem.

Acknowledgements. This work has been partially supported by: EU project H2020 SoBigData++ G.A. 871042; PNRR (Piano Nazionale di Ripresa e Resilienza) in the context of the research program 20224CZ5X4 PE6 PRIN 2022 "URBAI - Urban Artificial Intelligence" (CUP B53D23012770006), funded by the European Commission under the Next Generation EU programme; and by PNRR - M4C2 - Investimento 1.3, Partenariato Esteso PE00000013 - "FAIR - Future Artificial Intelligence Research" - Spoke 1 "Human-centered AI", funded by the European Commission under the NextGeneration EU programme.

References

1. Baccile, S., Cornacchia, G., Pappalardo, L.: Measuring the impact of road removal on vehicular CO2 emissions. In: EDBT/ICDT Workshops (2024)
2. Baur, A., Thess, M., Kleinschmit, B., Creutzig, F.: Urban climate change mitigation in Europe - looking at and beyond the role of population density. J. Urban Plann. Dev. **140** (2013). https://doi.org/10.1061/(ASCE)UP.1943-5444.0000165

3. Bivand, R., Wong, D.: Comparing implementations of global and local indicators of spatial association. TEST **27** (2018). https://doi.org/10.1007/s11749-018-0599-x
4. Brinchi, S., Fuschiotto, A., Gigli, R., Severini, G., Mancinelli, G., Ciavatta, G., Di Mambro, F.: Rapporto mobilità 2021. Tech. rep, Roma Servizi per la Mobilità Srl, Roma, Italy (2021)
5. Brunsdon, C., Fotheringham, A.S., Charlton, M.E.: Geographically weighted regression: a method for exploring spatial nonstationarity. Geogr. Anal. **28**(4), 281–298 (1996). https://doi.org/10.1111/j.1538-4632.1996.tb00936.x
6. Böhm, M., Nanni, M., Pappalardo, L.: Gross polluters and vehicle emissions reduction. Nat. Sustain., 1–9 (2022). https://doi.org/10.1038/s41893-022-00903-x
7. Castells-Quintana, D., Dienesch, E., Krause, M.: Air pollution in an urban world: a global view on density, cities and emissions. Ecol. Econ. **189**, 107153 (2021). https://doi.org/10.1016/j.ecolecon.2021.107153
8. Cervero, R., Kockelman, K.: Travel demand and the 3DS: density, diversity, and design. Transp. Res. Part D: Transp. Environ. **2**(3), 199–219 (1997). https://doi.org/10.1016/S1361-9209(97)00009-6
9. Cornacchia, G., Nanni, M., Pedreschi, D., Pappalardo, L.: Navigation services and urban sustainability. Fluctuation Noise Lett. **23**(3), 2450016-1–2450016-8 (2023). https://www.worldscientific.com/toc/fnl/23/03
10. Fragkias, M., Lobo, J., Strumsky, D., Seto, K.C.: Does size matter? Scaling of CO2 emissions and U.S. urban areas. PLOS ONE **8**(6), 1–8 (2013). https://doi.org/10.1371/journal.pone.0064727
11. Gately, C.K., Hutyra, L.R., Wing, I.S.: Cities, traffic, and CO2: a multidecadal assessment of trends, drivers, and scaling relationships. Proc. Natl. Acad. Sci. **112**(16), 4999–5004 (2015). https://doi.org/10.1073/pnas.1421723112
12. Graells-Garrido, E., Serra-Burriel, F., Rowe, F., Cucchietti, F.M., Reyes, P.: A city of cities: measuring how 15-minutes urban accessibility shapes human mobility in Barcelona. PLOS ONE **16**(5), 1–21 (2021). https://doi.org/10.1371/journal.pone.0250080
13. Gudipudi, R., Fluschnik, T., Ros, A.G.C., Walther, C., Kropp, J.P.: City density and CO2 efficiency. Energy Policy **91**, 352–361 (2016). https://doi.org/10.1016/j.enpol.2016.01.015
14. Hankey, S., Marshall, J.D.: Impacts of urban form on future US passenger-vehicle greenhouse gas emissions. Energy Policy **38**(9), 4880–4887 (2010). https://doi.org/10.1016/j.enpol.2009.07.005, special Section on Carbon Emissions and Carbon Management in Cities with Regular Papers
15. He, L., Liu, Y., He, P., Zhou, H.: Relationship between air pollution and urban forms: evidence from prefecture-level cities of the Yangtze River Basin. Int. J. Environ. Res. Public Health **16**(18) (2019). https://doi.org/10.3390/ijerph16183459
16. INRIX: INRIX 2022 traffic scorecard report (2022). https://inrix.com/scorecard/. Accessed 16 Jan 2023
17. Lee, S., Lee, B.: Comparing the impacts of local land use and urban spatial structure on household VMT and GHG emissions. J. Transport Geogr. **84**, 102694 (2020). https://doi.org/10.1016/j.jtrangeo.2020.102694
18. Lelo, K., Monni, S., Tomassi, F.: Socio-spatial inequalities and urban transformation. The case of Rome districts. Socio-Econ. Plann. Sci. **68**, 100696 (2019). https://doi.org/10.1016/j.seps.2019.03.002
19. Menard, S.: Applied Logistic Regression Analysis. SAGE Publications Inc, Thousand Oaks, California (2002)

20. Mohajeri, N., Gudmundsson, A., French, J.R.: CO2 emissions in relation to street-network configuration and city size. Transp. Res. Part D: Transp. Environ. **35**, 116–129 (2015). https://doi.org/10.1016/j.trd.2014.11.025
21. Moran, P.A.P.: Notes on continuous stochastic phenomena. Biometrika **37**(1-2), 17–23 (1950). https://doi.org/10.1093/biomet/37.1-2.17
22. Oliveira, E., Andrade, J., Makse, H.: Large cities are less green. Sci. Rep. **4**, 4235 (2014). https://doi.org/10.1038/srep04235
23. OpenStreetMap contributors: Planet dump (2017). https://planet.osm.org, https://www.openstreetmap.org
24. Pappalardo, L., Manley, E., Sekara, V., Alessandretti, L.: Future directions in human mobility science. Nat. Comput. Sci. **3**, 1–13 (2023)
25. Pedreschi, D., et al.: Human-AI coevolution (2024)
26. Roma Capitale – Open Data: Esercizi commerciali presenti nel territorio di Roma Capitale alla data dell'01/02/2018. https://dati.comune.roma.it/catalog/dataset/d148/resource/da49d933-c541-4466-920e-2cc24aec0258 (2018). licence: CC BY 4.0
27. Shahneh, M.R., Oymak, S., Magdy, A.: A-GWR: fast and accurate geospatial inference via augmented geographically weighted regression. In: Proceedings of the 29th International Conference on Advances in Geographic Information Systems, p. 564–575. SIGSPATIAL 2021, Association for Computing Machinery, New York, NY, USA (2021). https://doi.org/10.1145/3474717.3484260
28. Sun, H., Li, M., Xue, Y.: Examining the factors influencing transport sector CO2 emissions and their efficiency in central China. Sustainability **11**(17) (2019). https://doi.org/10.3390/su11174712
29. Susilo, Y., Maat, K.: The influence of built environment to the trends in commuting journeys in the Netherlands. Transportation **34**, 589–609 (2007). https://doi.org/10.1007/s11116-007-9129-5
30. Valentin Ribeiro, H., Rybski, D., Kropp, J.: Effects of changing population or density on urban carbon dioxide emissions. Nat. Commun. **10**, 3204 (2019). https://doi.org/10.1038/s41467-019-11184-y
31. Veneri, P.: Urban polycentricity and the costs of commuting: evidence from Italian metropolitan areas. Growth Chang. **41**(3), 403–429 (2010). https://doi.org/10.1111/j.1468-2257.2010.00531.x
32. Wang, S., Liu, X., Zhou, C., Hu, J., Ou, J.: Examining the impacts of socioeconomic factors, urban form, and transportation networks on CO2 emissions in China's megacities. Appl. Energy **185**, 189–200 (2017). https://doi.org/10.1016/j.apenergy.2016.10.052
33. Wang, Y., Hayashi, Y., Chen, J., Li, Q.: Changing urban form and transport CO2 emissions: an empirical analysis of Beijing. China. Sustain. **6**(7), 4558–4579 (2014). https://doi.org/10.3390/su6074558
34. Yeo, I., Johnson, R.A.: A new family of power transformations to improve normality or symmetry. Biometrika **87**(4), 954–959 (2000). https://doi.org/10.1093/biomet/87.4.954
35. Zhang, H., Peng, J., Wang, R., Zhang, J., Yu, D.: Spatial planning factors that influence CO2 emissions: a systematic literature review. Urban Climate **36**, 100809 (2021). https://doi.org/10.1016/j.uclim.2021.100809

Interpretable Machine Learning for Oral Lesion Diagnosis Through Prototypical Instances Identification

Alessio Cascione[1], Mattia Setzu[1], Federico A. Galatolo[1], Mario G.C.A. Cimino[1], and Riccardo Guidotti[1,2(✉)]

[1] University of Pisa, Largo Bruno Pontecorvo 3, Pisa, PI 56127, Italy
a.cascione@studenti.unipi.it,
{mattia.setzu,federico.galatolo,mario.cimino,riccardo.guidotti}@unipi.it
[2] KDD Lab, ISTI-CNR, Via G. Moruzzi 1, Pisa, PI 56124, Italy
riccardo.guidotti@isti.cnr.it

Abstract. Decision-making processes in healthcare can be highly complex and challenging. Machine Learning tools offer significant potential to assist in these processes. However, many current methodologies rely on complex models that are not easily interpretable by experts. This underscores the need to develop interpretable models that can provide meaningful support in clinical decision-making. When approaching such tasks, humans typically compare the situation at hand to a few key examples and representative cases imprinted in their memory. Using an approach which selects such exemplary cases and grounds its predictions on them could contribute to obtaining high-performing interpretable solutions to such problems. To this end, we evaluate PIVOTTREE, an interpretable prototype selection model, on an oral lesion detection problem. We demonstrate the efficacy of using such method in terms of performance and offer a qualitative and quantitative comparison between exemplary cases and ground-truth prototypes selected by experts.

Keywords: Interpretable Machine Learning · Explainable AI · Instance-based Approach · Pivotal Instances · Transparent Model · Dental Health AI · Oral Disease Prediction

1 Introduction

One of the sectors that has significantly benefited from the application of Machine Learning (ML) tools is healthcare [8,21]. However, although the models employed to solve diagnostic tasks are powerful in terms of predictive capability, their reliance on complex architectures often makes it difficult for experts and users to understand their reasoning. Moreover, the "cognitive process" employed by these models is frequently not comparable to how humans reason to solve the same tasks [49]. Given the pivotal role of these tools as decision-support systems for practitioners in healthcare, explaining and interpreting their predictions has become crucial and is the focus of active research in Explainable AI (XAI) [2].

As humans, our cognitive processes and mental models frequently depend on case-based reasoning [37], where past exemplary cases are stored in memory and retrieved to solve specific tasks. This type of reasoning is so deeply embedded in us that even young children can recognize and interact with unfamiliar objects they have never encountered before, provided these objects resemble something they already know [42]. Moreover, this ability extends across various modalities: we identify authors by their writing style, recognize relatives by shared facial features, and classify music genres based on similarities to familiar tracks [23]. In the healthcare sector, practitioners frequently diagnose or identify new conditions by referencing past case reports [20,38]. Additionally, the experiments detailed in [11] demonstrated that pattern recognition, grounded in examples gained through experience, is the diagnostic strategy with the highest likelihood of success.

Given these premises, a promising approach to designing inherently interpretable ML models for the healthcare sector is to explore the intuitive notion of similarity between *discriminative* and *descriptive* instances. The underlying assumption is that grounding a model's predictions on the similarity between test instances and exemplar cases would yield a naturally interpretable and trustworthy tool for medical experts and end-users alike. In this paper, we present a case study with an interpretable similarity-based model for decision-making applied to a specific medical context, i.e., for an oral lesion prediction task.

In particular, we study PIVOTTREE [7], a hierarchical and interpretable case-based model inspired by Decision Tree (DT) [6]. By design, PIVOTTREE can be used both as a *prediction* and *selection* model. As a selection model, PIVOTTREE identifies a set of training exemplary cases named *pivots*; as a predictive model, PIVOTTREE leverages the identified pivots to build a similarity-based DT, routing instances through its structure and yielding a prediction, and an associated explanation. Unlike traditional DTs, the resulting explanation is not a set of rules having features as conditions, but rules using a set of pivots to which the instance to predict is compared. Like distance-based models, PIVOTTREE allows to select exemplary instances in order to encode the data in a similarity space that enables case-based reasoning. Finally, PIVOTTREE is a *data-agnostic* model, which can be applied to different data modalities, jointly solving both pivot selection and prediction tasks. Given its modality agnosticism, PIVOTTREE represents an advancement over traditional DTs. As shown in [7], the case-based model learned by PIVOTTREE offers interpretability even in domains like images, text, and time series, where conventional interpretable models often underperform and lack clarity. Furthermore, unlike conventional distance-based predictive models such as k-Nearest Neighbors (KNN) [15], PIVOTTREE introduces a hierarchical structure to guide similarity-based predictions.

Figure 1 provides an example of PIVOTTREE on the `breast cancer` dataset[1], wherein cell nuclei are classified according to their characteristics computed from a digitized image of a fine needle aspirate of a breast mass. Starting from a dataset of instances, PIVOTTREE identifies a set of two pivots (Fig. 1 *(a)*) in this

---

[1] https://archive.ics.uci.edu/dataset/17/breast+cancer+wisconsin+diagnostic.

case belonging to the two distinct classes *Benign* and *Malignant*. Said *pivots* are used to learn a case-based model wherein novel instances are represented in terms of their similarity to the induced pivots (Fig. 1 *(b)*). Building on pivot selection, PIVOTTREE then learns a hierarchy of pivots wherein instances are classified. This hierarchy takes the form of a Decision Tree (Fig. 1 *(c)*): novel instances navigate the tree, percolating towards pivots to which they are more similar or dissimilar, and landing into a classification leaf. In the example, given a test instance x: if its similarity to *pivot 0* is lower than 3.61 (following the right branch), then x is classified as a *Benign*, i.e., x is far away from the *Malignant pivot 0* (see Fig. 1 *(b)*). Instead, following the left branch, if x's similarity to *pivot 1* is higher than 0.39 (left branch), then x is still classified as *Benign* as it is very similar to the *Benign pivot 1*, otherwise x is classified as *Malignant* as it is sufficiently similar to the *Malignant pivot 0*. In contrast, a traditional Decision Tree would model the decision boundary with feature-based rules, e.g., "if *mean concave points* < 2.4 then *Benign* else if *mean symmetry* < 1.7 then *Malignant*". However, traditional DTs *(i)* can only model axis-parallel splits, and *(ii)* cannot be employed on data types with features without clear semantics such as medical images. Hence, improving on traditional DTs, the case-based model learned by PIVOTTREE can provide interpretability even in domains such as images, text, and time series, by exploiting a suitable data transformation.

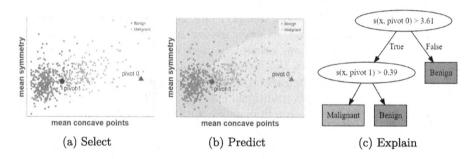

Fig. 1. PIVOTTREE as *(a)* selector, *(b)* interpretable model, *(c)* Decision Tree.

In this paper we demonstrate that PIVOTTREE represents a promisingly effective approach for *interpretability of oral lesion detection*, and we compare its selected pivots with instances identified as representative by domain experts. After an initial review of the literature concerning XAI in the healthcare sector, and prototype-based approach for explainability in Sect. 2, in Sect. 3 we summarize the PIVOTTREE method. Then, in Sect. 4 we report the experimental results on the oral lesion diagnostic problem. Finally, Sect. 5 completes our contribution and discusses future research directions.

2 Related Work

The wide use of explainability techniques for the medical field has been extensively reviewed in previous work [3,16]. ML [12], and specifically case-based reasoning, already finds application in the medical domain, where interpretable and uninterpretable models [5,10] already tackle a variety of tasks, including breast cancer prediction [28,41,48], melanoma detection [17,31–33] and Covid-19 detection [39]. The latter, in particular, introduces two-level interpretations: prototypes are also defined *contrastively*, i.e., both highly similar and highly dissimilar prototypes are provided, and they are also accompanied by heatmaps indicating regions of higher importance. These approaches integrate the discovery of prototypes directly into the model, which often uses similarity-based scoring function to perform predictions. Other examples include [44], which combines knowledge distillation with heterogeneous prototype selection for mammograms, building on [9], and [24] which leverage prototype learning for Autism spectrum disorder detection from fMRI images. In [25] besides prototypes criticism are also identified, i.e., instances representatives of some parts of the input space where prototypical examples do not provide good explanations.

Focusing on oral cancer detection, a relevant example is [46], which proposes an end-to-end, two-stage model for oral lesion detection and classification. This model leverages YOLOv5l [22] for detection and EfficientNet-B4 [43] for classification, making it suitable for deployment as a mobile application. In [27], the authors fine-tune a Single Shot Multibox Detector (SSD) [29] to identify the presence and location of oral disease. Finally, in [50] a self-supervised pre-training strategy is defined, followed by a semi-supervised learning approach on epithelial regions for carcinoma detection. A case-based approach specifically for oral lesion is offered in [13], which works with tabular descriptors by physicians. More at large, and aside from case-based interpretations, interpretability in the medical sector has been gaining attention for quite some years [34]. However, all the aforementioned works offer black-box models for oral lesion detection and classification. On the other hand, in terms of interpretability tools for oral cancer detection, only a handful of proposals are currently in place. In [1] an ensemble approach for oral cancer prediction using tabular data, which by design relays on SHAP values [4] for explainability, is discussed. In [14] an approach using gradient-weighted class activation mapping is presented and [40] provides visual explanations leveraging attention mechanisms also adding expert knowledge by incorporating manually edited attention maps in order to update classification results. Differently from the literature presented so far, to the best of our knowledge, our study is the first inquiring on explainability through prototypes for the oral lesion detection problem using a data-agnostic model.

3 Pivot Tree in a Nutshell

In this section, we present the main characteristics of PIVOTTREE. For more detailed information and extensive benchmarking, we refer readers to [7].

Given a set of n instances represented as real-valued m-dimensional feature vectors[2] in $\mathbb{R}^m$, and a set of class labels $C = \{1,\ldots,c\}$, in case-based reasoning, the objective is to learn a function $f : \mathbb{R}^m \to C$ approximating the underlying classification function, with f being defined as a function of k exemplary cases named *pivots*. Similarity-based case-based models define f on a similarity space $\mathcal{S}$, often inversely denoted as "distance space", induced by a similarity function $s : \mathbb{R}^m \times \mathbb{R}^m \to \mathbb{R}$ quantifying the similarity of instances [36]. Given a training set $\langle X, Y \rangle$, and a similarity function s, our objective is to learn a function $\pi : \mathbb{R}^{n \times m} \to \mathbb{R}^{k \times m}$ that selects a set $P \subseteq X$ of k pivots maximizing the performance of f. The instances in X are mapped into Z through $\mathcal{S}$, wherein they are represented in terms of their similarity to the pivots P. f is then trained on $\langle Z, Y \rangle$, and at inference time, instances are first mapped through the similarity space, before being fed to f.

Aiming for transparency of the case-based predictive model f, our objective is to employ as an interpretable model f Decision Tree classifiers (DT) or k-Nearest Neighbors approaches [18] (kNN). When f is implemented with a DT, split conditions will be of the form $s(x, p_i) \geq \beta$, i.e., "if the similarity between instance x and pivot p_i is greater or equal then β, then ...", allowing to easily understand the logic condition by inspecting x and p_i for every condition in the rule. On the other hand, when f is implemented as a kNN, every decision will be based on the similarity with a few neighbors derived from the pivot set P. A human user just needs to inspect x and the similarities with the pivots P and the instances in the neighborhood. When the number of pivots is kept small, the interpretability of both methods increases, limiting the expressiveness. Vice versa, using a selection model π that returns a large number k of pivots can increase the performance at the cost of interpretability. Our proposal aims to balance these two aspects by allowing the selection of a small number of pivots that still guarantee comparable performance to interpretable predictive models.

PIVOTTREE implements the selection function π, and leverages existing interpretable models to implement f. Much like Decision Tree induction algorithms [6], PIVOTTREE greedily learns a hierarchy of nodes wherein pivots lie. Node splits are selected so that the downstream performance of f is maximized, i.e., the split is chosen to maximize the information gain of the node. Notably, PIVOTTREE does not operate directly on the data, but rather on the induced similarities, thus the split is chosen among a set of candidates defining lower or higher similarities to a set of candidate pivots: the traditional "$x_i \leq \alpha$" split is replaced by a similarity rule of the form $s(x_i, p_i) \leq \alpha$ thresholding the similarity of instances to pivots. The training data is then routed according to the split, and the operation repeats recursively. Pivots come in two families: *discriminative*, which guide instance routing, and *descriptive*, which instead describe the node. The former are selected to maximize the performance, while the latter are selected to maximize similarity to the other instances percolating the node.

---

[2] For the sake of simplicity, we consistently treat data instances as real-valued vectors. Any data transformation employed in the experimental section to maintain coherence with this assumption will be specified when needed.

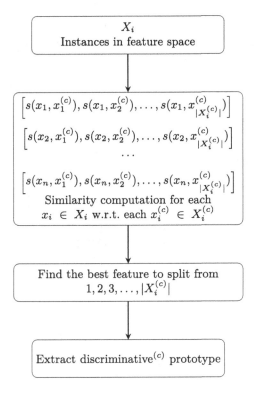

Fig. 2. PIVOTTREE workflow for selection of discriminative pivots of class c for a non-terminal node with X_i training instances. With $x^{(c)}$ we indicate an instance of class c and in analogous way with $discriminative^{(c)}$ the fact we are referring to a class c pivot.

Figure 2 displays the selection process for the *discriminative* pivots of class c in a node: choosing the best splitting feature in the similarity space implies finding the c-instance which best separates the current training data when instance similarity is taken into account, i.e., the *discriminative* pivot of such class.

In a sense, the *descriptive* and *discriminative* pivots extracted by PIVOT-TREE can be associated with the prototypical examples and criticisms identified by [25]. However, their usage is markedly different. By design, PIVOTTREE is a data-agnostic model that leverages the concept of similarity to conduct both selection and prediction tasks simultaneously. While some data types, e.g., relational data, are more amenable than others, e.g., images or text, to similarity computation, with our contribution, we aim to address all data types as one. By decoupling similarity computation and object representation, PIVOTTREE can be applied to any data type supporting a mapping to $\mathbb{R}^m$, i.e., text through language model embedding, images through vision models, graphs through graph representation models, etc. In the healthcare sector, this approach can be highly beneficial due to the heterogeneous nature of the data types involved in diagnostic processes: sequential data like EEG/ECG signals, text-based clinical reports, and medical images of lesions can all be processed using a unified PIVOTTREE framework by transforming each data type into an appropriate vector representation. This integrated approach can improve diagnostic explainability by allowing for a comprehensive analysis of multimodal healthcare data. In the following experiments, we focus specifically on images, particularly on oral lesion images, using embeddings provided by a pre-trained deep learning model.

4 Experiments

In this section, we evaluate the performance of PIVOTTREE[3] (PTC) on the DoctOral-AI dataset[4]. Our objective is to demonstrate that PIVOTTREE is an accurate predictor and selector tool for the task and show how comparable the learned pivots are to ground-truth cases deemed prototypical by expert doctors.

Classification Models. We refer to PIVOTTREE used as Classification model with PTC. We use P to denote the set of pivots identified by PIVOTTREE, and O to denote the set of ground-truth prototypes. DT_P and KNN_P refer to DT and KNN models, respectively, trained in the similarity space obtained by computing the similarity between each instance and every pivot in P. Similarly, DT_O and KNN_O are trained in the similarity space derived from the ground-truth prototypes in O. As further baselines, we compare PIVOTTREE with KNN and DT directly trained on feature space. Finally, as deep learning (DL) model we rely on the Detectron2 (D2) model [47] fine-tuned on the DoctOral-AI dataset. We report the performance of D2 to observe the loss in accuracy at the cost of interpretability. A comparison on DoctOral-AI w.r.t other DL architectures is also offered in [35].

Experimental Setting. We evaluated the predictive performance of the aforementioned models by measuring Balanced Accuracy and F1-score, Precision and Recall by computing the metric for each label and reporting the unweighted mean. In line with [7], for PIVOTTREE hyperparameter selection[5], both as a predictor and a selector, we aim to maintain a low number of pivots and an interpretable classifier structure. Therefore, the optimal *maxdepth* is searched within the interval $[2, 4]$. When using PIVOTTREE as a selector, we assess the performance of using different pivot types – *discriminative*, *descriptive*, both, and using only those considered as splitting pivots – to identify which combination achieves the best selection performance when paired with DT or KNN. The best performance for KNN_P are obtained with $maxdepth = 3$, while for PTC and DT_P with $maxdepth = 4$. Leveraging both *discriminative* and *descriptive* pivots consistently yields better results. Finally, for the baseline DT and KNN the best performance is achieved with $maxdepth = 4$ and $k = 5$, respectively, both in

---

[3] A Python implementation, along with experimental details, is available at https://github.com/acascione/PivotTree_DoctOral.
[4] https://mlpi.ing.unipi.it/doctoralai/.
[5] For every tree, we set 3 as min nbr. of instances a node must have to be considered leaf, and 5 as the min nbr. of instances a node must have to perform a split.

the original space and in the similarity feature space. As distance function, we always adopt the Euclidean distance.

Dataset and Embedding Model. The DoctOral-AI dataset comprises 535 images of varying sizes, which define a multiclassification oral lesion detection task with classes *neoplastic* (31.58%), *aphthous* (32.52%), and *traumatic* (35.88%). Neoplastic ulcers typically exhibit the loss of epithelial layers, with raised, poorly defined margins. The base of these ulcers is often grayish, yellowish, or whitish, presenting a crater-like or raised appearance, generally composed of necrotic tissue with a granular texture. Aphthous ulcers, on the other hand, are characterized by the loss of epithelial layers and have flat, erythematous (red) margins with a grayish-yellow base, surrounded by red mucosa. Traumatic ulcers can feature raised or flat margins, bordered by a whitish or reddish rim, with a crater-like base in shades of white, gray, and yellow. Over time, the edges of chronic traumatic ulcers may harden and thicken. This detailed categorization is crucial for accurate diagnosis and treatment in clinical settings. The dataset is divided into 70% development and 30% testing, the former further divided on a 80%/20% split for training and validation. We embed images with a Detectron2 (D2) [47] CNN architecture fine-tuned on the dataset[6]. We resized each image into an 800 × 800 format. Then relevant feature maps are selected from the D2's backbone output and passed to the D2's region of interest pooling layer. Finally, a pooling layer and a flattening layer map the feature maps to a 256-dimensional embedding. We also report the performance of D2 to observe the loss in accuracy at the cost of interpretability.

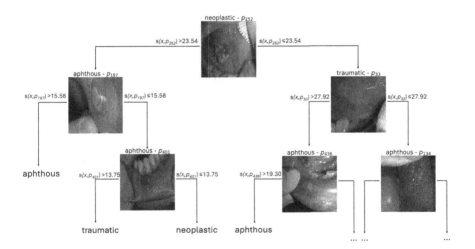

Fig. 3. Partial visual depiction of best PTC configuration on the test set. Branches are labeled with similarity threshold values used for prediction.

---

[6] We offer details regarding the training process in https://github.com/galatolofederico/oral-lesions-detection.

Qualitative Results. Fig. 3 depicts a visual representation of PTC decision rules and splitting pivots associated with the initial nodes[7]. Given a hypothetical instance x to predict, the predictive reasoning employed by the trained model proceeds as follows: x is first compared to p_{252}, a *neoplastic* instance. If the similarity between x and p_{252} is sufficiently high, then x traverses the left branch and is compared to the *aphthous* pivot p_{197}. If x is sufficiently similar to p_{197}, the model concludes the prediction and assigns x to the *aphthous* class. Otherwise, an additional comparison with p_{401} is performed, leading to a final classification as either *neoplastic* or *traumatic*. We underline that the path leading to *traumatic* decision lacks pivots belonging to such class. This suggests that the model can effectively perform comparisons with pivots belonging to other classes to exclude their possibility for x, thereby assigning x to the remaining class by exclusion[8]. On the other hand, if the initial comparison identifies x as dissimilar from the *neoplastic* p_{252}, the model then compares it to the *aphthous* p_{33} and applies analogous reasoning for subsequent comparisons.

Table 1. Mean predictive performance and number of pivots. Best performer in **bold**, second best performer in *italic*, third best performed underlined.

Model	Bal. Acc.	F1-score	Precision	Recall	Nbr. Pivots
D2	**0.859**	**0.854**	**0.854**	**0.858**	-
PTC	*0.834*	*0.832*	*0.839*	*0.834*	9
DT$_P$	<u>0.833</u>	<u>0.830</u>	<u>0.830</u>	<u>0.833</u>	47
KNN$_P$	0.811	0.807	0.810	0.811	**5**
DT$_O$	0.739	0.734	0.742	0.740	9
KNN$_O$	0.801	0.795	0.798	0.801	9
DT	0.770	0.766	0.772	0.770	-
KNN	0.809	0.808	0.811	0.810	-

Quantitative Results. Table 1 reports the mean predictive performance, and the number of pivots of the various predictive models[9]. D2 has the highest performance, at the cost of being not interpretable. However, a not markedly inferior performance is achieved by PIVOTTREE predictor, i.e., PTC, that only requires 9 pivots (6 of which are shown in Fig. 3). The third best performer is PIVOTTREE used as selector for a DT, i.e., DT$_P$. Unfortunately, such performance is accompanied by high complexity, as DT$_P$ requires 47 pivots. Finally,

---

[7] The actual trained tree has a *maxdepth* of 4. For visualization purposes, we limit the visualization to the initial nodes.

[8] We intend to fix this (possible) issue by extending PIVOTTREE with Proximity Trees [30] to compare the test x against two pivots instead of only one.

[9] For DT and PIVOTTREE selector/predictor models we trained each best configuration with 50 different random states. Since the standard deviation of the values resulted to be negligible, we report only the average result.

KNN$_P$, i.e., PIVOTTREE used as selector for a KNN is the predictor requiring the smallest number of pivots. Overall, PIVOTTREE both employed as selector and predictor leads to competitive results compared to D2. We underline how PTC has the best trade-off between accuracy and complexity, showing competitive results with respect to the fine-tuned D2 but providing an interpretable predictor through its pivot structure, and the low number of pivots adopted.

Remarkably, selecting the set of pivots P through PIVOTTREE leads to a KNN and a DT which are better than those resulting using the ground-truth prototypes, especially for the DT case, underlying that those instances which for humans are clear examples, perhaps didactic examples, of certain cases, are not necessarily the best ones to discriminate through an automatic AI system.

Finally, we remark that the performance of any PIVOTTREE-based model is better than those of the KNN and DT classifiers directly trained on embeddings.

Pivot-Prototypes Comparison. We provide here a quantitative comparison in terms of similarities between the pivots selected through PTC P with the ground-truth prototypes O. In particular, we consider as similarity measures the Euclidean distance on the D2 embeddings, and the Structural Similarity (SSIM) [45] on the original images. For the latter, we first resize the images regions of interest to 300×300 pixels. SSIM identifies changes in structural information by capturing the inter-dependencies among similar pixels, especially when they are spatially close. In Figs. 4 and 5 we report two heatmaps highlighting the similarities between the PIVOTTREE pivots (rows) and ground-truth prototypes (columns), on Euclidean and SSIM similarity, respectively. Darker colors indicate higher similarity. For the similarity comparison through Euclidean distance, we specify that the average distance between each pair of instances in the DoctOral-AI training set is 26.90 ± 6.48. When examining the average distance between pivot and ground-truth pairs w.r.t. each class in the heatmap, we find the following values: 23.93 for *neoplastic*, 24.65 for *aphthous*, and 24.60 for *traumatic*. This shows how the mean pairwise distances within individual classes are generally close to the overall mean pairwise distance. Pivots and ground-truth prototypes tend to not present robust similarities. Furthermore, we notice how for pivots p_{403} and p_{238}, both members of *aphthous* class, the most similar ground-truth prototypes belong to a different class. On the other hand, for the other pivots, the closest ground-truth counterpart is consistently one of the same class, sometimes with a very high similarity: some examples are p_{134} with o_{382} and p_{403} and o_{223}. A different tendency can be observed in Fig. 5 when using SSIM: the average SSIM with respect to each class is 0.46 for *neoplastic*, 0.70 for *aphthous*, and 0.57 for *traumatic*, with a mean similarity in the overall training set of 0.58 ± 0.10. This highlights a notably high internal similarity for the *aphthous* class. As evident from Fig. 5, the highest similarity is always observed when comparing pivots with the *aphthous* ground-truth prototypes, differently from Fig. 4 which shows higher variability across classes more oriented towards the right matching. This comparison corroborates the idea of relying on the Euclidean distance on the D2 embedding space for PIVOTTREE.

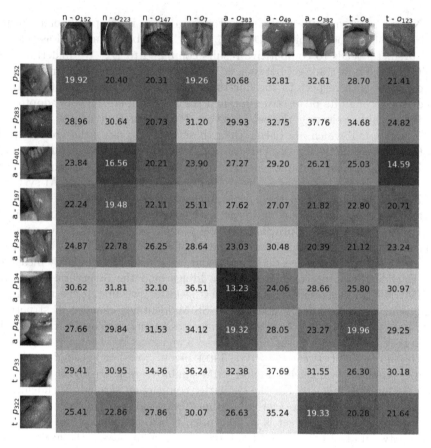

Fig. 4. PIVOTTREE pivots (rows) and ground-truth prototypes (columns) comparison as Euclidean distances on D2 embedding. The darker the color the more similar are a pivot and a ground truth prototype. The first letter identifies the class of the instances: neoplastic, aphthous, and traumatic.

Furthermore, we evaluate how pivots extracted using PTC group instances together compared to ground-truth prototypes. We partition dataset instances with Voronoi partition, each instance associated to the prototype closest to it in the D2 embedding space. In Fig. 6, we compare group size and entropy calculated w.r.t. the target variable Y. In PTC groups, we highlight how *aphthous* pivots tend to aggregate the majority of instances, alongside the *traumatic* pivot p_{322}. *Aphthous* pivots exhibit higher entropy levels and form less pure groups, except for the smaller, entirely pure group centered around p_{134}. Conversely, the *neoplastic* instance p_{252} significantly captures a substantial percentage of its class, paralleled by p_{33} which similarly captures *traumatic* instances effectively. On the other hand, regarding ground-truth prototypes, most instances group around o_8, o_{382}, and o_{233}, representing *traumatic*, *aphthous*, and *neoplastic* classes, respectively. Many *neoplastic* instances are well-represented by o_{152},

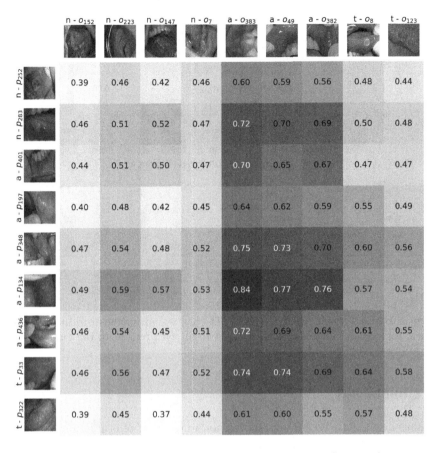

Fig. 5. PIVOTTREE pivots (rows) and ground-truth prototypes (columns) comparison as SSIM on raw regions of interest. Same rules from Fig. 4 apply.

and similarly for *aphthous* instances with o_{383}. Instances of pure or almost pure groups are observed for *neoplastic* and *aphthous* classes, whereas a highly pure group for *traumatic* instances is lacking in this scenario. Only a single entirely pure group for *neoplastic* instances is found for ground-truth prototypes, whereas PTC pivots are able to isolate two entirely pure groups around p_{33} and p_{134}.

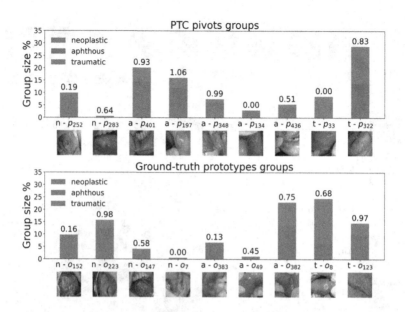

Fig. 6. Group sizes in percentage using PTC pivots and ground-truth prototypes as centers, respectively. Entropy values of each group with respect to the target variable are indicated by values above each bar.

5 Conclusion

We have discussed PIVOTTREE application in the case of oral lesion prediction, showing its superiority as a predictor with respect to other simple interpretable models and as selector when paired with such simple models trained on the similarity space induced by the selected pivots. Furthermore, we have compared expert-selected prototypes with PIVOTTREE-selected pivots, highlighting how a strong similarity can be observed in some of the pairs. Given its flexibility, PIVOTTREE lends itself to be applied for several other diagnostic task in the healthcare sector. Future investigations include testing PIVOTTREE on medical data of different modalities (time-series, text reports, tabular data) in order to assess its performance, comparing it against neural prototype-based approaches for medical data as explored in [26,39] and evaluating the interpretability of identified pivots through human subjects. Additional analysis could investigate the trade-off between performance and explainability by evaluating how PIVOTTREE compares to competing post-hoc explainers. Moreover, future research could focus on developing specialized interpretability metrics for PIVOTTREE and other case-based models, as this study primarily relied on depth and the number of pivots to assess interpretability and complexity. Furthermore, other splitting strategies could be analyzed, one being a direct comparison between pairs of pivots as shown in PROXIMITYTREE models [30] or attempting to generate instead of select the PIVOTTREE model [19].

Acknowledgments. This work has been partially supported by the European Community Horizon 2020 programme under the funding schemes ERC-2018-ADG G.A. 834756 "XAI: Science and technology for the eXplanation of AI decision making", "INFRAIA-01-2018-2019 - Integrating Activities for Advanced Communities", G.A. 871042, "SoBigData++: European Integrated Infrastructure for Social Mining and Big Data Analytics", by the European Commission under the NextGeneration EU programme - National Recovery and Resilience Plan (Piano Nazionale di Ripresa e Resilienza, PNRR) - Project: "SoBigData.it - Strengthening the Italian RI for Social Mining and Big Data Analytics" - Prot. IR0000013 - Avviso n. 3264 del 28/12/2021, and M4C2 - Investimento 1.3, Partenariato Esteso PE00000013 - "FAIR - Future Artificial Intelligence Research" - Spoke 1 "Human-centered AI", M4 C2, Investment 1.5 "Creating and strengthening of innovation ecosystems", building "territorial R&D leaders", project "THE - Tuscany Health Ecosystem", Spoke 6 "Precision Medicine and Personalized Healthcare", by the Italian Project Fondo Italiano per la Scienza FIS00001966 MIMOSA, by the "Reasoning" project, PRIN 2020 LS Programme, Project number 2493 04-11-2021, by the Italian Ministry of Education and Research (MIUR) in the framework of the FoReLab project (Departments of Excellence), by the European Union, Next Generation EU, within the PRIN 2022 framework project PIANO (Personalized Interventions Against Online Toxicity) under CUP B53D23013290006.

Disclosure of Interests. The authors have no competing interests to declare that are relevant to the content of this article.

References

1. Adeoye, J., et al.: Explainable ensemble learning model improves identification of candidates for oral cancer screening. Oral Oncol. **136**, 106278 (2023)
2. Ali, S., et al.: The enlightening role of explainable artificial intelligence in medical & healthcare domains: a systematic literature review. Comput. Biol. Medicine **166**, 107555 (2023)
3. Band, S.S., et al.: Application of explainable artificial intelligence in medical health: a systematic review of interpretability methods. Inf. Med. Unlocked **40**, 101286 (2023)
4. Baptista, M.L., et al.: Relation between prognostics predictor evaluation metrics and local interpretability SHAP values. Artif. Intell. **306**, 103667 (2022)
5. Bichindaritz, I., Marling, C.: Case-based reasoning in the health sciences: what's next? Artif. Intell. Med. **36**(2), 127–135 (2006)
6. Breiman, L., et al.: Classification and Regression Trees. Wadsworth (1984)
7. Cascione, A., et al.: Data-agnostic pivotal instances selection for decision-making models. In: Bifet, A., Davis, J., Krilavičius, T., Kull, M., Ntoutsi, E., Žliobaitė, I. (eds.) ECML/PKDD, vol. 14941, pp. 367–386. Springer, Cham (2024). https://doi.org/10.1007/978-3-031-70341-6_22
8. Celard, P., et al.: A survey on deep learning applied to medical images: from simple artificial neural networks to generative models. Neural Comput. Appl. **35**(3), 2291–2323 (2023)
9. Chen, C., et al.: This looks like that: deep learning for interpretable image recognition. In: NeurIPS, pp. 8928–8939 (2019)
10. Choudhury, N., Begum, S.A.: A survey on case-based reasoning in medicine. IJACSA **7**(8), 136–144 (2016)

11. Coderre, S., et al.: Diagnostic reasoning strategies and diagnostic success. Med. Educ. **37**(8), 695–703 (2003)
12. Dixit, S., et al.: A current review of machine learning and deep learning models in oral cancer diagnosis: recent technologies, open challenges, and future research directions. Diagnostics **13**(7), 1353 (2023)
13. Ehtesham, H., et al.: Developing a new intelligent system for the diagnosis of oral medicine with case-based reasoning approach. Oral Dis. **25**(6), 1555–1563 (2019)
14. Figueroa, K.C., et al.: Interpretable deep learning approach for oral cancer classification using guided attention inference network. JBO **27**(1), 015001 (2022)
15. Fix, E.: Discriminatory analysis: nonparametric discrimination, consistency properties, vol. 1. USAF school of Aviation Medicine (1985)
16. Frasca, M., et al.: Explainable and interpretable artificial intelligence in medicine: a systematic bibliometric review. Discov. Artif. Intell. **4**(1) (2024). https://doi.org/10.1007/s44163-024-00114-7
17. Grignaffini, F., et al.: Machine learning approaches for skin cancer classification from dermoscopic images: a systematic review. Algorithms **15**(11), 438 (2022)
18. Guidotti, R., et al.: A survey of methods for explaining black box models. ACM Comput. Surv. **51**(5), 93:1–93:42 (2019)
19. Guidotti, R., et al.: Generative model for decision trees. In: AAAI, pp. 21116–21124. AAAI Press (2024)
20. Harasym, P.H., et al.: Current trends in developing medical students' critical thinking abilities. KJMS **24**(7), 341–355 (2008)
21. Javaid, M., et al.: Significance of machine learning in healthcare: features, pillars and applications. Int. J. Intell. Networks **3**, 58–73 (2022)
22. Jocher, G.: YOLOv5 by Ultralytics. https://github.com/ultralytics/yolov5
23. Johnson-Laird, P.N.: Mental models and human reasoning. Proc. Natl. Acad. Sci. **107**(43), 18243–18250 (2010)
24. Kang, E., et al.: Prototype learning of inter-network connectivity for ASD diagnosis and personalized analysis. In: Wang, L., Dou, Q., Fletcher, P.T., Speidel, S., Li, S. (eds.) MICCAI. LNCS, vol. 13433, pp. 334–343. Springer (2022). https://doi.org/10.1007/978-3-031-16437-8_32
25. Kim, B., et al.: Examples are not enough, learn to criticize! criticism for interpretability. In: NIPS, pp. 2280–2288 (2016)
26. Kim, E., et al.: XProtoNet: diagnosis in chest radiography with global and local explanations. In: CVPR, pp. 15719–15728. Computer Vision Foundation / IEEE (2021)
27. Kouketsu, A., et al.: Detection of oral cancer and oral potentially malignant disorders using artificial intelligence-based image analysis. Head Neck **46**, 2253–2260 (2024)
28. Lamy, J., et al.: Explainable artificial intelligence for breast cancer: a visual case-based reasoning approach. Artif. Intell. Medicine **94**, 42–53 (2019)
29. Liu, W., et al.: SSD: single shot multibox detector. In: Leibe, B., Matas, J., Sebe, N., Welling, M. (eds.) ECCV 2016. LNCS, vol. 9905, pp. 21–37. Springer, Cham (2016). https://doi.org/10.1007/978-3-319-46448-0_2
30. Lucas, B., et al.: Proximity forest: an effective and scalable distance-based classifier for time series. Data Min. Knowl. Discov. **33**(3), 607–635 (2019)
31. Metta, C., et al.: Exemplars and counterexemplars explanations for image classifiers, targeting skin lesion labeling. In: ISCC, pp. 1–7. IEEE (2021)
32. Metta, C., et al.: Improving trust and confidence in medical skin lesion diagnosis through explainable deep learning. JDSA, 1–13 (2023). https://doi.org/10.1007/s41060-023-00401-z

33. Metta, C., et al.: Advancing dermatological diagnostics: interpretable AI for enhanced skin lesion classification. Diagnostics **14**(7), 753 (2024)
34. Panigutti, C., et al.: Doctor XAI: an ontology-based approach to black-box sequential data classification explanations. In: FAT*, pp. 629–639. ACM (2020)
35. Schank, R.C., Abelson, R.P.: Knowledge and Memory: The Real Story, pp. 1–85. Psychology Press (2014)
36. Pekalska, E., Duin, R.P.W.: The Dissimilarity Representation for Pattern Recognition - Foundations and Applications, Series in Machine Perception and Artificial Intelligence, vol. 64. WorldScientific (2005)
37. Schank, R.C., Abelson, R.P.: Knowledge and Memory: The Real Story, pp. 1–85. Psychology Press (2014)
38. Shin, H.S.: Reasoning processes in clinical reasoning: from the perspective of cognitive psychology. KJME **31**(4), 299 (2019)
39. Singh, G., Yow, K.C.: An interpretable deep learning model for Covid-19 detection with chest x-ray images. IEEE Access **9**, 85198–85208 (2021)
40. Song, B., et al.: Interpretable and reliable oral cancer classifier with attention mechanism and expert knowledge embedding via attention map. Cancers **15**(5), 1421 (2023)
41. Song, B., et al.: Classification of mobile-based oral cancer images using the vision transformer and the SWIN transformer. Cancers **16**(5), 987 (2024)
42. Spelke, E.S.: What babies know: Core Knowledge and Composition Volume 1, vol. 1. Oxford University Press (2022)
43. Tan, M., Le, Q.V.: Efficientnet: Rethinking model scaling for convolutional neural networks. In: ICML. Proceedings of Machine Learning Research, vol. 97, pp. 6105–6114. PMLR (2019)
44. Wang, C., et al.: Knowledge distillation to ensemble global and interpretable prototype-based mammogram classification models. n: Wang, L., Dou, Q., Fletcher, P.T., Speidel, S., Li, S. (eds.) MICCAI. LNVCS, vol. 13433, pp. 14–24. Springer, Cham (2022). https://doi.org/10.1007/978-3-031-16437-8_2
45. Wang, Z., et al.: Image quality assessment: from error visibility to structural similarity. IEEE Trans. Image Process. **13**(4), 600–612 (2004)
46. Welikala, R.A., et al.: Automated detection and classification of oral lesions using deep learning for early detection of oral cancer. IEEE Access **8**, 132677–132693 (2020)
47. Wu, Y., et al.: Detectron2 (2019). https://github.com/facebookresearch/detectron2
48. Yagin, B., et al.: Cancer metastasis prediction and genomic biomarker identification through machine learning and explainable artificial intelligence in breast cancer research. Diagnostics **13**(21), 3314 (2023)
49. Yang, G., et al.: Unbox the black-box for the medical explainable AI via multimodal and multi-centre data fusion: a mini-review, two showcases and beyond. Inf. Fusion **77**, 29–52 (2022)
50. Zhou, J., et al.: A pathology-based diagnosis and prognosis intelligent system for oral squamous cell carcinoma using semi-supervised learning. ESWA **254**, 124242 (2024)

Ensemble Counterfactual Explanations for Churn Analysis

Samuele Tonati[1,2](✉)[iD], Marzio Di Vece[1,3][iD], Roberto Pellungrini[1][iD], and Fosca Giannotti[1][iD]

[1] Scuola Normale Superiore, Piazza dei Cavalieri 7, 56126 Pisa, Italy
{samuele.tonati,marzio.divece,roberto.pellungrini,fosca.giannotti}@sns.it
[2] University of Pisa, Lungarno Antonio Pacinotti 43, 56126 Pisa, Italy
[3] IMT School for Advanced Studies, Piazza San Francesco 19, 55100 Lucca, Italy

Abstract. Counterfactual explanations play a crucial role in interpreting and understanding the decision-making process of complex machine learning models, offering insights into why a particular prediction was made and how it could be altered. However, individual counterfactual explanations generated by different methods may vary significantly in terms of their quality, diversity, and coherence to the black-box prediction. This is especially important in financial applications such as churn analysis, where customer retention officers could explore different approaches and solutions with the clients to prevent churning. The officer's capability to modify and explore different explanations is pivotal to his ability to provide feasible solutions. To address this challenge, we propose an evaluation framework through the implementation of an ensemble approach that combines state-of-the-art counterfactual generation methods and a linear combination score of desired properties to select the most appropriate explanation. We conduct our experiments on three publicly available churn datasets in different domains. Our experimental results demonstrate that the ensemble of counterfactual explanations provides more diverse and comprehensive insights into model behavior compared to individual methods alone that suffer from specific weaknesses. By aggregating, evaluating, and selecting multiple explanations, our approach enhances the diversity of the explanation, highlights common patterns, and mitigates the limitations of any single method, offering to the user the ability to tweak the explanation properties to their needs.

Keywords: Explainable AI · Counterfactual Explanations · Churn Analysis

1 Introduction

The recent surge of interest in Machine Learning (ML) and Artificial Intelligence (AI) has led to the development of a multitude of models aimed at decision-making across various sectors, including healthcare, financial systems, and criminal justice. Although it may seem logical to favor more accurate models when evaluating different options, the emphasis on accuracy has resulted in unintended consequences. ML Developers frequently prioritize higher accuracy, often at the expense of interpretability, making their models increasingly complex and difficult to understand. This lack of explainability becomes a significant concern when models are entrusted with making critical decisions that affect people's well-being. In order to address these concerns, the concept of Explainable AI (XAI) has emerged as a promising solution. XAI addresses the aforementioned challenges by offering explanations to make the inner workings of AI models interpretable and easy to understand, as AI-generated prescriptions are often unintelligible to humans. Counterfactual Explanations are a ubiquitous form of explanation, which aim to provide a contrastive way of explaining the decisions of a model. They also align with the requirements specified by the GDPR and the AI Act, in particular concerning the right to explanation and transparency. Counterfactual Explanations indicate what would need to change in the input data to alter the model's output, thereby offering a clear and intuitive understanding of the decision-making process and are found to be successful in many practical domains [1,2]. This is especially true for applications such as churn analysis. In such a scenario we have a churn officer relying on a machine learning model to understand which clients are more likely to churn. Here, understanding why certain clients are churning and being able to provide feasible alternatives to prevent churn is the key objective [3]. Therefore, explainable AI techniques are fundamental in giving some insight into what actions the officer could employ to engage the customer and prevent churn [4]. While there are many techniques to generate counterfactual explanations, each methodology has its own specific mechanisms that may provide counterfactuals with specific characteristics. Moreover, there is no accepted methodology in the literature to evaluate counterfactual explanations agnostically w.r.t. the application domain, therefore, the counterfactual explanations obtained with a certain methodology may have properties that are not entirely helpful in the context of churn analysis. In this paper, we propose a novel evaluation framework for counterfactual explanations for churn analysis. We rely on an ensemble of counterfactual explanations generated with diverse techniques and a flexible linear combination of metrics as an evaluation function to select the best counterfactuals. This function allows the officer to modify the counterfactual ensemble with little computational overhead, selecting the counterfactuals that could give the best chances of retention. We evaluate our approach on three publicly available datasets for churn prediction.

2 Related Works

Churn analysis is a crucial application of ML in various industries, particularly in telecommunications, finance, and subscription-based services. Churn refers to the phenomenon where customers stop using a company's services or products, leading to a loss of revenue [5-7]. Accurately predicting churn is essential for businesses to implement effective retention strategies and reduce customer attrition. Traditional ML models used for churn prediction focus on identifying patterns and factors that contribute to a customer's likelihood of leaving and are designed often as a combination of unsupervised and supervised techniques [8–11]. These models often rely on large datasets encompassing customer behavior, transaction history, demographics, and other relevant attributes. While such models can achieve high accuracy in predicting churn, their complexity and opaqueness in their decision-making processes poses challenges for decision-makers who need to understand the underlying reasons for customer attrition (i.e. the loss of customers by a business).

Counterfactual explanations are a distinct category of post-hoc local explanation methods that describe how to alter the input to achieve a desired outcome from the model. According to a commonly accepted definition in literature, these modifications should be minimal and closely resemble the original instance being explained. This similarity highlights how counterfactual explanations share many traits with adversarial attacks, as both aim to flip the model's prediction, albeit with different goals. The key distinguishing factor of counterfactual explanations lies in their desired properties, which are intended to provide informative value to the user. These properties typically include proximity to the original instance, minimality, actionability, and diversity among the others [2].

Counterfactual explanations align well with the goal of personalized customer experiences [12,13]. By understanding the specific factors that influence each customer's decision to stay or leave, businesses can tailor their engagement strategies to meet individual needs, thus enhancing customer satisfaction and loyalty. Moreover, counterfactual explanations have yet to be applied to the case of churn in the literature. Existing works have instead mainly focused on XAI techniques based on feature importance [4,14,15].

3 Problem Definition

Churn prevention encompasses all the actions that a company puts into place to prevent the loss of customers. The first and most important part of any churn prevention strategy lies in detecting which customers will likely interrupt their relationship with the company, given their current status. This task can be modeled as a binary classification task [3] where a machine learning model is used to predict which customers are the ones likely to churn. A churn officer has then the duty of interacting with these customers in order to find possible actions to prevent them from churning.

In this context, counterfactual explanations appear to be extremely useful from a business' perspective: firstly, counterfactual explanations help identify the minimal changes needed to retain a customer; for example, if the model indicates that a customer is likely to churn, a counterfactual explanation might reveal that offering a small discount, offering a particular product or improving service quality could change the prediction to retention. This allows businesses to implement precise interventions that are cost-effective and efficient. Secondly, the interpretability offered by counterfactual explanations builds trust among business stakeholders. Unlike explanatory methods that might provide abstract or general insights, such as in the case of global explanations, counterfactuals show specific scenarios and outcomes, making it easier for non-technical stakeholders to understand and trust the model's recommendations. This trust is vital in securing support and commitment from stakeholders for data-driven strategies and ensuring their successful implementation. However, depending on what counterfactual method one chooses, the explanations obtained may rely on specific optimization strategies and therefore explore only some particular aspect of the importance of a churn officer. Counterfactual explanations have not been explored as a solution to this kind of problem [4,14,15]. To tackle similar problems, Guidotti et al. [16] proposed an ensemble method that leverages the strengths of multiple counterfactual explainers to cover a set of desirable properties, such as minimality, actionability, stability, diversity, plausibility, and discriminative power. Their approach demonstrates the efficacy of boosting weak explainers into a powerful ensemble that is both model-agnostic and data-agnostic, capable of handling various data types including tabular data, images, and time series. Building upon this idea, we propose an ensemble approach that operates ex-post as an evaluation and selection mechanism. Our method is designed to identify the optimal set of counterfactual examples by employing a linear combination score of various metrics, that reflect on the possible aspects that a churn officer would explore in a churn prediction model. In contrast to the ensemble proposed by [16], which combines results through a diversity-driven selection function, our framework introduces a more nuanced evaluation score. This approach not only refines the selection process but also ensures that the chosen counterfactuals align closely with the desired properties - thereby improving the interpretability and reliability of the explanations provided - and that can be aptly tweaked by practitioners to give more emphasis to a specific metric. In the context of a binary classification task, given a model function $f : X \subset \mathbb{R}^n \to \{0, 1\}$ that maps instances $x \in X$ with features in $\mathbb{R}^n$ to predicted class labels (0 or 1), a counterfactual explanation aims to identify a modified version of an original instance x such that the modified instance x' leads to a different model prediction. Therefore, given an original instance ($x \in \mathbb{R}^n$), a predictive model ($f : \mathbb{R}^n \to \mathbb{R}$), the model's prediction for the original instance ($y_M = f(x)$) and a target prediction ($y_T \neq y_M$), The goal is to find a counterfactual instance ($x' \in \mathbb{R}^n$) such that $f(x') = y_T$ while ensuring that (x') is similar to the original instance (x). This similarity can be a function of the different metrics evaluating the relationships between these two instances.

Let's consider the case in which the similarity function is represented by a distance metric $d(x, x')$, typically chosen as the (L_1) or (L_2) norm. The problem of finding a counterfactual explanation can be solved by finding $\min_{x'} d(x', x)$ subject to the constrain $f(x') = y_T$ where $d(x', x)$ is defined as $d(x', x) = \|x' - x\|_p$ with ($\|\cdot\|_p$) denoting the (L_p)-norm, commonly (L_1) or (L_2)-norm.

Other than distance metrics, the optimization process may also incorporate constraints on the perturbation ($\delta_{CF} = x' - x$) to ensure sparsity or adherence to feature-specific constraints: $x' = x + \delta_{CF}$.

Our proposal is to score the counterfactual explanations produced by an ensemble of counterfactual methods using evaluation metrics that align with desired properties in the context of Churn analysis. The pseudocode of our approach is given in Algorithm 1.

Algorithm 1. k-CEM: k-Counterfactual Ensemble Method

1: **Input:** X_{test} - test set, E - set of CF explanations, M - trained model
2: **Output:** C - ensemble of top counterfactual explanations
3: $C \leftarrow \emptyset$ ▷ Initialize result set
4: **procedure** ENSEMBLECFEXPLANATIONS
5:     Load X_{test}, E, and M
6:     **for** each $e \in E$ **do**
7:         $C \leftarrow$ Extract subset of X_{test} corresponding to e
8:         $\hat{y}_{orig} \leftarrow M(C_{subset})$
9:         $\hat{y}_{cf} \leftarrow M(e)$
10:     **end for**
11:     Retain CFs where $\hat{y}_{test} \neq \hat{y}_{ensemble}$
12:     **for** each CF $c \in C$ **do**
13:         Calculate and normalize metrics: $d_{prox}(c)$, $s(c)$, $p(c)$, $d_{div}(c)$
14:         Define weights $w_{prox}, w_{spars}, w_{plaus}, w_{div}$
15:         Compute score: $score(c) \leftarrow w_{prox} \cdot d_{prox}(c) + w_{spars} \cdot s(c) + w_{plaus} \cdot p(c) + w_{div} \cdot (1 - d_{div}(c))$
16:     **end for**
17:     Sort C by $score$ in ascending order
18:     Group C by original instance index i
19:     **for** each group g_i **do**
20:         Select top k CFs from g_i based on $score$
21:     **end for**
22:     Return the top CFs DataFrame C
23: **end procedure**

As described in the pseudocode our method takes in input test data, counterfactual explanations, and a trained model. For each explanation method we retain only valid explanations and put them in an ensemble set C (lines 6–11). Then, we characterize each element of the set using relevant metrics which are first normalized and then incorporated into the score function with user-defined weights (lines 12–16). Explanations are then sorted by their score and the top k are selected for each reference instance in the test set (lines 17–22).

3.1 Counterfactuals Methods

For the implementation of our Counterfactual Ensemble Method, we choose four different counterfactual generation methods that, in our opinion, condense the most diverse approaches to synthetic counterfactual explanation generation.

- **DiCE** perturbs input features within model decision boundaries, leveraging a genetic algorithm to create multiple instances leading to different predictions [17]. It generates diverse counterfactual examples solving an optimization problem that balances properties of proximity and diversity.
- **Growing Spheres** (GS) uses a sphere-growing algorithm to iteratively explore the feature space around a given instance [18]. In our approach, we slightly modify GS to return the best k instances instead of just one, ranking them based on L_2 proximity to the original instance.
- **CFRL** is a model-agnostic counterfactual generation method that uses reinforcement learning [19] to train a generative model to produce counterfactual explanations.
- **T-LACE** is a counterfactual explanation method that constructs a transparent latent space using a linear transformation where also the original prediction of the model is added, ensuring that similar records in the latent space have similar features and predictions [20]. Counterfactuals are then searched in the latent space decomposing contributions from each feature to identify a prediction direction.

3.2 Evaluation of Counterfactual Explanations

The scoring function of the counterfactual ensemble is the core of our approach, and it is based on properties that reflect upon possible questions that a churn officer would need to answer in order to prevent a client from churning. Here, we present the measures we chose and highlight their purpose in the context of churn analysis.

Proximity Measures. *How minimal are the changes required to retain potentially churning customers? Proximity measures indicate close counterfactuals.* We choose an average proximity measure using a geometric mean that combines various normalized proximity measures. The geometric mean prevents skewing by outliers, ensuring equal contribution from all proximity measures. The individual proximity metrics we use are:

Euclidean Distance (L_2 norm) measures the overall difference in feature values:

$$\text{Proximity}_{L2} = \sqrt{\frac{m-h}{m} \sum_{i \in \text{cont}} (x'_i - x_i)^2 + \frac{h}{m} \sum_{j \in \text{cat}} \delta(x'_j, x_j)}$$

Manhattan Distance (L_1 norm) measures the sum of absolute differences:

$$\text{Proximity}_{L1} = \frac{m-h}{m} \sum_{i \in \text{cont}} |x'_i - x_i| + \frac{h}{m} \sum_{j \in \text{cat}} \delta(x'_j, x_j)$$

Maximum Absolute Difference L_∞ norm measures the maximum element-wise absolute difference:

$$\text{Proximity}_{L_\infty} = \max\left(\frac{m-h}{m}\max_{i\in\text{cont}}|x'_i - x_i|, \frac{h}{m}\max_{j\in\text{cat}}\delta(x'_j, x_j)\right)$$

Here, m is the total number of features, h the number of categorical features, cont continuous features, cat categorical features, and $\delta(x'_j, x_j)$ is 1 if $x'_j \neq x_j$ and 0 otherwise (Hamming distance).

Plausibility Measure. *Is the counterfactual explanation similar to a non-churning customer in the data and thus justifiable to the customer? Plausibility indicates counterfactuals that have close examples in the original dataset.* The plausibility measure assesses the degree of plausibility or soundness of the counterfactual instances (X') with instances in the original dataset to explain.

Specifically, it calculates the minimum distance of each $x' \in X'$ from its closest instance in the original data.

To compute the plausibility measure we build a KDTree on the X_{test} dataset to efficiently find the nearest neighbors, then we query the KDTree to find the nearest neighbor in X_{test} for each instance in the set of counterfactual instances and we calculate the distance between each x' and its closest instance in X_{test}. The use of a KDTree for computing the plausibility measure is motivated by the need to efficiently find the nearest neighbors of counterfactual instances within the dataset X_{test}. KDTree provides logarithmic search time complexity for nearest neighbor queries, making it more scalable compared to linear search methods, which have linear time complexity. Efficiency is crucial when the number of instances in X_{test} is large, a common situation for real-world applications like churn analysis.

Plausibility is represented as the euclidean distance of x' from its closest instance in the X_{test} population.

Sparsity Measure. *Does the counterfactual explanation modify as few features as possible, thus making the required changes easier for the churn officer to propose? Sparsity indicates counterfactuals that touch the least amount of features.* Sparsity is computed as the fraction of differing features to the total number of features n:

$$\text{Sparsity} = \frac{\sum_{i=1}^{n}(x'_i \neq x_i)}{n} \qquad (1)$$

Diversity Measure. *The counterfactuals produced do provide different courses of action for the churn officer? Diversity indicates that the produced explanations have enough variety for the churn officer to act upon.* The diversity measure quantifies the dissimilarity or variation within groups defined by the generation source. It is calculated as the mean of distances between pairs of instances within each group.

$$\text{Diversity} = \frac{1}{N}\sum_{i=1}^{N}\frac{1}{n_i(n_i-1)}\sum_{j\neq k}d(x_j, x_k) \qquad (2)$$

where N is the total number of groups - i.e. sets of counterfactuals for a given instance to explain -, n_i is the number of instances in group i, and $d(x_j, x_k)$ is the distance between instance j and instance k within the same group.

Evaluation Score. Our evaluation score is computed by combining multiple measures using a weighted linear combination. This approach aims to synthesize different aspects of the counterfactual explanations into a single score, allowing for comprehensive evaluation and comparison. The first three metrics need to be minimized-i.e., the lower, the better, whereas, since diversity needs to be maximized, we minimize its complement.

$$\text{Evaluation Score} = \begin{array}{l} + w_1 \times \text{Proximity}_{\text{avg;N}} \\ + w_2 \times \text{Sparsity} \\ + w_3 \times \text{Plausibility}_{\text{N}} \\ + w_4 \times (1 - \text{Diversity}_{\text{N}}) \end{array} \qquad (3)$$

Each weight represents the relative importance assigned to its corresponding measure in the overall evaluation. The selection of top counterfactual examples involves sorting the instances based on their linear combination scores in ascending order. Instances with lower scores are prioritized as they represent better adherence to the optimization objectives defined by the weighted combination of measures. By leveraging this approach, churn experts can efficiently identify and retrieve the most relevant and effective counterfactual explanations for each individual instance, in any scenario.

4 Experiments

4.1 Introduction

For our experiments, we used three public datasets specifically focused on the churn classification problem. The "Churn for Bank Customers" dataset[1] contains 10,000 records with 14 features aimed at predicting whether a customer has exited the bank (0.20 ratio of churners). The "Credit Card Bank Churn", dataset[2] includes 10,000 credit card user records with 18 features to predict if a customer will stop using the bank's credit card services (0.19 ratio of churners). The "Iranian Churn Dataset"[3] contains 3,150 records and provides telecommunications customer data from Iran with 13 features used to analyze churn behavior (0.18 ratio of churners) in the telecom industry.

---

[1] https://www.kaggle.com/datasets/mathchi/churn-for-bank-customers.
[2] https://www.kaggle.com/datasets/anwarsan/credit-card-bank-churn.
[3] https://archive.ics.uci.edu/dataset/563/iranian+churn+dataset.

We start by identifying which model to explain. We compare the performance of Light Gradient Boosting Machine (LightGBM)[21], XGBoost [22], and Random Forest classifiers [23]. Our focus is on the explanation methodology, more so than on the task, we therefore chose models that are commonly used for the task and easily applied [3]. To fine-tune the models, we conducted a randomized grid search with 5-fold cross-validation, optimizing for the ROC AUC score and we accounted for class imbalance penalizing errors on the minority class proportionally during training.

In Table 1 the comparison is displayed across datasets for F1-Score and Matthews Correlation Coefficient (MCC) measures. The LightGBM outperforms or at least performs as good as the XGBoost in the datasets under analysis while the Random Forest slightly underperforms. These results lead us to establish the LightGBM as the model of interest for the following counterfactual explanations. Statistics related to counterfactual explanations for the XGBoost model are similar and will be omitted to enhance clarity.

Table 1. Model Performance on Churn Datasets. LightGBM and XGBoost display similar performances while Random Forest slightly underperforms.

Metric	Bank Churn			Card Churn			Iranian Churn		
	2000 instances			2026 instances			630 instances		
	LGB	XGB	RF	LGB	XGB	RF	LGB	XGB	RF
F1 Score	0.60	0.59	0.57	0.87	0.86	0.80	0.90	0.88	0.88
MCC	0.52	0.48	0.50	0.79	0.78	0.73	0.87	0.74	0.85

4.2 Exploring Weight Combinations for Counterfactual Evaluation

We explore different weight combinations of our scoring function for evaluating counterfactual explanations. Specifically, we take into consideration the possibility of assigning equal weights on the four metrics (0.25 for each metric), or the possibility of focusing on one of them by imposing a "Higher Weight" (HW) on it, i.e. a weight of 0.5, while imposing for the others a weight of 0.1667.

For each weight combination, we calculate the score for each counterfactual example and select the top $k = 5$ examples for each algorithm. The counts of top counterfactuals selected in the ensemble method for each algorithm against the weight combinations are plotted in the figure below.

We can see in Fig. 1 that for the Bank Churn and the Iranian Churn datasets, the counterfactuals produced using the DiCE method are predominantly selected, regardless of the imposed imbalance on the score function. Contrarily, the T-LACE method is predominantly selected in the generation of counterfactuals for the Credit Card Churn dataset. The quality of counterfactuals across methods is, hence, strongly dataset-dependent. This shortcoming can be avoided by the proposed ensemble approach, being capable of selecting

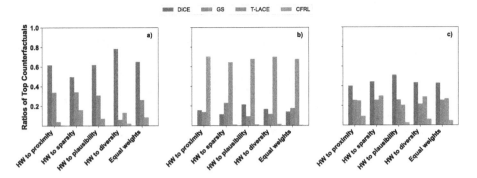

Fig. 1. Ratio of top counterfactuals by weight combinations that are selected by the ensemble for each of the three datasets: **a)** the Churn for Bank Customers, **b)** the Credit Card Bank Churn and **c)** the Iranian Churn dataset. In the x-labels, *HW* stands for "higher weight" in the score function (default value 0.5). While the ensemble selects a relatively large amount of counterfactuals originating from the DiCE method for **a)** and **c)**, it selects predominantly counterfactuals from T-LACE for Credit Card Churn. The ensemble method, hence, displays a high degree of adaptability to the dataset at hand.

at the individual-instance level, the best-performing counterfactual explanations according to the score function imposed by the user (following her preferences).

Next, we explore different weight combinations and predicted probability thresholds for evaluating counterfactual explanations. As for weight combinations, we refer to the ones introduced in Fig. 1, as for predicted probability threshold, instead, we refer to 3 different classes of thresholds for the predicted probability that the counterfactual instance has the opposite class of the reference instance according to the black-box model. The thresholds considered are 0.5, 0.7, and 0.9. For each combination of weights and thresholds, we calculate the scoring function for each counterfactual example and select the top 5 counterfactuals based on their scoring. The counts of top counterfactuals for each algorithm are then plotted against the weight combinations and predicted probability thresholds, as shown in Fig. 2.

As we can see from Fig. 2, in the Bank Churn (a) and Iranian Churn datasets (c), GS and CFRL algorithms tend to be selected, together with the predominant DiCE, for counterfactuals with lower confidence, i.e. prediction probability $\in (0.5, 0.7)$. Increasing the prediction confidence of the top counterfactuals offered by GS and CFRL reduce in favor of counterfactuals generated by DiCE. Instead, in the Credit Card Churn dataset, T-LACE is predominant across probability thresholds, while GS and DiCE offer a number of counterfactuals that fluctuate regardless of the given threshold. This shows how it is possible to explore the confidence of the underlying model by simply modifying the scoring function.

The plots in Fig. 3 offer insights into the performance of counterfactual explanations generated from different sources. We observe a peculiar trait in our ensemble: prediction probabilities are more spread-out, with respect to the sin-

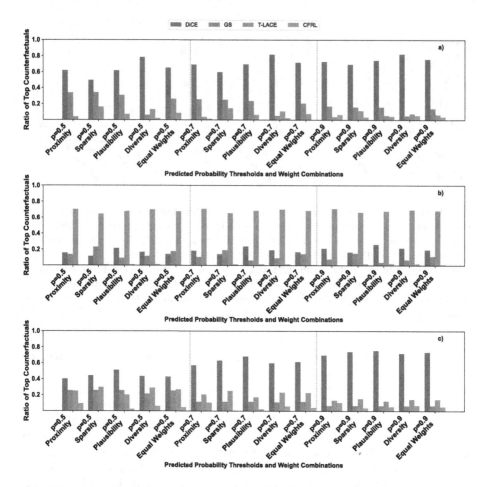

Fig. 2. Ratio of top counterfactuals by probability threshold and weight combinations for each of the three datasets: **a)** Churn for Bank Customers, **b)** Credit Card Bank Churn and **c)** Iranian Churn. In the x-labels, different thresholds are depicted stratified by the imposed coefficient on the score function. In the Credit Card Churn dataset, T-LACE is the predominant method across thresholds. In the other datasets, the predominant method is DiCE. However, note that in both datasets the distance between the ratio of top counterfactuals between DiCE and the other methods increases by increasing the threshold, in other words the selection of counterfactuals across the different methods tends to be more consistent when prediction confidence is lower.

gle methods. This is intriguing because it provides a set of explanations that can cover all the ranges of churning probability (i.e. <0.7: low churn probability, <0.9: mid churn probability, and >0.9: high churn probability), offering a possibility for market segmentation and diverse intervention strategies. A customer retention specialist might get a more exhaustive explanation with diversity in prediction probabilities rather than a focus only on labels with high confidence.

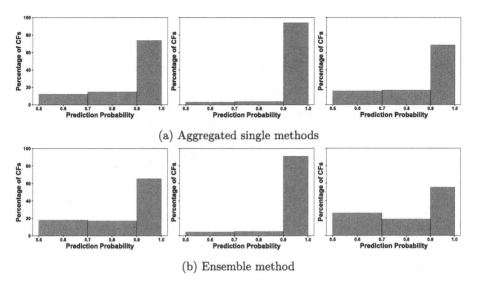

Fig. 3. Comparison of prediction probability of the single methods aggregated together (a) with the ensemble method (b) for the different datasets, namely Bank Churn (on the left), Credit Card Churn (on the center), and Iranian Churn (on the right). The prediction probability given by the application of the black-box on the counterfactual instances and the percentage of CFs with those probabilities is displayed respectively on the x-axis and the y-axis. Plots of the ensemble in (b) show higher diversity in probabilities with respect to single methods in (a). The increased diversity in prediction probabilities allows practitioners to better segment counterfactuals, and hence individuals, based on different confidence levels.

4.3 Analysis of Features Change Ratios

The plot in Fig. 4 displays a heatmap of Kendall's Tau correlations between the SHAP features rank and the ranks of the most changed features from the ensemble method, listed in decreasing order. SHAP values measure the contribution of each feature to the prediction for an individual instance, based on the concept of Shapley values [24]. These values help to understand the influence of each feature on the model's output.

The average change ratios of the features, on the other hand, indicate the proportion of counterfactual instances where a feature value differs from the original instance. This metric helps to identify which features are most frequently altered in the generated counterfactuals, suggesting their importance in driving changes in predictions. This visualization provides insights into how well the ranks of the SHAP values align with the ranks of the most changed features.

As illustrated in Fig. 4, methods like T-LACE and DiCE exhibit inconsistent correlations across different datasets which suggests they might be sensitive to specific dataset characteristics. On the other hand, GS is more consistent but uncorrelated and CFRL shows consistently negative correlations. The ensemble in all its variations frequently shows stronger and more consistent correlations,

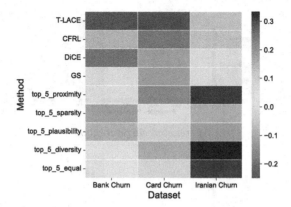

Fig. 4. Kendall-Tau heatmap for SHAP rank correlations, comparing the single methods to the ensemble with different weights configurations across the datasets. Results show that the ensemble selects diverse and more consistent CFs that have on average higher correlation to SHAP features rank.

indicating its robustness in aligning counterfactual changes with SHAP feature importance. The varying correlations suggest that different methods for generating counterfactual explanations may prioritize different features compared to SHAP. Methods with higher positive correlations - as in the case of the ensemble - can be considered more interpretable and aligned with SHAP's feature importance, making them preferable in scenarios where feature importance is a key part for effective actionability of the explanations.

4.4 Example of the Explanation

In Table 2 the output of the ensemble method is displayed and compared with a chosen original instance to be explained. Only the changes to the features made by the generation methods to flip the decision boundary of the black-box are reported. In this example, the original observation is a churning customer, and the ensemble offers a diverse set of counterfactual instances to explain alternative scenarios where the customer would not be predicted as churning. The user can explore different options, a more sparse alternative as in the case of CF_1, a more plausible (i.e. more similar to an actual example in the data) option as in CF_2, a low effort option as in CF_3 or a more diverse one as in CF_5.

Table 2. Comparison of the original instance with the set of top 5 counterfactual instances. In the chosen example from Bank Churn Dataset, CF_1 and CF_3 are selected from GS, CF_2 and CF_4 from DiCE and CF_5 from T-LACE.

\multicolumn{9}{c}{Original Instance}

CreditScore	Gender	Age	Tenure	Balance	NumOfProducts	HasCrCard	IsActiveMember	EstimatedSalary	Churn
650	1	30	6	0	1	0	0	67997	1

Counterfactual Instances

CF_1: Tenure +1 → Churn = 0
CF_2: CreditScore -157, NumOfProducts +1, EstimatedSalary -2181 → Churn = 0
CF_3: HasCrCard +1 → Churn = 0
CF_4: CreditScore -300, NumOfProducts +1, EstimatedSalary -2807 → Churn = 0
CF_5: Tenure +3, EstimatedSalary -67997 → Churn = 0

5 Conclusions

In this paper we focused on a counterfactual ensemble strategy for churn analysis, leveraging an evaluation score specifically suited to enhance the counterfactual explanation's utility for churn officers and experts. Our method is applicable to any type of classifier, regardless of its specific characteristics, as it only requires the classifier's prediction during the counterfactual generation phase. The ensemble method leverages the strengths of multiple counterfactual generation techniques, and the evaluation function ensures that useful properties are prioritized for the churn officer. This results in a more diverse set of explanations that offer a broader perspective on the local decision boundary of the black-box and multiple courses of intervention. The proposed evaluation scoring function allows for a multifaceted evaluation of counterfactual explanations based on multiple criteria which we believe is particularly useful in the context of churn analysis, where counterfactual explanations can provide actionable insights that can help businesses devise targeted intervention strategies. For example, understanding the minimal changes needed to retain a customer can lead to cost-effective and efficient retention strategies. The ability to tweak the evaluation parameters based on specific needs allows practitioners to prioritize certain aspects of the explanations, such as minimizing changes or ensuring high plausibility, making the approach highly adaptable to various scenarios. The experiments conducted on three publicly available churn datasets from different domains (banking, credit card services, and telecommunications) have shown the applicability of our evaluation strategy, and how easily it can be piloted by churn officers to explore churn prediction machine learning models. While the current implementation of our work has shown promising results, there are several areas for future research and improvement. Conducting user studies to evaluate the business utility and user satisfaction of the generated counterfactual explanations would provide valuable insights for refining the approach. Moreover, the computational costs associated with counterfactuals generation must be carefully considered when dealing with very large datasets, as it may hinder efficiency in real-time churn analysis scenarios. Additionally, exploring the integration with other explainability techniques, such as global explanations and feature importance measures, could provide a more comprehensive toolkit for understanding model behavior.

Acknowledgements. SoBigData.it receives funding from European Union - NextGenerationEU - National Recovery and Resilience Plan (Piano Nazionale di Ripresa e Resilienza, PNRR) - Project: "SoBigData.it - Strengthening the Italian RI for Social Mining and Big Data Analytics" - Prot. IR0000013 - Avviso n. 3264 del 28/12/2021. This work has been also supported by the PNRR-M4C2-Investimento 1.3, Partenariato Esteso PE00000013-"FAIR-Future Artificial Intelligence Research"-Spoke 1 "Human-centered AI", funded by the European Commission under the NextGeneration EU programme. MDV also acknowledges support by the European Community programme under the funding schemes: ERC-2018-ADG G.A. 834756 "XAI: Science and technology for the eXplanation of AI decision making." This work was also funded by the European Union under Grant Agreement no. 101120763 - TANGO. Views and opinions expressed are however those of the author(s) only and do not necessarily reflect those of the European Union or the European Health and Digital Executive Agency (HaDEA). Neither the European Union nor the granting authority can be held responsible for them.

References

1. Stepin, I., Alonso, J.M., Catalá, A., Pereira-Fariña, M.: A survey of contrastive and counterfactual explanation generation methods for explainable artificial intelligence. IEEE Access **9**, 11974–12001 (2021)
2. Guidotti, R.: Counterfactual explanations and how to find them: literature review and benchmarking. Data Mining and Knowledge Discovery (2022)
3. Geiler, L., Affeldt, S., Nadif, M.: A survey on machine learning methods for churn prediction. Int. J. Data Sci. Anal. **14**(3), 217–242 (2022)
4. Joy, U.G., Hoque, K.E., Uddin, M.N., Chowdhury, L., Park, S.-B.: A big data-driven hybrid model for enhancing streaming service customer retention through churn prediction integrated with explainable AI. IEEE Access **12**, 69130–69150 (2024)
5. Chen, W.: Customer churn analysis for telecom operators based on SVM. In: Proceedings of the 3rd International Conference on Signal and Information Processing, Networking and Computers (ICSINC), vol. 473, pp. 327–332. Springer (2017)
6. Mishra, A., Reddy, U.S.: A comparative study of customer churn prediction in telecom industry using ensemble based classifiers. In: International Conference on Inventive Computing and Informatics (ICICI), pp. 721–725. IEEE (2017)
7. Petkovski, A.J., Stojkoska, B.L.R., Trivodaliev, K.V., Kalajdziski, S.A.: Analysis of churn prediction: a case study on telecommunication services in Macedonia. In: 2016 24th Telecommunications Forum (TELFOR), pp. 1–4. IEEE (2016)
8. Burez, J., Van den Poel, D.: Handling class imbalance in customer churn prediction. Expert Syst. Appl. **36**(3), 4626–4636 (2009)
9. Maldonado, S., López, J., Vairetti, C.: Profit-based churn prediction based on minimax probability machines. Eur. J. Oper. Res. **284**(1), 273–284 (2020)
10. De Bock, K.W., De Caigny, A.: Spline-rule ensemble classifiers with structured sparsity regularization for interpretable customer churn modeling. Decis. Support Syst. **150**, 113523 (2021)

11. Adhikary, D.D., Gupta, D.: Applying over 100 classifiers for churn prediction in telecom companies. Multimed. Tools Appl. **80**, 1–22 (2020)
12. Lemon, K.N., Verhoef, P.C.: Understanding customer experience throughout the customer journey. J Mark. **80**(6), 69–96 (2016)
13. Luo, X., Kumar, V.: Operational efficiency and customer retention in outsourced customer service operations. J. Mark. Res. **50**(2), 264–278 (2013)
14. Theodoridis, G., Tsadiras, A.: Applying machine learning techniques to predict and explain subscriber churn of an online drug information platform. Neural Comput. Appl. **34**(22), 19501–19514 (2022)
15. Tao, J., et al.: Explainable AI for cheating detection and churn prediction in online games. IEEE Trans. Games **15**(2), 242–251 (2023)
16. Guidotti, R., Ruggieri, S.: Ensemble of counterfactual explainers. In: DS 2021. LNCS, vol. 12986, pp. 358–368. Springer (2021)
17. Sharma, S., Henderson, J., Ghosh, J.: CERTIFAI: counterfactual explanations for robustness, transparency, interpretability, and fairness of artificial intelligence models (2019)
18. Laugel, T., Lesot, M.-J., Marsala, C., Renard, X., Detyniecki, M.: The dangers of post-hoc interpretability: unjustified counterfactual explanations (2019)
19. Samoilescu, R.-F., Van Looveren, A., Klaise, J.: Model-agnostic and scalable counterfactual explanations via reinforcement learning. CoRR, abs/2106.02597 (2021)
20. Bodria, F., Guidotti, R., Giannotti, F., Pedreschi, D.: Transparent latent space counterfactual explanations for tabular data. In: DSAA, pp. 1–10. IEEE (2022)
21. Ke, G., et al.:. LightGBM: a highly efficient gradient boosting decision tree. In: NIPS, pp. 3146–3154 (2017)
22. Chen, T., Guestrin, C.: XGBoost: a scalable tree boosting system. In: KDD, pp. 785–794. ACM (2016)
23. Breiman, L.: Random forests. Mach. Learn. **45**(1), 5–32 (2001)
24. Lundberg, S.M., Lee, S.-I.: A unified approach to interpreting model predictions. In: NIPS, pp. 4765–4774 (2017)

Open Access This chapter is licensed under the terms of the Creative Commons Attribution 4.0 International License (http://creativecommons.org/licenses/by/4.0/), which permits use, sharing, adaptation, distribution and reproduction in any medium or format, as long as you give appropriate credit to the original author(s) and the source, provide a link to the Creative Commons license and indicate if changes were made.

The images or other third party material in this chapter are included in the chapter's Creative Commons license, unless indicated otherwise in a credit line to the material. If material is not included in the chapter's Creative Commons license and your intended use is not permitted by statutory regulation or exceeds the permitted use, you will need to obtain permission directly from the copyright holder.

This Sounds Like That: Explainable Audio Classification via Prototypical Parts

Andrea Fedele[1,2]($^{\boxtimes}$)[iD], Riccardo Guidotti[1,2][iD], and Dino Pedreschi[1,2][iD]

[1] University of Pisa, Pisa, Italy
andrea.fedele@phd.unipi.it, {riccardo.guidotti,dino.pedreschi}@unipi.it
[2] ISTI, National Research Council, Pisa, Italy
{andrea.fedele,riccardo.guidotti,dino.pedreschi}@isti.cnr.it

Abstract. The demand for understanding machine learning models has led to the development of interpretable-by-design models that provide both outcomes and explanations. In this paper, we extend the concept of Prototypical Part Networks to the audio domain with `SonicProtoPNet`. This model enables a "this sounds like that" reasoning for audio classification, where a test instance audio is classified based on *prototypical parts* that most resemble specific areas of specific training instances. Quantitative results from genre and environmental sound classification, as well as musical instrument recognition tasks, demonstrate satisfactory per formance using the Log-Mel transformation of the audio input signal, further supported by backbone pre-training on image-input data. Furthermore, we introduce a high-quality back-soundification method for the learned sonic prototypes, facilitating intuitive interpretation of classification decisions through auditory inspection.

Keywords: Explainable Artificial Intelligence · Explainable Audio Classification · Part Prototypical Interpretability

1 Introduction

The advancements in Deep Learning (DL) over the past five to ten years have significantly increased the applicability of this technology across various fields of our society. Indeed, DL technologies have become essential for Artificial Intelligence (AI) product development due to their remarkable performance in various tasks and domains. However, as DL models grow in complexity, interpretability often decreases [2]. This leads to opaque decision-making processes that cannot be easily audited or corrected for biases or errors [11]. The AI community has long been aware of this challenge, leading to the growing of eXplainable Artificial Intelligence (XAI) [4]. Research in this field aims to achieve the key goal of making these sophisticated models clear and transparent to a diverse range of users, including stakeholders, developers, and third parties or end-users.

Hence, interpretability is in the eye of the beholder, and various forms of explainer and explanations have been explored [1,4,11]. Many techniques, categorized as *post-hoc* explainers, elucidate machine learning models after the training phase [2,6,11]. However, it is important to note that such techniques commonly reason through surrogate models, which may fail to accurately elucidate the model's actual reasoning process, leaving the relevance of the original black box undisclosed [10]. In contrast, promising approaches of interpretability *by-design*, or *ex-ante* explainability, involve developing architectures that not only provide outcomes but also offer insight into the model's reasoning process to achieve those outcomes [2,6,11]. In particular, ProtoPNet [5], has demonstrated the ability to maintain comparable performance to black-box counterparts while adding the valuable aspect of transparency across computer vision tasks. ProtoPNet learns prototypes specific to each class during training and utilize a straightforward classification process based on these prototypes, resembling a *"this looks like that"* reasoning. ProtoPNet not only provides the classification outcome itself, but also identifies *prototypical parts* of the test image that *"looks like"* parts of the training samples considered discriminative for that particular class.

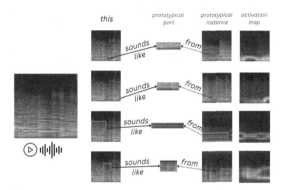

Fig. 1. SonicProtoPNet explanation process visualized as Log-Mel Spectrograms for an audio of a *piano* recording alongside the learned prototypical parts used for classification. The leftmost element is the test instance. Then, from left to right: *(i)* same test instance with bounding boxes, *(ii)* prototypical parts, *(iii)* prototypical instances from the training set, *(iv)* activation maps. Code for reproducibility and playable prototypes examples: https://github.com/andreafedele/SonicProtoPNet/

However, as of our current knowledge, the applicability of ProtoPNet in the audio domain remains unexplored. In order to fill this research vacuum, we introduce SonicProtoPNet, an interpretable-by-design network for audio classification tasks. SonicProtoPNet aims to emulate a *"this sounds like that"* reasoning. An example of audio classification along with SonicProtoPNet explanation is presented in Fig. 1. Our contribution can be summarized as: *(i)* adaptation of ProtoPNet towards Log-Mel Spectrogram representation of audio recordings, *(ii)* experimentation across genre classification, musical instrument recognition, and environmental sound classification tasks, *(iii)* assessment of image-based pre-training's impact on Log-Mel Spectrogram classification performance, and *(iv)* introduction of high-quality soundification for the *prototypical parts*, enabling auditory inspection of discriminative excerpts within audio recordings.

The rest of paper is organized as follows. In Sect. 2, we elaborate on pertinent studies. Section 3 details background information about `ProtoPNet` and the Log-Mel Spectrogram audio representation, while Sect. 4 details the problem addressed and our `SonicProtoPNet` proposed solution. The outcomes of our experiments are articulated and examined in Sect. 5. Lastly, Sect. 6 encapsulates our contributions and proposes potential avenues for future research.

2 Related Works

In recent years, `ProtoPNet` [5] has emerged as interpretable-by-design alternatives to conventional Convolutional Neural Networks CNNs [12], as it offers comparable performance but add value by learning class-specific prototypes that contribute to transparent predictions. Various adaptation followed modifying such classification process: in `ProtoTree` [20] a binary decision tree is used, while in `ProtoPool` [21] a soft assignment of prototypes to classes during training is learned, and these are transformed to hard assignments at test time. `ProtoConcepts` [15] extends the original work from a *"this looks like that"* to a *"this looks like those"* kind of reasoning to improve the assignment of parts capturing concepts in a test image to those of several training images.

Research on interpretability in audio signals lags behind that in image or textual domains [16], but several post-hoc techniques have been explored. LRP has been applied to audio digit recognition [3], and it has been used on CNNs processing 1-Dimensional waveforms [17]. `Sound-LIME`, an extension of LIME [27], perturbs individual segments of the input waveform to obtain explanations [18]. Similarly, `SINEX`, a post-hoc siamese network explainer, uses a perturbation-based method to explain outcomes on audio spectrograms [7]. Lastly, `AudioSLIME` generates interpretations encompassing playable audio objects.

In recent times, there has been a growing interest in exploring prototypical explanations within the audio domain [13,14,22,23,26,31]. In [22], `APNet`, an interpretable sound classification system extending the prototypical image classification network of [13], is introduced. `APNet` evaluates input similarity with learned prototypes using a frequency-dependent similarity measure tailored for audio data, allowing auditory inspection of the prototypes. `APNet` is further extended in [31] to handle multi-label polyphonic settings, providing explanations with local prototypes and associated attention maps. Another variant of `APNet` is presented in [26], utilizing cosine similarity between prototypes and feature maps, with a focus on exact matching between prototypes and training samples. [14] introduces an approach for representing audio clips using spectral prototypes and audio-specific transformations, enhancing prototype reconstruction by incorporating audio-specific transformations: gain, pitch, low-frequency filters, and high-frequency filters. In [23], an interpretable method is proposed for audio separation or denoising, focusing on regularized intermediate embeddings from Non-Negative Matrix Factorization. `DFT-LRP` in [8] investigates how input representations affect event classification tasks in audio.

In our study, we extend the scope of interpretable prototypical-based audio classification beyond monophonic sound tasks examined in [22], applying it to tasks like genre, environmental sound, and musical instrument recognition. We exclude evaluations from [31] and [26] as they address different problems; the former focuses on multi-label tasks, while the latter projects prototypes onto specific training samples, explaining them as specific instances. Unlike our aim, [14] focuses on audio prototype construction, and [23] uses NMF dictionaries for explanations, while we focus on identification and selection. Similar to [8], we explore how audio transformations affect predictive performance within the audio domain, but with an interpretable by-design approach rather than a post-hoc one. It is crucial to note that the discussed works differ from ours significantly as they do not aim to identify prototypical *parts*, which pinpoint sonic time-frequency segments of a test instance resembling those from training recordings.

3 Background

We offer here background information to ensure readers have a comprehensive understanding of the subsequent sections.

ProtoPNet Architecture. We provide here a summary of ProtoPNet and refer the reader to [5] for details, as our proposal is based on the same architectural structure. ProtoPnet consists of three main components: *(i)* a convolutional network f, *(ii)* a prototypical layer g_p, and *(iii)* a fully connected layer h. The role of f is to extract embeddings $z = f(x)$ from input x, with parameters w_{conv}. The resulting dimensions of z in the embedding space are denoted as $H \times W \times D$. The prototypical layer g_p learns m prototypes $P = \{p_j\}_{j=1}^{m}$, each with dimensions $H_1 \times W_1 \times D$ ($H_1 \leq H$) and ($W_1 \leq W$). Each jth prototype unit g_{p_j} measures the L^2 distance between a prototype p_j and the patches of z with the same shape, i.e., overlaps of p_j on z. These distances are converted to similarity scores, forming an activation map where each value represent how strongly a prototypical part is present in the given input. The up-sampled activation map in the input domain, is a heat-map representing which part of the input image is most similar to the learnt prototype. Each g_{p_j} unit's activation map is reduced using global max pooling to a single similarity score, indicating how strongly a prototypical part is present in *some* patch of the input matrix. The fully connected layer h, with weights w_h and no bias, holds the prototype-class activation weights and it is optimized as the last step of the training procedure.

The training phase of ProtoPNet includes three main steps. First, Stochastic Gradient Descent is applied to all layers except the last one to learn a latent space. This space clusters important patches corresponding to a given input around semantically similar prototypes of the input's true class, while simultaneously ensuring clear separation between clusters centered at prototypes of other classes. During this step, w_{conv} of f and prototypical layer g_p parameters are jointly optimized, while w_h parameters remain fixed. Within the optimization problem, the clustering cost encourages each training sample's latent patch to be close to *at least* one prototype of its own class, while the separation cost ensures distance from prototypes of other classes. Additionally, w_h weights are initialized such that for each connection between the jth output of the prototype unit and the logit of class k, $w_h^{(k,j)}$ is set to 1 if $p_j \in P_k$, and -0.5 otherwise ($p_j \notin P_k$). The positive association between a class k prototype and its corresponding logit raises the predicted probability of class k, while a non-class k prototype decreases it. In the second step, each prototype p_j is projected onto the nearest training patch from the same class. Finally, a convex optimization of the last layer's weights w_h is conducted to adjust the layer connections. Specifically, for entries $w_h^{(k,j)}$ where $p_j \notin P_k$, the model strives for sparsity[1], aiming for $w_h^{(k,j)} \approx 0$. The classification of a given test instance involves comparing the latent features $z = f(x)$ with the prototypes learned during training. For each class k, the network measures the similarity between its latent patch representation and every learned prototype p_j of class k. This comparison generates a map of similarity scores for each prototype. Finally, each similarity score is weighted by the prototype-class connection weights (w_h), determining the overall contribution of each prototype to the classification of the input instance's specific class.

The learned prototypes exist within the embedding space and, to identify the patch of the input space corresponding to a prototype p_j, the input x is fed into the network, and the activation map generated by the prototype unit g_{p_j} is upsampled to match the size of the input sample before applying max-pooling. This upsampling process produces a heatmap highlighting which areas of the input x bear the closest resemblance to the prototype p_j, with the highly activated region in the heatmap corresponding to the most activated patch of x. The visualization of prototype p_j involves identifying the smallest rectangular area within x, containing pixels with activation values surpassing the 95th percentile in the upsampled activation map generated by g_{p_j}.

Log-Mel Spectrogram Representation. Recent research utilizes Log-Mel Spectrogram transformation for audio data due to its compatibility with CNNs and the high-performance results it achieves [19,25,28,30]. Following an initial transformation of the time dimension using Short-Time Fourier Transformation (STFT), each time frame of the frequency spectrum undergoes conversion to the mel scale, which aligns with human sound perception. The frequency spectrum is then represented on a logarithmic scale to accentuate minor changes in amplitude. Finally, the log-amplitude spectra of individual time frames are aggregated over time, forming a two-dimensional matrix. Formally, this matrix is represented as $\mathbb{R}^{p \times q}$, where q denotes the length of the audio track, and p represents the number of observed mel-frequencies. Within each tuple (i, j) of the matrix, at time i and mel-frequency j, the magnitude of amplitude of the original recording is expressed in decibels (dB).

---

[1] Sparsity reduces reliance on negative reasoning, thus avoiding reasoning like "this sample is of class k because *it is not* of class k'".

4 SonicProtoPNet

We define here the `SonicProtoPNet` adaptation for processing audio data and we discuss the prototypes visualization and soundification process.

Network Adaptation. The `ProtoPNet` architecture, as presented in the previous section, cannot be directly applied to audio tracks input data in waveform. Such data, in `SonicProtoPNet`, have to be transformed into Log-Mel Spectrograms, which in turn require technical adjustments to ensure compatibility with the f CNN component. As such, the first adaptation involves introducing an audio to Log-Mel Spectrograms transformer ($x = audio2lms_\Theta(t)$) capable of converting an audio track t into a Log-Mel Spectrogram x. We highlight that the $audio2lms_\Theta$ implementation accommodates parametric values for the conversion Θ, which can be easily adjusted within the project settings.

Once loaded and converted, the input matrix from audio recordings differs from that of `ProtopNet`[2]. In our case, it has dimensions $F \times T \times C$, where F and T denote the y-axis (log-mel frequencies) and x-axis (time) of the original audio, respectively, and C represents the number of channels. For stereo audio recordings, $C = 2$, while mono recordings have a single channel with $C = 1$. This dimensional difference affects the usage of the pre-trained f CNN used in `ProtoPNet`, where architectures pre-trained on image-data, such as VGG and ResNet, are employed. Thus, for the `SonicProtoPNet` architecture, to not discard the pre-trained weights w_{conv} of f, we averaged the 3-channel kernel on the first layer, as it is the only layer affected by dimensionality compatibility, allowing our architecture to work with a single channel to model the Log Mel Spectrogram of mono audio recordings. This step plays a crucial role in the overall performance of `SonicProtoPNet`, as discussed in Sect. 5.

`ProtoPNet` heavily relies on augmented samples during the training phase, which must be treated distinctly from the original ones. This distinction is necessary because the projection of prototypes and the subsequent recovery of the actual training sample, which is matched with the "like that" in the authors' motto, must be performed on real samples rather than augmented ones. The augmentation techniques used in [5] on image-input data include *rotation*, *skew*, and *shear*. In `SonicProtoPNet`, similarly to [14], we replace these operations with *noise injection*, *pitch shifting*, and *time stretching*. The final adaptation of the [5] network was conducted during the push/projection epochs within the training procedure. During this step, the content of the Log-Mel training spectrograms is appended to the saved output, in addition to the graphical representation (in image format) performed in [5]. This step is crucial as it enables the recovery of the training audio recordings for soundification purposes at prediction time.

At prediction time, given a test audio waveform track t, `SonicProtoPNet` first transforms it into a Log-Mel Spectrogram x using the $audio2lms_\Theta$ transformer. It then provides the predicted target label y, along with a set of m prototypes $p_1, \ldots, p_m$ for class y and their corresponding bounding boxes $bb_1, \ldots, bb_m$ i.e.,

---

[2] `ProtoPNet` inputs are RGB-images of dimension $W \times H \times 3$, representing width pixels, height pixels, and the three RGB channels respectively.

Algorithm 1: ProtoSoundify(t, bb_x, Θ_x)

Input : t - waveform audio track, x - Log-Mel Spectrogram,
p - prototype, bb_x - bounding boxes of p in x,
Θ_x - conversion parameters
Output: a - playable sonic prototypical part

1 $s \leftarrow \text{PwrSpect}(t, \Theta_x)$;   // to power spectrogram
2 $F \leftarrow \text{FBMatrix}(\Theta_{x_{n\_fft/2+1}}, \Theta_{x_{n\_mels}})$;   // get matrix
3 $bb_s \leftarrow map(F, bb_x)$;   // map bounding boxes in s
4 $s_p \leftarrow isolate(s, bb_x, bb_s)$;   // isolate prototype in s
5 $a \leftarrow \text{GriffinLim}(s_p, \Theta_x)$;   // convert to waveform
6 **return** a;   // return audible prototypical part
---

the prototype coordinates on the Log-Mel Spectrogram x. The set of returned m prototypes includes: *(i)* the top-n activated prototypes overall, chosen from all possible classes, and *(ii)* the top-n prototypes for each of the top k classes contributing to the final classification towards label y. It is important to note that prototypes contributing to a specific y label classification may not always come from training samples of that same label, and misclassification often occurs when the most influential prototypes come from a different label than y. Both n and k are adjustable parameters at prediction time. Additionally, for each prototypical part m_i in the x test instance, the network identifies the corresponding most triggered areas on some training instance e. The set $t_{e_1}, \ldots, t_{e_m}$ of the most activated training samples in waveform format for each sonic prototypical part is included in the same output bundle. This set is constructed by considering the smallest absolute distance between the Log-Mel Spectrogram representation of the training instance t – which is available due to the network adaptation described earlier that saves the Log-Mel Spectrogram content during the push/projection phases of the training procedure – and each training sample recording belonging to the prototype p class after the *audio2lms* conversion. This allows for the aural inspection of both the complete recordings and the prototypical sonic parts.

Prototypes Visualization and Soundification. While Log-Mel Spectrograms can indeed be visualized as images, as shown in Fig. 1, such representations may not convey the complete information. Indeed, while the RGB images adopted in [5] encapsulate the same semantic information as the network input, Log-Mel Spectrograms represent audio recordings that is something humans are naturally attuned to perceive through sound, rather than visual observation. Therefore, given the context of audio input, prototypical part explanation should extend beyond visual representations within the Log-Mel Spectrogram x.

We name *Sonic Prototypical Part* (**SPP**) an audible prototype a containing only information identified by bounding box coordinates bb_x onto the Log-Mel Spectrogram x. However, due to the lossy nature of the Log-Mel Spectrogram transformation, the standard process of soundification, which involves estimating the STFT in the normal frequency domain from the mel frequency domain, would yield poor audio quality. Therefore, to produce high-quality playable SPP a, we define the following steps summarized by Algorithm 1.

In particular, ProtoSoundify takes as input a test audio track in waveform t, the corresponding Log-Mel Spectrogram x, a prototype p returned by the SonicProtoPNet as well as the bounding boxes bb of p in x, and the conversion parameters Θ_x used by the *audio2lms* tranformer of SonicProtoPNet in the Log-Mel Spectrogram transformation of x. The SPP soundification starts with a Power Spectrogram transformation PwrSpect applied to the test audio t (line 1, Algorithm 1) with the same FFT size (n_ftt), window size, and hop length (hop_length) between STFT windows as used *audio2lms* for x transformation. We condense such parameters in Θ_x for readability purposes. The resulting Power Spectrogram s shares the time dimension with x, but its frequency dimension scale is different, not matching the Mel scale of x. Next, ProtoSoundify creates a triangular filter bank matrix F of size (n_freq, n_mels) using the same number of Mels and Mel scale as x, and $n\_fft/2 + 1$ frequencies as used in PwrSpect for s (line 2, Algorithm 1). This filter bank matrix F is then used to map the bounding box coordinates from the Mel-Frequency dimension of x to the Spectrogram Frequency scale of s (line 3, Algorithm 1), therefore computing bb_s. F plays is crucial, as it addresses the disparity between the frequency domains of x and s, which are not directly aligned. Specifically, while the frequencies in x are represented in mel frequencies, those in s are in power frequencies. Using bb_s, areas outside such bounding box in the frequency-axis of s can be filtered out by using the minimum value present in s, i.e., by simulating silence. For time-dimension filtering, bb_x coordinates are used, as both x and s share such dimension. Both filtering are included in line 4, Algorithm 1, returning s_p which is the Power Spectrogram representation of the p prototype filtered on bb_x time and bb_s frequencies bounding boxes from x. Finally, the playable SPP a is the result of s_p GriffinLim transformation [9] from Power Spectrogram prototype to the corresponding waveform in the audio domain (line 6, Algorithm 1).

5 Experiments

We present here the experiments conducted to assess the performance of SonicProtoPNet. First, we present the experimental setting, detailing datasets, augmentation techniques, Log-Mel Spectrogram transformation, and batch size used in SonicProtoPNet. Then, we discuss the baseline neighbors-based classifier and three distinct data pre-processing techniques. Finally, we present the experimental results of SonicProtoPNet.

Experimental Setting. We experimented with ESC50 [24], GTZAN [29], and *Medley-Solos-DB*[3] (MEDLEY) datasets, employed for environmental sound classification, genre classification, and instrument recognition tasks, respectively. ESC50 comprises 2000 environmental audio recordings, each consisting of forty

---
[3] DOI: https://zenodo.org/doi/10.5281/zenodo.1344102.

5-second-long clips per class, e.g., rain, helicopter, cat, snoring. GTZAN contains 1000 audio tracks lasting 30 s each, distributed across 10 different genres, with each genre represented by 100 tracks, e.g., blues, classical, jazz, rock. Each 30-second recording in GTZAN has been divided into six non-overlapping 5-second excerpts. MEDLEY is comprised of 21571 audio clips, each lasting approximately 3 s and featuring a single instrument among 8 different ones, such as violin, piano, trumpet, female singer, etc. All recordings in these datasets are in the WAV format and are mono signals, containing a single channel.

All audio files are converted into Log-Mel Spectrograms using a consistent n_fft value of 12288. We varied the hop_length and n_mels parameters depending on the dataset[4]. These parameter choices were made after preliminary experiments aimed at creating square-shaped matrices that meet the requirements of the f component, while retaining audio information. As a result, the dimensions of the Log-Mel Spectrograms for ESC50, GTZAN, and MEDLEY are as follows: $300 \times 300 \times 1$, $350 \times 350 \times 1$, and $240 \times 240 \times 1$, respectively. It is important to note that tuning the Log-Mel spectrogram parameters does not necessarily require users to have a deep understanding of each parameter. The process can be guided by preliminary combinations aimed at creating square-shaped matrices that are compatible with f component. During these preliminary steps, the quality of audio reconstruction for the selected parameters can be further verified using Algorithm 1. Therefore, specific parameter initializations for the Log-Mel transformation can quickly be tested and discarded if they fail to create square-shaped spectrograms or result in poor audio reconstruction quality.

In our experiments, we augmented each audio recording in the ESC50 dataset with 12 samples (4 per technique), in the GTZAN dataset with 3 samples (1 per technique), and in the MEDLEY dataset with 1 sample using a randomly selected technique[5]. We chose these numbers to maintain consistency with the cardinality of the original datasets. Larger datasets would result in significantly larger sets of augmented samples, which deviates from the ratios suggested in [5].

We created multiple versions of the datasets to experiment with varying numbers of classes. For ESC50, we generated 4 versions with 3, 5, 10, and 50 classes, respectively. Similarly, for GTZAN and MEDLEY, we created 3 versions each with 3, 5, and 10 classes for GTZAN, and 3, 5, and 8 classes for MEDLEY. In each version, the classes were randomly selected from the complete dataset. The train-test split for MEDLEY was maintained as provided in the metadata. For ESC50 and GTZAN, we randomly split the data into 75% for training and 25% for testing. This split was consistent across all versions of the datasets. We chose not to use

---

[4] For ESC50, the hop_length was set to 600 and n_mels to 300. For GTZAN, these values were adjusted to 700 and 350, respectively. Finally, for MEDLEY, the hop_length was set to 500 and n_mels to 240.

[5] Time Stretch technique: slows down the recording within the range of 80 to 90% of the original length; Noise Injection technique: varies from 0 to 10 dBs; Pitch Shifting technique: raises the pitch within a range of 7 semitones above the original recording.

the provided split metadata for ESC50 due to the potential usefulness of different takes of the same recording as augmented samples for training[6].

In line with [5], in all experiments with SonicProtoPNet, we set the number of prototypes to recover per class as 10. Additionally, the batch size for training, testing, and augmentation sets was set to 20, while the number of convex optimization steps for the last layer was set to 5. The push operation (projection of prototypes) was performed every 5 training epochs, starting after the first 5 warming epochs. The remaining parameters of the network were kept consistent with the original network, including learning rates, starting clustering, and separation coefficients[7]. Each SonicProtoPNet experiment utilized a VGG19 convolutional-based backbone implementing the function f, and was conducted both with and without the use of pre-trained weights on this network. It is worth noting that this network was pre-trained on the 1k-ImageNet dataset[8], which consists of quality-controlled and human-annotated images. In line with [5], the training procedure stops when the training accuracy converges, and the clustering cost becomes smaller than the separation cost within the Stochastic Gradient Descent optimization of all layers prior to the last one.

Baselines. We compared SonicProtoPNet against a k-Nearest Neighbors (k-NN) classifier, chosen for its close resemblance to the classification process of a given instance in SonicProtoPNet. Indeed, while SonicProtoPNet classifies an instance based on the most activated prototypical parts relative to the training sample and its class connection, k-NN assigns the class based on the majority class of the k closest training instances considering the whole object w.r.t a specified distance metric. Furthermore, we selected k-NN because it enables the identification of the neighbors of a given test instance, allowing the full audio recordings themselves to be played for aural inspection[9]. We followed two experimentation paths for k-NN. The first path considers $k = 3$, creating a simple aural inspection scenario where only a very limited number of recordings need to be listened to in order to better understand the classification outcome. The second path follows the rationale of the SonicProtoPNet experimental training phase, where 10 prototypes are assigned to each class. In this case, we refer to it as z-NN, where z is the result of the number of classes in a given dataset multiplied by 10. Although aural inspection might be harder in this scenario, our aim is to mimic the SonicProtoPNet training procedure to compare the explanatory power of the two algorithm alternatives while keeping the number of total prototypes per class identical. Following suggestions outlined in [8], three different representations of the audio recordings were created for k-NN. The first transformation, cent, represents the audio file using spectral centroids of the magnitude spec-

---

[6] The code provided in the repository allows replication of the same dataset configurations used in our experiments, starting from the original dataset repositories.

[7] SonicProtoPNet also uses two additional 1×1 convolutional layers following f, with ReLU for all layers except the last one, which utilized a sigmoid activation.

[8] ImageNet: https://www.image-net.org/.

[9] k-NN does not highlight specific sonic prototypical parts but instead presents the complete audio recordings of the identified neighbors themselves.

trogram, normalizing each frame and extracting the mean centroids. The second transformation, spect, generates a Mel Spectrogram for all STFT frames and computes the mean of each column to create a tabular representation. The final transformation, comb, combines spect, the column mean of a Chromagram representation, and the column mean of the Mel Frequency Cepstral Coefficients (MFCC) representation for all STFT frames. While chromagram representation shows the distribution of energy in different pitch classes, i.e., musical notes, over time, MFCC aims to capture spectral characteristics of audio signals. We use these transformations[10] to maintain the spectral information kept in the input data of SonicProtoPNet and to assess the impact of additional information on a tabular feature-based dataset within the k-NN classifier[11].

Table 1. Test Set Accuracy. Best result is highlighted in bold, best result runner up in italic. Bottom row average and standard deviation of the accuracy for each method.

dataset	classes	3-NN			z-NN			SonicProtoPNet	
		cent	spect	comb	cent	spect	comb	$w^{conv}_{non\text{-}pre}$	w^{conv}_{pre}
ESC50	3	.66	.63	.73	.56	.63	.56	**.86**	*.80*
	5	.60	.56	.68	.48	.28	.56	*.94*	**.96**
	10	.29	.34	.46	.24	.20	.33	*.56*	**.93**
	50	.10	.18	.26	.07	.05	.08	*.57*	**.75**
GTZAN	3	.46	.79	*.90*	.44	.66	.81	.75	**.95**
	5	.40	.79	**.90**	.38	.62	.79	.86	*.89*
	10	.23	.67	*.82*	.20	.42	.53	.59	**.86**
MEDLEY	3	.53	.51	*.58*	.56	.47	.54	.50	**.59**
	5	.32	.42	.52	.25	.45	.46	**.55**	*.53*
	8	.43	.34	*.52*	.42	.33	.46	.40	**.54**
	avg	.40	.52	.63	.36	.41	.51	*.66*	**.78**
	std	**.16**	.19	.20	**.16**	.19	.20	*.17*	**.16**

Quantitative Results. Table 1 presents the predictive accuracy for each dataset, along with the different representations used with the 3-NN and z-NN baseline. Additionally, experiments for SonicProtoPNet are conducted with and without pre-trained weights (w^{conv}_{pre} and $w^{conv}_{non\text{-}pre}$ respectively) for the f CNN-based VGG19 backbone. The quantitative results reveal interesting insights. Firstly, the choice of representation for audio recordings significantly influences the overall predictive capability of k-NN. Notably, the comb transformation, which combines Mel Spectrogram, Chromagram, and MFCC features, achieves

---

[10] Default parameters for the three different transformations are utilized via the Librosa Python library, with the exception of using 40 Mel Frequency Spectral Coefficients and 128 mels for spectrograms.
[11] All k-NN experiments utilize the Euclidean distance metric within the classifier.

superior performance across all datasets. Conversely, the spectral centroids time-series representation performs the worst. In general, the comb transformation combined with 3-NN, exhibits performance comparable to SonicProtoPNet $w_{\text{pre}}^{\text{conv}}$ for GTZAN and MEDLEY. Moreover, pre-training f weights enhances the performance of SonicProtoPNet in 8 out of the 10 experiments, making it the overall best-performing network. This underscores the importance of pre-trained networks on image data, even for Log-Mel Spectrogram matrix representation in the context of audio classification. However, it is important to note that significant improvements (e.g., at least 20% points) are observed in only 3 of these 8 cases. These are the only instances where the pre-training procedure would be truly beneficial for overall performance and could influence the choice of one model over another in real-world applications. Therefore, reliance on such pre-trained architectures is not always necessary, as reasonable performance can still be achieved without them. Generally, tasks become more challenging with an increasing number of classes for both 3-NN, z-NN and SonicProtoPNet. Notably, pre-training is not always beneficial, as $w_{\text{non-pre}}^{\text{conv}}$ achieves the best performance on ESC050 3-class and MEDLEY 5-class datasets. z-NN experimental results demonstrate that using a number of neighbours resembling the class-prototypes learned by SonicProtoPNet, results in generally worse performance compared to the 3-NN alternative. Therefore, in k-NN, using a smaller number of neighbors for classification leads to better performance and an easier subsequent aural inspection analysis phase. The training times[12] for SonicProtoPNet vary depending on the dataset cardinality[13].

Qualitative Results. Figure 2 illustrates an example of local correct classification based on part-prototypical reasoning in SonicProtoPNet, using a *piano* audio recording from the MEDLEY 10-class dataset[14]. The classification outcome y correctly predicts the piano class label. The figure consists of three rows, each representing a different prototype contribution to the classification. In this example, all top three contributing prototypes correspond to the *piano* class. Even though all top three prototypes correspond to the piano class, we notice that each prototype correlates to a different training audio, with variations in frequency and time coordinates. For instance, the most strongly connected prototype is located in the lower spectrum of frequencies in both the test and training samples, approximately 1.5 s long. Auditory inspection confirms their frequency isolation similarity, further enhanced by the fact that the two notes played (E^b

---

[12] Experiments run using a *Quadro RTX 6000* GPU unit.

[13] Minimum, maximum, and mean hour times as follows: (0.2, 3, 0.9) for ESC50, (0.6, 2.2, 1.2) for GTZAN, and (0.5, 1, 1.5) for MEDLEY when using $w_{\text{pre}}^{\text{conv}}$. Same training times are observed in $w_{\text{non-pre}}^{\text{conv}}$ SonicProtoPNet, as, in both cases, training stops early due to convergence and the clustering cost being smaller than the separation cost verification condition. Only in ESC50_{10}, ESC50_{50}, GTZAN_5 and GTZAN_{10}, the training epochs achieve convergence twice the time reported for $w_{\text{pre}}^{\text{conv}}$.

[14] In the same manner as the isolation of prototypes from the prototypical instances, the parts bounded by bb_x in the test instances are also isolated, but not depicted in Figure for improved readability.

and A^b) form a perfect fourth interval. Similarly, the second row prototype is also situated in the lower frequency spectrum for both test and training audio recordings, suggesting a similar action of playing exactly four different notes in a comparable time interval. The third training audio plays a specific note throughout the recording, with test and train prototypes located at different frequency heights within the spectrogram. Auditory inspection in this case suggests that resonating harmonics of the train instances may be considered as the root for the test audio's different-note playing recording.

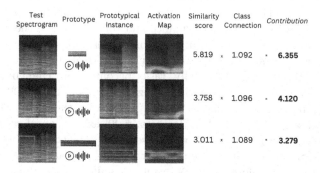

Fig. 2. SonicProtoPNet prediction and explanation process on Log-Mel Spectrograms of the audio of the *piano* recording presented in Fig. 1, reporting the top-3 activated prototypes for classification, all contributing to the *piano* class. Each row illustrates, from left to right: *(i)* the test instance spectrogram with bounding boxes highlighting the activated prototypical parts, *(ii)* the playable prototypical parts, *(iii)* the prototypical instances in the training set, *(iv)* the activation maps indicating the similarity of each prototypical part, *(v)* the similarity between prototypical and test instance, *(vi)* the prototype class contribution, and *(vii)* the final contribution for the *piano* class.

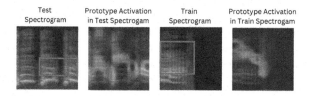

Fig. 3. SonicProtoPNet prototype parts for a *laughing* class ("ah ah ah") *test spectrogram*, along with the most activated prototypical instance from the training set, i.e., *train spectrogram*. The activation maps for both spectrograms, second and fourth plots, visually convey more information than the bounding boxes alone. The highest attention is clearly focused on the laughing instance transients, with less emphasis on the space in-between.

Aural inspection of GTZAN and MEDLEY dataset, suggest patterns similar to those highlighted in the previous example, although it is important to recall that they do not fully describe the overall reasoning of the network since they are local prediction. However, being local prediction based on the fixed and learned 10 prototypes, they do indicate tendencies that occur with a certain frequency. For instance, the GTZAN music genres correctly identify the blues genre, matching SPPs on elements that typically play solos or lead parts in the same frequency range, e.g., electric guitars and keyboards. Nevertheless, through the usage of SonicProtoPNet and by listening its audible prototypical parts, we realized that voice melodies within the songs tend to lead to misclassification when they are triggered as prototypical parts for guitar solos. Therefore, SPPs are a valuable tool for raising user awareness, especially in cases where networks only achieve satisfactory performance. Another pattern suggested by local analysis on the MEDLEY dataset indicates that environmental recordings where different events happen within the same excerpt, e.g., instances of "ah ah ah" laughter, isolate the spectrogram exactly on the same number of such instances in both test and training spectrograms. In such cases, the attention map adds value via graphical visualization, as it clearly shows that most attention is given to the audio transient of each instance. An example of such behavior is depicted in Fig. 3. Misclassification on these types of environmental sounds, however, is misleading when the prototype considers only a limited amount of frequencies. For example, the laughing class is commonly misclassified with tooth-brushing audios, as it also exhibits different event instances, e.g., toothbrush movements, that, if filtered in the higher spectrum of frequencies, might fail to capture semantic information for correct classification. In such cases, human aural inspection might also fail to distinguish between the two classes[15].

6 Conclusions

We introduced SonicProtoPNet, an interpretable-by-design network for audio classification, facilitating "this sounds like that" reasoning through playable prototypical audios for auditory inspection of discriminative aspects captured by the model. Using Log-Mel Spectrogram representation of audio inputs, experiments on genre, environmental sound, and musical instrument recognition demonstrate the effectiveness and benefits of our approach based on prototypical parts over naive methods, enabling reproduction of entire audio data. We addressed the importance of recovering playable sonic prototypical parts with a novel approach involving mapping Mel scale frequencies to Power spectrogram frequencies using triangular bank filters. Experimental results suggest that pre-trained weights on common CNN networks, like VGG19, can benefit Log-Mel Spectrogram representations of audio, and that interpretability can be enhanced by playable prototypes derived from visual representations like attention maps. Future research involves experimenting with SonicProtoPNet on other audio domain classification tasks and datasets, investigating combinations of Spectrograms, Chro-

---

[15] The examples discussed, along with many others for each of the different datasets, are available in the code repository, where sonic prototypical parts of the audio recordings can be played and inspected.

mograms, and MFCCs representations. Another direction is assessing other pretrained networks besides VGG19, and exploring the impact of pre-training on Log Mel Spectrograms instead of images. Finally, conducting a human-driven study to evaluate the classification behavior and validate the alignment with human reasoning is a promising avenue for future research.

Acknowledgments. This work has been partially supported by the European Community H2020 programme under the funding schemes ERC-2018-ADG G.A. 834756 "XAI: Science and technology for the eXplanation of AI decision making" (https://xai-project.eu/), "INFRAIA-01-2018-2019 - Integrating Activities for Advanced Communities", G.A. 871042, "SoBigData++: European Integrated Infrastructure for Social Mining and Big Data Analytics" (http://www.sobigdata.eu), G.A. 952026 HumanE AI Net (https://www.humane-ai.eu/), G.A. 952215 TAILOR (https://tailor-network.eu/), G.A. 101120763 TANGO, by the European Commission under the NextGeneration EU programme - National Recovery and Resilience Plan (Piano Nazionale di Ripresa e Resilienza, PNRR) - Project: "SoBigData.it - Strengthening the Italian RI for Social Mining and Big Data Analytics" - Prot. IR0000013 - Avviso n. 3264 del 28/12/2021, and M4C2 - Investimento 1.3, Partenariato Esteso PE00000013 - "FAIR - Future Artificial Intelligence Research" - Spoke 1 "Human-centered AI", and by the Italian Project Fondo Italiano per la Scienza FIS00001966 MIMOSA.

Disclosure of Interests. The authors have no competing interests to declare that are relevant to the content of this article.

References

1. Ali, S., et al.: Explainable artificial intelligence (XAI): what we know and what is left to attain trustworthy artificial intelligence. Inf. Fusion **99**, 101805 (2023)
2. Arrieta, A.B., et al.: Explainable artificial intelligence (XAI): concepts, taxonomies, opportunities and challenges toward responsible AI. Inf. Fusion **58**, 82–115 (2020)
3. Becker, S., et al.: Audiomnist: exploring explainable artificial intelligence for audio analysis on a simple benchmark. J. Frankl. Inst. **361**(1), 418–428 (2024)
4. Bodria, F., et al.: Benchmarking and survey of explanation methods for black box models. Data Min. Knowl. Discov. **37**(5), 1719–1778 (2023)
5. Chen, C., et al.: This looks like that: deep learning for interpretable image recognition. In: NeurIPS, pp. 8928–8939 (2019)
6. Dosilovic, F.K., et al.: Explainable artificial intelligence: a survey. In: MIPRO, pp. 210–215. IEEE (2018)
7. Fedele, A., et al.: Explaining siamese networks in few-shot learning for audio data. In: Pascal, P., Ienco, D. (eds.) Discovery Science. DS 2022. LNCS, vol. 13601, pp. 509–524. Springer, Cham (2022). https://doi.org/10.1007/978-3-031-18840-4_36
8. Frommholz, A., et al.: XAI-based comparison of input representations for audio event classification. CoRR **abs/2304.14019** (2023)
9. Griffin, D.W., Lim, J.S.: Signal estimation from modified short-time fourier transform. In: ICASSP, pp. 804–807. IEEE (1983)
10. Guidotti, R.: Evaluating local explanation methods on ground truth. Artif. Intell. **291**, 103428 (2021)
11. Guidotti, R., et al.: A survey of methods for explaining black box models. ACM CSUR **51**(5), 93:1–93:42 (2019)

12. Khan, A., et al.: A survey of the recent architectures of deep convolutional neural networks. Artif. Intell. Rev. **53**(8), 5455–5516 (2020)
13. Li, O., et al.: Deep learning for case-based reasoning through prototypes: a neural network that explains its predictions. In: AAAI, pp. 3530–3537. AAAI Press (2018)
14. Loiseau, R., et al.: A model you can hear: audio identification with playable prototypes. In: ISMIR, pp. 694–700 (2022)
15. Ma, C., et al.: This looks like those: illuminating prototypical concepts using multiple visualizations. In: NeurIPS (2023)
16. Mehrish, A., et al.: A review of deep learning techniques for speech processing. Inf. Fusion **99**, 101869 (2023)
17. Milosheski, L., et al.: XAI for self-supervised clustering of wireless spectrum activity. CoRR **abs/2305.10060** (2023)
18. Mishra, S., et al.: Local interpretable model-agnostic explanations for music content analysis. In: ISMIR, pp. 537–543 (2017)
19. N., S.B., et al.: Speech task based automatic classification of ALS and Parkinson's disease and their severity using log mel spectrograms. In: SPCOM, pp. 1–5. IEEE (2020)
20. Nauta, M., et al.: Neural prototype trees for interpretable fine-grained image recognition. In: CVPR, pp. 14933–14943. Computer Vision Foundation/IEEE (2021)
21. Nauta, M., et al.: This looks like that, because ... explaining prototypes for interpretable image recognition. In: Kamp, M., et al. (eds.) Machine Learning and Principles and Practice of Knowledge Discovery in Databases. ECML PKDD 2021. CCIS, vol. 1524, pp. 441–456. Springer, Cham (2021). https://doi.org/10.1007/978-3-030-93736-2_34
22. Pablo, Z., et al.: An interpretable deep learning model for automatic sound classification. Electronics **10**(7), 850 (2021)
23. Parekh, J., et al.: Tackling interpretability in audio classification networks with non-negative matrix factorization. IEEE ACM Trans. Audio Speech Lang. Process. **32**, 1392–1405 (2024)
24. Piczak, K.J.: ESC: dataset for environmental sound classification. In: ACM Multimedia, pp. 1015–1018. ACM (2015)
25. Purwins, H., et al.: Deep learning for audio signal processing. IEEE J. Sel. Top. Signal Process. **13**(2), 206–219 (2019)
26. Ren, Z., et al.: Prototype learning for interpretable respiratory sound analysis. In: ICASSP, pp. 9087–9091. IEEE (2022)
27. Ribeiro, M.T., et al.: Why should I trust you?: explaining the predictions of any classifier. In: KDD, pp. 1135–1144. ACM (2016)
28. Seo, S., et al.: Convolutional neural networks using log mel-spectrogram separation for audio event classification with unknown devices. J. Web Eng. **21**(2) (2022)
29. Tzanetakis, G., Cook, P.R.: Musical genre classification of audio signals. IEEE Trans. Speech Audio Process. **10**(5), 293–302 (2002)
30. Zhou, Q., et al.: Cough recognition based on mel-spectrogram and convolutional neural network. Front. Robot. AI **8**, 580080 (2021)
31. Zinemanas, P., et al.: Toward interpretable polyphonic sound event detection with attention maps based on local prototypes. In: DCASE, pp. 50–54 (2021)

TETRA: TExtual TRust Analyzer for a Gricean Approach to Social Networks

Federico Mazzoni[1](✉)[iD], Simona Mazzarino[2][iD], and Giulio Rossetti[3][iD]

[1] Department of Computer Science, University of Pisa, 56127 Pisa, Italy
federico.mazzoni@phd.unipi.it
[2] Clearbox AI, 10129 Turin, Italy
simona@clearbox.ai
[3] National Research Council, Institute of Information Science and Technologies A. Faedo (ISTI), 56127 Pisa, Italy
giulio.rossetti@isti.cnr.it

Abstract. Social ties in human relationships are often based on *trust* between peers. Depending on the context, trust can assume different forms, and can be computed and quantified in different ways. In this work, we introduce TETRA, a framework based on the theory of *cooperative principle* by the linguistic Paul Grice, employing state-of-the-art NLP techniques to assign three different trust scores to a sentence, focusing on *relation between sentences* (Relation), *information density* (Quantity), and *politeness* (Manner). Furthermore, we employ the framework to analyze a network of Reddit users in order to identify how trust scores can be leveraged to get a better insight into human relationships, assuming that the trust scores computed by TETRA can be applied to the network's edges, or averaged to assign a score to the network's nodes. Our experiments showed that trust scores computed by TETRA can be employed to cluster the network's nodes, can successfully validate another independent network trust model and can be used to gather interesting insights in the context of Social Balance Theory.

Keywords: Conversational Trust · Cooperative Principle · Natural Language Processing · Network Analysis

1 Introduction

Trust is a key component of human relations and can generally be seen as a measure of confidence that another entity will behave in a certain way [31]. The concept of trust assumes a relationship between two entities: a *trustor*, who gives trust, and a *trustee*, who receives trust. The entities might be single persons, groups, or firms [6]. These definitions apply to many fields, including psychology [19], economy [22], sociology [8]. The linguistic field has shown an interest in defining how language can convey trust [36], whereas the network analysis field has been focused either on a topological trust or on the interaction between the nodes [30], in order to understand additional information about the nodes and/or the edges, potentially helping users navigate the network to avoid bad-faith agents. However, to the best of our knowledge, these two fields have rarely intersected, and network analyses based on *external* linguistic trust metric are rare. With this work, we first propose a framework called TETRA (*TExtual TRust Analyser*), based on the theory of *cooperative principles* from the linguistic Paul Grice. Provided a sentence (such as the comment associated with a network's edge), TETRA outputs three trust scores, each representing one key different maxim of communication. We then employed those scores to perform the analysis of a network of Reddit users. Our experiments focused on how the trust scores can describe the various nodes, if it is possible to employ them to validate an independent interaction-based trust model, and their potential use within the context of the Social Balance Theory framework.

The rest of this work is organized as follows. Section 2 sums up relevant literature on the main topics surrounding this work, both from a network science and a linguistic viewpoint. In Sect. 3, we focus in detail on the TETRA framework, how it diverges from the main competitor and how it computes the three trust scores. In Sect. 4, we describe the results obtained by employing TETRA on a network of Reddit users by showcasing a set of three experiments. Finally, Sect. 5 reports the conclusions and potential future works.

2 Related Works

2.1 Trust in Computer Science

In computer science, trust has been classified as *user trust* and *system trust* [30]. The former involves the user-machine interaction, and it is defined as *the expectation that a device or system will faithfully behave in a particular manner to fulfil its intended purpose* [37]. The latter is closer to the purpose of this article, as it defines a form of relational trust between users mediated by a machine, e.g. in social networks [30]. The authors of [30] further classify user trust models into two macrocategories, namely *Network Structure Models* (or *Graph-Based Models*) and *Interaction-Based Models*.

The former assumes that *trust* can be computed w.r.t. how people (represented as nodes) are related. For example, areas of the networks with a higher density of connections (e.g., high interconnectedness of nodes) are assumed to hold more trust [2]. In this model, *trust* is a property of the nodes, i.e., some people are more trustworthy than others. Trust ratings are often assigned to users by other users, and depending on the context, each user can have multiple ratings [13]. TidalTrust, proposed in [12], infers a trust relationship between two nodes, assuming that neighbours with higher trust ratings agree about the trustworthiness of a third party. The results are employed to improve user interfaces and user experiences, e.g., by sorting emails using the inferred trust rating. In [24], the strengths of friendships and the extent of the trusted social network for each user are computed following how each user rated their connections. The social network is then used to compute the effective trust flow for users who are not directly within the social neighbourhood. As the ratings and the connections can change over time, the proposed model dynamically computes trust values in social networks.

On the other hand, interaction models focus on the actual interactions between the users. For example, Adali et al.'s model identifies two kinds of trust: *conversation trust*, and *propagation trust* [1]. For the former, if two users often have long conversations, they equally trust each other. The latter assumes that if the user Alph propagates a message received from the user Brittany to a third user Charlie, then Alph trusts Brittany (although not necessarily vice-versa). The model does not take into account the content of the conversations or the shared messages. The STrust model computes the overall *social trust* of a user following their *popularity trust* and their *engagement trust*. The former deals with the number of followers, positive feedback and views, whereas the latter measures how much the user is involved in the community, i.e., member's trust towards the community. It is calculated by considering how frequently a member is connected, how many other users they follow, and how often they read posts and provide comments. The STrust model aims to promote positive engagement [27].

The authors of [34] propose a hybrid model, where social trust is computed following *explicit trust* and *implicit trust*. Whenever two users engage with each other, they share their friend lists and store them as friendship graphs. Trust is determined based on the friendship graph proportionally to the distance (direct links have value 1). Implicit trust is based on the frequency and duration of the conversation between the users and on the similarity of their friends. Explicit trust captures topological information from the network, whereas implicit trust focuses on the relation between the users.

2.2 Trust in Linguistic

As pointed out by Ricardo Court [4,5], trust is inherently linked to communication, both verbal and non-verbal. Focusing on the verbal aspect, specifically written communication, some examples of trust being embedded within language itself can already be seen, for instance, in historical studies of the language used by merchants in pre-modern times. In fact, precise linguistic structures were employed to reassure clients and maintain good relations among merchants [36]. It can, therefore, be argued that there are some rules that help create effective communication, thereby generating trust between the two parties.

Some of those rules are explored in [25], which also goes into the domain of paraverbal (e.g., tone voice) and non-verbal (e.g., gestures) communication. In terms of pure linguistic content, it is noted that the occurrence of personal pronouns, the use of metaphors, and the attempt to make the audience identify with the speakers can help gain trust *rationally*, but it is argued that trust is often born from an *emotional* status. In [23], the author proposes a Machine Learning model to estimate trust from a conversation based both on linguistic and acoustic cues. An interesting case study is proposed in [18], which shows how linguistic trust rules can be born in modern times by focusing on how people with different mother tongues relate in an international working environment. While English is used as the *lingua franca*, the work notes that new words or syntactic structures are sometimes created between non-native speakers, increasing their common trust.

While not directly acknowledging trust between speakers, Paul Grice's concept of *cooperative principle* is close, and consistent with the idea of rationally defining trust through a set of rules. According to Grice, two speakers must follow four maxims to communicate efficiently [14,15]. Those are the maxims of *Quality* (telling the truth), *Quantity* (be informative), *Relation* (be on topic), and *Manner* (be perspicuous). Following Grice's, other maxims have been proposed, e.g., the maxim of *Politeness* [28].

Combined with a game-theoretical framework, Grice's maxims have been employed to study and simulate the speaker's rational behavior [9], or to to simulate agents' reasoning about each other's beliefs in maximizing joint utility [35]. An early computational linguistic model was also inspired by Grice to generate meaningful text [7]. However, Grice's work has also been criticized. Despite the maxims assuming the form of a well-defined taxonomy, in practice, they are often too vague or overlapping when used to analyze real-world scenarios, leading to multiple interpretations (or the lack thereof) from different scholars. Nonetheless, journalists do tend to follow the maxims while writing news to establish a relationship with their audience [21].

The applied linguistic field has proposed different approaches to quantify the maxims, employing computational linguistics techniques. For example, the authors of [33] focus on the maxim of quantity for short texts by linking informativeness with syntactic cohesion. On the other hand, traditional NLP techniques to quantify and formalize most of the maxims in order to rank community questions and answers in [10].

3 Methodology

We designed our framework TETRA[1] following two principles:

- given a sentence, the framework should provide an independent trust score for each of Grice's maxims in the $[0, 1]$ range;
- the framework should be as general purpose as possible.

To the best of our knowledge, the closer existing work is [10] by Freihat et al., which has been a good reference point. However, the two approaches diverge in significant ways – for once, Freihat et al. do not follow our first point, mixing the different maxims and ultimately providing a single score. Below, we detail how we dealt with each maxim.

---

[1] The library is available on: https://github.com/simonamazzarino/TETRA.

3.1 Quality

Like Freihat et al., we decided to exclude the maxim of Quality, focusing our framework on the *extensional* use of the language (what does the sentence mean) rather than the *intentional* one (what does the speaker mean). From a linguistic point of view, it has been noted that the maxim of Quality does not follow the same principles as the others, to the point of having been "de-maximized" [16].

3.2 Relation

Grice "expect[s] a partner's contribution to be appropriate to the immediate needs at each stage of the transaction" [14]. In practice, this formulation is extremely vague. In [10], it is assumed that the presence of domain-specific terms and imperative and expression of politeness can be a good proxy. However, the former point violates our second design principle, requiring ad-hoc dictionaries, while the latter might be seen as arbitrary and can bring to an overlap with the maxim of Manner. Freihat et al. also compute the cosine similarity by employing Word2Vec and Brown and Clark embeddings.

TETRA follows a similar approach, using Sentence-BERT instead [29]. Consistently with the scope of our framework, Sentence-BERT vectorizes entire sentences (or text) instead of words. By virtue of being trained on a large corpus of text, it also has a built-in context awareness. Note that computing the Relation score requires the user to provide two sentences as input: a "primary" $sent_1$, and a secondary $sent_2$, to which the main sentence is replying. Sentence-BERT then computes the respective embeddings emb_1 and emb_2, and the final Relation score r is given by:

$$r = \frac{cos\_sim(emb_1, emb_2) + 1}{2} \quad (1)$$

In other words, the TETRA's Relation score is the embeddings' cosine similarity normalized in the in the $[0, 1]$ range.

For example, assume "Who was the first man on the moon?" as the secondary sentence, and two potential primary sentences: "It was Neil Armstrong", and "It was William Shakespeare". TETRA assigns the former a score of 0.80, while the latter a score of 0.64, as "Neil Armstrong" is expected to be found in the context of "moon" more than "William Shakespeare".

3.3 Quantity

Grice provides a two-fold definition of the maxim of Quantity: be as informative as possible and avoid redundancy. The work of Freihat et al. focuses on the idea of being informative, checking the presence of named entities (NE), references, currency, and numbers. Our approach is similarly focused on entities, but it also takes into account the broader concept of *lexical density* – a text is semantically denser the greater the ratio of its content words over the sum of all its the words (tokens) [3,20].

First of all, TETRA employs Presidio, the SDK developed by Microsoft [26], to count the named entities of a given sentence (date, location, person, NRP[2] and phone number), obtaining their number w_NE. Each NE is then replaced by a filler word[3]. Subsequently, TETRA counts the number of all words (w_all) and the number of all content words (w_cont), in both cases, including the filler NE words. While computing the information density, compared to the traditional approach, TETRA sum to w_cont the number of NE words multiplied by their ratio over all the content words – i.e., each NE is counted both as part of w_cont (with weight 1), and as part of w_NE (with weight w_ne/w_cont).

Moreover, like for the maxim of Relation, TETRA can also take as input both a primary and a secondary sentence (though in this case, it is optional). TETRA checks if the secondary sentence includes an interrogative pronoun among "Who", "Where", and "When", and if the primary sentence provides a direct answer in the form of at least a person, a location, or a time. While this does not prove if the information provided by the primary sentence is *redundant*, it checks if it is *needed* for an adequate reply. TETRA then counts the number of matches with an internal counter k, which is then used to give a small reward (one for each pronoun-NE match; k range is thus $[0, 3]$. If a secondary sentence is not provided, k defaults to 0).

The sentence's Quantity score q is then given by:

$$q = \frac{w\_cont + w\_ne * w\_ne/w\_cont + k/w\_cont}{w\_all} \quad (2)$$

For example, TETRA assigns a Quantity score of 0.666 to the sentence "It was Neil Armstrong", but while paired to the secondary sentence "Who was the first man on the moon?" the score increases to 1 (as "Neil Armstrong" is a person, replying to "Who"). On the other hand, a very long sentence like "Neil Armstrong was the very first man who set foot on the moon in 1969, under Nixon's presidency" would see the score drop to 0.480 (in this case, the "Who" boost would be negligible: 0.486) due to the lower information density. Overall, we believe our approach to the maxim of Quantity manages to capture the two concepts described by Grice.

3.4 Manner

With the maxim of Manner, our work diverges the most from Grice's original intent. Grice posits a simple supermaxim ("Be perspicuos", or clear), which is then explored with four more specific maxims: avoid ambiguity, avoid expression of obscurity, be brief, be orderly. Freihat et al. redefine the maxim by focusing on being positive and avoiding negative, ironic and insulting expressions, detecting them through the use of sentiment lists. Conceptually, this approach turns the

---

[2] A person's Nationality, Religious or Political group.
[3] This is done to count multi-word name entities as a single word (e.g., "Neil Armstrong", "United States").

maxim of Manner close to the maxim of Politeness proposed by [28] and also brought to overlaps with their approach to the maxim of Relation.

TETRA follows the idea of focusing on politeness, but instead of employing a sentiment list, it leverages the open Python library Detoxify [17] to compute the *risk of toxicity* t in the $[0, 1]$ range. The final Manner score m is thus:

$$m = 1 - t \tag{3}$$

As Detoxify supports multiple languages and multiple potential dimensions to focus on (toxicity – which TETRA defaults on –, severe toxicity, obscenity, threat, insult, identity attack, sexually explicit content), we believe it fully follows our second design principle and provides a good choice to compute the Manner score.

4 Experiments

We analyzed the Reddit Mental Health network, which includes 91,659 posts (i.e., opening comments) and 516,691 comments shared on the r/socialanxiety subreddit by 123,883 users over a 5-year period from 01/01/2018 to 31/12/2022. We employed TETRA to assign a Relation, Quantity and Manner score (or weight) to each edge, i.e., to each comment (but not to the posts, as they are not directed towards a defined entity), resulting in average values of 0.67, 0.52, and 0.89 with a standard deviation of 0.09, 0.15, and 0.26. The very high values of Manner can be explained by the kind of discussion held in the mental health subreddit – a place users go to request help and kindness. We focused our experiments in three directions: ramification of the metrics on the nodes, comparison with Adali et al.'s trust model [1], and trust analysis employing Social Balance Theory.

4.1 Metrics and Nodes

While we posited trust as an edge property, it is possible to assign a trust score to each node as well. As each edge (with its three weights related to the Relation, Quantity, and Manner dimensions) is directed, each node has an amount of "exiting" trust – while playing the role of the trustee – and "entering" trust while playing the role of the trustor (in [27], it is effectively used the terminology "social capital"). The network includes 103,892 trustors and 72,750 trustees. 61,149 users cover both roles.

For each node, we computed the average entering and exiting Relation, Quantity and Manner scores. Afterwards, we computed the nodes' betweenness centrality, assuming a directed graph. The scores were subsequently scaled in a $[0, 1]$ range. A simple statistical analysis employing Spearman's coefficient shows a (weak) relationship only between Relation and Quantity, revealing that the various metrics describe different nuances, both for the trustors and trustees (Table 1 reports the trustees' scores). By analyzing people playing both the roles of trustors and trustees, we found a positive correlation between the same metric for the different roles (0.11 for Quantity and Manner, 0.20 for Relation) and, interestingly, a negative correlation for Quantity and Relation (–0.16, both for Trustor's Quantity - Trustee's Relation and vice-versa).

Table 1. Correlation between nodes' dimensions

	Relation	Quantity	Manner	Centrality
Relation	1	–0.170	–0.033	–0.004
Quantity	–0.170	1	–0.033	–0.001
Manner	–0.033	–0.033	1	0
Centrality	–0.004	–0.001	0	1

To identify the prototypical users of our network in terms of trust, we performed an X-Means clustering 100 times on the three trust metrics, assuming a maximum of 100 clustering. For both trustees and trustors, X-Means found 100 clusters to be the ideal divisions. The bigger ones share average Relation and Quantity values – thus confirming a relationship between the two – and very high Manner values. This can be seen as the ideal representation of the network's users. Interestingly, the ideal divisions feature only a handful of very small clusters with all the metrics close to 1. Again, this applies both to the trustors and the trustees.

Due to the lack of a proper correlation between all three metrics and the disparity between Relation and Quantity's average value on the one hand and Manner's on the other, we restrained from combining the three trust scores in our further experiments – although, in other contexts, this could be a viable option, providing the need of a single semantic trust score.

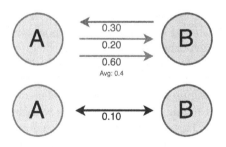

Fig. 1. Trust difference score

4.2 Comparison with Adali et al.'s Model

While dealing with conversational trust, Adali et al.'s model effectively blur the line between trustors and trustees – it is assumed that both participants hold

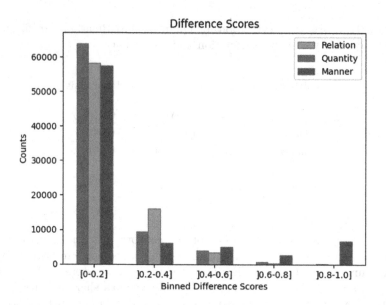

Fig. 2. Distribution of the difference scores

the same trust for each other. Moreover, longer and more frequent conversations are seen as a proxy for stronger trust. In order to verify if these assumptions hold true from a semantic point of view, we computed the *trust difference scores*. Trust difference scores are the absolute value of the difference of the average trust score of each couple of nodes. For example, if Alph has an average Quantity score of 0.40 with its interaction with Brittany, and Brittany has an average Quantity score with its interaction with Alph of 0.30, the Quantity difference score is 0.10 (see Fig. 1).

For Adali et al.'s first assumption to hold true, the difference scores of the three trust dimensions should be as close as 0 as possible – implying both ends of a conversation give each other as much trust as they receive. Our experiments show that average difference scores are indeed low (0.12, 0.14, and 0.17, respectively, for Relation, Quantity, and Manner). Figure 2 shows the distribution, using a binning with a 0.2 interval (in the Social Balance Theory section, we will return to the abnormal number of Manner nodes with high difference scores – i.e., a low mutual trust). However, we did not notice a correlation between very low difference scores and frequent conversations – in fact, more sparse conversations sometimes resulted in scores extremely close to 0. This might be due to the fact that, from a semantic point of view, longer conversations can get more nuanced or heated.

4.3 Trust and Social Balance Theory

Social Balance Theory is based on the principle that relationships involve triangles, or triads, can be either balanced or unbalanced. Each edge of the triangle has a sign, either positive or negative. Following the most common interpretation of the Social Balance Theory, a triangle is balanced if there is an even number of negative edges[4]; otherwise, it is unbalanced, and the relationship might not last long [32]. In the following, we use the triangle nomenclature from Fig. 3, which shows the four potential triangles.

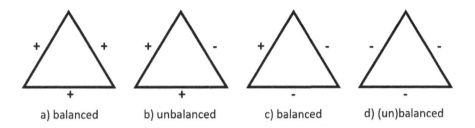

Fig. 3. Triangles in Social Balance Theory. Image source: [32]

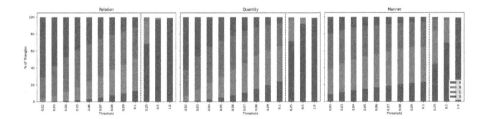

Fig. 4. Distribution of triangles with different thresholds

To apply Social Balance Theory, we first turned our directed graph into a signed undirected graph. As in the previous sections, we focused on the *couples of nodes interacting with each other*. For example, in a network with Alph, Brittany, and Charlie, if Alph has interacted with Charlie, but Charlie has not interacted with Alph or Brittany, Charlie is removed from the network. Self-loops were also removed. This resulted in a network with 51,366 nodes, 74,637 edges, and 2,829 unique triangles.

In order to provide a sign to each edge, we employed the trust difference scores computed before (therefore, each edge effectively has three signs). As the ideal difference score is 0, we selected a low threshold (0.02), and assigned a positive sign to scores under that threshold. Increasing the threshold obviously increases the number of A and B triangles (i.e., triangles with the most positive signs), as can be seen in Fig. 4, although the results become more arbitrary. In

---

[4] The Weak Balance Theory considers the triangle with three negative signs to be balanced as well, but this is beside the point of our article.

Table 2. Triangles overview (Threshold: 0.02)

		Relation				Quantity				Manner			
		A	B	C	D	A	B	C	D	A	B	C	D
All Triangles		2	80	726	2021	14	174	877	1764	245	842	1197	545
Top Nodes	Trustee	1	9	117	342	3	33	155	394	19	29	24	6
	Trustor	1	13	201	570	4	46	240	687	17	33	27	9
Bottom Nodes	Trustee	0	10	109	317	2	14	100	222	68	318	528	247
	Trustor	1	10	112	393	1	35	159	267	48	232	373	169

the first row of Table 2, we can notice a high density of unbalanced D triangles for Relation and Quantity, meaning that most people interacting with each other have a low mutual trust score. However, C triangles are also relatively common, implying that out of three persons, one is trusted by the other two. Consistently with its higher average trust scores, A, B and C triangles are more common for Manner.

Subsequently, for each trust dimension, we selected the top and bottom 25% trustors and trustees in order to analyze the kind of user participating in balanced and unbalanced triangles. The later rows of Table 2 report how many triangles involved *at least* one top or bottom node. Overall, the results are consistent with the general ones. Focusing on Relation and Quantity's top nodes, we can notice that trustors are more commonly involved in C and D triangles than trustees. This might represent the tendency of trustors to give more trust than the one received in the short-term (as D is unbalanced), and to get themselves involved in a context of (potentially initial) low mutual trust but establishing long-term relationships (as C is balanced). On the other hand, we notice a more even distribution of triangles for Relation and Quantity's bottom nodes. As for Manner, top nodes are very sparse (perhaps suggesting that very polite people tend to be less active), whereas there is a very high number of bottom nodes participating in B and C triangles. We already noticed those users while discussing Fig. 2. They might be trolls, or people providing provocative content. In particular, C triangles might depict a couple of trolls trusting each other and harassing another user.

Overall, our experiments show that trust analysis can easily be integrated into the existing Social Balance Theory framework to provide useful insights, although with the major drawbacks of the arbitrary sign threshold. Given the promising results, further research could focus on finding a heuristic for the threshold.

5 Conclusion

In this work, we presented TETRA, a new framework to capture the Relation, Quantity and Manner scores, as defined by Paul Grice. Following the semantic of each maxim, TETRA employs different NLP techniques alongside existing state-of-the-art models (Sentence-BERT, Presidio, Detoxify) with more granular results than the only available alternative. However, by its nature, TETRA may have limitations when dealing with texts written in an ironic manner or containing idioms. TETRA was extensively tested on a large corpus of messages shared on Reddit, discovering that the semantic trust scores can be used to gain better insights into a network. Experimental results prove that the semantic scores can define the network's users and are consistent with some of the assumptions of unrelated, interaction-based trust models. Moreover, they provided a good fit for the Social Balance Theory paradigm, showing interesting ramifications while creating triangles starting from the trust scores.

Future experiments could focus on communities and echo chambers, analyzing the density of trustworthy nodes or the average trust scores of the edges. It could be particularly interesting to focus on trust in dynamic communities, identifying users gaining or losing trust in the long term. We believe trust dynamics could also lead to different community events – e.g., a split led to a group of trustworthy nodes. The focus on the nodes' trustworthiness could also play a role in studying opinion dynamics, as nodes with high trust scores might spread information better.

As for TETRA potential improvements, future works could experiment with the missing maxim of Quality. The context of networks provides key information about each person – namely, their past history. Analyzing past messages can lead to good stance detection, and the simple (yet hard to conceptualize) Grice's maxim of "tell the truth" could be redefined as "be consistent with your past". In other words, TETRA should be able to understand to what extent a person's current stance is consistent with their past stance(s) – with the caveat that a person can obviously change opinion over time.

With that said, we believe that employing TETRA in its current state can already achieve good results in real use cases, for example, for recommender systems prioritizing users of certain kinds (i.e., polite users with high Manner scores) or helping navigating reviews on online shopping sites. Trust scores also naturally lead to moderation, and our experiments with the Social Balance Theory already hint at the potential of troll detection. However, a more "soft" approach might be offering personal *suggestions* rather than *moderation*, e.g., suggesting a user with a low Quantity score to write briefer or someone with a middle Manner score to slightly rephrase part of the text.

Social networks aside, we believe TETRA has a good potential in human-machine interaction, both for studying the extent to which the machine can understand Grice's maxims and for quantifying the degree of trustworthiness between the parts. Notably, LLMs already employ semantic similarity with RAG techniques [11].

Acknowledgment. This work is supported by (i) the European Union - Horizon 2020 Program under the scheme "INFRAIA-01-2018-2019 - Integrating Activities for Advanced Communities", Grant Agreement n.871042, "SoBigData++: European Integrated Infrastructure for Social Mining and Big Data Analytics" (http://www.sobigdata.eu); (ii) SoBigData.it which receives funding from the European Union

- NextGenerationEU - National Recovery and Resilience Plan (Piano Nazionale di Ripresa e Resilienza, PNRR) - Project: "SoBigData.it - Strengthening the Italian RI for Social Mining and Big Data Analytics" - Prot. IR0000013 - Avviso n. 3264 del 28/12/2021; (iii) EU NextGenerationEU programme under the funding schemes PNRR-PE-AI FAIR (Future Artificial Intelligence Research); (iv) TANGO G.A. 101120763.

References

1. Adali, S., et al.: Measuring behavioral trust in social networks. In: 2010 IEEE International Conference on Intelligence and Security Informatics, pp. 150–152. IEEE (2010)
2. Buskens, V.: The social structure of trust. Soc. Netw. **20**(3), 265–289 (1998)
3. Castello, E.: Text Complexity and Reading Comprehension Tests, vol. 85. Peter Lang, New York (2008)
4. Court, R.: Januensis ergo mercator: trust and enforcement in the business correspondence of the brignole family. Sixt. Century J. **35**(4), 987–1003 (2004)
5. Court, R.: The language of trust: reputation and the spread and maintenance of social norms in sixteenth century genoese trade. Rime **1**, 77–96 (2008)
6. Currall, S.C., Inkpen, A.C.: On the complexity of organizational trust: a multi-level co-evolutionary perspective and guidelines for future research. Handb. Trust Res. 235–246 (2006)
7. Dale, R., Reiter, E.: Computational interpretations of the gricean maxims in the generation of referring expressions. Cogn. Sci. **19**(2), 233–263 (1995)
8. Dumouchel, P.: Trust as an action. Eur. J. Sociol./Archives européennes de sociologie **46**(3), 417–428 (2005)
9. Franke, M.: Quantity implicatures, exhaustive interpretation, and rational conversation. Semant. Pragmat. **4**, 1 (2011)
10. Freihat, A.A.A.K., Mohammed R, H.Q., Giunchiglia, F., et al.: Using grice maxims in ranking community question answers. In: eKNOW 2018 Tenth International Conference on Information, Process, and Knowledge Management, pp. 38–43. IARIA XPS Press, 2018
11. Gao, Y., et al.: Retrieval-augmented generation for large language models: a survey. arXiv preprint arXiv:2312.10997 (2023)
12. Golbeck, J.: Personalizing applications through integration of inferred trust values in semantic web-based social networks. In: Semantic Network Analysis Workshop at the 4th International Semantic Web Conference, vol. 16, p. 30. Publishing, 2005
13. Golbeck, J., Parsia, B., Hendler, J.: Trust networks on the semantic web. In: Klusch, M., Omicini, A., Ossowski, S., Laamanen, H. (eds.) CIA 2003. LNCS (LNAI), vol. 2782, pp. 238–249. Springer, Heidelberg (2003). https://doi.org/10.1007/978-3-540-45217-1_18
14. Grice, H.P.: Logic and conversation. In: Speech Acts, pp. 41–58. Brill, 1975
15. Grice, P.: Studies in the Way of Words. Harvard University Press, Cambridge (1991)
16. Gunlogson, C., Fine, A.B.: De-maxim-izing quality. Working Papers in the Language Sciences at the University of Rochester, June 2011
17. Hanu, L.: Unitary team. Detoxify. Github, 2020. https://github.com/unitaryai/detoxify

18. Henderson, J.K., Louhiala-Salminen, L.: Does language affect trust in global professional contexts? perceptions of international business professionals. J. Rhetor. Prof. Commun. Glob. **2**(1), 2 (2011)
19. Heyns, M., Rothmann, S.: Trust profiles: associations with psychological need satisfaction, work engagement, and intention to leave. Front. Psychol. **12**, 563542 (2021)
20. Johansson, V.: Lexical diversity and lexical density in speech and writing: a developmental perspective. Working papers/Lund University, Department of Linguistics and Phonetics, vol. 53, pp. 61–79 (2008)
21. Kheirabadi, R., Aghagolzadeh, F.: Grice's cooperative maxims as linguistic criteria for news selectivity. Theory Pract. Lang. Stud. **2**(3), 547 (2012)
22. Korczynski, M.: The political economy of trust. J. Manag. Stud. **37**(1), no–no (2000)
23. Li, M., Erickson, I.M., Cross, E.V., Lee, J.D.: It's not only what you say, but also how you say it: machine learning approach to estimate trust from conversation. Hum. Factors **66**(6), 1724–1741 (2024)
24. Maheswaran, M., Tang, H.C., Ghunaim, A.: Towards a gravity-based trust model for social networking systems. In: 27th International Conference on Distributed Computing Systems Workshops (ICDCSW'07), p. 24. IEEE (2007)
25. Meck, A.-M.: Communication and trust: a linguistic analysis. Media Trust in a Digital World: Communication at Crossroads, pp. 81–96, 2019
26. Mendels, O., Peled, C., Levy, N. V., Rosenthal, T., Lahiani, L., et al.: Microsoft presidio: context aware, pluggable and customizable pii anonymization service for text and images, 2018
27. Nepal, S., Sherchan, W., Paris, C.: Strust: a trust model for social networks. In: 2011 IEEE 10th International Conference on Trust, Security and Privacy in Computing and Communications, pp. 841–846. IEEE (2011)
28. Pfister, J.: Is there a need for a maxim of politeness? J. Pragmat. **42**(5), 1266–1282 (2010)
29. Reimers, N., Gurevych, I.: Sentence-bert: sentence embeddings using siamese bert-networks. *arXiv preprint*arXiv:1908.10084, 2019
30. Sherchan, W., Nepal, S., Paris, C.: A survey of trust in social networks. ACM Comput. Surv. (CSUR) **45**(4), 1–33 (2013)
31. Singh, S., Bawa, S.: A privacy, trust and policy based authorization framework for services in distributed environments. Int. J. Comput. Sci. **2**(2), 85–92 (2007)
32. Teixeira, A.S., Santos, F.C., Francisco, A.P.: Emergence of social balance in signed networks. In: Gonçalves, B., Menezes, R., Sinatra, R., Zlatic, V. (eds.) CompleNet 2017. SPC, pp. 185–192. Springer, Cham (2017). https://doi.org/10.1007/978-3-319-54241-6_16
33. Tewari, M., Bensch, S., Hellstrom, T., Richter, K.-F.: Modelling grice's maxim of quantity as informativeness for short text. In: ICLLL 2020: The 10th International Conference in Languages, Literature, and Linguistics, Japan, 6–8 November 2020, pp. 1–7 (2020)
34. Trifunovic, S., Legendre, F., Anastasiades, C.: Social trust in opportunistic networks. In: 2010 INFOCOM IEEE Conference on Computer Communications Workshops, pp. 1–6. IEEE (2010)
35. Vogel, A., Bodoia, M., Potts, C., Jurafsky, D.: Emergence of Gricean maxims from multi-agent decision theory. In: Proceedings of the 2013 Conference of the North American Chapter of the Association for Computational Linguistics: Human Language Technologies, pp. 1072–1081 (2013)

36. Wubs-Mrozewicz, J.: The concept of language of trust and trustworthiness:(why) history matters. J. Trust Res. **10**(1), 91–107 (2020)
37. Yao, J., Chen, S., Nepal, S., Levy, D., Zic, J.: Truststore: making amazon s3 trustworthy with services composition. In: 2010 10th IEEE/ACM International Conference on Cluster, Cloud and Grid Computing, pp. 600–605. IEEE (2010)

SoBigData++: Societal Debates and Misinformation Analysis

What's Real News Today? A Multimodal, Continual-Learning Approach for Detecting Fake News Over Time

Luca Maiano[1,2], Martina Evangelisti[1], Silvia Bianchini[1], and Aris Anagnostopoulos[1](✉)

[1] Department of Computer, Control and Management Engineering, Sapienza University, 00185 Rome, Italy
{maiano,aris}@diag.uniroma1.it,
{evangelisti.1796480,bianchini.1796898}@studenti.uniroma1.it
[2] Ubiquitous, 00185 Rome, Italy
http://ubiquitous.green

Abstract. Multimodal fake news detectors are typically trained to work on fixed distributions, making them hardly applicable to ever-changing events. Although it is possible to apply transfer learning to retrain a model on the most recent facts, it will tend to lose its ability to recognize old contents. We mitigate this problem by considering news as a stream of data that becomes available over time and by introducing a continual-learning solution that learns from new events as they become available. Our solution maintains good performance on previously known tasks without limiting the applicability of this solution to older news, leading to a substantial gain of +9.22% accuracy on average compared to transfer learning and a +3.65% increase in F1 score over the ideal scenario where you train the model on all data in one session. Besides this, we introduce the Tri-Encoder, a state-of-the-art multimodal model that allows the cross-attention mechanism between images and texts to be applied.

Keywords: Fake news · Continual learning · Multimodal learning

1 Introduction

The massive adoption of social networks has made them a very effective tool for spreading false content. Fake news stories often spread faster and with a higher frequency than the real ones [1], but, more importantly, the more a user is exposed to the same content, the more she tends to perceive it as trustworthy [1]. This fact can have a more profound effect than one may expect. An example of this is the 2016 presidential election in the United States. Snopes[1] identified 529 social-media rumors about Donald Trump and Hillary Clinton that could have influenced the election outcome.

---
[1] https://www.snopes.com – Fact-checking website and reference source for urban legends, folklore, myths, rumors, and misinformation.

There are many challenges to face to counter this phenomenon. First, the most influential fake news contain both texts and images. For example, tweets with images obtain 18% more clicks, 89% more likes, and 150% more retweets than tweets with text-only content [30]. A similar trend takes place on Facebook, where the 87% of the posted photos have been liked, clicked, or shared [30]. Because of this fact, recent studies have analyzed the semantics of multimodal content to classify the news as real or fals with three main approaches: (1) *early-fusion* methods [24,31] learn low-level features from different modalities that are immediately fused, and fed into a single prediction model, (2) *late-fusion* models [3] fuse unimodal decisions with some mechanisms such as averaging and voting, and (3) *hybrid-fusion* [8] combines the early fusion and late fusion. Using VisualBERT [16], MMBT, and ViLBERT, Dimitrov et al. [8] evaluated several fusion techniques (such as early-fusion, late-fusion and self-supervised models) for propaganda identification. According to their research, self-supervised joint learning models, and in particular VisualBERT, outperform other fusion techniques.

Although this type of approach seems to attain high levels of accuracy in most of the studies, its applicability in real scenarios is still somewhat limited [12,17,22]. Indeed, most state-of-the-art approaches apply these techniques in a static setting, where the training and test data belong to a *fixed distribution*, known at design time. This assumption, however, does not reflect the ever-changing nature of news [1] being spread online based on recent events. Some studies proposed to tackle this problem from a different perspective, by analyzing the propagation of news or the communities and the users' reactions to such content [19]. However, these interactions can sometimes be complex to capture because they require monitoring of the entire network, something that is not always feasible. Therefore, analyzing the content stream remains the most accessible way. Motivated by this discussion, in this work, we propose to model the latest news flow as an *incremental task*, where data arrive sequentially in batches, and each batch corresponds to some new events that we want to learn to classify. Our contributions can be summarized as follows. (1) We introduce a multimodal architecture (the *Tri-Encoder*) for fake-news classification based on the analysis of texts and images. (2) Next, we apply a *continual-learning strategy*, which allows to continually learn to classify new topics without losing the ability to classify previously known ones. The proposed solution allows not only to maintain good performance over time, but it even improves compared to the ideal case in which all the topics are immediately available in the first training session. (3) Finally, we perform various measurements and comparison of our approach with others.

The remainder of this paper is organized as follows. In Sect. 2, we introduce our proposed method. Section 3 present our experiments. We conclude in Sect. 4.

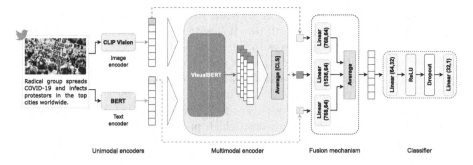

Fig. 1. Overview of the proposed *Tri-Encoder* multimodal architecture.

2 Methodology

Our model aims at learning the discriminable feature representations for fake-news detection in a way that can constantly adapt to the most recent events. We focus on the news spread on Twitter, but the same framework can be extended to other social networks. Formally, given a tweet $X = \{T, V\}$ comprising textual (T) and visual (V) information, our goal is to learn a target function $g(X, \theta) = Y$ that predicts whether the post is a fake ($Y = 0$) or true ($Y = 1$) content by examining the textual and visual information, as well as the semantic relationship between the two types of information. To this aim, we model news as a stream of unknown distributions $\mathcal{D} = \{\mathcal{D}^1, \ldots, \mathcal{D}^n\}$ over $X \times Y$, with X and Y input and output random variables, respectively. At time step i, the model learns a new function $f_i^{CL} = g(X, \theta^i)$ by updating its current parameters θ^{i-1} on a new fact $\mathcal{D}^i$ by training it on a training set $\mathcal{D}_{\text{train}}^i$ and testing it on a test set $\mathcal{D}_{\text{test}}^i$. The objective of the continual-learning algorithm is to minimize the loss $\mathcal{L}_D$ over the entire stream of data $\mathcal{D}$:

$$\mathcal{L}_D\left(f_n^{CL}, n\right) = \frac{1}{\sum_{i=1}^n |\mathcal{D}_{\text{test}}^i|} \sum_{i=1}^n \mathcal{L}_{\text{fact}}\left(f_n^{CL}, \mathcal{D}_{\text{test}}^i\right) \quad (1)$$

$$\mathcal{L}_{\text{fact}}\left(f_n^{CL}, \mathcal{D}_{\text{test}}^i\right) = \sum_{j=1}^{|\mathcal{D}_{\text{test}}^i|} \mathcal{L}_{\text{cls}}\left(f_n^{CL}(x_j^i), y_j^i\right) \quad (2)$$

where the loss $\mathcal{L}_{\text{cls}}\left(f_n^{CL}(x_j^i), y_j^i\right)$ represents the binary cross entropy loss.

2.1 The *Tri-Encoder* Model Architecture

The Tri-Encoder model architecture is shown in Fig. 1. The model involves an *image encoder* and a *text encoder* to obtain unimodal image and text representations, and a *multimodal encoder* to fuse and align the image and text representations for multimodal reasoning.

Text Encoder. Given the text of a tweet, we first tokenize and embed it in a list of word vectors using WordPiece [27] with a vocabulary of 30,000 tokens and append two special characters to the input: the class token [CLS], which is appended in front of each input example, and the separator token [SEP]. Then, we apply a transformer model over the word vectors to encode them into a list of N_T hidden state vectors $h_T \in \mathbb{R}^H$, including $h_{\text{CLS},T}$ for the text classification token. In all our experiments, we use the bidirectional BERT-base [7] model with 12 layers and 12 attention heads, which produces 768-dimensional hidden vectors. In the training phase, all weights are frozen except for the last two layers.

Image Encoder. For the image encoder, we use the pretrained CLIP's [20] visual feature extractor. Given an input image, we split it into 32 × 32 patches, which are then linearly embedded and fed into a ViT-B/32 [9] transformer model along with positional embeddings and an extra image classification token [CLS]. Similarly to the text encoder, the image-encoder output is a list of N_V image hidden state vectors $h_V \in \mathbb{R}^H$ ($H = 768$), each corresponding to an image patch, plus an additional $h_{\text{CLS},V}$ for the image classification token. Similarly to the text encoder, all weights are frozen except for the last two layers during training.

Multimodal Encoder. We use an additional transformer model for learning a joint contextualized representation of the image and text hidden states. Specifically, we apply the VisualBERT [16] model that is pretrained on the Visual Commonsense Reasoning dataset [29]. The model consists of a stack of transformer layers that align the regions of the input image with the textual input through self-attention. Compared to a simple concatenation of the two unimodal embeddings, this configuration allows *cross-attention* between the projected unimodal image and text representations and fuses the two modalities. This encoder takes as input the visual (h_V) and textual (h_T) hidden representations extracted from the unimodal models and produces $N_T + N_V + 2$ multimodal hidden state vectors $h_M \in \mathbb{R}^H$ ($H = 768$), where N_T and N_V are the numbers of text tokens and image patches, respectively, and the two additional vectors are the special [CLS] and [SEP] tokens. The output of the last layer may not always be the best representation of the input when fine tuning for downstream tasks. Previous studies proved that for pretrained language models, the most transferable contextualized representations of input text tend to occur in the middle layers, whereas the top layers specialize in language modeling [5]. Therefore, inspired by the same considerations, we average the penultimate last three layers' output and concatenate the averaged hidden state vector with the [CLS] hidden state vector of the output layer, producing a 1536-dimensional output h_{MM}. We validate this choice in Sect. 3.1.

Fusion Mechanism. In the fusion step, the visual, textual, and the multimodal [CLS] feature vectors $h_{T,\text{CLS}}$, $h_{V,\text{CLS}}$, and $h_{MM,\text{CLS}}$ are all projected onto a 64-dimensional subspace through a linear layer, producing the corresponding $h'_{T,\text{CLS}}$, $h'_{V,\text{CLS}}$, and $h'_{M,\text{CLS}}$ vectors. Finally, we calculate a weighted average of these

vectors
$$h_{TVM} = \text{avg}(w_T h'_{T,\text{CLS}} + w_V h'_{V,\text{CLS}} + w_M h'_{M,\text{CLS}}) \qquad (3)$$
where w_T and w_V are fixed to 0.25, and $w_M = 0.5$.

Classifier. The final step of the Tri-Encoder is the classification step. The classifier is composed of two linear layers generating a 32-dimensional and a 1-dimensional outputs, and are separated by the rectified linear unit (ReLU) and dropout operations. A sigmoid activation function follows the output of the last layer:
$$\sigma(x) = \frac{1}{1 + exp(-x)}$$
Values below a threshold τ are predicted *false*. Experimentally, we found $\tau = 0.46$ as the optimal value.

2.2 Continual-Learning Strategy

Now, we propose a continuous learning strategy that allows the Tri-Encoder to update its knowledge on the latest news as they become available. To this end, it is essential to avoid catastrophic forgetting [6], as we want the model to continue to classify previous content accurately. A naive idea might be to retrain the model on an ever-growing set of training data; however, such an approach can become prohibitively expensive as the volume of data grows over time. On the other hand, the model should not overfit to a new event because it would cause it to loose its previous skills.

We choose to adopt a *knowledge distillation* approach. We choose this simple regularization method since it allows maintaining the model size fixed as the data size increases and it does not require to store the previous data in *memory*. Distillation techniques were introduced by Hinton et al. [11] as a means to transfer knowledge from a neural network T (the *teacher*) to a neural network S (the *student*). The key idea behind knowledge distillation is that soft probabilities predicted by a network of trained "teachers" contain much more information about a data point than a simple class label. For example, if multiple classes are assigned high probabilities for an image, this could mean that the image must be close to a decision boundary between those classes. Forcing a student to mimic these probabilities should then cause the student network to absorb some of this knowledge that the teacher discovered, above and beyond the information in training labels alone. To implement this strategy, we modify the classification loss $\mathcal{L}_{\text{cls}}$ in Eq. 2 by adding a regularization factor

$$\mathcal{L}'_{\text{cls}} = \alpha \mathcal{L}_{kd} + \beta \mathcal{L}_{\text{cls}}(f_n^{CL}(x_j^i), y_j^i), \qquad (4)$$

where α and β are experimentally set to 0.5 and 0.6, respectively, and $\mathcal{L}_{kd}$ is the mean squared error (MSE) loss that measures the squared L2 norm between the teacher and the student outputs.

3 Experiments

In this section, we validate the model and the proposed continual-learning solution. All models were trained with the Adam optimizer, a learning rate fixed to $3e^{-5}$ and a batch size of 32 samples. For the experiments reported in Sects. 3.1, and 3.2, we train the models for 10 epochs, while for the other experiments we train for 5 epochs only. For evaluating our experiments, we chose three commonly used datasets: (1) *MediaEval Verifying Multimedia Use benchmark* [2], (2) *PolitiFact*, and (3) *GossipCop* [21].

3.1 Ablation Study

To evaluate the design choices of our model, we now analyze several possible multimodal variants. We consider three baseline models, which are combined with the following three feature extractors for the images: ResNet50 *(R)*, CLIP Vision *(C)*, and ViT *(V)*.

Table 1. Multimodal methods performances on the MediaEval [2] dataset.

Model	All news		Fake News			Real News		
	Acc/F1	F1-macro	Prec.	Rec.	F1	Prec.	Rec.	F1
SE(R)	0.6347	0.6338	0.7180	0.5984	0.6528	0.5584	0.6837	0.6147
SE(C)	0.7586	0.7520	0.7820	0.8031	0.7924	0.7250	0.6987	0.7119
DE(R)	0.7094	0.6711	0.6844	**0.9158**	0.7834	**0.7921**	0.4316	0.5587
DE(C)	0.7058	0.7024	0.7623	0.7079	0.7341	0.6413	0.7029	0.6707
VB(R)-b	0.7613	0.7564	0.7948	0.7873	0.7910	0.7172	0.7264	0.7218
VB(R)-cat	0.7513	0.7460	0.7846	0.7809	0.7828	0.7070	0.7115	0.7092
VB(R)-avg	0.7367	0.7313	0.7728	0.7666	0.7697	0.6892	0.6965	0.6928
VB(C)-b	0.7522	0.7382	0.7479	0.8571	0.7988	0.7606	0.6111	0.6777
VB(C)-cat	0.7358	0.7337	0.8003	0.7190	0.7575	0.6672	0.7585	0.71
VB(C)-avg	**0.7978**	**0.7934**	**0.8248**	0.8222	**0.8235**	0.7617	0.7649	**0.7633**
VB(V)-b	0.6766	0.6766	0.7956	0.5873	0.6757	0.5892	**0.7970**	0.6775
VB(V)-cat	0.6493	0.6485	0.7351	0.6079	0.6655	0.5719	0.7051	0.6315
VB(V)-avg	0.7167	0.7115	0.7593	0.7412	0.7502	0.6625	0.6837	0.6729

Simple-Encoder (SE). This model is based on the simple concatenation of the features extracted from images and texts. For text, we use BERT, taking the [CLS] representation for the last hidden state. The unimodal models are fed into a linear layer with an output size of 512 and concatenated, producing a 1024-dimensional vector that is passed through two linear layers of 1024×32 and 32×1 dimensions separated by a ReLU function, a dropout layer (set to 0.4), and a sigmoid activation function.

Dual-Encoder (DE). This architecture has been inspired by the double visual textual transformer model (DVTT) [18]. Each modality is conditioned by the other, enriching the text with visual information from the text encoder and vice-versa. The textual representation is taken from the last hidden state of the BERT model. Similarly, when CLIP Vision is employed as a visual feature extractor, the image representation comes from the last hidden state of the transformer encoder. In the case of ResNet50, feature maps are extracted from the second last layer, and a $(6,6)$ pooling is applied. Finally, we concatenate the [CLS] tokens from both transformers, obtaining a 1024-dimensional embedding.

VisualBERT (VB). This model combines image regions and language with a transformer, allowing self-attention to discover implicit alignments between language and vision. It is pretrained on visual-reasoning tasks. We consider the following three variants with all the backbones:

- *base (b):* the representation of the [CLS] token representation from the last hidden state is fed into the classifier;
- *concatenation (cat):* the [CLS] token representations from the last four layers are concatenated before classification;
- *average (avg):* the [CLS] token representations from the penultimate three layers are averaged and concatenated with the last [CLS] token before classification.

Table 1 summarizes all the experiments. The results show a consistent advantage of the *VB(C)-avg* configuration over the others, which is the same configuration used for our Tri-Encoder. Besides that, we can observe that using CLIP Vision for the visual component achieves superior performance in all the configurations. As the original model is trained in a multimodal setting, we make the hypothesis that this is because of the fact that it manages to extract features more aligned with the textual component. Regarding the compared architectures, VisualBERT achieves, on average, superior performance compared to the Simple-Encoder and Dual-Encoder.

3.2 Fake-News Detection Performance

To validate the proposed Tri-Encoder, we propose a comparison with state-of-the-art methods. In particular, we validate the model's performance against (1) well-known unimodal deep-learning architectures and (2) multimodal solutions designed for fake-news detection. All the models have been trained on MediaEval for 10 epochs with a batch size of 32, a learning rate of 3e-05, and the Adam [14] optimizer.

Table 2 reports the results of this first experiment. We can notice that our Tri-Encoder architecture outperforms all the other methods, followed by CALM [28], which achieves comparable performance. In the next section we study the architectural choices that led to the proposed Tri-Encoder. We can generally observe that multimodal models perform better than unimodal ones, confirming the additive contribution of the images to an accurate classification. We can also notice

Table 2. Fake news detection performance on the MediaEval [2] dataset.

Model	F1-micro	F1-macro
BERT [7]	0.6247	0.6238
ResNet50 [10]	0.7021	0.6962
VGG19 [23]	0.6275	0.6273
CLIP Vision [20]	0.7440	0.7353
VisualBERT [16]	0.7978	0.7934
MVAE [13]	0.745	0.744
EANN [25]	0.715	0.719
EANN- [25]	0.648	0.6385
SpotFake [24]	0.778	0.760
MFN [3]	0.808	0.785
MCAN [26]	0.809	0.808
CALM [28]	0.845	0.839
CAFE [4]	0.806	0.805
Simple-Encoder (ours)	0.7586	0.7520
Dual-Encoder (ours)	0.7058	0.7024
Tri-Encoder (ours)	**0.851**	**0.845**

that for the unimodal architectures, models that analyze images outperform BERT, a model based on text. A possible explanation for this could be that in the MediaEval dataset, many fake images have been manipulated in a way that makes the detection of such manipulation highly accurate by the image classifiers.

3.3 Robustness to Incremental Topics

Whereas the previous results demonstrate the effectiveness of the proposed method on a task, in this section, we evaluate the model's performance on new tasks in a continual-learning scenario. Specifically, we evaluate the performance of knowledge distillation (*KD*) compared to two other strategies: the transfer learning (*TL*) and the elastic weight consolidation (*EWC*) [15], which can be seen as an improvement of the L2-regularization.

Static Training Sessions. Before illustrating the results on the continual-learning setting, let's evaluate the performance of the models when the Tri-Encoder is trained from scratch on *all datasets simultaneously*. This allows us to evaluate the performance of the continual learner in the ideal scenario where the data are immediately available in the first training session. In Table 4, we report the results in terms of F1 score when the training set is balanced or unbalanced between the three datasets. As expected, when the data are unbalanced, the F1 score on the MediaEval dataset is higher than the others, having three times the number of samples that the other slices have. By balancing the data, the overall accuracy doesn't change much, but the distribution between the different portions is more even. We can also notice that the model's performance on MediaEval drops slightly compared to the case in which the model is trained only on this dataset (see Table 2). This could be justified by the fact that when the model is trained on all datasets simultaneously, the broader distribution of facts present in all datasets leads the model to converge into a region where it minimizes errors on all topics, but which leads to a slight performance drop on the MediaEval topics.

Table 3. Transfer learning performance in terms of F1 score of the model trained on $T1$, followed by TL to $T2$, followed by TL to $T3$. The rows indicate the dataset on which we perform the TL (or the initial training for row $T1$), and the columns indicate the dataset on which we perform evaluation.

Task	MediaEval	PolitiFact	GossipCop
$T1$	0.8515	-	-
$T2$	0.7312	0.7932	-
$T3$	0.7218	0.5224	0.7117

a $T1 =$ MediaEval, $T2 =$ PolitiFact, $T3 =$ GossipCop

Task	MediaEval	GossipCop	PolitiFact
$T1$	0.8515	-	-
$T2$	0.6860	0.7259	-
$T3$	0.6466	0.5123	0.7351

b $T1 =$ MediaEval, $T2 =$ GossipCop, $T3 =$ PolitiFact

Table 4. Results of the model trained from scratch on all three datasets available at once. *Balanced* indicates that in the training dataset, we balance a number of samples for each fact.

Tested dataset	Not Balanced		Balanced	
	F1-micro	F1-macro	F1-micro	F1-macro
MediaEval	0.7810	0.7762	0.7105	0.6855
PolitiFact	0.6251	0.6232	0.6529	0.6519
GossipCop	0.6781	0.6776	0.6855	0.6848
All	0.6984	0.6978	0.6966	0.6895

Furthermore, we evaluate the model's performance in a scenario where we apply *transfer learning*. Starting from MediaEval as the first task ($T1$), in

Table 3a we see the performance after applying transfer learning to $T2 =$ PolitiFact and to $T3 =$ GossipCop, and in Table 3a we see the performance after applying transfer learning to $T2 =$ GossipCop and to $T3 =$ PolitiFact. We can see that in both cases, the model suffers from catastrophic forgetting. Indeed, as we train it on new datasets, it becomes less accurate on previously seen ones. In the following section we see how our incremental-learning strategy manages to reduce this problem.

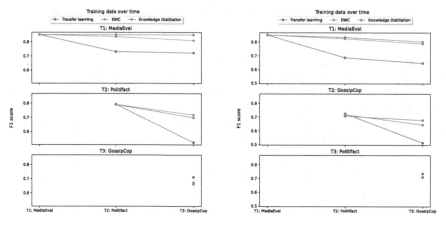

(a) $T_1 =$ MediaEval, $T_2 =$ PolitiFact, $T_3 =$ GossipCop

(b) $T_1 =$ MediaEval, $T_2 =$ GossipCop, $T_3 =$ PolitiFact

Fig. 2. F1 score of the Tri-Encoder over all tasks during time. In each of figures (a) and (b), the first plot shows the F1 score evaluated on dataset $T1$ (MediaEval for both of them), the second one the F1 score for $T2$ (PolitiFact for (a), GossipCop for (b)), and the third one for $T3$ (GossipCop for (a), PolitiFact for (b)).

Continual-Training Sessions. We now evaluate the robustness of the model through several continual-learning sessions. To do this, in the first training session $T1$, we train the model on MediaEval. We chose this dataset for the first session as it is the largest among those considered and it allows a first training phase of the Tri-Encoder without causing overfitting. In subsequent training sessions, we expose the model to the new facts in GossipCop and PolitiFact. To this end, we introduce two more training sessions, namely $T2$ and $T3$. The goal of the continual learner is to learn new tasks without encountering catastrophic forgetting of the previous ones. To validate this approach, we compare the performance of knowledge distillation (KD) with respect to EWC and transfer learning.

In Fig. 2, we show the performance of all the strategies in terms of F1 score. In Fig. 2(a) we train first on MediaEval ($T1$), and then on PolitiFact ($T2$) and GossipCop ($T3$). In Fig. 2(b) we switch GossipCop ($T2$) and PolitiFact ($T3$). From both figures, we can see that the performance of the knowledge-distillation approach remains more or less constant on $T1$ during all the training sessions. For concreteness let us look at Fig. 2(a). EWC's performance on this task is more or less comparable, although it suffers a more pronounced drop in F1 score when we switch from PolitiFact ($T2$) to GossipCop ($T3$). In the case of transfer learning, we notice a substantial drop in performance with the arrival of new tasks. This is absolutely justifiable because, in transfer learning, we do not impose to the model to perform the well also in the previous tasks. In the second plot, where we evaluate with respect to the dataset $T2 =$ PolitiFact, we can observe a similar behavior. Knowledge distillation and EWC have more or less similar performance, with a small drop (about 10%) in F1 score from $T2$ to $T3$, and with a high drop in the case of transfer learning. Finally, in the last training session, transfer learning outperforms the other strategies, whereas knowledge distillation and EWC again have comparable performance. The results generally suggest greater robustness of continual learning methods compared to transfer learning. Although knowledge distillation and EWC obtain comparable performance in all training sessions, the former is more robust on the oldest task ($T1$), guaranteeing superior stability on all learning sessions.

Table 5 shows the average accuracy and forgetting of all methods on the three tasks after the last training session. In all settings, transfer learning attains the worst results regarding average accuracy and forgetting. As for EWC and knowledge distillation, these achieve comparable accuracy values with a slight advantage of knowledge distillation. In terms of forgetting, however, knowledge distillation achieves the best performance, confirming the considerations made in the previous section. The only case that the continual-learning approaches give an inferior score compared to transfer learning is in the evaluation of the third training session ($T3$), but even there the difference is small (see the two bottom plots in Fig. 2).

Table 5. Average accuracy (ACC) and forgetting (BWT) of the continual-leaning approaches on the three datasets. For ACC a higher value is better, for BWT a lower value is better.

Method	ACC	BWT
Transfer learning	0.6520	0.2002
EWC	0.7294	0.0576
Knowledge distillation	**0.7401**	**0.0475**

a $T1$: MediaEval, $T2$: PolitiFact, $T3$: GossipCop.

Method	ACC	BWT
Transfer learning	0.6314	0.2092
EWC	0.7151	0.0681
Knowledge distillation	**0.7277**	**0.0399**

b $T1$: MediaEval, $T2$: GossipCop, $T3$: PolitiFact.

For a more detailed report of the performance of the three strategies after the last training session, we report the results in terms of F1 score in Table 6. For each task, we report the best results in bold. We also mark with ∗ the strategy that achieves the best performance on a given task. As also mentioned in the

Table 6. F1 score of the Tri-Encoder over all tasks after the last training session. Bold values indicate the best performance on a task. Tasks marked with ∗ indicate the learning strategy that performed best on those specific tasks.

Training	Task	All news		Fake News			Real News		
		Acc/F1	F1-macro	Prec.	Rec.	F1	Prec.	Rec.	F1
TL	T1	0.722	0.721	0.839	0.656	0.736	0.621	**0.818**	0.706
	T2	0.522	0.521	0.522	0.478	0.499	0.522	0.566	0.543
	T3∗	**0.712**	**0.711**	**0.703**	**0.786**	**0.695**	**0.719**	**0.735**	**0.727**
EWC	T1	0.810	0.796	0.801	0.903	0.849	0.828	0.675	0.744
	T2∗	**0.717**	**0.717**	0.729	**0.687**	**0.707**	**0.706**	0.747	**0.744**
	T3	0.661	0.658	0.661	0.595	0.627	0.661	0.721	0.689
KD	T1∗	**0.849**	**0.841**	**0.845**	**0.913**	**0.878**	**0.857**	0.758	**0.804**
	T2	0.698	0.693	**0.758**	0.578	0.656	0.661	**0.817**	0.731
	T3	0.673	0.673	0.643	0.709	0.674	0.706	0.639	0.671

a T1: MediaEval, T2: PolitiFact, T3: GossipCop.

Training	Task	All news		Fake News			Real News		
		Acc/F1	F1-macro	Prec.	Rec.	F1	Prec.	Rec.	F1
TL	T1	0.647	0.584	0.650	0.873	0.745	0.633	0.318	0.423
	T2	0.512	0.428	0.495	**0.937**	0.647	0.683	0.123	0.208
	T3∗	**0.735**	**0.734**	0.702	**0.813**	**0.754**	**0.78**	0.658	0.714
EWC	T1	0.788	0.771	0.777	**0.902**	**0.835**	**0.814**	0.624	0.707
	T2	0.644	0.635	0.589	0.845	**0.694**	**0.765**	0.461	0.575
	T3	0.712	0.708	**0.773**	0.597	0.674	0.674	**0.826**	**0.742**
KD	T1∗	**0.802**	**0.791**	**0.810**	0.868	0.838	0.787	**0.705**	**0.744**
	T2∗	**0.677**	**0.677**	**0.636**	0.758	0.692	0.731	**0.604**	**0.661**
	T3	0.707	0.704	0.757	0.607	0.674	0.674	0.807	0.735

b T1: MediaEval, T2: GossipCop, T3: PolitiFact.

discussion of Fig. 2, transfer learning achieves the best performance only in the last task ($T3$). However, knowledge distillation is shown to be the most robust method on the first task ($T1$), followed by EWC. Compared to Table 4, we can see that although the performance degrades slightly on all tasks compared to standard training, continuous-learning strategies still achieve acceptable performance on all three tasks. Moreover, comparing the results with those of Table 4, we can even notice that with continual-learning strategies, we achieve a higher performance compared to training all three datasets in a single session.

It is interesting to note a detail that emerges both from the experiments presented in Table 5, as well as from those of Tables 6 and 3. We can observe a small difference in terms of performance in the order in which we train the model on the various tasks, which seems to suggest that it may have an effect at the model's capacity to generalize. Training on PolitiFact and then on GossipCop seems to improve performance in all experiments. This could be because the topics in GossipCop are very different from those in the other two datasets. Consequently, introducing this dataset in the second training session could have a negative effect on the third session. We leave the exploration of this phenomenon as future work.

4 Conclusion

In this work, we have introduced a content-based multimodal continual-learning strategy, which allows to learn from event streams as they become available over time. This strategy offers the advantage of being able to model the problem of false-content detection in a more realistic way than what is traditionally done by the state-of-the-art approaches, in which one works on fixed distributions. In addition to being more realistic, this paradigm shift still allows for satisfactory performance and leads to a more than 9% improvement of the average accuracy. We have shown that not only does the performance of the continual learner remains high as new tasks arrive, but it can even improve compared to training on a dataset obtained by concatenating different datasets.

The finding in this work creates a lot of interesting future directions. First of all, regarding the number of topics, it is necessary to understand how much every single topic can influence the performance of the learning strategy. Thus an extension of this study to other datasets and measurement of one topic at a time may provide useful insights. Furthermore, a higher number of training sessions may reveal more interesting patterns and may raise the question of whether there are some moments in time where complete retraining may be beneficial. Besides this, it would be interesting to test other incremental-learning techniques. Finally, it would be important to evaluate the possibility of integrating human feedback within the continuous-learning component, for example, by assessing the possibility of applying reinforcement learning.

Acknowledgements. Supported by the ERC Advanced Grant 788893 AMDROMA, the EC H2020RIA project "SoBigData++" (871042), the PNRR MUR project PE0000013-FAIR," the PNRR MUR project IR0000013-SoBigData.it, and the MUR PRIN project 2022EKNE5K "Learning in Markets and Society."

References

1. Alam, F., et al.: A survey on multimodal disinformation detection. In: Proceedings of the 29th International Conference on Computational Linguistics, pp. 6625–6643. International Committee on Computational Linguistics, Gyeongju, Republic of Korea (Oct 2022). https://aclanthology.org/2022.coling-1.576
2. Boididou, C., et al.: Verifying multimedia use at mediaeval 2015 in mediaeval benchmarking initiative for multimedia evaluation (Nov 2015)

3. Chen, J., Wu, Z., Yang, Z., Xie, H., Wang, F., Liu, W.: Multimodal fusion network with latent topic memory for rumor detection. In: 2021 IEEE International Conference on Multimedia and Expo, ICME 2021, pp. 1–6. Proceedings - IEEE International Conference on Multimedia and Expo, IEEE Computer Society, United States (Jun 2021). https://doi.org/10.1109/ICME51207.2021.9428404
4. Chen, Y., Li, D., Zhang, P., Sui, J., Lv, Q., Tun, L., Shang, L.: Cross-modal ambiguity learning for multimodal fake news detection. In: Proceedings of the ACM Web Conference 2022, pp. 2897–2905 (2022)
5. Dai, Z., Lai, G., Yang, Y., Le, Q.: Funnel-transformer: Filtering out sequential redundancy for efficient language processing. In: Larochelle, H., Ranzato, M., Hadsell, R., Balcan, M., Lin, H. (eds.) Advances in Neural Information Processing Systems, NIPS 2020, vol. 33, pp. 4271–4282. Curran Associates, Inc., Red Hook, NY, USA (2020), https://proceedings.neurips.cc/paper/2020/file/2cd2915e69546904e4e5d4a2ac9e1652-Paper.pdf
6. De Lange, M., et al.: A continual learning survey: defying forgetting in classification tasks. IEEE Trans. Pattern Anal. Mach. Intell. **44**(7), 3366–3385 (2022). https://doi.org/10.1109/TPAMI.2021.3057446
7. Devlin, J., Chang, M.W., Lee, K., Toutanova, K.: Bert: pre-training of deep bidirectional transformers for language understanding (2018). https://doi.org/10.48550/ARXIV.1810.04805
8. Dimitrov, D., et al.: SemEval-2021 task 6: Detection of persuasion techniques in texts and images. In: Proceedings of the 15th International Workshop on Semantic Evaluation (SemEval-2021). pp. 70–98. Association for Computational Linguistics, Online (Aug 2021). https://doi.org/10.18653/v1/2021.semeval-1.7, https://aclanthology.org/2021.semeval-1.7
9. Dosovitskiy, A., et al.: An image is worth 16x16 words: transformers for image recognition at scale (2020). https://doi.org/10.48550/ARXIV.2010.11929
10. He, K., Zhang, X., Ren, S., Sun, J.: Deep residual learning for image recognition. In: 2016 IEEE Conference on Computer Vision and Pattern Recognition (CVPR), pp. 770–778 (2016https://doi.org/10.1109/CVPR.2016.90
11. Hinton, G., Vinyals, O., Dean, J.: Distilling the knowledge in a neural network (2015). https://doi.org/10.48550/ARXIV.1503.02531
12. Horne, B.D., Nørregaard, J., Adali, S.: Robust fake news detection over time and attack. ACM Trans. Intell. Syst. Technol. **11**(1) (2019). https://doi.org/10.1145/3363818
13. Khattar, D., Goud, J.S., Gupta, M., Varma, V.: Mvae: Multimodal variational autoencoder for fake news detection. In: The World Wide Web Conference, WWW 2019, pp. 2915-2921. Association for Computing Machinery, New York (2019). https://doi.org/10.1145/3308558.3313552
14. Kingma, D.P., Ba, J.: Adam: A method for stochastic optimization (2014). https://doi.org/10.48550/ARXIV.1412.6980
15. Kirkpatrick, J., et al.: Overcoming catastrophic forgetting in neural networks. Proceedings of the National Academy of Sciences **114**(13), 3521–3526 (2017). https://doi.org/10.1073/pnas.1611835114
16. Li, L.H., Yatskar, M., Yin, D., Hsieh, C.J., Chang, K.W.: Visualbert: a simple and performant baseline for vision and language (2019). https://doi.org/10.48550/ARXIV.1908.03557
17. M. Silva, R., R. Pires, P., Almeida, T.A.: Incremental learning for fake news detection. J. Inform. Data Manag. **13**(6) (2023). https://doi.org/10.5753/jidm.2022.2542, https://sol.sbc.org.br/journals/index.php/jidm/article/view/2542

18. Messina, N., Falchi, F., Gennaro, C., Amato, G.: AIMH at SemEval-2021 task 6: multimodal classification using an ensemble of transformer models. In: Proceedings of the 15th International Workshop on Semantic Evaluation (SemEval-2021), pp. 1020–1026. Association for Computational Linguistics, Online (Aug 2021). https://doi.org/10.18653/v1/2021.semeval-1.140, https://aclanthology.org/2021.semeval-1.140
19. Monti, F., Frasca, F., Eynard, D., Mannion, D., Bronstein, M.M.: Fake news detection on social media using geometric deep learning (2019). https://doi.org/10.48550/ARXIV.1902.06673
20. Radford, A., et al.: Learning transferable visual models from natural language supervision (2021). https://doi.org/10.48550/ARXIV.2103.00020
21. Shu, K., Mahudeswaran, D., Wang, S., Lee, D., Liu, H.: Fakenewsnet: a data repository with news content, social context and dynamic information for studying fake news on social media. arXiv preprint arXiv:1809.01286 (2018)
22. Siciliano, F., Maiano, L., Papa, L., Baccini, F., Amerini, I., Silvestri, F.: Adversarial data poisoning for fake news detection: how to make a model misclassify a target news without modifying it (2024)
23. Simonyan, K., Zisserman, A.: Very deep convolutional networks for large-scale image recognition (2014). https://doi.org/10.48550/ARXIV.1409.1556
24. Singhal, S., Shah, R.R., Chakraborty, T., Kumaraguru, P., Satoh, S.: Spotfake: a multi-modal framework for fake news detection. In: 2019 IEEE Fifth International Conference on Multimedia Big Data (BigMM), pp. 39–47 (2019). https://doi.org/10.1109/BigMM.2019.00-44
25. Wang, Y., et al.: Eann: event adversarial neural networks for multi-modal fake news detection. In: Proceedings of the 24th ACM SIGKDD International Conference on Knowledge Discovery & Data Mining, KDD 2018, pp. 849-857. Association for Computing Machinery, New York (2018). https://doi.org/10.1145/3219819.3219903
26. Wu, Y., Zhan, P., Zhang, Y., Wang, L., Xu, Z.: Multimodal fusion with co-attention networks for fake news detection. In: Findings of the Association for Computational Linguistics: ACL-IJCNLP 2021, pp. 2560–2569. Association for Computational Linguistics, Online (Aug 2021). https://doi.org/10.18653/v1/2021.findings-acl.226, https://aclanthology.org/2021.findings-acl.226
27. Wu, Y., et al.: Google's neural machine translation system: bridging the gap between human and machine translation (2016). https://doi.org/10.48550/ARXIV.1609.08144
28. Wu, Z., Chen, J., Yang, Z., Xie, H., Wang, F.L., Liu, W.: Cross-modal attention network with orthogonal latent memory for rumor detection. In: Zhang, W., Zou, L., Maamar, Z., Chen, L. (eds.) WISE 2021. LNCS, vol. 13080, pp. 527–541. Springer, Cham (2021). https://doi.org/10.1007/978-3-030-90888-1_40
29. Zellers, R., Bisk, Y., Farhadi, A., Choi, Y.: From recognition to cognition: visual commonsense reasoning. In: The IEEE Conference on Computer Vision and Pattern Recognition (CVPR) (June 2019)
30. Zhang, D., et al.: Fauxbuster: a content-free fauxtography detector using social media comments. In: Proceedings of IEEE BigData 2018 (2018)
31. Zhou, X., Wu, J., Zafarani, R.: Safe: Similarity-aware multi-modal fake news detection (2020). https://doi.org/10.48550/ARXIV.2003.04981

Beyond the Horizon: Using Mixture of Experts for Domain Agnostic Fake News Detection

Carmela Comito(✉), Massimo Guarascio, Angelica Liguori, Giuseppe Manco, and Francesco Sergio Pisani

Institute for High Performance Computing and Networking (ICAR-CNR), Via Pietro Bucci 8/9C, 87036 Rende, Italy
{carmela.comito,massimo.guarascio,angelica.liguori,giuseppe.manco, francescosergio.pisani}@icar.cnr.it

Abstract. In recent years, social media have become one of the main means to quickly spread information worldwide, but this rapid dissemination also brings significant risks of misinformation and fake news, which can cause widespread confusion, erode public trust, and contribute to social and political instability. This scenario is further exacerbated by the fact that fake news can span various topics across different domains, making it impracticable for a single moderator to manage the massive quantity of data. The use of Machine Learning, particularly language models, is rising as an effective solution to mitigate the risk of misinformation. However, a single model cannot fully capture the complexity and variety of the information it needs to process, often failing to classify examples from new domains. In this work, the aforementioned challenges are addressed by leveraging a novel hierarchical deep-ensemble framework. This framework aims to integrate various domains to offer enhanced predictions for new ones. Specifically, the approach involves learning a distinct model for each domain and refining them through domain-specific adaptation procedures. The predictions of these refined models are hence blended using a Mixture of Experts approach, which allows for selecting the most reliable for predicting the new examples. The proposed approach is fully cross-domain and does not necessitate retraining or fine-tuning when encountering new domains, thus streamlining the adaptation process and ensuring scalability across diverse data landscapes. Experiments conducted on 5 real datasets demonstrate the robustness and effectiveness of our proposal.

Keywords: Cross-Domain Fake News Detection · Deep Ensemble Learning · Language Models · Mixture of Experts

---

All the authors equally contributed to the paper and are considered all first authors.

1 Introduction

In the modern landscape, the pervasive influence of social media has revolutionized the dissemination of information on a global scale. However, this unprecedented speed in spreading information also carries substantial risks, particularly in the form of misinformation and fake news. Such malicious content not only fosters widespread confusion but also undermines public trust in traditional sources of information, leading to an erosion of credibility and reliability. Moreover, the proliferation of misinformation poses a significant threat to societal harmony and political stability, as it can amplify polarization and incite unrest. Thus, while social media offer unparalleled connectivity and access to information, they rise new needs in terms of mechanisms for discerning real and fake information.

Indeed, nowadays, the rapid and widespread propagation of fake news has reached emergency levels, especially given the suspicion that recent critical societal events, such as the Brexit referendum and the 2016 United States election, were influenced by the deliberate spread of false or deceptive information.

This complex scenario is further exacerbated by the fact that fake news can span a wide range of topics across multiple domains, making it impractical for a single moderator to handle the large amount of data yielded. Since individual efforts alone are inadequate to address the pervasive issue of fake news, there has been growing interest in recent years in automatic tools that can assist and support moderators in their tasks.

In this context, verifying the veracity and authenticity of news is a critical challenge that can greatly benefit from recent advancements in Artificial Intelligence (AI) and Machine Learning (ML) [1]. Given that this task is time-consuming, expensive, and impractical for the vast amounts of data generated online, AI-based tools offer an effective solution by automating the identification of deceptive information, thereby reducing the reliance on specialized and trusted professionals. In particular, the automatic detection of fake news has garnered significant interest from the research community. Traditionally, this problem has been approached as a text classification task in the literature [11], aiming to distinguish between real and fake news.

However, learning reliable detection models able to identify fake news requires coping with different complex issues [5]. First, an effective solution should allow handling low-level raw data frequently affected by noise, as the channels used to spread fake news typically allow for sharing only short text (e.g., Twitter). Moreover, the number of labeled training instances is limited; the labeling phase is a difficult and time-consuming task manually performed by domain experts. Finally, fake news can concern different topics; therefore, the features leveraged to perform the prediction should be domain-independent to handle different topics.

Contribution. To address the aforementioned issues, we proposed a deep-ensemble framework for fake news detection tailored for cross-domain applications. The adoption of the Deep Learning paradigm [3] represents a natural solution to address the above issues, as DL techniques permit the learning of accurate classification models also from raw data (in our solution, the words composing the news) without requiring heavy intervention by data-science experts. Basically, these DL models are structured according to a hierarchical architecture (consisting of several layers of base computational units, i.e., the artificial neurons are stacked one upon the other), allowing for learning features at different abstraction levels to represent raw data.

In this work, we define $MoDA - FND$ (Mixture of Domain Agnostic Fake News Detectors), a hierarchical Deep-Ensemble approach for learning cross-domain fake news detectors from text data. Basically, the idea is to combine different base Language Models (LMs) learned using a combination of standard training stages and adaptation procedures performed over domains different from the ones where they were initially trained. These specialized models are, hence, blended by leveraging a Mixture-of-Experts scheme to yield accurate predictions. Experiments conducted on five real datasets and against different competitors demonstrate the effectiveness of our solution.

Organization of the Paper. The remainder of the paper is structured as follows. Section 2 defines the addressed problem. In Sect. 3, we survey recent works concerning the cross-domain fake news detection problem; Sect. 4 introduces our approach and details the devised deep ensemble strategy, and Sect. 5 showcases numerical results. Finally, Sect. 6 concludes the paper and outlines possible future research directions.

2 Problem Statement

This paper addresses the challenge of detecting fake news in new or emerging domains. The objective is to utilize knowledge from known domains to effectively identify fake news in unfamiliar contexts. We treat cross-domain fake news detection as a domain generalization problem, where a pre-trained model from a source domain is applied to classify instances in a distinct target domain. Specifically, in this work, we focused only on analyzing the text of the news to reveal its veracity.

Domain generalization (DG), i.e., out-of-distribution generalization, has attracted increasing interest in recent years. Domain generalization deals with a challenging setting where one or several different but related domain(s) are given, and the goal is to learn a model that can generalize to an unseen test domain. Great progress has been made in the area of domain generalization for years. There are many generalization-related research topics such as domain adaptation, meta-learning, transfer learning, covariate shift, and so on. Domain Adaptation (DA) aims to maximize the performance on a given target domain using existing training source domain(s). The difference between DA and DG is that DA has access to the target domain data while DG cannot see them during training. This makes DG more challenging than DA but more realistic and favorable in practical applications.

The proposed approach focuses on a domain invariant representation between source and target domains, allowing the classifier trained on the source domains to accurately classify samples from the target domain. The problem can be formally described as follows:

Given news from an unknown domain, D_T, we aim to learn a domain-independent news representation from a set of known different application domains (labeled source data) $\{D_1, D_2, ..., D_N\}$ that are available for training the classification model and testing on the D_T.

Let be $\{D_1, D_2, ..., D_N\}$ a set of fake news data collections from N different domains such that each dataset D_k consists of pairs of news content and its real label, i.e., $D_k = \{(x_i^k, y_i^k)\}_{i=1}^{|k|}$, where x_i^k is the textual content of the $i - th$ example from $k - th$ domain and the label $y_i^k \in \{0, 1\}$ (i.e., false or true).

Our goal is to learn a fake news classifier from the source data $\{D_1, D_2, ..., D_N\}$. Formally, let $D_{train} = \{D_1, D_2, ..., D_N\}$, we want to discover a function $f : D_{train} \to \{0, 1\}$ that can accurately predict the real label of any unseen news in the target domain D_T. In our framework, the function f takes the form of a *Domain-Agnostic Ensemble of Fake News Detectors* illustrated in the following section.

In the rest of the paper, we use x, y, $\hat{y}$ to represent news content, the true label, and the predicted one, respectively.

3 Related Work

In recent years, a multitude of studies have been conducted on the problem of fake news detection. These studies primarily utilize various supervised models that mainly analyze the text to identify false news. However, these cutting-edge detection techniques struggle significantly when applied to news from domains other than the ones they were trained on. For instance, models trained on political news may perform well within that specific context but exhibit poor performance when tasked with identifying fake news in other areas, such as healthcare or entertainment, particularly in domains that are rarely encountered during training. News content varies greatly across domains, with each domain having its own unique language and vocabulary, user engagement patterns, and dissemination methods. Furthermore, current models tend to favor features tied to specific events, leading to bias. To overcome these challenges, it is essential to develop models that can comprehend and integrate information across multiple domains.

To address these issues, it is imperative to design models capable of capturing and leveraging cross-domain knowledge. Cross-domain modeling entails constructing models that can acquire insights from data within one domain and effectively apply them to another distinct domain. Although a few studies have ventured into using cross-domain datasets for detecting fake news, this section reviews some of the most prominent methodologies proposed in the academic literature.

In the work by Wang et al. [24], the authors introduced an Event Adversarial Neural Network (EANN) designed to improve the detection of fake news by minimizing the influence of domain-specific features. This framework includes an event discriminator that operates in tandem with a multimodal fake news detector. The event discriminator identifies and discards features unique to specific events, retaining only the shared characteristics across various events. This approach enables the model to extract event-invariant features, enhancing its capability to detect fake news in newly emerging events by engaging in a minimax game with a multimodal feature extractor.

In Zhang et al. [25], the authors introduced BDANN, a BERT-based domain adaptation neural network for multimodal fake news detection. BDANN features a multimodal feature extractor, a domain classifier, and a fake news detector. The extractor uses BERT for text features and VGG-19 for image features, which are combined and fed to the detector. The domain classifier aligns features from different events into a common feature space, aiding domain adaptation. However, this method is limited as it only handles a fixed number of known events and does not consider the sequence or timing of events.

In Silva et al. [21] is presented a multimodal fake news detection framework designed for cross-domain data, leveraging both domain-specific and cross-domain information through two separate embedding spaces. The framework comprises two main parts: unsupervised domain embedding learning and supervised domain-agnostic news classification. The unsupervised component uses multimodal content, such as text and propagation networks, to create low-dimensional domain vectors. This content is represented as a heterogeneous network with nodes for both users and words in the news titles. The approach allows for selecting unlabelled news to train the fake news detection model, which performs well across multiple domains while reducing labeling costs.

In Li et al. [10] is proposed the Multi-source Domain Adaptation with Weak Supervision (MDA-WS) approach for detecting misinformation. This method leverages adversarial training to transfer knowledge from multiple source domains, ensuring consistent feature representation across domains. It trains fake news classifiers for these source domains and incorporates researchers' expertise through weak supervision in the target domain. Weak labels are assigned based on heuristic rules established by researchers. The effectiveness of MDA-WS hinges on the assumption of domain similarity between multiple sources and the target. However, significant domain differences can impede knowledge transfer, leading to less effective adaptation. Moreover, relying on heuristic rules for labeling unlabeled samples in the target domain introduces potential limitations. The accuracy of this approach heavily relies on the reliability of the chosen heuristics, which are based on domain-specific terms extracted using proprietary services. Inaccurate or insufficient heuristic rules may introduce bias or errors in labeling, thereby impacting the performance of fake news classifiers.

In Shu et al. [19] is introduced CrossFND, a deep learning framework for cross-domain multimodal fake news detection. This approach combines cross-domain knowledge transfer with within-domain analysis of news content, user comments, and user-news interactions. It focuses on learning unsupervised feature representations for domain adaptation. A key element is a domain classifier designed to identify the domain of news content by detecting feature differences across domains. By adding a small portion of the target dataset to the source domain dataset, the domain classifier-a three-layer neural network-is trained to distinguish domains. The model maximizes domain loss to force learning of domain-independent features, thereby making it harder for the classifier to detect the domain. Simultaneously, it minimizes the loss function of the fake news classifier to accurately identify fake news.

In Mosallanezhad et al. [13] is introduced REAL-FND, a novel multi-modal domain-adaptive approach that integrates user comments and interactions into fake news detection. Building upon CrossFND [19], REAL-FND enhances the domain classifier by replacing the Multilayer Perceptron (MLP) with a sophisticated architecture based on reinforcement learning. This adaptation ensures that the model effectively transforms learned representations from the source to the target domain, obscuring domain-specific features while preserving domain-invariant components.

Discussion. Despite the variety of approaches proposed for addressing fake news detection, many face significant challenges and limitations. They typically assume that news records from different domains arrive sequentially, which does not always reflect real-world data streams, and they require prior knowledge of the news record's domain, which may not always be available. Moreover, these methods often struggle with capturing newly emerging domains and adapting to temporal changes within domains since they typically require a substantial amount of labeled examples from the target domain to effectively train detection models. In contrast, our approach focuses on learning cross-domain news knowledge without relying on domain information. In summary, most state-of-the-art approaches rely heavily on supervised models. Our self-supervised learning, in contrast, leverages unlabeled data and pre-trained representations to enhance generalization across domains. Additionally, the cross-domain transfer learning approach enhances detection accuracy by transferring knowledge from related domains.

4 $MoDA - FND$: Detecting Cross-Domain Fake News

In this section, we present a comprehensive description of our Deep Learning-based approach capable of identifying malicious and misleading information across various domains. This problem is particularly relevant in scenarios where fake news emerges in new contexts with no prior data available. Existing methods are often tailored to specific events, making them ineffective for detecting fake news in emerging situations [7–9,15–17,22].

$MoDA - FND$ combines the capabilities of Language Models with the boosting offered by the Ensemble Learning scheme depicted in Fig. 1. Let $\{D_1, D_2, \ldots, D_N\}$ be N datasets extracted from different domains. Each dataset D_i is split into a pair $\langle D_i^{tr}, D_i^{ad} \rangle$. The first subset, D_i^{tr}, is used to train a base specialized fake news detector. The classifier utilizes a pre-trained language model from the BERT family as its backbone to extract features, which are then processed by one or more fully connected dense layers to make predictions.

Next, an adaptation procedure is performed for each model M_i against another domain D_j^{ad} with $j \neq i$ to obtain an adapted model M_i^j (this method is performed against all the other domains). Finally, the predictions from the obtained models are combined using an ensemble strategy es, which is a parameter of our solution. Specifically, the ensemble model is trained over a view containing data from all the domain seen.

The learning process described above is reported in the Pseudo Algorithm 1.

Algorithm 1. $MoDA - FND$ learning procedure

Require: $\{D_1, D_2, \ldots, D_N\}$: Datasets from different domains,
  LM: Pre-trained Language Model,
  es: Ensemble Strategy,
  b: adaptation_budget
Ensure: EM: Ensemble Model
1: $M_{list} \leftarrow \emptyset$ {Initialize empty list of models}
2: **for** $i = 1$ to N **do**
3:   $\langle D_i^{tr}, D_i^{ad} \rangle \leftarrow$ SPLITDATA(D_i, b)
4:   $M_i \leftarrow$ TRAINBASEMODEL(D_i^{tr}, LM)
5:   Add M_i to M_{list} {Store trained base model}
6: **end for**
7: **for** $i = 1$ to N **do**
8:   **for** $j = 1$ to N **do**
9:     **if** $j \neq i$ **then**
10:      $M_i^j \leftarrow$ ADAPTMODEL(M_i, D_j^{ad})
11:      Add M_i^j to M_{list} {Store adapted model}
12:    **end if**
13:   **end for**
14: **end for**
15: $D^{Ens} \leftarrow$ Merge$(\{D_1^{ad}, D_2^{ad}, \ldots, D_N^{ad}\})$ {Merge datasets}
16: $EM \leftarrow$ LEARNENSEMBLE(D^{ens}, M_{list}, es) {Learn a function to combine the predictions of the models}
17: **return** EM {Return ensemble model}

4.1 Base Model Architecture

As discussed above, the base models used in $MoDA - FND$ rely on a pre-trained Language Model belonging to the BERT family that represents the backbone of the detector. The BERT model (Bidirectional Encoder Representations from Transformers) [4], along with its variations, is a state-of-the-art neural architecture able to extract informative representations (*embeddings*) from textual data. In our approach, we utilize a pre-trained instance of DeBERTa (Decoding-enhanced BERT with disentangled attention) [6], illustrated in Fig. 2. DeBERTa builds upon the strengths of BERT and RoBERTa models by integrating two innovative techniques. The first technique employs a disentangled

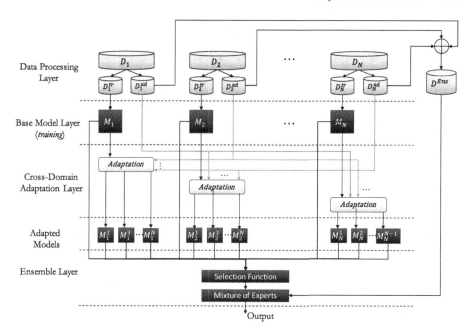

Fig. 1. $MoDA - FND$ solution approach.

attention mechanism, wherein each word is represented by two vectors, capturing both its content and position. By utilizing disentangled matrices based on content and relative positions, the model computes attention weights among words, enhancing its ability to understand context. The second technique involves an enhanced mask decoder, replacing the traditional output softmax layer, to more accurately predict masked tokens during model pretraining. These advancements have demonstrated substantial improvements in both the efficiency of model pretraining and the overall performance across various downstream tasks.

Similar to BERT, the pretraining process of DeBERTa follows two main steps: Word Masking and Next Sentence Prediction (NSP). In the Word Masking step, a certain percentage of words within a sentence are masked, and the model is trained to predict these masked tokens by considering the surrounding word context. This context includes terms preceding and following the masked word. Subsequently, the model undergoes further fine-tuning through the Next Sentence Prediction task, aimed at understanding relations among sentences. Notably, the DeBERTa instance we employ is pre-trained using a combination of Wikipedia and bookcorpus datasets [26].

The Language Model remains fixed in this setup (basically, it is freezed), while three trainable layers are added on top of the architecture to learn the fake news classification function. Basically, the Language Model acts like an embedder that allows to map the text data in a latent space e_x with which we feed the Multi-Layer Perceptron on top of the base learner.

Fig. 2. DeBERTa architecture.

4.2 Combining the Detectors: Mixture of Experts

The combination mechanism plays a pivotal role in augmenting the overall detection capabilities by harnessing the diverse expertise encapsulated within individual local models. Employing an ensemble strategy in the form of a Mixture of Experts (MoE) scheme [12], the combination mechanism orchestrates the fusion of predictions from the base classifiers. Its objective is to ensure that the final classification of input x is primarily influenced by a subset of base classifiers, ideally, those demonstrating higher expertise in the regions of the instance space relevant to x.

As illustrated in Fig. 3, this architecture includes a trainable Gate sub-net, essentially an MLP with $L = N \times N$ outputs, which computes a set of weights $z_1, \ldots, z_L$ for the base classifiers. These weights are then utilized to calculate the final fake news probability $\tilde{y}$ as an optimal convex combination of predictions $\tilde{y}^{(1)}, \ldots, \tilde{y}^{(L)}$ provided by the base classifiers for input x. Notably, differently from a standard MoE the gate input is the embedding e_x, computed using the shared backbone of the base models. Formally, the architecture computes the overall prediction for input x as:

$$\tilde{y} = \sum_{i=1}^{L} \tilde{y}_i \cdot z_i \qquad (1)$$

where, $\langle z_1, \ldots, z_L \rangle = MLP^L(e_x)$. We point out that the experts in Fig. 3 are depicted in a different color to emphasize that they are not further fine-tuned, i.e., their model weights are frozen. The only trainable component of the ensemble is the Gate and it takes the form of a neural model.

One of the risks in the learning process of the MoE is the increased utilization of a subset of experts. Consequently, the information regarding the specialization of experts in a particular domain may be significantly affected. To address this issue, a domain identification task has been introduced, allowing the Gate to preserve information on the specific characteristics of a domain and selecting the expert who best handles features compatible with that domain. The final objective optimized by the learning process is defined as follows:

$$\mathcal{L} = -(y\log(\tilde{y}) - (1-y)\log(1-\tilde{y})) - \sum_{c=1}^{N} y_{d,c}\log(\tilde{y}_{d,c}) \qquad (2)$$

where N is the number of source domains, the first two terms represent the Binary Cross Entropy computing the classification error of the MoE, the $\tilde{y}_{d,c}$ is the prediction score of the detected source domains by the Gate, and the last term is a regularisation component to preserve domain knowledge. In addition, a selection function based on the F1-score achieved by the adapted models across the various domains is used to select the top-K most reliable models to limit this choice of the MoE.

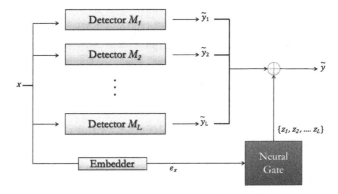

Fig. 3. Ensemble mechanism: Mixture of Experts.

5 Experimental Assessment

In this section, we evaluate the effectiveness of our solution. Specifically, first, we define the experimental setting (including datasets, parameters, metrics, and competitors), and then we discuss the obtained results.

5.1 Datasets and Experimental Setup

To assess the quality of the proposed framework, we used five benchmark datasets described below:

- **GossipCop**[1] comprising of news related to the entertainment industry, sourced from the FakeNewsNet data repository [18,20].
- **Politifact**[2] consisting of news related to the US political system, sourced from the FakeNewsNet data repository [18,20].

---
[1] https://www.gossipcop.com/.
[2] https://www.politifact.com/.

- **MD**[3] including news concerning the 2016 US election.
- **CoAID**[4] [2] consisting of news about the COVID-19 virus.
- **Liar**[5] [23] comprising short statements that have been collected from the fact-checking website PolitiFact and labeled by human.

In Table 1, we report dataset statistics, such as the total number of articles, the class distribution, the vocabulary size, and the word count distribution.

Table 1. Dataset statistics

Dataset	#Articles	#Real	#Fake	Vocabulary size	#Words per article			
					Avg.	Median	Q1	Q3
GossipCop	4,736	2,117	2,619	167,759	581	348	180	551
PolitiFact	818	370	448	69,796	1,470	336	108	882
MD	6,296	3,171	3,125	216,169	781	601	294	1,026
CoAID	4,439	4,023	416	15,738	56	74	14	80
Liar	12,685	9,183	3,502	24,549	18	17	12	22

We implemented $MoDA - FND$[6] using the PyTorch framework [14]. As explained in Sect. 4, the classifier consists of a pre-trained DeBERTa acting as the feature extractor and three fully connected dense layers that learn the fake news detection function. These three layers are instantiated with 256, 128, and 1 neurons. The first two layers have a `tanh` as activation function. Adam, with a learning rate of 1e−3, is used as optimizer. To evaluate our approach, we used the following evaluation protocol: we considered N datasets and analyzed different configurations by taking N-1 datasets as the source domain and the remaining one as the target domain. Each source domain D_i is split into D_i^{tr} and D_i^{ad}. First, we pre-trained each classifier M_i against its source dataset D_i^{tr} for 100 epochs. It is then adapted on the other source domains D_j^{ad} with $j \neq i$ for an additional 100 epochs. Next, the ensemble model is trained using D_i^{ad} from all the source datasets. Finally, the trained model is tested on the target dataset.

All the experiments were conducted on an NVidia DGX Station featuring 4 GPU V100 32 GB.

To evaluate the effectiveness of $MoDA - FND$, we performed a comparative evaluation against three competitors. Specifically:

- **MDA-WS** [10]: Multi-source Domain Adaptation with Weak Supervision (MDA-WS) employs adversarial training for learning domain invariant feature representation and a set of fake news classifiers for multi-source domains. For embedding, MDA-WS uses TextCNN and RoBERTa-base, with news content truncated to 300 tokens.

---

[3] https://github.com/miafranc/fakenews2020.
[4] https://github.com/cuilimeng/CoAID.
[5] https://huggingface.co/datasets/ucsbnlp/liar.
[6] https://github.com/massimo-guarascio/ds24.

- **EANN** [24]: is a state-of-the-art model for domain-agnostic fake news detection, designed to improve generalization by capturing common features across various events of news. In our experiments, we excluded the component that handles the image while keeping the rest of the architecture unchanged. The parameter settings are identical to those specified in the original work [24].
- **CrossFND** [19]: CrossFND is a deep learning-based model for detecting fake news across different domains using a multimodal approach. It consists of several components: (i) an encoder for news content, user comments, and user-news interactions to learn representations; (ii) a fusion component that integrate these representations for fake new detection; and (iii) a domain adaptive representation learning component. In our experiments, we excluded the parts dealing with news comments and user-news interactions, keeping the rest unchanged. This version of the framework is referred to as **CrossFND\A** in the original work [19]. The parameter settings remain the same as those in [19].

In addition, based on the same base learners used by $MoDA - FND$, we considered further three combination strategies, respectively named Max-Probability, Stacking, and Stacking HP. The first one is a non-trainable function that determines the final prediction by selecting the one with the highest confidence score among those produced by the base learners. The latter methods involve concatenating the output probability scores from the base learners, which are then used as input to a subsequent network that generates the final prediction. The key difference between these two strategies is that, in the second one, the output from the base learners is rounded to produce a binary response before being fed into the subsequent network.

Finally, to evaluate the effectiveness of the detectors, we adopt the $F1$-Score. The $F1$-Score is calculated as the harmonic mean of Precision and Recall, providing a single, concise measure of overall system performance. This choice simplifies the interpretation of the results, making it easier to compare the effectiveness of different models. In particular, we used a macro-avg approach to handle the unbalance of some datasets.

5.2 Numerical Results

Table 2. Performance results in terms of Macro-F1 varying different models and datasets.

	GossipCop	Politifact	MD	CoAID	Liar
CrossFND	0.4374	0.3717	0.5415	0.2259	0.4880
EANN	**0.5122**	0.3477	0.5503	0.5303	0.4199
MDA-WS	0.3089	0.3539	0.3317	0.4754	0.2174
$MoDA - FND$	0.4857	**0.4817**	**0.6474**	**0.5660**	**0.5045**

Here, we conducted two suites of experiments to validate our approach. In the first suite of experiments, we compared our approach, $MoDA - FND$, with other state-of-the-art methods for fake news detection across various datasets: GossipCop, Politifact, MD, CoAID, and Liar. As shown in Table 2, our approach outperforms the other approaches in most cases. Specifically, $MoDA - FND$ achieves the highest performance on Politifact, MD, CoAID, and Liar datasets, indicating its robustness and superior capability in detecting fake news across different domains. Although EANN performs better on GossipCop, $MoDA-FND$ consistently delivers strong results across all datasets, showcasing its overall effectiveness.

In the second suite of experiments, we focused on comparing our combination strategy based on the Mixture of Experts scheme with other ensemble strategies. Table 3 presents the results achieved in terms of $F1$. The combination strategy adopted by $MoDA-FND$, achieves the best performance on Politifact, MD, and CoAID datasets, demonstrating its superior ability to integrate the predictions from various base classifiers effectively. While the Stacking and Max-Probability methods slightly outperform $MoDA - FND$ on GossipCop and Liar, respectively, our approach remains highly competitive, validating the efficacy of the MoE-based ensemble strategy.

Consistently across both scenarios, $MoDA - FND$ performs slightly less effectively on GossipCop, likely due to the specific characteristics of the dataset, including the number of examples and the used terminology.

Table 3. Performance results in terms of Macro-F1 varying different combination strategies and datasets.

	GossipCop	Politifact	MD	CoAID	Liar
Max-Probability	0.4939	0.2896	0.4745	0.2548	**0.5215**
Stacking	**0.5042**	0.4201	0.4117	0.4423	0.5079
Stacking - HP	0.4990	0.4026	0.4980	0.5251	0.5077
$MoDA - FND$	0.4857	**0.4817**	**0.6474**	**0.5660**	0.5045

6 Conclusion and Future Work

In this work, we designed an ensemble framework for detecting fake news using the deep learning paradigm that effectively addresses a major issue in this field, i.e., recognizing misinformation across different domains by exploiting one learning model able to generalize through the domains. Here, we introduced $MoDA - FND$, a hierarchical Deep-Ensemble approach for training cross-domain fake news detectors using text data. Our approach involves combining diverse base Language Models, trained through standard training phases and adaptation procedures across various domains. These specialized models are then fused together using a Mixture-of-Experts scheme to generate precise predictions. Experimental evaluations conducted on five real datasets demonstrated the effectiveness of our approach.

As future work, we are interested in integrating additional sources of information, thus expanding our approach to a multi-modal framework. For instance, we aim to explore the incorporation of the social network of users who propagate misinformation and fake news, thereby enhancing the comprehensiveness and effectiveness of our detection system.

Acknowledgment. This work has been partially supported by: (i) European Union - NextGenerationEU - National Recovery and Resilience Plan (Piano Nazionale di Ripresa e Resilienza, PNRR) - Project: "SoBigData.it - Strengthening the Italian RI for Social Mining and Big Data Analytics" - Prot. IR0000013 - Avviso n. 3264 del 28/12/2021; (ii) project SERICS (PE00000014) under the NRRP MUR program funded by the EU - NGEU; (iii) Italian MUR, PRIN PNRR 2022 Project "Limiting MIsinformation spRead in online environments through multi-modal and cross-domain FAKe news detection (MIRFAK) ", Prot.: P2022C23K9, ERC field: PE6, funded by European Union - Next Generation EU.

References

1. Comito, C., Caroprese, L., Zumpano, E.: Multimodal fake news detection on social media: a survey of deep learning techniques. Soc. Netw. Anal. Min. **13**(1), 101 (2023). https://doi.org/10.1007/S13278-023-01104-W
2. Cui, L., Lee, D.: CoAID: COVID-19 healthcare misinformation dataset (2020)
3. Cun, Y.L., Bengio, Y., Hinton, G.: Deep learning. Nature **521**(7553), 436–444 (2015)
4. Devlin, J., Chang, M.W., Lee, K., Toutanova, K.: BERT: pre-training of deep bidirectional transformers for language understanding. In: NAACL-HLT, pp. 4171–4186 (2019). https://doi.org/10.18653/v1/N19-1423
5. Folino, F., Folino, G., Guarascio, M., Pontieri, L., Zicari, P.: Towards data- and compute-efficient fake-news detection: An approach combining active learning and pre-trained language models. SN Comput. Sci. **5**(5) (2024). https://doi.org/10.1007/s42979-024-02809-1
6. He, P., Liu, X., Gao, J., Chen, W.: DeBERTa: Decoding-enhanced BERT with disentangled attention (2020)
7. Jin, Z., Cao, J., Guo, H., Zhang, Y., Luo, J.: Multimodal fusion with recurrent neural networks for rumor detection on microblogs. In: Proceedings of the 25th ACM International Conference on Multimedia (2017)
8. Jing, Q., et al.: TransFake: multi-task transformer for multimodal enhanced fake news detection. In: IJCNN, pp. 1–8 (2021)
9. Kumari, R., Ekbal, A.: AMFB: attention based multimodal factorized bilinear pooling for multimodal fake news detection. Expert Syst. Appl. **184**, 115412 (2021)
10. Li, Y., Lee, K., Kordzadeh, N., Faber, B., Fiddes, C., Chen, E., Shu, K.: Multisource domain adaptation with weak supervision for early fake news detection. In: 2021 IEEE International Conference on Big Data, pp. 668–676 (2021). https://doi.org/10.1109/BigData52589.2021.9671592
11. Liu, C., et al.: A two-stage model based on BERT for short fake news detection. In: Knowledge Science, Engineering and Management, pp. 172–183 (2019)

12. Masoudnia, S., Ebrahimpour, R.: Mixture of experts: a literature survey. Artif. Intell. Rev. **42**(2), 275–293 (2014)
13. Mosallanezhad, A., Karami, M., Shu, K., Mancenido, M.V., Liu, H.: Domain adaptive fake news detection via reinforcement learning. In: Proceedings of the ACM Web Conference 2022, pp. 3632–3640 (2022). https://doi.org/10.1145/3485447.3512258
14. Paszke, A., et al.: PyTorch: an imperative style, high-performance deep learning library (2019)
15. Raj, C., Meel, P.: ARCNN framework for multimodal infodemic detection. Neural Netw. **146**, 36–68 (2022)
16. Sachan, T., Pinnaparaju, N., Gupta, M., Varma, V.: SCATE: shared cross attention transformer encoders for multimodal fake news detection. In: Proceedings of the 2021 IEEE/ACM ASONAM, pp. 399–406 (2021)
17. Shu, K., Cui, L., Wang, S., Lee, D., Liu, H.: Defend: explainable fake news detection. In: Proceedings of the 25th ACM SIGKDD International Conference on Knowledge Discovery and Data Mining, pp. 395–405 (2019)
18. Shu, K., Mahudeswaran, D., Wang, S., Lee, D., Liu, H.: FakeNewsNet: a data repository with news content, social context and dynamic information for studying fake news on social media. arXiv:1809.01286 (2018)
19. Shu, K., Mosallanezhad, A., Liu, H.: Cross-domain fake news detection on social media: a context-aware adversarial approach. In: Khosravy, M., Echizen, I., Babaguchi, N. (eds.) Frontiers in Fake Media Generation and Detection. Studies in Autonomic, Data-driven and Industrial Computing, pp. 215–232. Springer, Singapore (2022). https://doi.org/10.1007/978-981-19-1524-6_9
20. Shu, K., Sliva, A., Wang, S., Tang, J., Liu, H.: Fake news detection on social media: a data mining perspective. ACM SIGKDD Expl. New. **19**(1), 22–36 (2017)
21. Silva, A., Luo, L., Karunasekera, S., Leckie, C.: Embracing domain differences in fake news: cross-domain fake news detection using multi-modal data (2021). https://doi.org/10.48550/ARXIV.2102.06314
22. Wang, J., Mao, H., Li, H.: FMFN: fine-grained multimodal fusion networks for fake news detection. Appl. Sci. **12**(3) (2022)
23. Wang, W.Y.: "Liar, Liar pants on fire": a new benchmark dataset for fake news detection. In: Proceedings of the 55th Annual Meeting of the Association for Computational Linguistics (Volume 2: Short Papers), pp. 422–426 (2017). https://doi.org/10.18653/v1/P17-2067
24. Wang, Y., et al.: EANN: event adversarial neural networks for multi-modal fake news detection. In: Proceedings of the 24th ACM SIGKDD International Conference on Knowledge Discovery & Data Mining, pp. 849–857 (2018). https://doi.org/10.1145/3219819.3219903
25. Zhang, T., et al.: BDANN: BERT-based domain adaptation neural network for multi-modal fake news detection. In: 2020 International Joint Conference on Neural Networks (IJCNN), pp. 1–8 (2020). https://doi.org/10.1109/IJCNN48605.2020.9206973
26. Zhu, Y., et al.: Aligning books and movies: towards story-like visual explanations by watching movies and reading books. In: 2015 IEEE ICCV 2015, pp. 19–27 (2015). https://doi.org/10.1109/ICCV.2015.11

Quantifying Attraction to Extreme Opinions in Online Debates

Davide Perra[1], Andrea Failla[1,2(✉)], and Giulio Rossetti[2]

[1] Department of Computer Science, University of Pisa, Pisa 56127, Italy
d.perra@studenti.unipi.it, andrea.failla@phd.unipi.it
[2] National Research Council, Institute of Information Science and Technologies A. Faedo (ISTI), Pisa 56127, Italy
giulio.rossetti@isti.cnr.it

Abstract. Opinion polarization and political segregation are key societal concerns, especially on social media. Although these phenomena have been traditionally attributed to homophily—preference for like-minded individuals—recent work in social psychology suggests that acrophily—preference for extreme rather than moderate opinions—might play a role as well. In this work, we introduce a methodology to estimate the degree of preference for connecting with users who hold strong opinions on social media. Our framework is composed of four phases: (i) opinion estimation, (ii) opinion thresholding, (iii) network construction, and (iv) acrophily estimation. We apply it to study the climate change debate on Reddit and find that users show higher-than-expected acrophilic patterns, especially if they are climate skeptics or have extreme opinions. Acrophilic patterns are stable over time, while polarization gradually leaves space for pluralism.

Keywords: acrophily · opinion polarization · climate change · social media · social networks

1 Introduction

Online Social Platforms, such as X/Twitter and Reddit, are digital environments where individuals freely manifest their opinions and typically engage in discussions with peers. When it comes to debating hot-button issues such as climate change [23], minority discrimination [9], and abortion rights [2], users self-organize in opposed opinion-driven subgroups called Echo Chambers [2,17]. In these settings, users' beliefs are reinforced by exposure to like-minded peers, and opposing views are instead discredited. As a consequence, opportunities for change are scarce, and opinions might radicalize over time [12]. Human biases such as homophily [8]—preference for connecting with similar others—were shown to play a pivotal role in these dynamics. Moreover, recent work in social psychology has unveiled that acrophily—preference for like-minded peers with more extreme opinions (as opposed to more moderate)—also influences tie and opinion formation mechanisms [5], and that the combined effects of homophily and acrophily ultimately harshen polarization effects [6]. Therefore, measuring these effects becomes of utmost importance to understand (and consequently, reduce) the effects of these warning phenomena. Nonetheless, acrophily is a rather recent concept, and literature lacks a methodology to measure it in the wild. To bridge this gap, we introduce a framework to model, measure, and interpret attraction to extreme opinions in online debates. Our framework is flexible, and can be applied to discussions on any social media platform, as it leverages features that are common to virtually all platforms. In this work, we apply it to Reddit discussions on climate change and study the evolution of polarization and homophilic/acrophilic effects across time. Notably, this is the first research effort providing a standardized methodology to quantify acrophily in online social networks, and also to study its temporal evolution.

The remainder of this work is organized as follows. In Sect. 2, we provide an overview of related literature; Sect. 3 introduces the methodology and details its steps; subsequently, Sect. 4 describes the data collection process, and Sect. 5 introduces and discusses the experiments; finally, Sect. 6 concludes the work and suggests future research directions.

2 Related Work

This section discusses relevant literature related to the present work. Specifically, it focuses on acrophily and homophily, human biases that were claimed to drive social interactions online.

Acrophily. Acrophily is a social tie formation mechanism first introduced in [5], where it is defined as the tendency to affiliate with more extreme (as opposed to more moderate) like-minded peers. In this study, participants were asked to rate their emotions toward political policies in a controlled setting. After viewing peers' responses to the same policies, they selected which peers they wanted to affiliate with. The findings indicate that participants' choices were influenced by both homophily and acrophily. Moreover, the inclination toward acrophily was linked to the perception that more extreme expressions are seen as more representative of one's political group. More recent work has studied the presence of acrophily on Twitter (now X) [6]. The authors associated users with political leaning estimated from media consumption and then compared the probability of retweeting extreme users with controlled simulations. The findings suggest that liberals' social ties may be driven by both homophily and acrophily, while conservatives show extreme acrophilic preferences. Authors hint that research should move toward exploring acrophily on other platforms as well, a suggestion we address in the present work.

Homophily. Homophily is the principle that *similarity breeds connection* [8], and describes the tendency to associate with peers sharing similar tastes, beliefs and/or demographics. This phenomenon was observed in a variety of online and offline settings [1,10,18]. In the online domain, it was linked to the rise of (especially political) polarization [2,17,20]. Several methods exist to quantify homophilic tendencies in social networks. Newman's assortativity coefficient is the most popular measure, and quantifies attribute correlation between adjacent node pairs [10]. More recent work [13,15] has moved to study local patterns of homophily, to address the fact that, in large networks, homophilic and heterophilic wiring patterns may coexist. Attribute-driven wiring patterns were studied among higher-order interactions as well, by defining measures over simplicial complexes and hypergraphs [16,22].

3 Methodology

In this section, we introduce our analytical pipeline for estimating the preference for extreme opinions in online debates. Of the two works on acrophily to date, only one studies this phenomenon on social media data [6]. While relying on a sound methodology, and yielding interesting results, acrophily is studied on retweet networks, a choice that makes it impossible to generalize their strategy to other social platforms like Reddit. Instead, we tackle this issue by leveraging information on social interactions and semantics extracted from user-generated content—elements shared by any social platform. Our framework is composed of the following four phases: (i) Opinion estimation, (ii) Opinion thresholding, (iii) Network construction, and (iv) Acrophily estimation. These steps are described in detail below.

Opinion Estimation. This step deals with the task of identifying user ideology with respect to a controversial issue. Specifically, the aim is to understand (i) which of the two stances the user agrees with, and (ii) how strong their opinion is. Estimating user opinions is a challenging task because most users do not necessarily associate themselves with an explicit label, e.g., political ideology. Moreover, estimating the *intensity* of such opinion is paramount difficult. We model the problem of jointly predicting stance and opinion intensity as a text classification task. Informally, a classifier learns to discern between two stances (e.g., democrats vs. conservatives) based on annotated examples of each ideology (e.g., posts by prominent democrats/conservatives). The assumption here is that an individual's opinion is conveyed by the posts he/she produces within the debate, and that like-minded individuals share similar language to an extent.

As a proxy of opinion intensity, we use the model's class probability, a value in [0,1] identifying the model's confidence level on a particular prediction. Here, values closer to 1 reflect a stronger belief in one stance, and values closer to 0 indicate a stronger belief in the other. The intuition behind this choice is that strong opinions emerge more clearly from text, and thus are assigned scores closer to 0 and 1 than moderate opinions. After obtaining scores for each post, we compute the average score over a predefined time period (e.g., one month, semester, year, etc.) and assign the results to each user. Lastly, we rescale and center these values on zero by applying the following transformation:

$$y = 2x - 1, \tag{1}$$

where x is the average model score, and y is the resulting opinion value. After applying this function, the opinion sign is indicative of the stance (pro/against), while the (absolute) value is indicative of the opinion strength. For instance, in an abortion debate, stances may hold values in $[-1, 1]$ where 1 represents a strong "pro-abortion rights" ideology, -1 represents a strong "pro-life/anti-abortion" ideology, and 0 implies neutrality.

Opinion Thresholding. After estimating user ideology, a key step of this pipeline involves distinguishing between extreme vs. moderate opinion holders. This typically involves setting a threshold, so that opinions falling beyond it are considered strong, and otherwise are considered moderate. The choice of an appropriate threshold strictly depends on the data, and specifically on the platform and discussion topic. We suggest choosing thresholds depending on the opinion distribution and the topic of interest. Namely, on a controversial issue, one would expect a considerable fraction of debaters to have extreme opinions. However, other analysts might choose other approaches, such as setting symmetric thresholds (e.g., 0.7 and -0.7), or relying on external data such as survey data.

Network Construction. In order to study connections to users with extreme opinions, we must first build a model representing their social ties. Social topology on Online Social Networks typically comes in multiple layers [3]. Indeed, on most platforms, users can follow/befriend each other, and comment, share and like other people's content. Since we are interested in the dynamics of online debates, we will focus on networks emerging from conversations. We model an online debate as a directed graph $\mathcal{G} = (V, E)$, where V is the set of nodes representing social media users, and E is the set of edges that encode conversations between such users (e.g. comment/reply interactions). Each node is further enriched with information on its stance toward a specific topic on a binary spectrum (the result of the first step), and on whether this stance is extreme or not (the result of the second step).

Acrophily Estimation. This phase involves computing measures on the structure and semantics obtained from previous steps. To measure attraction to extreme opinions, we rely on the concept of acrophily [5,6] and formalize it as follows.

Definition 1. (Acrophily Index.) *Let $G = (V, E)$ be a node-attributed graph where V is the set of nodes, and E is the set of edges. Let each node $v \in V$ be enriched by a tuple (p_v, q_v) where p_v is a binary value describing user ideology among two opposing factions, and q_v is a binary value indicating whether user ideology is extreme (1) or moderate (0).*

Let S_v identify the set of successors of the node v, and $S_v^{same} \subset S_v$ its subset of nodes adhering to the same faction, namely, all successors s for which $p_v == p_s$ is true. Let M_v^{same} be the subset of S_v^{same} containing nodes with extreme ideologies, i.e., all successors $s \in S_v^{same}$ for which $q_s == 1$ is true. The Acrophily Index for node v is computed as:

$$Acrophily(v) = \frac{|M_v^{same}|}{|S_v^{same}|}. \qquad (2)$$

This function returns a value in $[0, 1]$, where 1 indicates strong acrophilic behavior (preference for like-minded peers with more extreme ideology), and 0 indicates the absence of such behavior. The measure is undefined for $|S_v^{same}| = 0$, and thus is computed only for nodes having at least one like-minded successor. The Acrophily Index above can be generalized to the whole graph by averaging over the number of non-null values.

4 Data Collection

We apply our framework to Reddit debates on climate change. In this section, we detail the data collection and preparation processes.

In this phase, our aim is to find text content that accurately represents each of the two opposing stances in the climate change debate, namely climate skeptics vs. those who believe climate change is genuine. To do so, we identify a collection of subreddits where discussions on this issue take place, and retrieve all their submissions[1] and comments missed from 2019 to 2022 via the Pushshift API[2]. Since we want to capture the controversial nature of the debate, we make sure to include subreddits that clearly relate to each of the ideological positions (e.g., r/climateskeptics, r/ClimateActionPlan). We rely on subreddit names and descriptions to identify relevant subreddits. Moreover, we sample 50 random submissions from each of the considered subreddits and manually verify adherence to the topic. Since a previous study suggests that many debates on climate change occur in general-purpose settings/subreddits [19], we also include subreddits with a broader scope (e.g., r/TrueAskReddit). In these cases, we filter out submissions that do not contain the word "climate" to ensure adherence to the topic of interest. This process resulted in 40,872 submissions and 661,024 comments. The final dataset contains both textual information—which is used to estimate the user ideology—and relational information—which is used to build the interaction network.

5 Attraction to Extreme Opinions in the Climate Change Debate

In the following, we apply our methodology to Reddit data on climate change debates. Our aim is twofold. First, we want to discover whether users show signs

---

[1] Reddit distinguishes between a "submission", i.e., a post that initiates a discussion, and a "comment", i.e., a post that replies to a submission or to another comment.
[2] https://api.pushshift.io/.

of attraction to extreme opinions, and whether this phenomenon occurs at a higher rate than expected. Then, we investigate whether opinions in climate change debates are polarized and how polarization evolves over time. All data produced during this study is anonymized and released in a dedicated Zenodo repository along with the code to reproduce the experiments (see [14]).

5.1 Opinion Estimation

In order to estimate user opinions on the climate change issue, we rely on BERT, a neural language model based on a bidirectional transformer architecture [21]. We finetune the BERT base uncased model on annotated examples extracted from our dataset. Specifically, we use submissions on stance-specific subreddits (e.g., r/climateskeptics), where users argue on their own view of the topic. We use submissions because they are usually longer pieces of text where opinions emerge more clearly. The filtered dataset contains 19K posts by climate skeptics—to which we assign label '1', and 28K by supporters, to which we assign label '0'. Text is then processed via a standard pipeline, including lowercasing, removing non-printable characters, XSLT tags, URLs, numbers, punctuation, extra spaces, and English stopwords. Then, we remove posts with less than 15 characters to ensure a minimum level of informativity. This leaves 5K skeptics submissions, and 9K supporters submissions. This data is split into train (70%), validation (10%), and test (20%) sets, and the former two are fed to the BERT model for the fine-tuning phase.

We experiment with several configurations of the batch size, number of epochs, and input length. The batch size refers to the number of training examples that are used by the model during a single weights update; the number of epochs is the amount of complete iterations across the entire training dataset throughout the learning process; finally, the input length is the maximum number of tokens allowed, after which the input is truncated. In the best configuration, we set these parameters at 22, 2, and 155, respectively. Performance metrics for the best model performances are shown in Table 1. The model is able to learn the patterns in the training data well, as all metrics are beyond 0.95 in this phase. Validation statistics are slightly lower. This indicates that the model generalizes well but is slightly less accurate on unseen data compared to the training data. The difference between training and validation metrics suggests

Table 1. BERT Performance Metrics

Metric	Training	Validation	Test
Accuracy	0.954	0.888	0.818
Precision	0.957	0.899	0.848
Recall	0.963	0.906	0.883
F1 Score	0.960	0.903	0.865

a small degree of overfitting. The model's performance further decreases when applied to completely unseen data, which is expected. Still, the metrics indicate that the model is suitable to be applied to new contexts as it reaches a respectable F1 score of over 0.85. Figure 1 displays a confusion matrix with the counts of true/false positives/negatives obtained on the test set. The confusion matrix does not suggest significant bias, given the good performance on both classes, and similar magnitudes of misclassification class-wise.

As a last step, we apply the model to unlabeled data from discussions on climate change in general subreddits. Since we want to investigate whether preference for extreme opinions changes over time, we partition the 2022 data by quarters. Moreover, we average the model's class probabilities for each user across each trimester. Finally, each value is rescaled and centered at 0 via Eq. 1 to obtain the final opinion scores. In summary, the output of this step is, for each user discussing climate change in a non-specialized scenario (e.g. general discussion of news or politics), a score that quantifies (i) their stance on climate change in a given quarter of 2022, and (ii) the intensity of this belief.

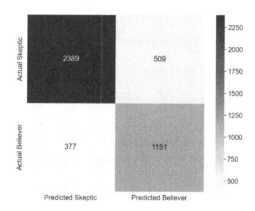

Fig. 1. Confusion Matrix of BERT performance on the test set

5.2 Opinion Thresholding

Figure 2 displays the Kernel Density Estimations of opinion distributions across the four timesteps. Opinions show bimodal distributions with peaks nearing extreme values, which is a typical sign of polarization [7,9]. However, as highlighted in the previous section, selecting appropriate thresholds that separate extreme from moderate users is a challenging task. Ideally, in a polarized scenario, we would expect to find several users with extreme opinions. We study the impact that cutting at different percentiles on both ends has on the volume of extreme users. We study the following percentiles: 5th, 10th, 15th, 20th, 25th for the left threshold; 65th, 70th, 75th, 80th, 85th, 90th, 95th percentiles for the right threshold. We choose to cut (i) at the percentile immediately after the peak on the left and (ii) at the percentile immediately before the peak on the right. The left and right thresholds for each snapshot are represented in Fig. 2 as red dashed lines, and their values are reported in grey.

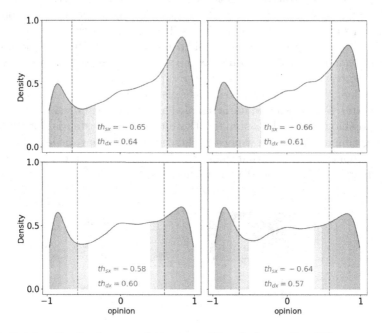

Fig. 2. Opinion distribution at each time step. Blue shades outline different percentiles, with darker ones indicating more restrictive thresholds. Dashed lines outline the thresholds used in this work. (Color figure online)

5.3 Network Construction

To account for the temporal evolution of user ideology and social topology, we build four networks from data, each encompassing three months of online interactions. Nodes are Reddit users, and directed edges refer to replies/comments to submissions or comments in the studied subreddits. Information on who replies to who is available in the Reddit post metadata. In accordance with the previous section, we enrich the nodes with information on (i) their estimated faction at that time step and (ii) whether their opinion is extreme or not. Summary statistics are reported in Table 2.

5.4 Acrophily Estimation

In the following, we compute Acrophily for all networks via Eq. 2. KDE plots are shown in Fig. 3. Distributions of acrophily values are almost identical across all

Table 2. Network statistics for each snapshot. $|V|$ and $|E|$ refer to the number of nodes and edges respectively. k_{avg} refers to the average degree. C_{max} refers to the number of nodes in the largest weakly connected component.

Quarter	$|V|$	$|E|$	k_{avg}	C_{max}	#supporters	#skeptics
Q1	23,253	57,128	4.91	21,069	8753	14,500
Q2	28,426	78,913	5.55	27,373	11,049	17,377
Q3	30,048	87,749	5.84	29,443	13,203	16,845
Q4	19,854	51,964	5.23	19,358	9,358	10,496

observation periods. Most users show an absence of acrophilic behaviors, as evident from the peak on the left. Still, some peaks can be observed at around 0.5 and 1, highlighting a considerable presence of acrophilic users as well. Overall, 25 to 30% of users in each timestamp show strong acrophilic tendencies (> 0.5).

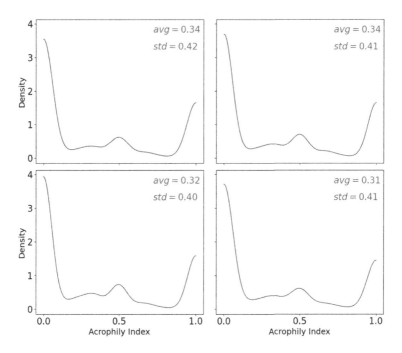

Fig. 3. KDE of acrophily values at each timestep. Mean and standard deviation values are reported in the top-right corner.

As shown in Fig. 4(a), acrophily shows some variation across time, although minor. In general, however, we find that (i) skeptical users show higher acrophily than supporters (with the exception of Q3, where values are comparable), and (ii) extreme users are more acrophilic than moderates. To understand whether this debate exhibits more acrophily than expected at random, we compare average acrophily values with null models. Specifically, we compare each value with those obtained from a directed configuration model [11] (DCM), which shuffles connections while preserving in and out-degree distributions. Each node in the DCM is assigned an opinion value drawn from the empirical opinion distribution of the corresponding time period. This process ensures that connections in the null model are random, while taking into account both the structural properties of the system and the opinion distribution. Figure 4(b) shows the ratios of real over expected acrophily values for different categories. In all timestamps, the climate change debaters show higher-than-expected acrophilic tendencies. This is true both on a global scale, where our data is up to 1.4 times more acrophilic than the DCM, and class-wise. Again, extreme users and skeptical users show the highest ratios with respect to the DCM (up to 1.6 and up to 1.7, respectively), suggesting that climate change debates on Reddit are particularly heated and at risk of polarization.

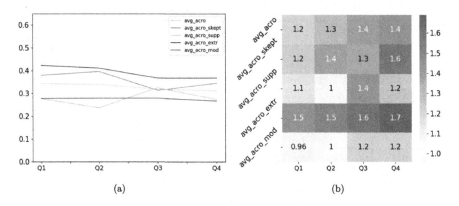

Fig. 4. Temporal trends of acrophily (a) and ratios over the directed configuration model (b)

To assess whether users are embedded in local echo chambers, we compare their opinions with those of the users they actively interact with [20]. Figure 5 displays this comparison across the four snapshots. We find that users are typically surrounded by peers with similar opinions (both in stance and magnitude), and especially so for extreme climate skeptics. Howvever, pluralism emerges as time goes by (as denoted by the brighter areas now aligning horizontally, see [20]), especially in the second half of the year, and opinions become more moderate. Indeed, the Pearson coefficient for user opinion and average successors' opinion lies around 0.4 in the first two quarters, but goes down to around 0.2 afterwards. Note that all coefficients are statistically significant ($pvalue < 0.01$). Interestingly, the highest pluralism is observed in the last quarter, concurrently with the United Nations Climate Change Conference (more commonly, COP27). We hypothesize that relevant events in the second part of 2022—such as COP27—might have induced Reddit users to engage with users with opposing views, with the aim of convincing and/or criticizing.

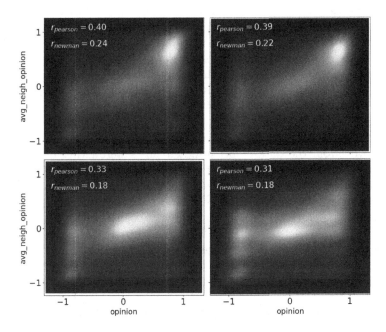

Fig. 5. Contour plot of user opinion (x-axis) against average opinions of users they interact with (y-axis). $r_{pearson}$ refers to the Pearson correlation coefficient between the two variables. r_{newman} refers to the Newman categorical assortativity coefficient for the stance label.

6 Conclusion

With the aim of better understanding polluting online dynamics, we have introduced a descriptive framework to measure attraction to extreme opinions within debates on social platforms. Contrary to previous work on acrophily—which relied on surveys and/or retweet-based Twitter simulations—our framework is scalable and platform-independent. This is due to the fact that it relies strictly on user-generated content and conversation structure, which are present in most—if not all—social networks. We apply our framework to Reddit debates on climate change, and study the evolution of acrophily during 2022. In our analysis, we discover varying degrees of acrophily. Moreover, we find that extreme opinion holders and climate skeptics are the categories that show the highest acrophilic tendencies, and do so at higher-than-expected rates. Users also show a degree of opinion polarization, especially in the first half of the year. Subsequently, polarization leaves space for pluralism, possibly tied to relevant real-world events.

Limitations. Our work is not free of limitations. First of all, due to the lack of ground truth data, there is no way to evaluate results in a sound manner. This is a common issue of data-driven works that estimate user opinions [9], which emphasizes the need for open, topic-specific datasets with this kind of information. Moreover, we use class probabilities of a deep learning model to estimate opinions, and leverage extreme values (i.e., close to 1 or 0) to distinguish between extreme vs. moderate stances. We do this assuming that strong opinions emerge more clearly from the text. Although reasonable, this might not always be true. Still, we believe it is the most systematic way of doing so at scale. Relatedly, to gather enough data for a reasonable user opinion classification we use large window sizes (three months per snapshot), which might overlook within-quarter opinion changes. Moreover, a small percentage of users were removed from the dataset becausetheir posts were either deleted or not sufficiently long/informative. This, of course, has an impact on the observed network topology, although minor. Finally, the validation process is based on a comparison of average acrophily values, which does not take into account the variability in acrophily across different network realisations generated by the null model. We also stress that our findings are strictly relative to the data and should not be generalized to larger populations without further statistical analyses.

Future Research. Future works might attempt to tackle the issues discussed above and/or apply this framework to other controversial debates, such as those related to vaccination, abortion rights, and more. Other platforms should be investigated as well, and especially recent/understudied ones such as the Twitter-like Bluesky [4] and Mastodon [24]. This would allow to understand whether different algorithmic choices and platform affordances may impact observed acrophilic tendencies. Finally, we plan to devise new measures to estimate acrophily that account for chance and class size imbalance, and to further study the combined effects of acrophily with other biases.

Acknowledgments. This work is supported by (i) the European Union—Horizon 2020 Program under the scheme "INFRAIA-01-2018-2019—Integrating Activities for Advanced Communities", Grant Agreement n.871042, "SoBigData++: European Integrated Infrastructure for Social Mining and Big Data Analytics" (http://www.sobigdata.eu); (ii) SoBigData.it which receives funding from the European Union—NextGenerationEU—National Recovery and Resilience Plan (Piano Nazionale di Ripresa e Resilienza, PNRR)—Project: "SoBigData.it—Strengthening the Italian RI for Social Mining and Big Data Analytics"—Prot. IR0000013—Avviso n. 3264 del 28/12/2021; (iii) EU NextGenerationEU programme under the funding schemes PNRR-PE-AI FAIR (Future Artificial Intelligence Research).

Disclosure of Interests. The authors have no competing interests to declare that are relevant to the content of this article.

References

1. Barberá, P., Jost, J.T., Nagler, J., Tucker, J.A., Bonneau, R.: Tweeting from left to right: is online political communication more than an echo chamber? Psychol. Sci. **26**(10), 1531–1542 (2015)
2. Cinelli, M., Morales, G.D.F., Galeazzi, A., Quattrociocchi, W., Starnini, M.: Echo chambers on social media: a comparative analysis. arXiv preprint arXiv:2004.09603 (2020)
3. Dickison, M.E., Magnani, M., Rossi, L.: Multilayer Social Networks. Cambridge University Press (2016)
4. Failla, A., Rossetti, G.: "I'm in the bluesky tonight": insights from a year worth of social data. arXiv preprint arXiv:2404.18984 (2024)
5. Goldenberg, A., et al.: Homophily and acrophily as drivers of political segregation. Nat. Hum. Behav. **7**(2), 219–230 (2023)
6. Goldenberg, A., et al.: Attraction to politically extreme users on social media (2023)
7. Liu, J., Huang, S., Aden, N.M., Johnson, N.F., Song, C.: Emergence of polarization in coevolving networks. Phys. Rev. Lett. **130**(3), 037401 (2023)
8. McPherson, M., Smith-Lovin, L., Cook, J.M.: Birds of a feather: homophily in social networks. Ann. Rev. Sociol. **27**(1), 415–444 (2001)
9. Morini, V., Pollacci, L., Rossetti, G.: Toward a standard approach for echo chamber detection: reddit case study. Appl. Sci. **11**(12), 5390 (2021)
10. Newman, M.E.: Mixing patterns in networks. Phys. Rev. E **67**(2), 026126 (2003)
11. Newman, M.E., Strogatz, S.H., Watts, D.J.: Random graphs with arbitrary degree distributions and their applications. Phys. Rev. E **64**(2), 026118 (2001)
12. Nguyen, C.T.: Echo chambers and epistemic bubbles. Episteme **17**(2), 141–161 (2020)
13. Peel, L., Delvenne, J.C., Lambiotte, R.: Multiscale mixing patterns in networks. Proc. Natl. Acad. Sci. **115**(16), 4057–4062 (2018)
14. Perra, D., Failla, A., Rossetti, G.: Reddit climate change debate dataset (2024). https://doi.org/10.5281/zenodo.13603527
15. Rossetti, G., Citraro, S., Milli, L.: Conformity: a path-aware homophily measure for node-attributed networks. IEEE Intell. Syst. **36**(1), 25–34 (2021)
16. Sarker, A., Northrup, N., Jadbabaie, A.: Generalizing homophily to simplicial complexes. In: International Conference on Complex Networks and Their Applications, pp. 311–323. Springer (2022)
17. Sunstein, C.R.: The Law of Group Polarization. University of Chicago Law School, John M. Olin Law and Economics Working Paper (91) (1999)
18. Traud, A.L., Mucha, P.J., Porter, M.A.: Social structure of Facebook networks. Phys. A **391**(16), 4165–4180 (2012)
19. Treen, K., Williams, H., O'Neill, S., Coan, T.G.: Discussion of climate change on reddit: polarized discourse or deliberative debate? Environ. Commun. **16**(5), 680–698 (2022)
20. Valensise, C.M., Cinelli, M., Quattrociocchi, W.: The drivers of online polarization: fitting models to data. Inf. Sci. **642**, 119152 (2023)
21. Vaswani, A., et al.: Attention is all you need. Adv. Neural Inf. Process. Syst. **30** (2017)
22. Veldt, N., Benson, A.R., Kleinberg, J.: Combinatorial characterizations and impossibilities for higher-order homophily. Sci. Adv. **9**(1), eabq3200 (2023)

23. Williams, H.T., McMurray, J.R., Kurz, T., Lambert, F.H.: Network analysis reveals open forums and echo chambers in social media discussions of climate change. Glob. Environ. Chang. **32**, 126–138 (2015)
24. Zignani, M., Gaito, S., Rossi, G.P.: Follow the "mastodon": structure and evolution of a decentralized online social network. In: Proceedings of the International AAAI Conference on Web and Social Media, vol. 12, pp. 541–550 (2018)

Structure-Attribute Similarity Interplay in Diffusion Dynamics on Social Networks

Salvatore Citraro[(✉)], Valentina Pansanella, and Giulio Rossetti

Institute of Information Science and Technologies, National Research Council,
Pisa, Italy
{Salvatore.Citraro,Valentina.Pansanella,Giulio.Rossetti}@isti.cnr.it

Abstract. Social interactions are shaped by homophily, the tendency for individuals to connect with others who share similar attributes. Exploring this phenomenon is crucial for understanding a wide spectrum of social behaviors, including the spread of misinformation and the dynamics of societal debates. In this study, we leverage a graph transformation strategy—which analyzes the interplay between individuals' personal preferences and their structural connections—to investigate mechanisms of opinion/information diffusion. Among these latter ones, we focus on the Deffuant-Weisbuch model to simulate opinion dynamics and the Independent Cascade model to simulate information spread. Our findings on real-world social networks suggest that emphasizing attribute similarities enhances graph cohesion, whereas forcing structural similarities leads to fragmentation. Moreover, we observe a trend towards consensus opinion formation when enhancing attribute similarities, and faster as well as complete coverage of information spread in the same setup. These results motivate the importance of considering both individual attributes and network structure in studying social dynamics.

Keywords: Homophily · Opinion Dynamics · Information Diffusion

1 Introduction

The Latin phrase *similia similibus solvuntur*[1], used in chemistry to describe how polar solvents dissolve polar solutes and non-polar solvents dissolve non-polar solutes, can indeed be adopted as a metaphor to describe human mixing dynamics as well. Social interactions are governed by an analogous principle of "similarity", known in sociology as *homophily* [21]. This property describes the tendency to bond together individuals with similar opinions, beliefs, and thoughts, but, eventually, segregate them from the rest of the population [34]. Interactions between individuals involve both influence and selection mechanisms [1], leading to correlations of people's preferences across connections. In the language of

---

[1] https://chempedia.info/info/similia_similibus_solvuntur/.

complex systems/networks, the tendencies of one "type" to connect to another "type" are referred to as mixing patterns [23]. Assortative mixing, namely the tendency to connect to the same type, is the quantifiable homophilic principle in the analysis of complex networks. Quantifying assortative patterns in social networks is crucial for understanding many group behaviors, such as the emergence of opinion clusters when people debate [13]. People, in fact, tend to create bubbles of like-minded peers willing to minimize disagreement and conflicts [25]. Accordingly, not all individuals influence others equally. Factors like similarity to peers could shape the extent of influence in opinion exchanges [8] and "word-of-mouth" information spread [18]. Assortative tendencies can reinforce existing beliefs and the spread of information, including fake news and misinformation.

This work explores how variations in people's preferences and connections can affect various social dynamics. To achieve this, we use a graph transformation strategy to assign weights to the strength of people's interactions in social networks. Weights help us study how the changes in opinions and the spread of information vary with connection strengths, by embedding such strengths into these dynamics. We claim that people in social networks assign different importance to their neighbors' opinions based on structural connections and homophilic tendencies [20,21], and we aim to quantify their varying degrees of importance. Our analysis of real-world datasets shows that structural and node-attribute similarities affect diffusion processes differently. For instance, information diffusion is faster when edges are weighted giving more importance to attribute similarities; in similar setups, this importance lead to the formation of one dominant opinion clusters, while mixed structure-attribute importance leads to polarized scenarios; moreover, structural importance only leads to fragmentation. All such behaviors can be related to the topology of the resulting directed, weighted graphs, where more disconnected components and fewer strong edges emerge when structural similarities are emphasized over attributive ones.

The rest of the paper is organized as follows. Section 2 reviews essential literature on graph transformations and diffusion dynamics in social networks. Section 3 describes our framework for graph transformation and its application to diffusion dynamics. Section 4 introduces an experimental setup testing our methodology on real-world social network data. Section 5 concludes with a discussion of future research directions and current limitations.

2 Related Work

In this section we review the significant literature related to the current work. Particularly, we discuss the relevance of using a graph transformation method within the context of node-attributed graph mining for community detection tasks. Then, we sum up the main literature on information spreading models and opinion dynamics in complex networks.

2.1 Structure-Attribute Fusion in Community Detection

The graph transformation strategy we adopted in this work has been widely used for community detection methods that leverage both links and node attributes/features to identify graph partitions [17]. The task, commonly referred to as node-attributed community detection [3], aims to find graph clusters by means of both structural and semantic similarities between nodes. One approach is to convert the original, unweighted graph into a weighted one, where the weights represent the similarity between node feature vectors. This allows a community detection algorithm that handles edge weights to work on the improved structure [37]. A more detailed strategy involves introducing a linear combination of structural and attribute similarities, such that weights reflect the strength of edges based on a fusion of structure and features [6], and, by adjusting the parameter of the linear combination, one can prioritize either structural or attributive information exclusively. The most comprehensive generalization of this weight-based approach is presented in [4], where the authors provide an analytical study of the mechanism and introduce an automatic parameter tuning scheme to achieve the desired balance between structure-attribute impact on the specific task of community detection. Their strategy involves first converting the attributes into an attributive graph, resulting in two graphs-structural and attributive. These graphs are then fused into a composite one encapsulating both structural and attributive information. The strategy we will explore in the current work involves using weights combining features and structure, similar to the approaches discussed so far. For the sake of clarity, there exist other strategies for fusing structure and attributes in the literature about node-attributed community detection. These methods include fusing structural components into the feature space then running a clustering method on the improved feature space [30], or using probabilistic approaches [24], or employing ensemble clustering techniques [10].

2.2 Information Diffusion Models

Numerous studies have aimed to model the propagation of ideas, influence, and information within social networks [14,18], similar to the modeling of computer viruses and human diseases [29]. In information diffusion models, nodes can decide to do or not do something, e.g., to choose or not choose to buy a product. In the Threshold model, for instance, a node's decision depends on the proportion of its neighbors who have already made the same choice: if this proportion exceeds a predefined threshold, the node adopts the behavior as well [14]. An important reference study for the current work is the Independent Cascade Model [18]. In this model, a bunch of initial active nodes tries to influence their neighbors with a certain probability. If successful, the neighbors become active in the next round; otherwise, no further attempts are made to convince an already targeted node. The success probability, thus, is crucial as it determines how influence spreads across the network. Various studies have investigated how this parameter influences diffusion dynamics [2], and studied methods to predict it [33]. Moreover, several extensions of the Independent Cascade model has been proposed in the literature, e.g., for exploring the role of community structure as either a barrier or an accelerator [22]. Among the several applications implying information diffusion models, the influence maximization problem is the most challenging one, which can be described as the task of identifying the optimal set of nodes to maximize influence spread in a network [2].

2.3 Opinion Dynamics Models

Opinion dynamics models typically study how a population of agents, each with their own opinions, interact and influence each other. The population is typically represented as a (directed/undirected) graph, where connections signify *potential* influence. Each agent holds an opinion which can be binary, discrete, or continuous. The dynamics of these models are governed by two rules: selecting interaction partners and updating opinions after interaction. Interactions can be synchronous (all agents update simultaneously) or asynchronous (sequential updates). The dynamics end when the population reaches equilibrium (consensus, polarization, or fragmentation) or meets a stopping condition. There exist plenty of works which aim to study opinion formation and evolution through social interactions [8,9,15]. In bounded confidence models, like the Deffuant-Weisbuch one [8], agents only influence others when their opinion distance is within a confidence threshold ϵ [8], resembling sociological concepts like homophily [21]. This parameter can be either homogeneous or heterogeneous among the population [19,28]. Moreover, in many opinion dynamics models, bounded confidence ones included, agents put a weight on their peers to identify how much they trust their opinions or, in other words, to say how much they influence them in forming/changing theirs. For instance, in the Deffuant-Weisbuch model [8], this is represented as an influence parameter, or convergence rate, μ, which can either be homogeneous across the whole population [8] or heterogeneous [16]. This latter scenario, however, is understudied in the opinion dynamics literature. Importantly, most opinion dynamics models are studied on undirected networks, assuming symmetry of relationships [35]. However, there are examples of directed networks employed [12] to resemble, for example, online social networks characteristics [7].

3 Methods

In this section we formalize our strategy for transforming a node-attributed graph into a directed, weighted one using asymmetric similarities for structure and attributes. We then illustrate how this transformation could be relevant in the contexts of 1) opinion dynamics, by modifying the parameter of influence of the Deffuant-Weisbuch model [8], and 2) information diffusion, by embedding the weights in the mechanisms of the Independent Cascade model [18].

3.1 Graph Transformation

We represent social networks as directed attributed graphs $\mathcal{G} = (\mathcal{V}, \mathcal{E}, \mathcal{A})$, where a node $i \in \mathcal{V}$ represents individuals, and an ordered pair $(i, j) \in \mathcal{E}$ represents the interaction between individual i and individual j. We also assume the presence of m binary node attributes, i.e., $\mathcal{A} \subseteq \{0, 1\}^{|\mathcal{V}| \times m}$. To include strengths, we build the asymmetric weighted adjacency matrix $\mathcal{W}_\alpha$ of $\mathcal{G}$ using the following function:

$$w_\alpha(i, j) = (1 - \alpha) \cdot w_\mathcal{E}(i, j) + \alpha \cdot w_\mathcal{A}(i, j), \tag{1}$$

where $w_\mathcal{E}(i, j)$ indicates a structural similarity measure, and $w_\mathcal{A}(i, j)$ an attribute vector similarity between nodes i and j. The parameter α controls the importance of the two similarities.

Accordingly, $\mathcal{W}$ is defined as follows:

$$\mathcal{W} = \begin{cases} w_\alpha(i, j) & \text{if } (i, j) \in \mathcal{E} \\ 0 \end{cases}. \tag{2}$$

Then, we use the following function to reduce edge reciprocities such that

$$\phi(\mathcal{W}) = \begin{cases} w_\alpha(i, j) & \text{if } w_\alpha(i, j) \geq w_\alpha(j, i) \\ 0 \end{cases}, \tag{3}$$

The $\phi(\mathcal{W})$ function allows us to consider only the strongest edge between two endpoints.

Our graph transformation strategy allows to define $\mathcal{G}_\alpha = (\mathcal{V}, \phi(\mathcal{E})) = (\mathcal{V}, \mathcal{E}_\alpha)$ as the graph where $\mathcal{V}$ is the set of $\mathcal{G}$'s original nodes, and $\mathcal{E}_\alpha$ is the set of weighted edges such that $(i, j) \in \mathcal{E}_\alpha$ with $u, v \in \mathcal{V}$ is a ordered pair of nodes. Note that, from Eq. 3, $\mathcal{G}_\alpha$ can be a pure bidirectional graph as $\mathcal{G}$ only when $w_\alpha(i, j) = w_\alpha(j, i)$ for all the edges in the graph, otherwise only the ordered pair with the highest weight between (i, j) and (j, i) is kept.

We use the following definition of Jaccard similarity for measuring the structural weight $w_\mathcal{E}(i, j) \in [0, 1]$:

$$w_\mathcal{E}(i, j) = \frac{|\Gamma(i) \cap \Gamma(j)|}{|\Gamma(i)| - 1}, \tag{4}$$

where $\Gamma(i)$ indicates the set of i's adjacent nodes, or i's open neighborhood, and $|\Gamma(i)|$, the cardinality of the set. Note that in a social network this quantity measures how many common friends i and j have compared to all i's friends.

We use the cosine similarity for measuring the attribute vector strength $w_\mathcal{A}(i, j) \in [0, 1]$ as follows:

$$w_\mathcal{A}(i, j) = \frac{1}{|\Gamma(i) \cap \Gamma[j]|} \sum_{k \in \Gamma(i) \cap \Gamma[j]} \frac{A_i \cdot A_k}{\|A_i\| \cdot \|A_k\|}, \tag{5}$$

where $\Gamma[j]$ indicates the set of j's adjacent nodes plus j itself, or j's closed neighborhood, namely we measure $w_\mathcal{A}(i, j)$ as the average cosine similarity between i's attribute vector and all k's attribute vectors belonging to the set of common neighbors of i and j, including j's attribute vector.

3.2 Weights in the Deffuant-Weisbuch Model

The Deffuant-Weisbuch model [8] is an opinion dynamics model where a population of N agents/nodes interacts in a pairwise mode for a number of time steps T, until convergence is reached or a stopping condition is met. Each agent/node i holds an opinion $x_i(t) \in [0, 1]$; at $t = 0$, opinions generally are uniformly distributed among nodes, although several different initial conditions could be investigated. At each time step $t \in T$, a pair of nodes (i, j) is randomly selected and the opinions are updated as follows:

$$x_i(t+1) = \begin{cases} x_i(t) + \mu_{ij}(x_j(t) - x_i(t)) & \text{if } ||x_i(t) - x_j(t)|| < \epsilon \\ x_i(t). \end{cases} \quad (6)$$

Here, parameter ϵ, homogeneous across the population, indicates the bounded confidence threshold, and $\mu_{ij} \in [0, 0.5]$ represents the influence that node j has on i. According to the Eq. 6, the node i changes its opinion after the interaction with the node j if and only if their opinion distance is below the bounded confidence threshold ϵ. Moreover, μ_{ij} controls the degree of influence if node distances fall within ϵ. Thus, when $\mu_{ij} = 0$, node i trusts so little node j that the former one will not change its opinion, even if their distance falls within the confidence bound. When $\mu_{ij} = 0.5$, node i averages its opinion with agent j's; if the parameter is symmetric, it means instantaneous agreement between the two nodes, both taking their average opinion.

In the current work, we aim to use the edge weights of the graph transformation to tune the parameter μ_{ij} so that $w_\alpha(i, j)$ represents the weight that node i poses on the opinion of node j when updating its own opinion. Hence, we make the parameter asymmetric, leading us to model a more complex agreement dynamics between two nodes. Moreover, it is normally not considered the case of influence ratios above 0.5, meaning that both nodes value more the other's opinion than their own and, at the extreme case of $\mu_{ij} = 1$, node i takes node j's opinion and vice versa. To avoid this "swapping" behavior, which we do not consider reasonable, all weights (only in the context of opinion dynamics) $w_\alpha(i, j)$ have been scaled to the interval $[0, 0.5]$.

3.3 Weights in the Independent Cascade Model

The Independent Cascade Model [18] is used to simulate the spread of influence or information through a network. The model starts with an initial set of active nodes, and each node has a chance to influence its neighbors after being activated. An active node i at time t has a single chance to activate each currently inactive neighbor j with a probability $p(i, j)$. If i succeeds with its neighbor j, the latter one will be active at time $t+1$; if i does not succeed with j, it can not make any further attempts to activate the latter one in next rounds.

The probability $p(i,j)$ is thus a fundamental parameter of the model guiding the dynamics of the diffusion. Here, we aim to use $w_\alpha(i,j)$ as the probability that node i has to activate node j. Note that we exploit a slightly different logic compared to the previous use of weights for node influence in opinion dynamics. Whereas in the directed and weighted version of the Deffuant-Weisbuch model node i updates its opinion with respect to its similarity with j (so, high the similarity, high the ability of j to influence i), in this version of the Independent Cascade, instead, we use similarities from i to j to activate the latter one. This way, we always expect that "clusters" of similar nodes (same neighbors, same attributes) influence each others with high probabilities, but we also enforce a dynamics where hubs are influenced only by those neighbors sharing high similarities with them; moreover, hubs have less power to spread information uniformly across all the neighbors.

4 Results

The experiments of this section are organized as follows: we first outline some properties of the graphs resulting from the transformation, testing it on a collection of well-known node-attributed social networks; next, we select one of these networks (specifically one that reflects the average properties of the collection), and use it to apply the weighted extensions of the chosen opinion/information diffusion models.

4.1 Data

Facebook100 [38] is a collection of 100 Facebook friendships networks among the first 100 U.S. colleges admitted to Facebook, created during the early history of the social network. Nodes are labeled with several attributes. For our experiments we use the following ones: gender {*male, female*}, college year, dormitory, and status {*undergraduate, graduate student, summer student, faculty, staff, alumni*}. We encode these attributes as one-hot vectors for measuring the attribute node similarities as described in Eq. 5. Since several missing values are present for situations in which individuals do not volunteer a characteristic [38], nodes with any missing attribute values were excluded from the analysis.

We use all 100 networks to study the graph transformation strategy properties. For the two diffusion models, we focus on the *Haverford* college network ($N = 1446$, $E = 59589$ before processing; $N = 980$, $E = 33524$ after removing nodes with missing values) as a case study.

4.2 Transformation Properties

First, we are interested in studying the properties of the resulting transformation of an undirected, node-attributed graph into a directed, weighted graph. From Eq. 3, we know we keep only the strongest weight between two reciprocal edges to determine edge directions. As a consequence, the graph connectivity could be

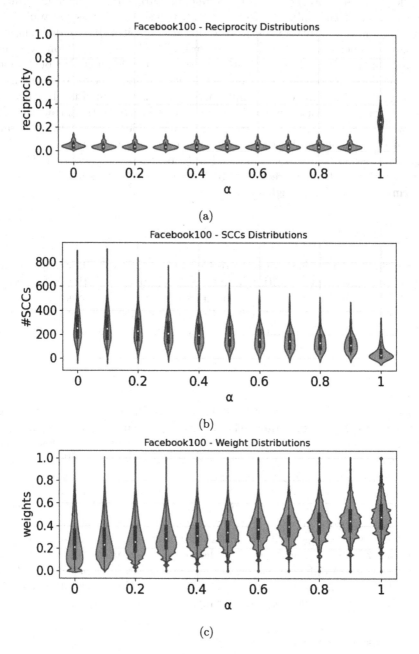

Fig. 1. Distributions of (a) reciprocity, (b) the number of Strongly Connected Components (SCCs), (c) edge weights across all 100 *Facebook* graphs as a function of α.

consistently affected by this strategy choice. Figure 1(a) shows the distribution of reciprocity across all *Facebook100* graphs as a function of the α parameter, where reciprocity is formally described as follows: $reciprocity = \frac{|\{(u,v) \in G | (v,u) \in G\}|}{|\{(u,v) \in G\}|}$. This measure indicates that it is rare for two connected nodes to have equal strengths, especially when structural similarities are considered. When the structure is "switched off" ($\alpha = 1$), reciprocity increases but never reaches total reciprocity.

Figure 1(b) shows the distribution of the number of Strongly Connected Components (SCCs) across all *Facebook100* graphs as a function of the α parameter. SCCs are subgraphs in which every node is reachable from every other node within the same subgraph, reflecting a high degree of mutual connectivity [5]. Here, the violin plots in the figure reveal some differences for low and high values of the parameter. A significant strength directed towards structural similarities (low values of α) can fragment the graphs and potentially restrict diffusion dynamics: SCCs distributions reveal long tails and the presence of many components across the networks. When forcing the impact of attribute similarities (high values of α), instead, the number of SCCs and variances in distributions are lower. This means that the graphs tend to form larger cohesive components, enhancing overall connectivity and reducing fragmentation.

To these findings are related the observations in Fig. 1(c) about edge weight distributions. Here, we observe again long tails for low values of α: structural similarities decrease the strength of connections for the majority of node interactions, indicating a few number of very strong connections. For high values of α, instead, connections tend to have, on average, stronger strengths, and the expected impact of a diffusion should be that information/influence propagates more efficiently throughout the overall graph.

4.3 Analysis on the Deffuant-Weisbuch Model

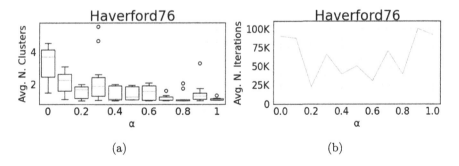

Fig. 2. Distribution of the number of opinion clusters at convergence (a) as a function of α for $\epsilon = 0.35$ and (b) average convergence time; results refer to 10 runs with 10^5 maximum iterations.

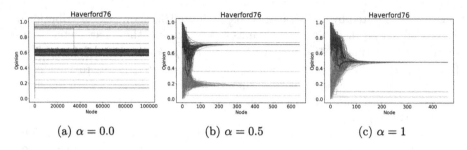

Fig. 3. Example of opinion evolution as a function of t for $\epsilon = 0.35$ and $\alpha = 0.0$ (a), $\alpha = 0.5$ (b) and $\alpha = 1.0$ (c); results refer to a single run with 10^5 maximum iterations.

The experimental setup for this analysis is as follows: we simulate the heterogeneous influence Deffuant-Weisbuch model on the α-varying transformed graphs from the Haverford college, varying the values of α in the range $[0, 1]$ with bins of 0.1. Since in the original graph the opinion values are absent, we set initial opinions to all nodes. Such opinions are uniformly distributed so that each agent i holds an opinion $x_i \in [0, 1]$. The results obtained from the model are averaged over 10 runs. We impose the simulations to stop when the population reaches an equilibrium, i.e., the cluster configuration will not change anymore, even if the agents keep exchanging opinions. Hence, we set an overall maximum number of iterations at 10^5 to account for situations where an equilibrium may never be reached. These choices were done to balance the need for statistical validity of the results and computational feasibility of the simulations. We chose to study $\epsilon = 0.35$ based on findings in the literature. While the universal threshold for consensus is theoretically $\epsilon = 0.5$ across various network topologies [11], recent studies [26–28,36] suggest that setting $\epsilon \approx 0.3$ typically guides populations towards consensus in the baseline model, i.e., when $mu_{i,j} = 0.5$ for all interactions. For our present work, we opt for $\epsilon = 0.35$ because it yields results qualitatively similar to $\epsilon = 0.3$, but with reduced sensitivity to parameter variations. Finally, μ_{ij} values vary as $w_\alpha(i, j)$ vary. Experiments are run using the NDlib package [32].

We are interested in whether the different values of α influence the final configuration, enhancing or reducing polarization. To explore the structure-attribute similarity interplay, we measure the number of clusters in the final opinion distribution. Figure 2(a) shows that, on average, the number of clusters decreases as α increases. We can relate this effect to the graph fragmentation (Fig. 1(b)) and to the positively skewed weights distribution, with a median μ_{ij} of 0.1 (Fig. 1(c)), for low values of α. This prevents social influence to lead the population steadily towards convergence around a single (or few) opinion value(s). What we observe, and we can see it more clearly in the example in Fig. 3(a) is an unstable dynamics for some node clusters and a fragmented final state. For high values of α, the number of clusters tends to decrease, converging towards one. The distribution of weights tends to be normally distributed with mean 0.5 as α tends to 1 (Fig. 1(c)): a higher percentage of nodes averages their opinions with those of their neighbors, bringing our setting closer to the baseline studied in the original model [8], and thus towards consensus. We can see from the examples in Fig. 3(b) and 3(c) that—despite maintaining an unstable dynamics—nodes tend to concentrate around increasingly fewer opinion values. Surprisingly, despite enhancing consensus, a higher α does not equally facilitate convergence. The presence of an heterogeneous distribution of μ_{ij} slows down the dynamic by making it unstable, as also emerges from Fig. 3. If we look at the distributions in Fig. 2 (b) we can see that the median value is mostly around the maximum iteration value (10^5), and in just few cases (α from 0.4 to 0.6 and 0.8) is around a few thousands iterations.

4.4 Analysis on the Independent Cascade Model

The experimental setup for this analysis is as follows: we simulate the weight-heterogeneous Independent Cascade model on the α-varying transformed graphs from the Haverford college, varying the values of α in the range $[0, 1]$ with bins of 0.1. Results are based on 5000 runs of the model, selecting, in each run, one random active starting node; 15 iterations per runs. The probabilities of activation $p(u, v)$ vary as $w_\alpha(i, j)$ vary. Experiments are run using the NDlib python package [32].

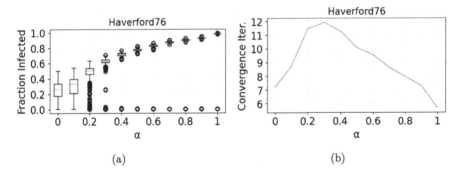

Fig. 4. (a) Fraction of infected nodes at the last iteration as a function of α; (b) Time to diffusion convergence (i.e., no more infected nodes at $t+1$) as a function of α; results refer to 5000 runs and 15 iterations.

To explore the structure-attribute similarity interplay, we measure the fraction of infected nodes, as this straightforwardly indicates the spread of influence/contagion within the networks. Figure 4 (a) shows that, on average, this number increases as α increases. We relate this observation to the distributions discussed in Fig. 1. A high number of disconnected components prevents the infection from spreading to the entire graph. This is a structural effect caused by weighing edges based solely on neighborhood similarities. Instead, including (even a low strength of) attribute similarities allows information to spread

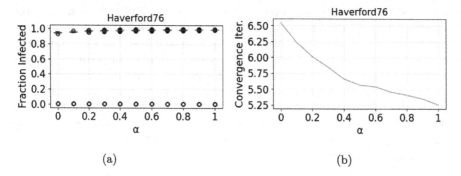

Fig. 5. (a) Fraction of infected nodes at the last iteration as a function of α; (b) Time to diffusion convergence (i.e., no more infected nodes at $t+1$) as a function of α; results refer to 5000 runs and 15 iterations using all the reciprocated edges in the transformed graph, i.e., no edge directions and both $w_\alpha(i,j)$ and $w_\alpha(j,i)$.

across almost the entire graph. This effect increases as α increases. Moreover, as suggested by Fig. 4 (b), the time to convergence decreases while increasing α. Convergence, here, is defined as the point at which no new nodes are infected at time $t+1$, i.e., the end of the diffusion. Hence, in addition to infect larger fractions of nodes, information spread is also faster, except for $0.2 \leq \alpha \leq 0.3$. At that range, the balance between structure-attribute similarities could create temporary bottlenecks or more isolated clusters that slow down the diffusion process before it resumes spreading efficiently at higher values of α. However, this effect disappears when using all the reciprocated edges, i.e., not applying Eq. 3 for imposing edge directionality, as illustrated in Fig. 5(b); in this regime, moreover, all nodes are infected for all values of α, Fig. 5 (a). Here, the constraints are relaxed, allowing for a more uniform and efficient, although less realistic, diffusion across the entire graph.

5 Discussion and Conclusion

Merging structural and node-attributive components of a network is a widely used approach in feature-rich community detection [3,17]. Among the techniques used in the latter task, the weight-based method, consisting in fusing the attributes into the structure, has been extensively adopted and analyzed [4]. In this work, we applied such a weight-based transformation beyond its original purpose/task of clustering, exploiting it in the contexts of opinion dynamics and information diffusion. These contexts are often used to simulate the spread of viral contents and misinformation through social networks, as well as the formation of opinions in societal debates. We acknowledge that these (eventually polluted) phenomena can be often facilitated by homophilic tendencies, that we indeed embed as α-varying edge weights. By studying such variations in weight-based extensions of the Deffuant-Weisbuch model [8] (for opinion dynamics) and the Independent Cascade model [18] (for information diffusion), we could gain insights into fostering more balanced and informed discussions within social networks. Taking into account this point of view, our results could suggest that mis/information diffusion and opinion formations in debates should be strongly influenced by the intrinsic attributes of nodes more than the structural component only, i.e., sharing common friends. Mixed importance can lead to polarization and to faster spread of information; gradually "switching off" structural similarities leads to reach consensus. Hence, a combination of both similarities can simulate more realistic dynamics of debates and diffusion of ideas.

Among the limitations of the current work, we acknowledge that we provide examples for only one college when studying diffusion processes, compared to the analysis of the whole collection when studying the properties of the graphs. Other colleges could manifest different dynamics [38]. Moreover, we used a feature space composed by gender, year, dormitory and status, but we did not analyze the specific impact of each attribute on the outcome of attribute similarities. For future research, we aim to address these limitations. Furthermore, we will explore additional factors that may influence opinion dynamics and information diffusion, e.g., heterogeneous values of the confidence bound ϵ [28], or node-based thresholds other than edge-based ones; an idea could be leveraging measures for assortative mixing estimation [31]. Regarding the graph transformations, we acknowledge the presence of many alternative methods for measuring the similarity with respect to the different components considering symmetry [3]. In the future, we aim to consider alternative strategies for asymmetric similarities as well. Our goal is to encourage researchers to focus on the complex interplay between individual attributes and network structure in shaping social dynamics.

Acknowledgments. This project was funded by SoBigData.it which receives funding from the European Union—NextGenerationEU—National Recovery and Resilience Plan (Piano Nazionale di Ripresa e Resilienza, PNRR)—Project: "SoBigData.it—Strengthening the Italian RI for Social Mining and Big Data Analytics"—Prot. IR0000013—Avviso n. 3264 del 28/12/2021.

Disclosure of Interests. The authors have no competing interests to declare that are relevant to the content of this article.

References

1. Aral, S., Muchnik, L., Sundararajan, A.: Distinguishing influence-based contagion from homophily-driven diffusion in dynamic networks. Proc. Natl. Acad. Sci. **106**(51), 21544–21549 (2009)
2. Chen, W., Wang, Y., Yang, S.: Efficient influence maximization in social networks. In: Proceedings of the 15th ACM SIGKDD International Conference on Knowledge Discovery and Data Mining, pp. 199–208 (2009)
3. Chunaev, P.: Community detection in node-attributed social networks: a survey. Comput. Sci. Rev. **37**, 100286 (2020)

4. Chunaev, P., Gradov, T., Bochenina, K.: The machinery of the weight-based fusion model for community detection in node-attributed social networks. Soc. Netw. Anal. Min. **11**, 1–20 (2021)
5. Cormen, T.H., Leiserson, C.E., Rivest, R.L., Stein, C.: Introduction to Algorithms. MIT Press (2022)
6. Dang, T., Viennet, E.: Community detection based on structural and attribute similarities. In: International Conference on Digital Society (ICDS), vol. 659, pp. 7–12 (2012)
7. Das, A., Gollapudi, S., Munagala, K.: Modeling opinion dynamics in social networks. In: Proceedings of the 7th ACM International Conference on Web Search and Data Mining, pp. 403–412 (2014)
8. Deffuant, G., Neau, D., Amblard, F., Weisbuch, G.: Mixing beliefs among interacting agents. Adv. Complex Syst. **3**(01–04), 87–98 (2000)
9. DeGroot, M.H.: Reaching a consensus. J. Am. Stat. Assoc. **69**(345), 118–121 (1974)
10. Elhadi, H., Agam, G.: Structure and attributes community detection: comparative analysis of composite, ensemble and selection methods. In: Proceedings of the 7th Workshop on Social Network Mining and Analysis, pp. 1–7 (2013)
11. Fortunato, S.:Universality of the threshold for complete consensus for the opinion dynamics of deffuant et al. Int. J. Mod. Phys. C **15**(09), 1301–1307 (2004)
12. Gandica, Y., del Castillo-Mussot, M., Vázquez, G.J., Rojas, S.: Continuous opinion model in small-world directed networks. Phys. A **389**(24), 5864–5870 (2010)
13. Gargiulo, F., Gandica, Y.: The role of homophily in the emergence of opinion controversies. arXiv preprint arXiv:1612.05483 (2016)
14. Granovetter, M.: Threshold models of collective behavior. Am. J. Sociol. **83**(6), 1420–1443 (1978)
15. Holley, R.A., Liggett, T.M.: Ergodic theorems for weakly interacting infinite systems and the voter model. In: The Annals of Probability, pp. 643–663 (1975)
16. Huang, C., Dai, Q., Han, W., Feng, Y., Cheng, H., Li, H.: Effects of heterogeneous convergence rate on consensus in opinion dynamics. Phys. A **499**, 428–435 (2018)
17. Interdonato, R., Atzmueller, M., Gaito, S., Kanawati, R., Largeron, C., Sala, A.: Feature-rich networks: going beyond complex network topologies. Appl. Netw. Sci. **4**(1), 1–13 (2019)
18. Kempe, D., Kleinberg, J., Tardos, É.: Maximizing the spread of influence through a social network. In: Proceedings of the Ninth ACM SIGKDD International Conference on Knowledge Discovery and Data Mining, pp. 137–146 (2003)
19. Li, G.J., Luo, J., Porter, M.A.: Bounded-confidence models of opinion dynamics with adaptive confidence bounds. arXiv preprint arXiv:2303.07563 (2023)
20. Liben-Nowell, D., Kleinberg, J.: The link prediction problem for social networks. In: Proceedings of the Twelfth International Conference on Information and Knowledge Management, pp. 556–559 (2003)
21. McPherson, M., Smith-Lovin, L., Cook, J.M.: Birds of a feather: homophily in social networks. Ann. Rev. Sociol. **27**(1), 415–444 (2001)
22. Milli, L., Rossetti, G.: Community-aware content diffusion: embeddedness and permeability. In: Complex Networks and Their Applications VIII: Volume 1 Proceedings of the Eighth International Conference on Complex Networks and Their Applications COMPLEX NETWORKS 2019 8, pp. 362–371. Springer (2020)
23. Newman, M.E.: Mixing patterns in networks. Phys. Rev. E **67**(2), 026126 (2003)
24. Newman, M.E., Clauset, A.: Structure and inference in annotated networks. Nat. Commun. **7**(1), 11863 (2016)
25. Nguyen, C.T.: Echo chambers and epistemic bubbles. Episteme **17**(2), 141–161 (2020)

26. Pansanella, V., Rossetti, G., Milli, L.: From mean-field to complex topologies: network effects on the algorithmic bias model. In: Complex Networks & Their Applications X: Volume 2, Proceedings of the Tenth International Conference on Complex Networks and Their Applications COMPLEX NETWORKS 2021 10, pp. 329–340. Springer (2022)
27. Pansanella, V., Rossetti, G., Milli, L.: Modeling algorithmic bias: simplicial complexes and evolving network topologies. Appl. Netw. Sci. **7**(1), 57 (2022)
28. Pansanella, V., Sîrbu, A., Kertesz, J., Rossetti, G.: Mass media impact on opinion evolution in biased digital environments: a bounded confidence model. Sci. Rep. **13**(1), 14600 (2023)
29. Pastor-Satorras, R., Castellano, C., Van Mieghem, P., Vespignani, A.: Epidemic processes in complex networks. Rev. Mod. Phys. **87**(3), 925 (2015)
30. Pizzuti, C., Socievole, A.: Multiobjective optimization and local merge for clustering attributed graphs. IEEE Trans. Cybernet. **50**(12), 4997–5009 (2019)
31. Rossetti, G., Citraro, S., Milli, L.: Conformity: a path-aware homophily measure for node-attributed networks. IEEE Intell. Syst. **36**(1), 25–34 (2021)
32. Rossetti, G., Milli, L., Rinzivillo, S.: Ndlib: a python library to model and analyze diffusion processes over complex networks. In: Companion Proceedings of the Web Conference 2018, pp. 183–186 (2018)
33. Saito, K., Nakano, R., Kimura, M.: Prediction of information diffusion probabilities for independent cascade model. In: International Conference on Knowledge-Based and Intelligent Information and Engineering Systems, pp. 67–75. Springer (2008)
34. Schelling, T.C.: Dynamic models of segregation. J. Math. Sociol. **1**(2), 143–186 (1971)
35. Sîrbu, A., Loreto, V., Servedio, V.D., Tria, F.: Opinion dynamics: models, extensions and external effects. In: Participatory Sensing, Opinions and Collective Awareness, pp. 363–401 (2017)
36. Sîrbu, A., Pedreschi, D., Giannotti, F., Kertész, J.: Algorithmic bias amplifies opinion fragmentation and polarization: a bounded confidence model. PLoS ONE **14**(3), e0213246 (2019)
37. Steinhaeuser, K., Chawla, N.V.: Identifying and evaluating community structure in complex networks. Pattern Recogn. Lett. **31**(5), 413–421 (2010)
38. Traud, A.L., Mucha, P.J., Porter, M.A.: Social structure of Facebook networks. Phys. A **391**(16), 4165–4180 (2012)

Author Index

A

Afshar, Bahar Emami II-115
Alfaro, Juan C. I-401
Alkhatib, Amr I-310
Alkhoury, Fouad II-3
Anagnostopoulos, Aris II-381
Andresini, Giuseppina II-183
Angileri, Flora I-325
Angiulli, Fabrizio I-19
Apicella, Andrea II-249
Appice, Annalisa II-183

B

Bastiaanssen, Patrick II-134
Baur, Lennart I-229
Beck, Florian II-50
Belaid, Mohamed Karim II-284
Bellamy, Hugo II-34
Bencini Farina, Antonio I-150
Bengs, Viktor I-401
Bianchi, Luigi Amedeo I-325
Bianchi, Mario I-150
Bianchini, Silvia II-381
Biondi, Elisabetta I-260
Birihanu, Ermiyas II-99
Biza, Konstantina II-65
Bocklandt, Sieben II-149
Bohm, Matteo II-303
Boldrini, Chiara I-260
Bos, Tychon II-134
Boström, Henrik I-310, II-19
Branco, Paula II-115

C

Cantini, Riccardo I-52
Cascione, Alessio II-316
Cekini, Kamer I-260
Cerqueira, Vitor I-135
Cerrato, Mattia I-229
Cimino, Mario G. C. A. II-316
Citraro, Salvatore II-425

Comito, Carmela II-396
Conti, Marco I-260
Cosentino, Cristian I-3
Cosenza, Giada I-52
Crombach, Anton I-369
Curk, Tomaž I-244

D

De Luca, Francesco I-19
De Raedt, Luc II-149
Decoupes, Rémy I-86
Derkinderen, Vincent II-149
Di Caro, Luigi II-231
Di Cecco, Antonio I-275
Di Mauro, Antonio I-116
Di Vece, Marzio II-332
Dost, Katharina II-215

E

Eugenie, Reynald I-339
Evangelisti, Martina II-381

F

Failla, Andrea II-411
Fantozzi, Marco I-275, I-325
Faraone, Renato I-325
Fassetti, Fabio I-19
Fedele, Andrea II-348
Ferrod, Roger II-231
Finkelstein, Edward I-229
Flesca, Sergio I-69
Fois, Andrea I-325
Fontana, Gianpietro II-183
Fontanesi, Michele I-295
Franzon, Lauri I-385
Fürnkranz, Johannes II-50

G

Galatolo, Federico A. II-316
Galfré, Silvia Giulia I-275

Galfrè, Silvia Giulia I-325
Gámez, José A. II-83
Geurts, Pierre II-19
Giannotti, Fosca II-332
Giugliano, Salvatore II-249
Glocker, Ben II-200
Guarascio, Massimo II-396
Guidotti, Riccardo I-150, II-316, II-348
Gündüz-Cüre, Merve I-3

H
Hartung, Lisa I-167
Henriksson, Aron I-36
Hess, Sibylle II-134
Hüllermeier, Eyke II-284
Huynh, Van Quoc Phuong II-50
Hvatov, Alexander I-213

I
Ienco, Dino II-231
Impedovo, Angelo I-116
Interdonato, Roberto I-86
Isgrò, Francesco II-249

J
Jukic, Selina I-229

K
Ketenci, Utku Gorkem II-115
Kimmig, Angelika II-149
King, Ross D. II-34
Köbschall, Kirsten I-167
Koloski, Boshko I-101
Koop, Simon I-198
Kori, Avinash II-200
Koumpanakis, Michail I-183
Kramer, Stefan I-167, I-229
Kurt, Tolga II-115

L
Lendák, Imre II-99
Liguori, Angelica II-396
Likas, Aristidis I-354
Lindgren, Tony I-36
Lombardi, Giulia I-325
Luu, Hoang Phuc Hau I-385

M
Maccagnola, Daniele I-150
Maiano, Luca II-381
Malerba, Donato II-183
Manco, Giuseppe II-396
Marozzo, Fabrizio I-3
Mazmanoglu, Hikmet II-115
Mazzarino, Simona II-364
Mazzoni, Federico II-364
Menkovski, Vlado I-198
Metta, Carlo I-275, I-325, II-167
Micheli, Alessio I-295
Monreale, Anna II-167
Morandin, Francesco I-275, I-325
Moreo, Alejandro II-267
Münzel, Lars I-229

N
Nanni, Mirco II-303
Navigli, Roberto I-101
Niederle, Jonas I-198
Nisticó, Simona I-19
Ntroumpogiannis, Antonios II-65

O
Orsino, Alessio I-52
Öztürk-Birim, Şule I-3

P
Pagès-Gallego, Marc I-198
Pansanella, Valentina II-425
Papini, Andrea I-275
Pappalardo, Luca II-303
Park, Sean II-215
Parton, Maurizio I-275, I-325
Passarella, Andrea I-260
Patron, Anri I-385
Paul, Felix Peter I-229
Pavesi, Daniele I-325
Pavlopoulos, John I-36, I-354
Pedreschi, Dino II-348
Peignier, Sergio I-369
Pellungrini, Roberto II-332
Pensa, Ruggero G. I-369
Perra, Davide II-411
Pfannes, Pascal I-229
Pinelli, Fabio II-167
Pisani, Francesco Sergio II-396

Author Index

Podda, Marco I-295
Pollak, Senja I-101
Pontieri, Luigi I-69
Prevete, Roberto II-249
Puerta, José M. II-83
Puolamäki, Kai I-385

R

Rabus, Maximilian II-284
Rahnama, Amir Hossein Akhavan II-19
Randl, Korbinian I-36
Recchia, Vito II-183
Reyes, Patricio II-303
Rigotti, Christophe I-369
Rinzivillo, Salvatore II-167
Rizzo, Giuseppe I-116
Roche, Mathieu I-86
Rohr, Benedikt I-229
Roque, Luis I-135
Rossetti, Giulio II-364, II-411, II-425
Russo, Fabio Michele II-167

S

Salvi, Michele I-325
Savvides, Rafael I-385
Scala, Francesco I-69
Schellenberg, Julius I-229
Schmitt, Nicholas I-229
Sebastiani, Fabrizio II-267
Setzu, Mattia II-316
Škrlj, Blaž I-101
Soares, Carlos I-135

Soullami, Ayyoub II-99
Špendl, Martin I-244
Spinnato, Francesco I-150
Stattner, Erick I-339

T

Talia, Domenico I-52
Teisseire, Maguelonne I-86
Thies, Santo M. A. R. I-401
Titov, Roman I-213
Tonati, Samuele II-332
Toni, Francesca II-200
Torrijos, Pablo II-83
Tortorella, Domenico I-295
Triantafillou, Sofia II-65
Tsamardinos, Ioannis II-65

V

Valentin, Sarah I-86
van Ree, Famke II-134
van Weezel, Thijs II-134
Vardakas, Georgios I-354
Vilalta, Ricardo I-183
Volpi, Lorenzo II-267

W

Welke, Pascal II-3
Wicker, Jörg II-215
Wolf, Philipp I-229

Z

Zupan, Blaž I-244

Printed in the United States
by Baker & Taylor Publisher Services